ANTS OF FLORIDA

ANTS OF FLORIDA
IDENTIFICATION AND
NATURAL HISTORY

MARK DEYRUP

Archbold Biological Station, Venus, Florida, USA

CRC Press
Taylor & Francis Group
Boca Raton London New York

CRC Press is an imprint of the
Taylor & Francis Group, an **informa** business

CRC Press
Taylor & Francis Group
6000 Broken Sound Parkway NW, Suite 300
Boca Raton, FL 33487-2742

First issued in paperback 2020

© 2017 by Taylor & Francis Group, LLC
CRC Press is an imprint of Taylor & Francis Group, an Informa business

No claim to original U.S. Government works

ISBN-13: 978-1-4987-5467-5 (hbk)
ISBN-13: 978-0-367-65836-6 (pbk)

Library of Congress Cataloging-in-Publication Data

Names: Deyrup, Mark, author.
Title: Ants of Florida : identification and natural history / Mark Deyrup.
Description: Boca Raton : Taylor & Francis, 2016. | Includes bibliographical references and index.
Identifiers: LCCN 2016012078 | ISBN 9781498754675
Subjects: LCSH: Ants--Florida.
Classification: LCC QL568.F7 D53 2016 | DDC 595.79/609759--dc23
LC record available at https://lccn.loc.gov/2016012078

Visit the Taylor & Francis Web site at
http://www.taylorandfrancis.com

and the CRC Press Web site at
http://www.crcpress.com

CONTENTS

ACKNOWLEDGMENTS

My wife Nancy has been a loving and patient partner in my three decades of work on Florida ants. She accompanied me as I scoured the state for ants; she labeled some 50,000 ant specimens for this project and compiled the plates for ant identification. Nancy and our three children, Ingrith, Leif, and Stephen, deserve a special award for accepting my long obsession with ants. Every local Florida trip, including swim meets and soccer games, somehow devolved into an occasion for ant collecting. Later in the careers of our offspring when I visited them at college and at military training camps, they continued undismayed by this public display of parental weirdness. Thanks!

Several myrmecologists became important ant-collecting companions, lending their knowledge and sharp eyes to many productive collecting expeditions. I am determined that the completion of this book will not mean the end of collaborations with my trusty friends Lloyd Davis, Stefan Cover, and Zachary Prusak. Edward O. Wilson sponsored and led several collecting trips, and has also consistently encouraged the Ants of Florida project. Thanks to all!

As mentioned above, I have been studying Florida ants for more than 30 years, with some distractions and detours along the way. Here, I acknowledge the work of myrmecologists, many of them friends or acquaintances, who have made timely contributions by revising groups of Florida ants, usually providing the Archbold Biological Station with authoritatively identified specimens after completing the taxonomic work. These 30-year colleagues include (in no particular order) James Trager (revisions of *Nylanderia, Dorymyrmex, Formica, Solenopsis,* and *Polyergus*), Phillip Ward (*Pseudomyrmex*), William Mackay (*Temnothorax*), William Mackay and José Pacheco (*Solenopsis*), Barry Bolton (*Strumigenys* and *Tetramorium*), Clifford Johnson (*Crematogaster* and *Dolichoderus*), Roy Snelling (*Camponotus*), Marcio Naves (*Pheidole*), C. Baroni Urbani and Maria De Andrade (*Proceratium*), and John LaPolla and Robert Kalla (*Nylanderia*). I also gratefully acknowledge Barry Bolton, whose catalog of ants came in time to tremendously simplify all taxonomic work, and Norman Johnson, whose Hymenoptera Name Serve website gives access to the original descriptions of ants.

During my three-decade exploration of Florida ants, my understanding of the natural history of many species has depended on the contemporaneous work of additional colleagues. The encyclopedic treatise *The Ants* by Bert Hölldobler and Edward O. Wilson provided a natural history foundation for all my work. A series of studies by Walter Tschinkel and his team of myrmecologists has revealed often unexpected discoveries in the lives of many ecologically important Florida ants. Recently, James Wetterer has been presenting new perspectives on a long series of exotic Florida ants.

The distribution maps of Florida ants benefited from three different levels of assistance. Lloyd Davis collected thousands of voucher specimens from north Florida and compiled collection information from specimens at the U.S. Museum of Natural History (Smithsonian). Clifford Johnson contributed a large collection of *Strumigenys*, including many specimens of rare species. The late Walter Suter allowed me to look through his Berlese funnel residues, which also yielded many new distributional records. The tabulation of the thousands of specimen records was done by Philipp Wiescher, who geo-referenced all the specimens preparatory to making distribution maps. The maps were designed and produced by Roberta Pickert, with recent updates by Vivienne Sclater.

Throughout my career in Florida entomology, the Archbold Biological Station has been my support, as a grouper might be supported by a gloriously diverse and beautiful coral reef. The unique natural and intellectual properties of the Archbold Biological Station could never persist in our turbulent society without the faithful guardianship of the Archbold board of directors and executive directors. I thank both the board and the executive directors James Layne, James Wolf, John Fitzpatrick, and Hilary Swain for their support of the Ants of Florida project and other projects in entomology at the Archbold Biological Station. I particularly thank John Fitzpatrick, an ornithologist, for the suggestion of an illustration-based ant guide loosely based on the kinds of field guides that have been so successful in ornithology.

ACKNOWLEDGMENTS

INTRODUCTION

Purpose: This book is for those Floridians who live in proximity to a diversity of interesting ant species. Actually, that is all of us. This book is an invitation to detach from personal servomechanisms, flip open the airlock, extend the landing ramp, and take a pioneering step onto the exciting Planet of the Ants. It is a world that is amazing, intricate, beautiful, savage, and almost unexplored.

ANTS ARE IN CHARGE
OF THE STATE OF FLORIDA

Florida is crawling with ants. A cookie crumb dropped on the ground quickly attracts ants in almost any terrestrial habitat. If the cookie crumb is placed on the trunk or branch of a tree, patrolling ants carry it off. If the bait is buried in a perforated vial, it is found by ants that are moving freely through Florida's sandy soil. This suggests that virtually every meter of the whole state is being scoured for food by scavenging ants, and this ceaseless foraging covers an area many times that of the surface area of Florida, because it includes the leaves and stems of plants, and the volume of soil around their roots.

Not one of Florida's many ant species, however, is living on cookies. Florida's ants make their living by exploitation of all kinds of ecological systems, both natural and artificial. Some ants are ecologically specialized; there is an ant that subsists on spider eggs and another that lives in hollow mangrove twigs. Some ants are generalists; there are ants that are equally at home in a forest or the edge of a parking lot, and ants that will attempt to grab any form of protein or sweets. Fortunately for us, we are very large as animals go, but if we were the size of beetles or caterpillars, we would realize the absolute necessity of discovering how to survive in a world ruled by ants.

Ants are a major force in Florida landscapes; this is one reason for learning about ants. A deeper reason might be curiosity about how the social behavior of a relatively small group of species has allowed them to dominate regions with a warm climate like that of Florida. Perhaps the study of ants may, in some people, provide a more basic reward: reassurance that the natural world around us can always astonish us with its intricate order.

BECOMING AN ANT EXPERT

Theoretically, it should not be very difficult to become an expert on Florida ants. The number of known species, 239, is not daunting compared with that of groups such as birds or butterflies. This book attempts of make identification relatively easy and gathers together the highly fragmented literature in species-by-species accounts of biology. The main problem is that ants are small, requiring examination through a microscope. This means that collecting ants is part of becoming an ant expert. On the good side, collecting insects can be an enjoyable challenge, and few people care if a small number of ants get collected. A single insecticide application on a lawn probably kills more ants than could be collected in a lifetime.

Ant experts are in more demand than one might expect. Most Floridians have ant stories or ant problems that they are happy to relate in hope of getting further information. There is often genuine curiosity behind the standard comment, "Interested in ants? Come to my house, you can have them all!"

LEARNING TO IDENTIFY FLORIDA ANTS

Illustrations

This book is intended to make it as easy as possible to identify and learn about Florida ants. It is designed for general naturalists as much as for entomologists or ant specialists. It obviously reflects modern field guides, which take advantage of the remarkable human ability to observe and remember small differences between species when standardized illustrations are provided in the form of "plates" of similar species. An important advantage of field guides is that the user's brain registers a large number of images while the user leafs through the plates, effortlessly increasing general expertise. The taxonomic keys that are traditional in entomological works are not included because it is unlikely that many users would prefer them to the "field guide" identification method. This book, of course, cannot aspire to be a field guide, as all ants are small and must be examined through a microscope to see any details. Eventually, one can become proficient at identifying many ants in the field, just as ornithologists can identify birds that are tiny dots against a cloud, but this is a skill not taught by books. Incidentally, the posture of the ants in the illustrations is not meant to be "natural." The appendages have been spread so they will not hide the body, and only appendages on the left side are included in the drawings.

Plates 1 and 2 are diagrams showing the general morphology of ants.

Diagnoses

Each species account begins with a small section on "Taxonomy and Similar Species." This section reminds the user of confusing similar species and confusing taxonomic uncertainties, if any.

COLLECTING FLORIDA ANTS

Many of the diagnostic features shown in the identification plates can only be observed by moving a specimen this way and that under a microscope. This requires collecting, killing, and preparing specimens of ants. I usually kill ants by putting them in a vial of alcohol or in the freezer. The dead specimens are glued (I use Elmer's Glue) to a little, narrow triangle of stiff paper (I make points from leftover strips of heavy, smooth drawing paper). It is possible to order a point punch that makes these paper triangles. There are hundreds of specimens of ants displayed on the AntWeb site, providing dorsal and lateral views of properly prepared ant specimens. The insect labels associated with each specimen were temporarily removed from each specimen and photographed separately. Labels from recently collected specimens show the kind of information that should be included on the label.

Not many people collect Florida ants, so it is currently easy to make significant contributions to knowledge of Florida ants, such as new county records or range extensions within a county. These new records or other natural history discoveries should be associated with a small number of representative "voucher specimens" that provide a permanent confirmation that the ant is the purported species. These voucher specimens should end up in a permanent insect collection. All the dots on the range maps of ants in this book are associated with a voucher specimen, usually in the collection of the Archbold Biological Station, occasionally in the Florida State Collection of Arthropods or in the U.S. National Museum.

DISTRIBUTION, NATURAL HISTORY, AND NAME DERIVATION

Each species has its own sections on distribution, natural history, and name derivation. Although there are maps showing known Florida distribution for most species, distribution outside of Florida is described, not shown in a map. The natural history section includes whatever is known to me about the biology of the species, widely varying from one species to another. The section on name derivation explains how the species got its name, or a suggested explanation if the reason for the name was not included by the author of the species.

This book is the result of years of study of Florida ants by the author and by many other entomologists. Knowing this, the reader may be surprised and disappointed by avowals of ignorance, for example, "knowledge of the distribution of this species remains incomplete," "the male of this species is unknown," "the function of these peculiar hairs is a complete mystery," and "this is apparently a species complex that includes one or more undescribed species." I apologize for any sense of dissatisfaction such phrases may impart. There is still an enormous amount of research to be done on Florida ants, much of it relatively simple and requiring little fancy equipment. This book is a guide to what is known about the ants of Florida, but it is also a guide to what is *not* known about the ants of Florida. The reader will soon see that there are enough mysteries to absorb the energies of many generations of inquisitive naturalists.

ABOUT THE AUTHOR

Mark Deyrup grew up in New York City, home to a surprising number of insect species, including ants, which find a living even in sooty areaways and cracks between slabs of concrete. Inspired by this limited but interesting fauna, he went to Cornell University for an undergraduate degree in entomology. This was followed by a Peace Corps stint in Ecuador, a hot spot of biodiversity. After the Peace Corps, he got a doctorate in forest entomology at the University of Washington, followed by several years teaching entomology at Purdue University. Since 1983, he has enjoyed an ideal entomologist's job, studying insect diversity and natural history at the Archbold Biological Station in Venus, Florida. He has published approximately 110 papers on subjects ranging from pygmy mole crickets that eat cyanobacteria beneath the surface of sandy areas, to a moth whose caterpillar hangs out in webs of social spiders. While working on the Ants of Florida project, he has described 12 new species of ants.

PART ONE

An Overview of the Ants of Florida

INTRODUCTION

The following sections are an informal compendium of all the patterns that I see in the ant fauna of Florida. This includes not only patterns in distribution but also ecological patterns, the kinds of relationships between ants and humans, and even the trends in the nomenclature and research dealing with Florida ants. For an introduction to ant morphology see Plates 1 and 2.

FLORIDA ANT STATISTICS

There are 239 species of ants known in Florida. The ants of the state are relatively well known, and the users of this book should expect to find accounts of almost any ant they find. Two lines of evidence, however, suggest that there are some more species to be discovered: (1) Additional species have been accumulating, although at a slow rate, right up to the time of writing these words (2015); (2) several species have been collected in Florida only once or twice, a circumstance that always hints that there are more species to be found. Species that are very localized or difficult to collect (imagine a species that lives in hollow twigs at the tops of trees in Hell's Half Acre in Jefferson County) are likely to have been overlooked. The real number of species of ants in Florida is probably close to 245 or 250, excluding any new exotic species that may invade in the future. Any enthusiastic and ingenious ant hunter is likely to be rewarded eventually with species that have never been reported in Florida. Some of these ants will be species unknown to science. The best area to find unreported native species is the northern tier of counties in the Peninsula and

all of the Panhandle, a poorly studied area with localized and endemic species of animals in many groups, from frogs to beetles. Newly introduced species could turn up almost anywhere but are most likely to be found near major ports.

With its 239 species, Florida is one of the most ant-rich states, a fact strangely absent from the promotional literature of the Florida Tourist Bureau. One site in Florida, the Archbold Biological Station in Highlands County, has 128 species of ants, the most ants known for any site in the United States. Florida undoubtedly has more species of ants than any other eastern state, largely because its unique fauna of tropical and subtropical species is combined with the more widespread fauna of the southern Atlantic Coastal Plain, with elements of the southern Appalachians thrown in. The only other eastern state that has been carefully surveyed is North Carolina (Carter 1962; Guénard et al. 2012), which has 192 species, a remarkably high diversity that is attributable to North Carolina's great diversity in topography and habitats.

As one goes north from the Carolinas, the ant fauna inevitably diminishes because ants are basically creatures of warm climates, even though there are some groups of ants, such as the genera *Formica*, *Lasius*, and *Myrmica*, whose rich (and confusing) proliferation of species is almost entirely restricted to temperate regions.

Large southwestern states have even more ant species than Florida. California has 270 species (Ward 2005), New Mexico has 239 species (Mackay and Mackay 2002), and Arizona probably has even more species than New Mexico (Hunt and Snelling 1975). West Texas, with an area approximately twice that of Florida, has 184 species listed

(Cokendolpher 1990), and we already know of many additional species from the pine woods of East Texas and from the Brownsville area. It may still be difficult, however, to find a southwestern site with more than the 128 species known from a relatively small Florida site, so Florida may be able, in some sense, to retain its proud claim to being the "antiest" state for some time to come.

NAMES OF FLORIDA ANTS

On the Joys and Frustrations of Knowing the Names of Ants

Many naturalists will use this book primarily for identifying ants. Biologists know that the purpose of the scientific name of an organism is to provide a unique pair of code words under which is indexed the information about that organism. Every correct identification is a key to a file box of knowledge, whether in somebody's head ("What do you know about *Temnothorax torrei*?"), or (formerly) in the indices of an abstracting journal, now replaced by electronic search engines, or the book you are holding in your hand. The knowledge-holding boxes, of course, are often empty. The enterprising Florida naturalist who encounters a small yellow ant, identifies it as *T. torrei*, and looks it up in the index of this book for the species account, still would not be able to find out how this species makes its living, because nobody knows. That too, however, is useful information. In the naturalist's head, a new, empty file box has been set up, ready to receive any observations on *T. torrei*.

Field biologists are aware of another phenomenon: the possession of a name sharpens the senses. The mechanism in our brain that sifts sensory input is connected to our information files, and shunts into a spotlight of greater awareness the organisms that we recognize. To know more is to see more. This discussion is all by way of explaining what might seem an inordinate preoccupation with names in this book, and among many biologists, who are willing to argue for hours over the application of a name. This obsession has an adaptive basis, although, like other adaptive obsessions, it can easily get out of hand.

Changing Names

The nomenclature of Florida ants is not quite settled. There are taxonomic problems about certain species in approximately 10 of the 49 genera of Florida ants. This may be frustrating, but it is no great cause for concern, as biologists have not been given any deadline for finalizing the names of ants, or any other organisms. The great majority of names used here should remain stable. A residue of species may have their names changed for two reasons.

Generic names may be changed when the higher classification of a group of ants is revised. The Florida species of *Conomyrma*, for example, were returned to the genus *Dorymyrmex* in Shattuck's 1992 revision of the Dolichoderinae. All such changes are made for the sake of the internal consistency in our classification of a group. There is no such thing as a biological genus concept, in the sense that there is a biological species concept. The use of the name *Conomyrma elegans* is not really wrong, because there could be no doubt which species was intended, and because a revision does not have the force of law, but most myrmecologists would agree that *Dorymyrmex elegans* is the current usage. Some names, *Colobopsis*, for example, are applied as a subgenus of *Camponotus* by some and as a full genus by others. This seems like a small difference in opinion that can eventually be resolved with a revision of the species groups of *Camponotus*, but the alleged misuse of a genus name can occasionally arouse as much pedantic petulance as the misidentification of a species.

Species names (or "specific epithets") can change when our concept of a species changes. The Florida subspecies of *Leptogenys elongata*, *Leptogenys elongata manni*, was raised to species level some years ago, so we now recognize two species, *L. elongata* and *L. manni* (Trager and Johnson 1988). Names can change on lists because the name was mistakenly applied. Old records of *Monomorium minimum* from Florida probably all refer to two other species, *Monomorium viridum* and *Monomorium trageri* (DuBois 1986). Species names can change for the trivial reason that there was an earlier, disused name that takes priority, but most species name changes reflect advances in our understanding of the status of the particular species.

There will be some species-level name changes in Florida ants. We are, for example, waiting for names for several species of *Brachymyrmex*. *Crematogaster ashmeadi* might be a complex of at least two species. Working out correct species names has a biological significance, because species, when looked at in a particular area and era, have an ecological and genetic integrity that is absent at the genus level and above.

This manual generally follows the subfamily classification outlined in Table 1 of Ward (2007) and the lists of genera from Barry Bolton's *A New General Catalogue of the Ants of the World* (1995, with online updates) and *Synopsis and Classification of Formicidae* (2003) with a few recent updates of particular genera.

On English Names for Florida Ants

Most species of insects do not have English names or "common names" of any sort. There is a good reason for this. There are hundreds of thousands of species of insects, and most of these species can only be identified by a few experts. Many naturalists are interested in ants, but the number who can identify most of the species they see is very small. The specialists who can identify species of the genus *Crematogaster*, to take a typically difficult genus of ants, are quite satisfied to have only the single Latinized name for each species. They certainly do not want to deal with a batch of vernacular names, when they already know species by their formal scientific names, such as *Crematogaster atkinsoni* and *Crematogaster cerasi*. In practice, almost all our discussions of such species are with each other, as nobody else seems very interested. The term *common name* seems absurd in such a context: fewer people would know the species by its common name than by its scientific name. To make matters worse, if one gives a species a common name, for example, calling *Odontomachus ruginodis* the "West Indian Snapping Ant," this name might be unacceptable in Spanish-speaking countries such as the Dominican Republic and Cuba, where this ant also occurs, so they might want their own name.

I personally have had a reverence for the system of binomial nomenclature since the age of 9 or 10, when I began to teach myself scientific names, which I thought of as the "real names" of animals and plants. I remember my surprise when I discovered that scientific names are occasionally changed. I am still quite conservative about nomenclature, and find myself resentful of the attitude among some modern systematists that names of genera are bits of falsifiable trivia that can be changed whenever a vagrant intellectual breeze rustles the topmost twigs of some computer-generated phylogenetic tree. Fortunately,

ants have not suffered much in this regard, and most of the recent changes, such as transferring the Florida species *Iridomyrmex pruinosus* to the genus *Forelius* (Shattuck 1992), are broadly based in ecology, morphology, and phylogeny.

Although I was teaching myself scientific names of insects with flash cards while standing in line at the school cafeteria at the age of 15, I would not hold others to this standard of oddness. The world is largely populated with people who are less peculiar, even as teenagers. Scientific names are not at all the currency of ordinary conversation. They are likely to be seen as meaningless concatenations of syllables, difficult to remember, embarrassing to try to pronounce. I spend some of my time talking to gardeners and other informally trained naturalists, and have become painfully familiar with the looks, ranging from befuddled to hostile, that I get when I start to toss around scientific names. If I say "Caribbean Trailing Ant," I get an entirely different reaction than if I say "*Monomorium ebeninum*," even though the two names are equally unfamiliar. Sometimes, I think that there might be some brainstem response to a person who seems to interlard their conversation with foreign words from an alien tribe. Sometimes, I think it is simple resentment that I am providing an unintelligible name that will be impossible to remember. It does not matter; it is undesirable to have anything, including a mass of insect names intoned in Latin, standing between the naturalist and nature.

I have therefore concluded that all the ants of Florida should have English as well as scientific names. I am encouraged by the fact that better entomologists than I have made the same decision. In 1953, Jaques Helfer provided English names for the many orthopteroids in his manual of the group. In Charles Covell's field guide to eastern moths (1984), a book that finally made moths easily accessible to a generation of naturalists, each moth has a common name. He claims that most of these names were taken from a preexisting list of common names (in Sutherland 1978), but the great majority of them cannot be found in that list, and are really from Holland's pioneering 1903 moth guide, or Covell invented them himself, with admirable aptness and brevity. Among guides to Florida insects, the books on dragonflies (Dunkle 1989), damselflies (Dunkle 1990), and grasshoppers (Capinera et al. 2001) include English names for all species. There are only a few available English names for Florida ants, so I have coined almost all the common names in this book. The Entomological Society of America has allocated to itself the authority to establish "approved" common names (Sutherland 1978), but this list is restricted to the species "commonly of concern to entomologists," which excludes almost all the ants of Florida. I doubt that I am the best person to make up names; many of the names in this book are more appropriate than concise—"the Wooly Pygmy Snapping Ant" is even clumsier than "*Strumigenys lanuginosa*." Perhaps these names will become streamlined with usage.

Subspecies Names for Florida Ants

Subspecific names have a dismal history in myrmecology. Specimens of ants that looked different might be given different subspecies names, even though the differences were related to caste, or to differences in the preservation of specimens. To make matters much worse, many ants were saddled with varietal names, so a specimen might have four or even five names attached to it. As Creighton (1950) pointed out, there was no simple way out of this "altogether horrendous maze of nomenclature." Some of these names applied to ants that should really be considered distinct species, some applied to ants that were legitimate distinctive geographic subspecies, and a large number of them applied to variants that had no geographic basis and were different castes, or specimens that were discolored in preservation, or, as Creighton discovered in a surprising number of cases, distinguished by characters that seemed to be completely imaginary. Many Florida ants were caught up in this mess, but, largely thanks to Creighton, we have relatively few such problems remaining. Fossilized vestiges of these problems, like taxonomic coprolites, can still be seen occasionally. The Florida species *Odontomachus ruginodis*, for example, was described by W. M. Wheeler in 1905 as *Odontomachus haematodis insularis ruginodis*. The last name, referring to a "variety," has no taxonomic standing, so *ruginodis* was not really the name of anything until 1935, when M. Smith used it as a subspecific name, *Odontomachus haematodis ruginodis*. This is why M. Smith, not Wheeler, is considered the author of *O. ruginodis* in the species account in this book. *Odontomachus ruginodis* was first recognized as a full species by E. O. Wilson in 1965. This little tangle and a few thousand similar cases are succinctly dissected in Barry Bolton's 1995 catalogue, a book that is a miracle of scholarship.

The flagrant abuse of the subspecies concept in ant taxonomy may have helped inspire Wilson and Brown's general attack on the practice of designating subspecies (1953). To this day, American myrmecologists often avoid using the surviving subspecific names, and almost nobody names new subspecies of ants. But although the logic of the 1953 attack on subspecies is as cogent as ever, and continues to appeal to each new generation of young taxonomists, the usefulness of the geographic subspecies category persistently overwhelms all theoretical objections, and seems set to outlive all its detractors. The young want definitions that always work, but eventually we must always come back to working definitions. Although I am not about to name new subspecies of Florida ants, I recognize the legitimacy of geographic subspecies, and deal with them as they come up in the species accounts.

One More Thing about Names: Author Names

Attached to the name of each ant is the name of the person who described it, or rarely, as in the case of *O. ruginodis*, the first person who used the name properly. When a species is transferred to a genus that is different from the genus in which it was first described, the author's name goes in parentheses. The species *Conomyrma elegans* Trager, when transferred to *Dorymyrmex*, became *D. elegans* (Trager). I generally follow Bolton's 1995 catalogue for author names. The use of an author name outside of a taxonomic treatise is, as the myrmecologist Bill Brown used to say, "An exercise in useless pedantry." In this book, author names appear only in one place, at the beginning of each species account.

ECOLOGICAL SIGNIFICANCE OF FLORIDA ANTS

Ants as Predators

Most species of ants are predatory, as well as scavengers, and ready to attack any inadequately defended arthropod. Adult ants are not, strictly speaking, carnivorous, since they feed mostly on fluids, either sweet fluids such as nectar, or the juices of prey. The predaceous behavior of ants is inspired by the appetites of larvae back in the nest, and the more food that can be found, the faster the nest grows. With this growth comes strength for nest defense and an ability to invest in the expensive business of producing queens and males that will disperse to new sites. Among more generalist predatory ants, there may be some targeting of the most available type of prey. This allows individual ants to become more efficient at finding and dealing with prey. This also makes these ants more important as regulators of ecosystems because they concentrate on species that are especially abundant.

Anti-Ant Defenses

Few terrestrial or arboreal arthropods would be able to survive in Florida without anti-ant defenses, either behavioral, morphological, or chemical. Scientists studying the defenses of Florida arthropods often use generalist predatory ants as prototypic archpredators. Florida Carpenter Ants (*Camponotus floridanus*) were used to test the effectiveness of the morphological defense of the Palmetto Tortoise Beetle (*Hemisphaerota cyanea*), which uses adhesive pads on its feet to clamp itself down on a leaf when attacked (Eisner 1972). "Gin traps," rows of snapping teeth on the edges of pupae of lady beetles and many other insects, were tested with the Red Imported Fire Ant (Eisner and Eisner 1992). Peculiar detachable bristles on the millipede *Polyxenus fasciculatus* were shown to have the ability to hogtie Ashmead's Acrobat Ant (*C. ashmeadi*), which is often found under the bark of pine trees, along with the millipede (Eisner et al. 1996). The Florida Harvester Ant (*Pogonomyrmex badius*), which is a predator as well as a seed eater, was photographed being blasted by a species of Florida bombardier beetle (Eisner 1972). A pyralid caterpillar (*Laetilia* sp.) that feeds on scales on lignum vitae trees on Lignum Vitae Key defends itself with regurgitates against Florida Carpenter Ants, which guard the scales (Eisner et al. 1972). The Slender Crazy Ant (*Paratrechina longicornis*) was used as a bioassay to test the chemical defenses of the caterpillar of the Palmetto Borer Moth (*Litoprosopus futilis*) that feeds on palmettos in Florida (Smedley et al. 1993).

It is an absurd understatement to say that we have just begun to discover arthropod defenses against ants. In the sand soils of Florida, which are permeable to ants (as shown by tests with buried baits), every species of soil-dwelling arthropod must have an anti-ant defense. The larvae of such arthropods, of which there are literally thousands of species, might be able to use active defenses, but the pupae must often depend on passive defenses, such as an impermeable cocoon or chemical deterrents. Consider the digger wasps in the families Sphecidae and Pompilidae: they stock their larval larders with paralyzed arthropods that would be ideal food for a great variety of scavenging and predatory ants. Although the underground groceries are safe from our eyes, they should not be at all safe from ants; it is like hiding hamburger in a hyena pit. We have to assume that these hundreds of species of digger wasps, and the sand-living bees as well, have potent defensive chemicals that can be smeared on the sides of the burrow to exclude ants from what would normally be a highly attractive resource. This is not quite hypothesis in vacuo: ant-repellent chemicals are known in social wasps (Jeanne 1970). Nonetheless, anybody who has ever dealt with an infestation of Pharaoh's Trailing Ants in their kitchen knows how persistent ants can be, even in the face of massive chemical assaults. An aggregation of sand wasp burrows can remain active for years, presumably protected by the minute amounts of whatever chemicals the wasps are able to produce from their various glands. Why haven't ants developed resistance to these defenses over the millions of years during which ants and digger wasps have been living side by side? The answer, as in the case of plant defenses against herbivores, is likely to be that there has been a long history of defense and counter-defense, culminating in a wide variety of defenses against ants. It would be gratifying to know just a few of them.

Ants as Specialized Predators

A portion of the diversity of life is derived from the fact that any general defense provides an exclusive resource to those species that are able to overcome that defense. This fuels evolutionary specializations and divergences, among the ants as in other groups. Hölldobler and Wilson (1990) have provided a list of specialized predators among the ants, including a number of species that occur in Florida. The largest group of species is among the dacetine ants: as far as is known, Florida species of *Strumigenys* are heavily dependent on Collembola (springtails) as prey. The relationship between the dacetine ants and the Collembola is discussed in detail later, but the short version is that the only thing as fast as the spring-loaded leaping appendage of a springtail is the spring-loaded impaling mandibles of a dacetine ant. Other specialized predators among Florida ants listed by Hölldobler and Wilson are *Stigmatomma pallipes*, which feeds primarily on geophilomorph centipedes, *Leptogenys manni*, which feeds on oniscid sow bugs, and army ants (*Neivamyrmex*), which attack nests of other ants. In addition, the members of the genera *Proceratium* and *Discothyrea*, including several Florida species, are all likely to be specialized predators of arthropod eggs (Brown 1979). The Japanese myrmecologist Masuko, one of the finest observers of the natural history of ants, has discovered (1994) that two Japanese species of *Myrmecina* are specialized predators of armored oribatid soil mites, and there is every reason to believe that this remarkable behavior occurs in other *Myrmecina*, including species found in Florida.

All the known examples of specialized predators in Florida belong to the subfamilies Ponerinae, Ecitoninae, and Myrmicinae, groups that retain the ancestral stinging ability of ants. It has been suggested that the possession of a sting may be a preadaptation to prey specialization because it allows elusive or well-defended prey to be subdued quickly. This explanation may well be true, but one can't help thinking that if specialized predation had been concentrated in the stingless subfamilies Dolichoderinae and Formicinae, we would happily explain the phenomenon by claiming that immobilizing chemical sprays are the only good way to deal with especially fast or fierce prey that are difficult or dangerous to seize and handle. One would also expect the greatest degree of specialization in the fastest-moving ants, which would have the best ability to encounter and bring back scarce or dispersed prey; this is not at all the case. Returning to reality, additional specialized predators among Florida ants are most likely to be found among the species of Ponerinae and Myrmicinae. It would be worth taking note of prey and middens of species such as *Ponera exotica*, *Hypoponera inexorata*, *Temnothorax smithi*, *Temnothorax bradleyi*, and also the species of *Proceratium*, *Discothyrea*, and *Strumigenys*, which may have interesting narrow specializations beyond the known predilections of their genera.

Most specialist predators among Florida ants are uncommon species compared to generalists that live in the same habitat, and their ecological effects are sure to be more subtle. In a simple specialist predator and host system, predators lag behind the populations of hosts and may accentuate population fluctuations, but as soon as one is dealing with a complex of specialist and generalist predators, the system loses much of its predictability, even in models (Hassell and Godfray 1992).

Cumulative Effects of Predatory Ants

The idea that ants control arthropod communities of Florida, and have helped shape the defensive adaptations of thousands of species, may seem extreme (or myrmecocentric), but it is supported by the devastation of Hawaiian arthropod communities when ants were introduced onto previously ant-free islands (Huddleston and Fluker 1968; Reimer 1994). This is not to say that ants alone rule Florida ecosystems; other groups such as nematodes and bacteria, beetles, and spiders are at least as important, and on a larger scale, humans have the ability to modify or destroy every habitat. It is, however, reasonable to believe that if Florida were divested of ants by means of some fantastic new insecticide, the whole state would suffer major biological perturbations. Since many species of arthropods, despite their defensive adaptations, have their populations repressed to some degree by ants, the most obvious effect would be the drastic increase of many other arthropods. On the scale of an individual plant, this has been shown by removing the extrafloral nectaries that certain plants use to attract predatory ants, and watching the resulting depredations by caterpillars (Beattie 1985).

Ants as Seed Dispersers in Florida

The dispersal of seeds by ants is a phenomenon that was discovered approximately a hundred years ago, but received little attention until quite recently. If early myrmecologists observed ants carrying seeds, they probably assumed that the seeds themselves would be consumed by the ants. Twenty-five years ago, only 300 plants were known to have ant-dispersed seeds; today, the number is approximately 10 times that (Westoby et al. 1991). Ants are induced to disperse seeds by a fleshy, usually lipid-rich structure (called an elaiosome) attached to the seed, or more rarely, an attractive coating on the seed. After this edible treat is removed, the seed is discarded, usually in a midden near the nest, where it may be buried by subsequent dumping. Seed dispersal by ants and other forms of seed dispersal have been compared by region and habitat (Willson et al. 1990), and it is clear that ants are major seed dispersers in only a few places: the fire- and drought-prone habitats of Australia, the droughty and nutrient-poor open shrub habitats of South Africa, and, to a much more limited extent, deciduous forests of eastern North America, where many spring-flowering herbs are dispersed by ants.

There are various theories to account for these patterns, and one might hope that these theories would tell us what to expect in Florida. Thompson (1981) suggested that spring-flowering woodland plants produce seeds at a time when the birds that are more normal seed dispersers are concentrating on collecting insects for their young, and less interested in fleshy fruits. We might, therefore, expect that there would be a good proportion of ant-dispersed seeds in the relatively small areas of hardwood forest in northern Florida where there are populations of spring-flowering herbs in genera that are often dispersed by ants, such as Erythronium, Uvularia, Trillium, and Viola (Thompson 1981). Ants in the genera Lasius, Myrmica, Temnothorax, Tapinoma, and especially Aphaenogaster, which are seed dispersers of violets in West Virginia (Culver and Beattie 1978), are also present in northern Florida. There are no published studies of seed dispersal by ants in northern Florida forests, but methods of Culver and Beattie (1978) could easily be adapted to northern Florida. The prevalence of ant-dispersed plants in open, scrubby, often fire-maintained habitats with nutrient-poor soils (Westoby et al. 1991) should make the heart of the Florida myrmecologist beat faster, because this habitat characterization fits most terrestrial habitats of Florida. With the abundance and diversity of ants available in these habitats, ant-dispersed plants should be common. Nobody has looked very carefully, but indications are not promising. The only widespread and diverse genus known to have elaiosomes is Polygala, although there are also a few species of open site Viola, mostly in northern Florida. There are probably many plants whose fresh seeds have never been examined for elaiosomes.

A particularly interesting case of seed dispersal by ants in tropical Florida was reported by Kaufmann et al. (1991), who studied seed dispersal in the exotic fig Ficus microcarpa. This species has a two-phase dispersal system: the fruits are eaten by birds, and then the ants collect the seeds from bird droppings. This disperses the seeds from their original clump, and if the ant species is arboreal, the seeds may be taken up into the nest where they can grow epiphytically and eventually take over the tree. There is no reason why this bird–ant two-step system could not be the dispersal mechanism of some other Florida plants that have berries and small seeds, and the methodology of Kaufmann et al., while reasonably demanding, would not be impossible to replicate. The discovery of a complex ecological interaction such as the two-phase dispersal of fig seeds normally leaves ecologists smiling fondly over the intricacies of nature, but in this case, there was no joy in Dade County: the tree is an exotic species of fig (F. microcarpa) that seems poised to invade remaining natural tropical hammocks of the area. Accomplice ants that transport the seeds are Paratrechina longicornis and Solenopsis invicta, themselves exotics.

Ants as Seed Harvesters in Florida

Several species of Florida ants harvest and consume seeds. The most obvious of these is the Florida harvester ant, P. badius, which gathers large quantities of seeds, although it is also a predator and scavenger (Ferster and Traniello 1995). At some sites, one may see streams of foragers collecting seeds of scrub rosemary (Ceratiola ericoides), and after fires in sand pine scrub, these ants collect by the thousands seeds of sand pines (Pinus clausa) that are released from their serotinous cones by fire. The seed hosts of this ant in many natural communities in Florida are unknown, much less the impact of these ants on plant associations in which they live. Western species of seed-harvesting ants can have pronounced and unexpected effects on plant associations (Hölldobler and Wilson 1990), but it is unlikely that Florida species have such strong effects, because intensive flowering in many plant species is triggered by fires, rather than by seasonal rains that produce more or less dependable seed crops. On the other hand, in habitats dominated by long-lived perennials, such as the open habitats of Florida uplands, seed predators could have major and generally ineluctable effects, especially through maintenance of patchy distributions of plant species, if the ability of a plant species to establish itself in a spot that is temporarily available for long-term establishment is determined by proximity to a nest of seed-harvesting ants.

Solenopsis geminata is another Florida species that harvests seeds on a large scale. The largest workers are morphologically specialized as seed grinders (Wilson 1978). In experimentally seeded plots at a site in Mexico, Risch and Carroll (1986) showed that this ant changed the composition of plants in study plots. These researchers also discovered complex foraging behavior in *S. geminata*. They showed, for example, that when ants encounter small numbers of less preferred seeds mixed in with large numbers of their favorite seeds, the ants are more likely to bring back the less preferred seeds than if they encountered the less preferred seeds alone. The reverse is also true: preferred seeds are safer among a large number of less preferred seeds. When this kind of complexity crops up in an ecological experiment that has been drastically simplified, it does not make one feel that it will be easy to pin down ecological effects of ants in natural systems.

Several other species of Florida ants regularly harvest seeds. I have observed seed harvesting in *Pheidole littoralis*, *Pheidole adrianoi*, and *Pheidole metallescens*, and it almost certainly occurs in *Pheidole carrolli*. The only nest of *Solenopsis globularia* that I have seen had a chamber packed with small seeds. In open sandy habitats, seed harvesting is a conspicuous phenomenon, and it should be easy to discover many new relationships between specific plants and seed predators.

Ants in the Diet

Generalist Ant Predators: We myrmecologists are so fond of expounding on the dominant role of ants in ecosystems that it is difficult for us to think of ants as victims. It is useful to remember that any group of superabundant organisms is likely to feature on the menu of numerous creatures. I have frequently seen ants in the stomach contents of Florida Scrub Lizards (*Sceloporus woodi*), Green Tree Frogs (*Hyla cinerea*), and Southern Toads (*Bufo terrestris*). Numerous insects and spiders eat ants opportunistically, including antlions (Myrmeleontidae), most species of which appear to be generalist predators. Flights of ants are a bonanza for general insectivores. Queen ants are especially choice, as they are usually plump with resources needed to found a colony, and generally unable to sting (except for some Ponerinae). After a night flight of ants, webs of night-weaving spiders (*Neoscona* spp.) may have bundles of alate *Pheidole* waiting to be stashed for later consumption. Florida Scrub-Jays (*Aphelocoma coerulescens*) and Blue Jays (*Cyanocitta cristata*) come to road shoulders to harvest queen fire ants running along the pavement. Tree swallows (*Iridoprocne bicolor*) swoop through early flights of fire ants in the spring and are probably responsible for the winged but gaster-less mutilated queens that one finds crawling over the ground. Cuban Brown Anoles (*Anolis sagrei*) perched on shrubs bordering my driveway engage in frenzied footraces across the paving to seize recently landed fire ant queens and males.

Specialized Predators: Florida ants also face specialized predators. Ants comprise approximately 50% of the diet of Flickers (*Colaptes auratus*) (Moore 1995), and a single species of small ant, *Crematogaster pinicola*, makes up approximately 40% of the diet of the Red-Cockaded Woodpecker (Hess and James 1998). The Eastern Narrow-Mouthed toad (*Gastrophryne carolinensis*) is an ant specialist: at the Archbold Biological Station, the stomachs of 146 Narrow-Mouthed Toads were found to contain a total of 4859 individual ants, representing 43 species and comprising 95% of all food items (Deyrup et al. 2013). Specialized myrmecophilous insects may appear as commensals in Florida ant nests. These include Syrphidae of the genera *Microdon* and *Rhopalogaster*, an undescribed species of Histeridae in the genus *Terapus* associated with *Pheidole morrisi*, Staphylinidae in the genus *Adranes*, and others.

FLORIDA ANTS AS PESTS OF HUMANS

The great majority of Florida's ant species do not bite or sting humans, and conduct their affairs in complete obscurity. The red imported fire ant (*S. invicta*) is the most extreme exception to this rule, and earns more attention than all other species combined. For many Floridians, the word *ant* is synonymous with fire ant, and any ant these Floridians see is considered a noxious pest. There is an increasing tendency for Floridians to label as pests any arthropods that occur in their yard or garden, certainly an untraditional and paranoid view of the outdoors. Some pest control operators and other professional entomologists seem to encourage this view, perhaps unintentionally. The idea that outdoor Florida is packed coast to coast with hostile and even dangerous ants and other arthropods is not only misleading, but unhealthy, as it needlessly raises our anxiety levels, drives us indoors, and encourages us to permeate our living space with insecticides.

Out of almost 240 species of ants in Florida, only 7 can legitimately be considered outdoor pests. These species are *Solenopsis invicta* (the worst, by far), *S. geminata* (similar to *S. invicta*, but much less common), *Wasmannia auropunctata* (a minute stinging species, most common in south Florida), *Pseudomyrmex gracilis* (a casual stinging species that falls out of trees). *Hypoponera punctatissima* (there are scattered reports of many people being stung by alate queens when there are large flights of this species in south and central Florida), *Camponotus pennsylvanicus* (makes galleries in live trees in north Florida), and *C. floridanus* (establishes large, well-defended nests in outdoor installations such as pump housings and master switch boxes). These pest species are discussed in detail in their species accounts. Innocuous outdoor "yard ants," such as species of the genera *Dorymyrmex*, *Crematogaster*, *Cardiocondyla*, *Cyphomyrmex*, and *Trachymyrmex*, should not be considered pests just because they presume to live near our homes (Plates 1 and 2).

Species that achieve pest status by regularly invading buildings are more numerous, including approximately 17 species. These species are *Monomorium pharaonis* (the most persistent and widespread species), *Monomorium floricola* (a south Florida species, not very common or persistent indoors), *Tapinoma melanocephalum*, *Brachymyrmex patagonicus*, *P. longicornis* and *Nylanderia fulva* (common species that persistently enter buildings, although the nests are usually outdoors), *Nylanderia bourbonica* and *Nylanderia steinheili* (species that regularly but infrequently enter houses, *S. invicta* (most likely to invade buildings when nests are flooded outdoors), *W. auropunctata* (relatively infrequent in buildings, but very irritating when it does move in), *Linepithema humile* and *Pheidole megacephala* (there are large populations of these invasive species in a few sites in Florida), *Technomyrmex difficilis* (a recent arrival that is proving to be even more of a household problem than *Paratrechina longicornis*), *Camponotus floridanus* and *C. inaequalis* (two of the species most often brought in for identification in south Florida; often nest in wall voids, behind cabinets, and in electric appliances), and *C. pennsylvanicus* (our only ant that does much excavating of sound timber; it is a relatively uncommon problem in northern Florida). The pest status of these species is discussed in their species accounts.

If doors and windows are left open, or if the structure of the house has large cracks, various species of ants, and many other arthropods, are likely to wander in and out. Males and queens of many species are attracted to lights, and I have had some unusual ants, suspected of being termites, brought in for identification. When outdoor ants invade buildings, a small amount of detective work will usually reveal their mode of entry, and they can be excluded by a physical barrier or by a band of insecticide on the outside where they enter the building. Fumigating the building or attempting to eliminate all ants within foraging distance of the house is an expensive, sometimes unhealthy, last resort. These drastic treatments are profitable to exterminators, who are also the authorities who provide recommendations for methods of pest control. It is useful to remember that there can be a conflict of interest in these recommendations, even though most exterminators are undoubtedly responsible and well-trained individuals. When I get an inquiry about an invasion of ants in a building, I always recommend that the owner do a little research on their own to discover the identity of the ants and where they are coming from, before yelling for an exterminator. Since these ants are almost never a threat to human health or to the structure of the building, there is little to lose by doing a personal investigation of the problem. A good example would be ants invading a school building, a situation that could evoke a reflexive fear of lawsuits. Pesticide treatment is more likely to cause actual health problems than an ant infestation. A biology class assigned to analyze the case could get a more valuable lesson in biology and behavior than they would derive from any textbook, and possibly save the school a pile of money.

CONSERVATION OF FLORIDA ANTS

A solicitude for rare ants is unusual. Most people feel that there are far too many kinds of ants that are far too numerous. There are good reasons, however, why we should be concerned if species of ants are becoming rare or endangered. Even rare species may be ecologically important if they are rare because they are restricted to scarce habitats. A series of species, for example, seems to be restricted to the remaining scrub ecosystems in south and central Florida (*Odontomachus relictus*, *Dorymyrmex elegans*, and *D. flavopectus*) and are likely to be important members of these ecosystems (Deyrup 1989). There is the argument that we don't know whether rare species might be "good for something" because no one has looked, and we will never know if the species disappear before they have been examined, and reexamined with the technology of the future (Eisner 1991). Associated with the rare ant *O. relictus* is the rare firefly *Pleotomodes needhami*, and we know that fireflies are hotbeds of unusual and potentially useful chemicals (Eisner 1991), and ants themselves, for that matter, have more glands than a wildlife photographer has pockets (Hölldobler and Wilson 1990). We should be modest in our judgment of which species are useful and attractive, and which species are useless and obnoxious, remembering that most hawks and owls were considered vermin in northern Europe and the United States until quite recently. Future generations may well have a more encompassing and sophisticated appreciation of biological diversity that is prevalent at the moment. Though few people (if any) expound on the clean, racehorse lines of *D. elegans*, or revel in the mysterious shifting patterns of ornate scales in the genus *Strumigenys*, it would be wrong to deny future generations those potential pleasures, or the intellectual and practical insights they might derive from them.

Although many of Florida's ants are rare in collections, there is little evidence that most of these species are actually becoming rare and endangered. These species, as well as most native Florida species of plants and animals, are becoming less abundant as native habitats are being remodeled by the relentless influx of humans into the state, but this does not mean that populations of these species have been reduced to the point of concern. Our clearest examples of rare ants are those that seem to be dependent on a habitat that occupies a very small area in the state, and that area is steadily shrinking. These include the three species mentioned above that occur in Florida scrub habitat, and *Trachymyrmex jamaicensis*, which lives in hammocks in the Florida Keys, but is common only on Elliott Key. *Formica subsericea*, which had an isolated Florida population far south of its normal range, in an unusual habitat (Wilson and Francoeur 1974), has not been seen since it was first reported. The Florida slave-making ant *Polyergus oligergus* (King and Trager 2007; Trager 2013; Trager and Johnson 1985) appears to be both isolated and restricted in distribution. As we get more records and more ecological information, the number of rare and endangered species that we can recognize is sure to increase.

In addition to species that are endangered by habitat conversion, there are probably species endangered by the direct or indirect effects of introduced species. In the dark interstitial mazes of the soil, and in hollow twigs swaying in the tree tops, ant empires rise and fall in pitched mandible-to-mandible battles while we pass by oblivious. These replacements of one species by another should also become more evident as we accumulate enough records to track changes in the ant fauna.

Ants are, to say the least, unlikely flagships for conservation projects. In some cases, a list of all the rare and endangered species of a site is useful to establish the significance of the site; any rare ants should be on the list. For the most part, ants would be well advised to remain discretely in the background, allowing more showy or cuddly species to make public appearances and beg for money. There are situations, however, in which a nice illustration of an ant could be used to symbolize the specialness of the site, reminding the public that any really unusual place is likely to harbor a host of strange and interesting-looking tiny creatures, whose habits are usually a mystery.

EXOTIC ANTS IN FLORIDA

As a young boy growing up on the lower East Side of New York City, on the way to school I used to walk up part of 14th Street, in those days a haunt of innocuous sleaze. Next to a movie theater that showed films that were considered racy ("Bridget Bardot in 'Girl in a Bikini,' Mon.–Sat.") was a place with darkened plate glass windows that advertised "Exotic Dancers Nightly." I had no real idea of what an exotic dancer might be, but I somehow knew not to ask my parents. I also knew intuitively that exotic dancers were less virtuous than nonexotic dancers, but at least as interesting. This is a good attitude for approaching exotic ants. They are not welcome members in natural communities, but they are extremely interesting. When I have found a new exotic ant in Florida, part of me exclaims "another new and exciting ant in Florida," even while my more responsible persona is decrying "another new, evil exotic." We should all fight the introduction of more invasive exotic species of ants and other organisms, but it is a bit hypocritical to consider exotics, such as the red imported fire ant, as evil organisms. These exotics are only manifestations or side effects of our own unparalleled ability to invade and disrupt natural ecosystems.

Exotic ants have moved into Florida in such large numbers that one might suspect them of having read the enticing promotional literature provided by state agencies. With 57 species and counting, Florida has the dubious distinction of having the largest number of introduced ants of any state. The percentage of exotic ant species is also high, approximately 24%, as compared, for example, with North Carolina, with 16 species, or 8% (Guénard et al. 2012). The state with the largest proportion of exotic ants is Hawaii, which has no native ants and a 100% exotic ant fauna. The large number of exotic ant species in Florida is attributable to the warm climate of the state. The large percentage of exotic ant species in Florida is attributable to the isolation of the state with respect to the huge Neotropical ant fauna. Neotropical species suited to Florida, with the exception of some Caribbean species, have been forced to wait for transport by humans.

Exotic species are those that were brought to Florida by humans after the arrival of Europeans. There are no known examples of ants intentionally imported into Florida. The great majority probably arrived with plants in containers, or in plant material, or as stowaways in ships. The recent massive increase in air traffic increases the chance of species arriving as air passengers from almost any part of the world. Included as exotic ants are some species, such as *Pheidole obscurithorax*, that were brought to adjacent states and seem to have spread into Florida by their own powers.

Exotic ants in Florida are covered in their respective species accounts. A long discussion of exotic ants in Florida and their effects and origins is provided by Deyrup et al. (2000). This paper is available online. The checklist of Florida ants at the end of the Species Accounts section and before the Literature Cited designates which species were imported from the New World tropics and subtropics (33 species) and which species were from the Old World tropics and subtropics (24 species).

MAPS OF FLORIDA ANTS

The Quality of the Data—Not Very Good

Ant species are not uniformly distributed over Florida. For example, exotic species as a group are concentrated in the southern part of the peninsula,

and there are other distributional patterns, but these patterns are often lacking in detail. It might seem simple to stack up the distribution maps of a series of species to give us all kinds of brilliant ideas on the ecological factors that determine the distribution of various groups of ants in Florida. This book includes a distribution map for most species, and the sum of the collection sites (which appear as large dots) is more than 11,000. When one looks at a series of these maps, with the same state and county outlines, the same size dots, most of the collections and identifications by the same person, a fantasy can easily begin to build that these maps are equivalent and comparable. Unfortunately, as soon as organisms are transformed into dots, they become susceptible to being improperly juggled. A look at the distribution maps for one genus, *Hypoponera*, is depressingly instructive.

Hypoponera opacior (see map for this species in the map section) is a very common ant throughout the state, although it is never seen crawling about in the open. Gather a few liters of leaf litter from a moist or dry site from Pensacola to the Florida Keys, and *H. opacior* will be there in your Berlese funnel extract. This species is often absent from wet areas, but large gaps in the distribution map are not attributable to huge wetlands, but rather to a lack of sampling. Likewise, clusters of dots indicate favorite collecting areas. This species, because it is so common, indicates immediately how much remains to be done before it can be said that Florida has been adequately surveyed for ants. Large areas of the state have hardly been touched. The grouping of the dots is an artifact of collecting. There is no way to recognize these problems by looking at the map; one needs the information that this species turns up almost every time it is sought. Given this information, one might feel justified in shading the entire map of Florida to indicate the ubiquitous nature of *H. opacior*.

Hypoponera inexorata (see map for this species in the map section) is also widely distributed and also occurs (usually with *H. opacior*) in upland and mesic sites. This species, however, is much more rarely collected: during the entire study, approximately 90 specimens were collected, while the specimens of *H. opacior* must number in the tens of thousands. The enormous gaps on the map of *H. inexorata* might reflect the rarity of the species throughout its range, or they might reflect the patchy nature of some unknown habitat requirement, or they might reflect some problem with the sampling method. The latter is probably the case; in most areas, this seems to be a subterranean species that seldom comes up into the leaf litter, so that one finds specimens by persistence in the more heavily sampled sites, and occasionally by luck on the less heavily sampled sites. We do not know nearly enough to consider shading the whole map because this species really might be absent from the western part of the peninsula and the northern edge of the Panhandle.

Hypoponera opaciceps (see map for this species in the map section) is almost always found in open, wet areas, such as marsh tussocks and the open edges of lakes. It is as widespread and dependably common as *H. opacior*, but much less commonly collected because the wet substrate it inhabits is harder to run through a litter extraction funnel, and there is less incentive to look for ants in soil of wet areas because there are far fewer species than in better-drained leaf litter where *H. opacior* occurs.

Hypoponera punctatissima (see map for this species in the map section) is a tropical exotic that prefers wet disturbed areas, such as heavily watered lawns, mulched shrubbery, seasonally wet pastures, and edges of ditches. It also lives in marsh tussocks, but cannot be found dependably in these natural habitats. The map of this species is particularly deceptive. *Hypoponera punctatissima* is probably present on every acre of the huge expanse of wet pasture, sugar cane fields, sod farms, and disturbed Everglades south of Lake Okeechobee, but since that area is generally inaccessible and faunistically boring, there is a big gap on the map. In contrast, the blank areas in the northern part of the map probably represent a genuine sporadic distribution of *H. punctatissima*, but there may be a sampling deficiency in north Florida as well. The northernmost part of Florida has such a diversity of undersampled natural habitats that one is less inclined to spend time sampling lawns and foundation plantings in order to get records of weedy exotic ants.

So what does the publication of a locality map accomplish, beyond immortalizing a series of biases and fortuitous events? As in many other scientific studies, positive data are more useful than negative data. A site record is unequivocal, while the absence of a site record could be explained in several ways. Over a period, the failure to find a species in a well-collected area (such as Alachua County), using methods that work in other areas, begins to give some reality to the blank spots on the map. Even now, it appears that there are some recognizable broad patterns in the distribution of ants in Florida. One of the most valuable contributions of these maps is that they frequently enable naturalists to identify any range extensions.

Spotting Future Biogeographic Changes: The nonmethodical nature of most ant collecting should not prevent recognition of some future changes in the fauna. If some species becomes so locally dominant that it could not have been missed even in a cursory review of the fauna, it is safe to assume that such a species was previously rare or absent. I have applied this kind of evidence to suggest that species such as *Cyphomyrmex rimosus* and *Brachymyrmex patagonicus* are new to Florida, even though they were both present during the current survey. I am also hoping that my records of litter samples may be used to show future changes. Although the selection of sites for sampling has not been methodical, the amount of litter and the extraction technique have been consistent. I have taken thousands of samples and recorded the species of ants in each one. It should be possible to use these records to show gross changes, if any, in the litter fauna a few decades hence.

PART TWO

Species Accounts

SUBFAMILY AMBLYOPONINAE

Genus *Stigmatomma*

Saw-Toothed Ants

When one first looks into the face of a worker *Stigmatomma* (pronounced **stig' ma tō" ma**), one can only be grateful that these ants are not the size of terriers. The long jaws, set with double rows of sharp, triangular teeth, are fearsome weapons. The prey of at least some species is relatively large and well-defended centipedes (Brown 1960a; Gotwald and Levieux 1972). Alex Wild provides a lively and profusely illustrated journal of a captive colony of *Stigmatomma oregonensis*, a species that appears closely related to the Floridian *Stigmatomma pallipes* (Notes from Underground, online, May 23, 2005). Workers in this colony killed centipedes by stinging and then carried the prey back to the nest, where they were repeatedly bitten, but not dismembered. Workers consumed hemolymph oozing from the centipedes, and eventually piled larvae on the prebitten centipedes. Queens and, less frequently, workers fed on larval hemolymph, puncturing the larvae in specific areas of the body. The larvae developed scars from this treatment, but otherwise seemed unharmed. This nonlethal cannibalism increased when the colony ran out of centipedes. Similar behavior occurs in the Japanese species *Strumigenys silvestrii* (Masuko 1986). The hemolymph-drinking behavior of *Stigmatomma* has inspired some myrmecologists to use the English name "vampire ants." The appropriateness of this name, however, may be questioned in light of the recent spate of interest in the natural history of vampires. Vampirism specialists, whose work may be scanned in the remarkable array of books currently appearing in the "Paranormal Romance" section of a large-chain bookstore, seem to disagree on whether vampire reproduction can even produce babies, and I know of no suggestion in the literature that vampires support themselves by feeding on the blood of their offspring.

The genus *Stigmatomma* was recently revived when the genus *Amblyopone* was divided into three genera; most of the species previously in *Amblyopone* have been transferred to *Stigmatomma*, which now includes 66 species (Yoshimura and Fisher 2012). *Stigmatomma* occurs worldwide except for arctic and subarctic areas (Brown 1960a). Most species live in moist habitats and forage underground; species in semiarid regions tend to remain deep in the ground (Brown 1960a).

Name Derivation: Apparently derived from *stigmata* (Greek), meaning "spot or dot," and *omma* (Greek), meaning "eye," probably referring to the tiny eyes of *Stigmatomma* workers.

Stigmatomma pallipes (Haldeman)— Common Saw-Toothed Ant (Plate 3)

Taxonomy and Similar Species: *Stigmatomma pallipes* varies in size and color in such a way as to give the impression that there is more than one species involved. Usually, larger, blackish brown specimens are found in moist habitats, and smaller, reddish brown specimens are found in dry habitats, as if there were two species with different habitat preferences. The smaller and lighter forms were first considered a subspecies before they were regarded as a full species, *Amblyopone subterranea* (Brown 1949a). These forms, however, are not distinguished by any known structural features, and one can put together a series of specimens that connects their extremes. These forms were eventually synonymized (Brown 1960a), but now and then I haul out a batch of specimens and check them again for structural differences. One of the problems with this and other species that vary in color is that sculptures of the integument and hairs show up better in darker individuals, increasing the apparent differences between light and dark individuals. The lingering question of whether there could be more than one species lumped under *pallipes* might someday be resolved by genetic analysis, if such analysis becomes as convenient and routine as its practitioners claim.

Distribution: Quebec, west to Wisconsin, south through Florida (Smith 1979), then west and south, possibly in a series of disjunct populations, into the Coastal Range of California (Ward 1988). In Florida, it is known throughout the state, but with many gaps that may reflect the cryptic and subterranean habits of the species in Florida.

Natural History: This species has been studied by several able myrmecologists, whose work has been summarized and augmented by Traniello (1982). The diet consists of fairly large, elongate arthropods, especially centipedes, but also beetle larvae (there are no records of millipedes as prey). The prey is seized, stung, and carried back to the nest. Workers may cooperate in subduing prey, but only one carries it back. Back at the nest, the prey is not dismembered, but the larvae are carried to it. Workers do not regurgitate food for larvae or for each other. The queen is said to squeeze larvae until a drop of clear liquid is regurgitated; this might be reexamined in the light of the non-lethal cannibalism seen in *S. oregonensis*. Workers and queens generally show few behavioral differences, except that mature queens do not forage. Workers do not change tasks as they mature. The genesis of colonies is not completely understood, but circumstantial evidence suggests that groups of 9–16 workers form small colonies with one or more queens, and these colonies, in turn, fission after the emergence of alates and workers in late summer. Within an area, workers from apparently isolated colonies are compatible. Alates leave the colony to mate, but usually return to the nest after mating. Alate females probably occasionally disperse and find new colonies removed from the parental population. In Florida, I often collect

males in Malaise traps, but I have only captured a few winged females.

In the northern part of its range, the habitat of *pallipes* is usually moist woodland with a heavy canopy. It is sometimes difficult to believe that the same species could be in a cool, Appalachian rhododendron forest, and also in the dry scrub forests of Florida, but it is not so unusual for a species of ant to occur in different habitats in different parts of its range.

Name Derivation: From *palleo* (Latin), meaning "to be pale," and *pes* (Latin), meaning "foot." There is little contrast between the color of the feet and the body in Florida specimens, but some dark, northern individuals are more obviously pale-footed.

Genus *Prionopelta*

Minute Three-Toothed Hunter Ants

The genus *Prionopelta* (pronounced **pry' on ō pel" tuh**) includes 13 species widely distributed over the tropics, including Central and South America, Africa, Madagascar, New Guinea, and New Caledonia (Bolton 1995). These are small yellowish ants that probably never emerge in the open. Sifted from leaf litter, they could easily be mistaken in the field for one of the yellow, subterranean *Solenopsis*, but they move about with a slightly more undulating or squirming gait. Under the microscope, members of this genus display elongated, three-toothed jaws, gradually clubbed antennae, and a broad attachment between the petiole and the gaster. This last feature is shared with the genus *Stigmatomma*, and the two genera are now grouped together in the subfamily Amblyoponinae (Bolton 2003). There is one species in Florida.

Name Derivation: From *prion* (Greek), meaning "a saw," and *pelta* (Greek), meaning "a small shield." This refers to the expanded, serrate clypeus, seen in frontal view.

Prionopelta antillana Forel—Caribbean Minute Three-Toothed Hunter Ant (Plate 3)

Taxonomy and Similar Species: *Prionopelta antillana* is an unmistakable species, thanks to its unusual mandibles, tiny eyes set far back on the head, and broadly attached petiole. It appears to be closely related to the mainland tropical species *Prionopelta amabilis*, and Brown (1960a) has suggested the two forms might be one variable species.

Distribution: *Prionopelta antillana* occurs in the Lesser Antilles and in northern South America (Brown 1960a). In Florida, it is known from a few sites in Marion County and one site in adjacent Sumter County. It has not been found in south Florida, where one might think that the climate might be more suitable. This species is small and cryptic, but it should have appeared in some of the thousands of litter samples that were collected in south Florida in the course of the Florida ant survey. My impression is that this species was imported directly into Marion County, and has spread slowly ever since. Eventually, it should occupy southern Florida. Although Marion County is not a logical point of entry for tropical ants, the city of Silver Springs has long promoted an image of tropical lushness. There is a botanical garden, which might be a possible source of exotic ants, and movies supposedly set in the tropics used the clear springs for photogenic and relatively safe and convenient locales. The most famous of these were the Johnny Weissmuller Tarzan movies shot in Silver Springs between 1932 and 1942, in which aquatic adventures in Silver Springs are jarringly intermixed with old footage from Africa and elsewhere. Perhaps tropical plants in containers were brought in for some scenes.

Natural History: Collected from soil and leaf litter in mesic hammocks and in sand pine scrub. A 2004 expedition to hunt for this species near the Big Scrub Campground in Ocala National Forest produced hundreds of workers from a dense population found under low, evergreen scrub oaks. No males or queens were seen. Colonies found on the island of Dominica in 2006 were in a very different situation: the chambers of the nest were in lumps of clay near a stream in a dense tropical forest. The biology of *antillana* has not been investigated, but the similar species *amabilis* has been studied by Hölldobler and Wilson (1986a). Colonies are in rotten wood, and a colony may be dispersed into more than one piece of wood. In feeding tests, the favored prey was Diplura of the family Campodeidae. Workers use fragments of old pupal cocoons to "wallpaper" the pupal galleries, probably to reduce moisture. Workers produce edible eggs and present them to the queen. Foraging workers show a "foot-dragging" behavior, and in a later paper, Hölldobler et al. (1992) show that a trail pheromone is exuded from a previously unknown gland in the hind basitarsi. The male is known from collections, but apparently has not been described.

Name Derivation: From the Antilles; the type is from St. Vincent in the Lesser Antilles.

SUBFAMILY PROCERATIINAE

Genus *Discothyrea*

Pygmy Egg-Eating Ants

The species of *Discothyrea* (pronounced **dis' kō thir" ē uh**) are tiny roly-poly ants with an endearing infantile appearance. They may be recognized by their relatively huge antennal club and enlarged second gastral segment, with the terminal gastral segments tucked underneath. There are 27 known species (Bolton 1995), distributed from North America to Brazil, and from Africa and Taiwan to Australia and New Zealand (Brown 1958a). There is one species in the United States. Colonies are small (10–20 workers) and usually found in rotten wood, in leaf litter, or under stones (Brown 1958a). Like the related genus *Proceratium*, *Discothyrea* species seem to feed primarily on arthropod eggs (Brown 1958b, 1979). Queens of one African species even found their nests within spider egg sacs, which provide both food and lodging for the first generation of workers (Dejean and Dejean 1998).

Name Derivation: From *diskos* (Greek), meaning "a flat circular plate," and *thyreos* (Greek), meaning "shield." This refers to the peculiar plate (fused frontal lobes) that sticks out from between the antennae; this plate is somewhat rectangular in most specimens, but is described as semicircular in the type specimen.

Discothyrea testacea Roger— Southern Pygmy Egg-Eating Ant (Plate 4)

Taxonomy and Similar Species: There are no similar species in the United States. There is some variation in the shape and development of the plate between the antennae. The male is distinguished by its small size (under 2 mm), projecting clypeus, and characteristic wing venation. The latter has a small stigma and a large, complete, apically rounded marginal cell. This combination of features is not known for other Florida ants, so it should be possible to recognize the species from males taken in flight traps. The tiny size of the males, however, will still demand a sharp-eyed myrmecologist going through the trap sample.

Distribution: North Carolina through Florida, west into Oklahoma. Through most of its range, it is known from widely separated sites, probably because these ants are so small and difficult to find. The species was named in 1863 (Roger 1863) on the basis of a worker and queen found somewhere in "North America," and apparently not seen again until 1948 (Smith and Wing 1954). This helps explain the absence of D. testacea from Creighton's 1950 manual of ants. In Florida, persistent Berlese sampling at any upland or mesic site eventually seems to produce specimens.

Natural History: In Florida, this species occurs in both xeric and mesic habitats, including oak scrub, high pine, flatwoods, and temperate hardwood forest. One specimen was found in a tropical hammock in the Florida Keys. The diet is probably the eggs of other arthropods, especially spiders, but this awaits confirmation. A male, apparently the only one known, was found in a small colony in a hollow acorn buried in the leaf litter in Florida scrub habitat.

Name Derivation: From testaceus (Latin), meaning "having a shell," probably referring to the first two gastral tergites, which conceal and protect the terminal abdominal segments.

Genus Proceratium

Egg-Eating Ants

Workers and queens of Proceratium (pronounced **prō ser ā" shum**) have the end of the gaster tucked under the first two segments, like that of Discothyrea, but lack the enormously enlarged antennal club. The genus, which is widespread in temperate and tropical areas, includes 78 species (some are fossil species), 45 of which were added in an exhaustive revision of the genus (Baroni Urbani and De Andrade 2003). Florida has six species, the largest number known for any state, although adjacent states probably have the same six species. This does not mean one can waltz down to Florida for one-stop shopping for Proceratium, as several species are extremely difficult to find, and none are a dime a dozen.

Almost all species of Proceratium spend their lives in concealment, usually in rotten wood or deep humus, but sometimes under stones in colder climates. Captive colonies may accept, usually reluctantly, various types of insect prey, but the main diet seems to be the eggs of arthropods, especially spiders (Brown 1958b, 1979). The peculiarly modified gaster seems to have a defensive function. The hypertrophied second gastral tergite conceals the terminal gastral segments, it presents a hardened surface to enemies approaching from the rear, and it is equipped with a gland that probably produces defensive chemicals (Baroni Urbani and De Andrade 2003). Although unusual in ants, some insects in other groups have convergently evolved similar adaptations to provide rigid armor for the abdomen while maintaining flexibility for the terminal abdominal segments. Parasitic wasps of the genera Chelonus and Ascogaster (Braconidae) are good examples. The anteriorly directed terminal segments of the gaster have also been seen helping to manipulate and transport spider eggs (Brown 1979), but one would not claim that the abdominal modifications are a necessary adaptation for this purpose, as ants in general have no difficulty dealing with small round objects. Proceratium colonies are small, usually with 10–50 workers (Brown 1958a). In the North Temperate Zone, nuptial flights are usually in late summer (Brown 1958a).

There is often considerable variation in size, sculpture, and color within a species of Proceratium, and it is easy to look at a series of one of these species and imagine that there are additional undescribed species. With large series, these forms can usually be seen to intergrade. The revision by Baroni Urbani and De Andrade (2003) seems to have cleared up these problems in the Nearctic fauna.

Male Proceratium can be collected in Malaise traps or other flight traps, which may reveal the presence of a species whose colonies are difficult to find. At the Archbold Biological Station (Highlands County), pergandei is known only from males, although the ants of this site have been studied for more than 20 years.

Name Derivation: From the prefix pro- (Latin), meaning "in front," and keration (Greek), meaning "a small, horn-like projection," possibly referring to the projecting frontal carinae in the type species, silaceum.

Proceratium chickasaw De Andrade— Chickasaw Egg-Eating Ant (Plate 5)

Taxonomy and Similar Species: This species is similar to the commoner and more widespread pergandei, but the rearward extension of the second gastral segment is longer and narrower, in lateral view as long as the first tergite; the second gastral segment is covered with slender hairs that stand up on the surface and bend back at the ends, rather than being covered with uniformly decumbent hairs as in *pergandei*.

Distribution: Known from northwest Florida, west into Mississippi, and north into Tennessee (Baroni Urbani and De Andrade 2003). There are only few Florida collections, all from Liberty County.

Natural History: Florida specimens were collected from leaf litter in deep ravines along the Apalachicola River. These ravines are refuges for certain plants and animals of the southern Appalachian Mountains. It appears that no colonies have been found, which suggests that nests may be in the soil, rather than in rotten wood, where they would be more easily found. Males and queens are unknown.

Name Derivation: Named for the Chickasaw Indians of southeastern North America.

Proceratium crassicorne Emery—
Emery's Egg-Eating Ant (Plate 6)

Taxonomy and Similar Species: *Proceratium crassicorne* is strikingly similar to *silaceum*, and the two species were synonymized by Creighton (1950) in a detailed and somewhat caustic account, not really softened by phrases such as, "I have no wish to seem unduly harsh...." In the recent revision of the genus (Baroni Urbani and De Andrade 2003), *crassicorne* is reinstated on the basis of a series of small, but consistent morphological differences, which do not occur together in nest series, and on the fact that the two forms often occur in the same area, or even at the same site. The two species are most easily distinguished, in my experience, by a difference in the density of the hairs on the gastral tergites, especially the second. In *crassicorne*, the small punctures from which the hairs emerge are close together, only one or two puncture-widths apart, when seen under a diffuse light, such as that of a fluorescent bulb. The second tergite of *silaceum* is sparsely punctate, with most punctures separated by several puncture-widths. When Creighton synonymized the two species, he included a commentary on the illusory nature of supposed differences in the shape of the petiole. Despite this, there really does seem to be a small but consistent difference in the shape of the petiole, with *crassicorne* less squared off above than *silaceum*, and the petiole of *crassicorne* less shining at the base of the petiolar scale, the basal carina of the petiolar scale less developed, and the petiole more pubescent, especially the anterior face near the base. The frontal carinae of *crassicorne* project more above the surface of the head and are less expanded than those of *silaceum*. Several series of specimens of *silaceum* and *crassicorne* that I have seen from Arkansas suggest that in the western part of their ranges, the two species diverge in size, with *silaceum* larger than *crassicorne*.

Distribution: New York into north Florida, west into Missouri and Arkansas. In Florida, it is known from only two sites, one in Liberty County and one in Santa Rosa County.

Natural History: The habitat of *crassicorne* is hardwood forest, including ravines of the Florida Panhandle. It is not known whether there are any habitat differences between *crassicorne* and *silaceum*. The diet is probably arthropod eggs, especially those of spiders, but it is not clear which of the dietary records for *silaceum* (Brown 1958a,b) apply to that species and which apply to *crassicorne*. It would be necessary to look at voucher specimens. Males and queens are described and illustrated by Baroni Urbani and De Andrade (2003).

Name Derivation: From *crassus* (Latin), meaning "thick" or "stout," and *cornu* (Latin), meaning "horn." This refers to the thick antennal scapes in frontal view.

Proceratium creek De Andrade—
Pocket Gopher Egg-Eating Ant (Plate 5)

Taxonomy and Similar Species: This species has a low, rounded petiole, similar to that of *pergandei*, but the antennal scapes are relatively longer, exceeding the posterior border of the head when the scapes are laid back against the head. The known workers are noticeably larger than those of *pergandei*, but since only about a dozen workers are known, there could be more variation than is known to date. Any specimen that appears to be an unusually large *pergandei* could be checked to see if it is this rare species, but so far no specimens have shown up in casual ant collecting.

Distribution: Known only from Thomas County, Georgia; Marion County, Arkansas; and Lafayette County, Florida; a male that is believed to represent this species was collected in a Malaise trap in Leon County, Florida.

Natural History: This remarkable ant causes one to wonder what other rare ants might still be undiscovered in the soil beneath our feet. All but one of the specimens were collected in pitfall traps placed in the burrow systems of pocket gophers (Geomys pinetis) by Paul Skelley and Peter Kovarik in a study emphasizing the specialized scarabs that live with pocket gophers. Trapping in these burrow systems is not a simple affair: the gopher must be trapped and removed first, or it will fill the pitfall traps with sand. The trap is then set in the burrow system, and the excavation into the burrow is carefully covered with a board and soil (Skelley and Woodruff 1991). Although they can defend themselves with sharp teeth, pocket gophers are vegetarians and probably not a threat to insects that share their tunnels. It is not known whether creek is restricted to pocket gopher burrow systems, or whether they occasionally wander in there.

Name Derivation: Named for the Creek Indians of southeastern North America.

Proceratium croceum (Roger)— Angulate-Petiole Egg-Eating Ant (Plate 6)

Taxonomy and Similar Species: Proceratium croceum is similar to crassicorne and silaceum, but has a more squared-off petiole in lateral view and (in the worker) a slight concavity in the mesosomal profile when viewed from the side. It is also larger (worker is approximately 3 mm long) than Florida specimens of crassicorne and silaceum (workers are approximately 2.5 mm).

Distribution: Virginia into north Florida, west into Illinois, southwest into Oklahoma and Texas (Smith 1979). In Florida croceum is known from a small number of sites in the northern part of the state, from Levy County north and west.

Natural History: This species lives in rotten logs and stumps in shady areas. Colonies are sometimes found beneath the bark of decaying pine logs. Cole (1940) found a colony of approximately 30 workers in a small nest in firm wood near the middle of a rotten log; he points out how easy it would be to miss the nests of this species unless one carefully dissects entire logs. Van Pelt (1958) found several nests in rotten pine logs. Haskins (1930) fed a captive colony on larvae and pupae of other ants, but the croceum larvae died before becoming full grown. It is probable, considering the diet of other species (Brown 1958a,b), that the normal diet is arthropod eggs. Males and queens are described and illustrated by Baroni Urbani and De Andrade (2003).

Name Derivation: From krokos (Greek), meaning "crocus," a spring flower well known for its orange anthers used in cooking, referring to the orange color of this ant.

Proceratium pergandei (Emery)— Pergande's Egg-Eating Ant (Plate 5)

Taxonomy and Similar Species: Proceratium pergandei belongs to a trio of Florida species (including pergandei, chickasaw, and creek) that have a low rounded petiole. Originally, species with this character state were placed in the genus Sysphingta, also spelled Sysphincta, which was synonymized with Proceratium (Brown 1958a). The antennal scapes of pergandei do not reach the posterior border of the head when laid back against the head (unlike creek), and the second tergite does not protrude rearward for a distance as great as the length of the first gastral tergite (unlike chickasaw).

Distribution: Massachusetts south into Florida, west into Colorado and eastern Texas (Baroni Urbani and De Andrade 2003). In Florida, it is known from Highlands County (males only) northward and west through the Panhandle.

Natural History: Nests have been found under stones (Wesson and Wesson 1940) and in dead wood (Brown 1979). In the sandy soils of Florida, it is probable that the nests are normally subterranean, as the species is known almost entirely from strays taken in leaf litter. A small colony (eight specimens) was found in a rotten pine stump below the surface of the ground. Paul Skelley collected several specimens from traps set up in burrow systems of pocket gophers (G. pinetis). Wesson and Wesson (1940) offered living and dead insects to a captive colony, but the only food accepted was the contents of the gasters of a few dead ants. Brown (1979) found a colony feeding on and storing spider eggs in a nest in rotten wood. The queen and male are described and illustrated by Baroni Urbani and De Andrade (2003).

Name Derivation: Named by Carlo Emery for Theodore Pergande (1840–1916), a German immigrant and Civil War veteran who provided biological observations and curatorial services for his ambitious, articulate, and exploitative boss, Charles Riley of the U.S. Department of Agriculture (Mallis 1971). Pergande is highly regarded among American myrmecologists for the quality of his specimens. He sent many species of ants to Emery and Forel for identification and description. William Wheeler regularly corresponded with Pergande to obtain specimens and get details on the biology of ants.

Proceratium silaceum Roger— Common Egg-Eating Ant (Plate 6)

Taxonomy and Similar Species: *Proceratium silaceum* is very similar to *crassicorne*; the distinctions are discussed under that species. To make matters more confusing, *silaceum* is variable in size, color, and surface sculpture. Color is probably related to time since emergence of the adult, with recently emerged adults a golden yellow color. Some other poneromorph ants also seem to adopt their mature color slowly, and, unlike young adult myrmicines and dolichoderines, these callows may even be found foraging away from the nest.

Distribution: Southern Ontario, south into Florida, west into Illinois and Oklahoma, and south into the pine forests of eastern Texas. Records from Mexico and Central America have referred to the species *mancum* (Baroni Urbani and De Andrade 2003). In Florida, it is known from scattered sites from Highlands County northward and west through the Panhandle. With enough effort, it could probably be found in most mixed pine and hardwood forested areas through the northern half of the state.

Natural History: The habitat of *silaceum* is pine forests or mixed pine and hardwood forests. Nests are usually in rotten wood, but outside of Florida, nests have also been found under stones (Brown 1958a). Workers can also be found in litter samples, although they are never abundant. One Florida colony was found under bark flakes at the base of a living pine tree, and another was found between layers of asphalt roofing material that had been illegally dumped in a pine flatwoods site. Some colonies have been found with arthropod eggs stored in chambers, and egg-storing behavior with spider eggs also occurs in the laboratory (Brown 1958a). When supplied with spider eggs, captive colonies ignored other prey, including the eggs, larvae, and pupae of other ants (Brown 1958a). When provided with spider eggs, "the reflexed gastric tip is used to tuck the slippery eggs forward toward the mandibles when the eggs are being carried by the ants (Brown 1958b). The male and queen of *silaceum* are described and illustrated by Baroni Urbani and De Andrade (2003).

Name Derivation: From *sil* (Latin), a kind of yellowish earth or yellow ochre, hence *silaceus*, or "ochre colored."

SUBFAMILY PONERINAE

Genus *Anochetus*

Lesser Snapping Ants

The genus *Anochetus* (usually pronounced **an ō kē" tus**) occurs throughout the tropics. It includes at least 97 species (Bolton 1995). This genus is closely related to *Odontomachus*, and members of the two genera are often similar in appearance. There can be no confusion between the genera in Florida, as we have only one species, *Anochetus mayri*, whose small size (approximately 4 mm) and double-pointed petiole immediately distinguish it from Florida *Odontomachus*, which are larger (9 mm or more) with a single petiolar spine.

Like *Odontomachus*, *Anochetus* has a snapping mechanism that can close the jaws with great force and speed, disabling small prey. Foraging is usually nocturnal, and prey is captured by ambush or slow stalking (Brown 1978). Nests are usually in hollow twigs in leaf litter, or in small cavities in soil, but there are some arboreal species (Brown 1978). *Anochetus kempfi*, a Caribbean species studied by Torres et al. (2000a), shows a number of interesting biological traits, including flightless queens, colonies that apparently reproduce by fission, and workers that attack males and some supernumerary queens. The eggs of this species are not left in chambers but are carried about by workers until they hatch.

Name Derivation: From the prefix *an-* (Greek), meaning "not" or "lacking," and *ochetos* (Greek), meaning "aqueduct" or "ditch," probably referring to the absence of a furrow or elongate depression found in some other ants, such as the furrows on the underside of the head of some species of *Odontomachus*.

Anochetus mayri Emery— Mayr's Lesser Snapping Ant (Plate 7)

Taxonomy and Similar Species: In the Neotropics, there appears to be a confusing array of species similar to *A. mayri*, some of which may be geographic variants of that species (Brown 1978). There is some evidence that males may provide useful characters for separating the species in this complex (Brown 1978). The Florida population is introduced, most probably from the West Indies (Deyrup et al. 2000), but could have come from some other region. Florida specimens appear identical to specimens I have seen from Puerto Rico and St. John (the type locality is St. Thomas). In Florida and elsewhere in the mainland United States, there is no danger of confusing *mayri* with any other species, unless additional *Anochetus* species are introduced.

Distribution: The geographic range of *mayri* will be uncertain until the taxonomy of the complex has been resolved. It clearly occurs in the Greater Antilles and nearby islands, and may range through southern Mexico into Bolivia as well (Brown 1978). In Florida, it was first discovered in 1987 in Miami-Dade County, and rediscovered in 2002 in Palm Beach County (Deyrup 2002). It seems likely that the species will expand its range in southern Florida, as Miami-Dade and Palm Beach counties have numerous nurseries for tropical and subtropical plants, as well as a population of mobile humans who take their landscape plants to new residences. It seems unlikely that this species could survive in colder parts of the state.

Natural History: The natural history of this species is largely unknown. Perhaps the presence of a population in Florida, a state well endowed with entomologists (although never too many), will lead to some research on the biology of the species. At the West Palm Beach site, numerous specimens, along with a variety of native and introduced ant species, were found by sifting deep litter beneath mature slash pines and oaks (Deyrup 2002). I have not seen males from the Florida population.

Name Derivation: Named in honor of Gustav Mayr (1830–1908), an enormously productive Austrian myrmecologist who described hundreds of species of ants, including many from North America. Gustav Mayr is also responsible for establishing approximately 48 generic ant names still in use today. *Anochetus* is one of these names.

Genus *Cryptopone*

Pit-Jawed Ants

The genus *Cryptopone* (in my experience usually pronounced **Krypt" ō pōne**, but also pronounced **Krypt' ō pō" nē**) is a small, but widely distributed, genus of approximately 15 species. These ants are similar in structure to some species of *Ponera* and *Hypoponera*, but have a large shallow pit at the bases of the jaws. This pit (whose function is unknown) also occurs in some other genera of ponerine ants. The legs are short, and the middle tibiae have strong bristles on their outer faces. These bristles are probably used for digging and are apparently associated with unusual anterior mobility of the middle legs, which are often rotated forward in

preserved specimens. Similar bristles and spines occur on the middle tibiae of many fossorial Bethylidae, Tiphiidae, and even sand-swimming bees of the genus *Perdita*. One would expect such bristles to be more widespread in ants. It seems possible that while most digging ants employ their jaws to excavate passageways, *Cryptopone* species use their legs as well to force their way through rotten wood in search of prey. This is speculation, not based on observations of behavior.

Name Derivation: From *kryptos* (Greek), meaning "hidden," and *pone*, in this case a contraction of "*Ponera*." This genus was erected by Carlo Emery, who, as usual, provided no derivation for the name. His main concern seems to have been to distinguish *Cryptopone* from *Ponera*, and the "*crypto*" probably does not mean "hidden," but rather refers to the "crypt," sometimes used anatomically to mean a "pit" or "recess," at the bases of the jaws.

Cryptopone gilva (Roger)—Southern Pit-Jawed Ant (Plate 9)

Taxonomy and Similar Species: *Cryptopone gilva* is distinguished from somewhat similar species of *Ponera*, *Hypoponera*, and *Pachycondyla* by the stout bristles on the outer face of its middle tibiae and the pit at the base of the jaws; both these features are difficult to see without diffuse illumination, such as that from a fluorescent bulb. In the field, this species can, with practice, be identified by a combination of color (reddish brown) and size (bigger than *Ponera* and *Hypoponera*, smaller than *Pachycondyla*). This species appears under the name *Euponera gilva* in Creighton (1950).

Distribution: North Carolina (Carter 1962) south into Florida, west into Texas (Smith 1979), and south into Costa Rica. Presumably, there are populations somewhere between Texas and Costa Rica. In Florida, this species occurs from the northern border south into Highlands County.

Natural History: *Cryptopone gilva* lives in rotten logs in mesic to wet woodlands (Haskins 1931; Smith 1934). Haskins, in his study of the biology of *C. gilva* (1931), found that the brood is kept in large chambers, the eggs, larvae, and pupae usually separated, with some additional sorting of larvae by size. Larvae are active and "athletic," responding to food before it has been presented by workers. Larvae feed on bits of arthropods; adults (in the laboratory) also consume sweets. Workers cover mature larvae with soil, which is removed after the cocoon has been constructed. Larvae are not covered with soil form naked pupae, which are usually destroyed by the workers. Workers help young adults to emerge from their cocoons and appear to obtain some exudate from the newly emerged callows. The larval period (in the laboratory) is about three to four weeks, and the pupal period is about a month. Mature queens and males attempt to leave the nest within two weeks of emerging from their cocoons, so it is possible that there are no strongly synchronized mating flights. Colonies collected in the field may have several dealate queens, and Haskins believed that dispersal in this species might be by both fission of large colonies and solitary nest founding by young queens. Haskins also performed a series of interesting experiments testing the sensitivity of this species to various stimuli, but it is difficult for me to relate these to the field natural history of the species.

Name Derivation: From *gilva* (Latin), meaning "pale yellow." This species is normally orange brown, and the type specimens may have been callow or discolored.

Genus *Gnamptogenys*

Grooved Ants

The combination of the conspicuous grooves on the head and body, and the one-segmented petiole separates this genus from all other Florida genera. *Gnamptogenys* (usually pronounced **Namp toj" en ēs**, but sometimes with the accent on the third syllable) is a genus of 99 species that ranges through most of the tropics, with the notable exception of Africa (Bolton 1995). As is often the case among the ponerine ants, some species seem to be specialized predators. A single introduced species occurs in Florida.

Name Derivation: From *gnamptos* (Greek), meaning "curved," and *genys* (Greek), meaning "jaw" or "cheek," referring to the strongly curved jaws, seen in lateral view.

Gnamptogenys triangularis (Mayr)— Spine-Thighed Grooved Ant (Plate 4)

Taxonomy and Similar Species: Longitudinal grooves, covering the head and body, including the gaster, distinguish this species from other Florida ants. There is a sharp, curved spine on the hind coxa. The name *Gnamptogenys aculeaticoxae* has been applied to the Florida population, but this name has been synonymized with *G. triangularis* (Lattke 1995).

Distribution: Native distribution is from Panama into Bolivia (Brown 1958a). In the United States, this species is known from several sites in Florida and Alabama. In Florida, it is known from only a few sites, and never seems abundant. At this point, it is one of Florida's rarer exotics.

Natural History: This remarkable species has been studied in Venezuela by Lattke (1990). He found nests, containing approximately 80–120 workers, in fallen logs and branches. In four nests that he examined, the only prey remains were those of millipedes, whose disarticulated segmental rings surrounded the area near the nest entrance. Larvae were found with their heads thrust into the body of a dead millipede. Many millipedes secrete powerful defensive secretions, including cyanide (Eisner et al. 2005). Lattke found that both adults and larvae of *triangularis* are resistant to cyanide, surviving for more than three hours in a potassium cyanide killing jar that dispatched other ants in less than five minutes. The male has been described under the name *aculeaticoxae* (Santschi 1921).

Name Derivation: From *triangulus* (Latin), triangular; perhaps this refers to the outline of the mandibles in frontal view, or the frontal view of the head.

Genus *Hypoponera*

Mini-Ponerine Ants

The genus *Hypoponera* (usually pronounced **Hī pō pō" ne ruh**) occurs in tropical and warm temperate regions. There are 132 described species (Bolton 1995), but some of these are probably synonyms, and there must be a batch of undescribed species. Almost all species have cryptic habits, and there are probably subterranean species that have never been seen. In the mainland tropics, there are numerous well-marked species, but it is difficult to put names on specimens because the genus has never been reviewed. In a recent version of his website on the ants of Costa Rica, Longino lists nine undescribed species from that country alone. Apparently, there is a revision of New World species about to begin. There are only a few species in Florida, but there is a good chance that additional species will be imported. There are, for example, two species found in disturbed areas in the Bahamas that would undoubtedly thrive in the Florida Keys. One of these, tentatively identified as *Hypoponera parva*, is so small and cryptic that it could be established for a long time in some restricted locale before anybody noticed. While there are many exotic *Hypoponera* that could probably live in Florida, there is no reason to believe that there are native species that remain undetected. Even subterranean species such as *inexorata* produce alates that are caught in flight traps, and I know of no unassigned alates.

Species of *Hypoponera* look similar to *Ponera*, but lack a translucent "porthole" found on the ventral part of the petiole of *Ponera*. This structure is inconveniently located, as it is normally hidden by the middle coxae. It is usually necessary to use a fine pin to raise the gaster slightly and reposition the hind legs; since the articulations are ball-and-socket joints, it is usually possible to make these adjustments to dry specimens without causing damage. Eventually, one becomes accustomed to more subtle and convenient differences between Florida species of *Hypoponera* and *Ponera*. The *Hypoponera* species have relatively longer legs, higher petioles, and less heavily clubbed antennae than the *Ponera*. Some large tropical species of *Hypoponera* are similar in general appearance to small *Pachycondyla*, but this is not a Florida problem.

In contrast to *Ponera*, queens of which seldom (if ever) appear in flight traps, queens of Florida *Hypoponera* fly readily. There are also flightless queens in the Florida species *opaciceps*, *opacior*, and *punctatissima*. The introduced species *punctatissima* has peculiar, large-headed ergatoid males that can occur in two sizes. Flightless males may not be unusual in the genus: two Japanese species each have two forms of males and two forms of queens, with various behavioral permutations (Yamauchi et al. 1996, 2001).

Another interesting feature in certain Japanese species is trophallaxis (passing of liquid food from one adult to another), not otherwise known in the ponerines (Hashimoto et al. 1995). Larval *Hypoponera* resemble *Ponera* in having door knob–shaped tubercles that stick the larvae to the walls of nest chambers; the morphology and function of these have been studied by Peters and Hölldobler (1992).

My reading of the literature on *Hypoponera* gives me the impression that members of this genus are considered general predators feeding on small subterranean organisms. I have not come across any specifics on how any species actually makes its living. The microhabitats favored by three common Florida species (*opaciceps*, *opacior*, and *punctatissima*) are consistent with a somewhat specialized diet on small fly larvae. This is just a guess.

Name Derivation: From the prefix *hypo-* (Greek), meaning in this case "less than," and *ponera* (Greek), referring to the genus *Ponera* and roughly translating as "little *Ponera*." The word *ponera* (Greek) means "toilsome," but might be indirectly derived from *ponikos* (Greek), meaning "hardworking," referring to the general reputation of ants.

Hypoponera inexorata (Wheeler)— Orange Mini-Ponerine (Plate 10)

Taxonomy and Similar Species: This species is distinguished by the combination of orange brown coloration and the shape of its petiole that tapers from top to bottom in lateral view. *Hypoponera punctatissima* may be orange brown, but has a lower and less tapering petiole; it is also conspicuously smaller when the two species are compared side by side. The mandibles of *inexorata* in frontal view are less convex along their outer edges than those of other Florida species, and are often described as "sinuate." This feature is also best appreciated when there are specimens of other species for comparison, at least when one is dealing with Florida specimens. Wheeler's drawing of one of the type series from Texas has curves that would probably meet the approval of Mae West. Male *inexorata* in Florida are large (more than 3 mm long), with well-pigmented wing veins and round eyes.

Distribution: South Carolina through peninsular Florida, west into Arizona, and south into Central America (Smith 1979). In Florida, *inexorata* is known from scattered localities around the state, reflecting its subterranean habits and consequent difficulty in collecting specimens. It is probably completely absent from extensive lowland areas.

Natural History: In Florida, the habitat of *inexorata* is dry coastal hammocks, Florida scrub, and sandhill. Specimens are found under objects on the sand, or occasionally extracted in ones and twos from leaf litter. In other parts of its range, it can be found under rocks, which provide a window into subterranean habitats not found in Florida. I have never seen an actual colony, and there is no evidence that this species has ergatoid males or flightless queens. I once observed a mating flight in scrub habitat on a June morning.

Name Derivation: From *in* (Latin), meaning "in," *ex* (Latin), meaning "out," and *ora* (Latin), meaning "edge," meaning "in and out edge," referring to the sinuate outer edges of the mandibles in frontal view.

Hypoponera opaciceps (Mayr)— Granulate Mini-Ponerine (Plate 10)

Taxonomy and Similar Species: The nearly parallel front and rear faces of the worker petiole (in lateral view) and nonshining integument (thanks to minute, dense granulation) distinguish this species from other Florida *Hypoponera*. Wingless individuals with large eyes, probably ergatoid queens, occasionally occur in this species. There are six subspecies (Bolton 1995), most of which are probably synonyms or separate species. Males have pale, almost transparent wing veins and slightly elongate eyes.

Distribution: South Carolina through Florida, west to Colorado; Central and South America, West Indies, Southeast Asia, and Polynesia (Smith 1979). Reported from Nevada by Wheeler and Wheeler (1986) and from Arizona by Mackay and Mackay (2002). It was probably transported by commerce from the Neotropics to Pacific tropical areas (Wilson and Taylor 1967). It undoubtedly occurs in southern California and New Mexico in moist disturbed areas such as golf courses, edges of ditches, and low spots in lawns. This species is sometimes found in soil of potted plants in Florida and is presumably dispersed widely by the nursery trade. At this time, there is no good reason to believe that *opaciceps* is not native to Florida. It is common in wracks above high tide lines along Florida beaches, and it is easy to imagine that this species could have arrived by island-hopping through the Caribbean or by following the coast around the Gulf of Mexico. In Florida, *opaciceps* occurs throughout the state.

Natural History: In Florida, *opaciceps* forms small colonies in open or partially shaded areas. It is most common in wet or periodically flooded areas, where it tends to replace *opacior* in fallen logs, grass tussocks, and accumulations of organic matter (Van Pelt 1958). In wet pastures in Highlands County, it may coexist with high populations of *Solenopsis invicta*. It often occurs under piles of seaweed on coastal beaches and under heaps of detritus left by high water along lakes. It is not clear how *opaciceps* is able to burrow through mud without becoming dirty, but its dense covering of fine pubescence might have something to do with this trick, as in mole crickets (Gryllotalpidae) and the unrelated pygmy mole crickets (Tridactylidae).

It is a rapid colonizer of disturbed areas such as lawns and ditches, and is even able to move into areas that have recently been subjected to volcanic eruptions in Hawaii (Fellers and Fellers 1982). In southern Florida, alates may be found in flight in afternoon and evening any time of year. Alate queens can deliver a noticeable sting when they become trapped under clothing or stuck to the skin by sweat. Whatever water-repellent qualities this species may have, they are evidently not proof against human perspiration.

Name Derivation: From *opacus* (Latin), meaning "dark," and *-ceps* (Latin), meaning "head," possibly referring to the dark coloration of the head, although the entire body is a similar shade, but more likely meaning "nonshining," referring to the conspicuously microgranulate surface of the head.

Hypoponera opacior (Forel)— Common Mini-Ponerine (Plate 10)

Taxonomy and Similar Species: Worker and queen *opacior* have the petiole in lateral view strongly narrowed above, like a narrow triangle, not more or less parallel sided as in *opaciceps*. This species is seldom orange like *inexorata*, and the outer borders of the jaws are smoothly convex, not slightly concave as in *inexorata*. This species is usually dark brown, but reddish brown, probably callow individuals can be found, and could be confused with *punctatissima*. The mesopleuron of *opacior* is slightly granulate, while that of *punctatissima* is smooth and shining. The petiole of *opacior* is more elevated than that of *punctatissima*, but this difference is best seen by comparing specimens of the two species. Queen *punctatissima* are dark brown, like those of *opacior*, but have a smooth and shining mesopleuron. Male *opacior* have light brown wing veins, unlike the transparent wing veins of *opaciceps*. Male *inexorata* have brown veins, but are at least 3 mm long, larger than male *opacior*. Wingless females with large eyes, probably ergatoid queens, occasionally occur in colonies of *opacior*.

There is some variation in size and color, as in many other ponerine ants, but I have not found any structural differences associated with this variation. I have looked at specimens from much of the range in the United States, and several Caribbean islands, and have a feeling that there may be two or more species included under the name *opacior*. There appear to be some small geographic differences in size and color, suggesting that this species has not recently moved into an enlarged range, but occurs as various separate lineages. If this is the case, it would not be surprising if some populations had diverged into separate species.

Distribution: Virginia south through Florida and the West Indies, west into Oregon, south through Central and South America into Chile and Argentina (Smith 1979), assuming it is one species throughout this huge range. This species is not, apparently, known from the Old World tropics and subtropics, and there is no reason to consider it a tramp species. In some areas, especially in arid climates, the distribution of opacior may be localized and discontinuous: in a survey of the ants of Nevada, Wheeler and Wheeler (1986) found opacior only once. In Florida, opacior can be found wherever there is cursory sampling of ants in leaf litter.

Natural History: This is the most abundant species of Hypoponera in Florida, occurring in a variety of mesic and upland habitats, both open and shaded, including fields, Florida scrub, sandhills, flatwoods, and hardwood hammocks. In wet areas, it is usually replaced by opaciceps and punctatissima. This species forms small colonies in soil, in rotten logs, and in accumulations of leaf litter. Nests are often in hollow nuts or small hollow twigs. In south Florida, I have found alates in flight traps from May through December. Alate queens are able to sting when trapped against the skin; I have been stung when running, the queens caught in sweat on my arms and neck.

Name Derivation: From opacior (Latin), meaning "darker," or "more opaque," probably relative to some other lighter or more shining species of Hypoponera or Ponera, such as punctatissima.

Hypoponera punctatissima (Roger)— Pantropical Mini-Ponerine (Plate 11)

Taxonomy and Similar Species: The shiny mesopleural area of punctatissima distinguishes it from reddish (callow) specimens of opacior. It is much smaller than inexorata and lacks the large punctures on the front of the head of inexorata. This species is somewhat similar to Ponera exotica but the latter species has conspicuous punctures on the head, short antennae with a relatively large antennal club, and a portholelike structure on the subpetiolar flange. According to Taylor (1968), H. punctatissima is a synonym of Ponera ergatandria, a name that appears in Creighton's manual (1950) and Smith's 1936 catalog.

Like many tramp species, punctatissima has accumulated a number of names, listed in Bolton (1995), probably because it has been collected in all sorts of out-of-the-way places, and because it is not a particularly distinctive species to begin with. It is possible, however, that myrmecologists have been flummoxed by another factor. Hypoponera punctatissima apparently only produces flightless males, which presumably have a strong tendency to mate with their sisters. This could, in theory, complicate taxonomy. If noticeable divergences arise in a lineage with sibling mating, it would be possible to have consistent sympatric forms that appear to be separate species. This is most likely to be seen in a tramp sib-mating species that has been imported two or more times from separate parts of its range. One might say that these forms have some of the characteristics of separate species, since they are sympatric and reproductively isolated. In such cases, entomologists tend to recognize distinct "species" of sib-mating or parthenogenetic lineages if they show distinctive morphological differences and if they also show ecological differences. Such "species by analogy" appear to be numerous in the Scolytinae, especially in the genera Hypothenemus and Xyleborus. In Hawaii, punctatissima has a light form found in dry areas and a dark form in wet areas. This suggests two functional species, but it is also possible that the color differences reflect direct environmental effects, which I think that I have seen in some other ants, such as Ponera pennsylvanica and Stigmatomma pallipes. Van Pelt (1958) also reports two forms of punctatissima, but one of these is probably Ponera exotica.

Distribution: Pantropical; in North America, colonies seem able to survive outdoors in Florida and southern Texas to California. It is of Old World origin (Delabie and Bland 2002). It is also able to survive in warm microhabitats around and under buildings and in warm, composting organic matter; this has allowed it to move into cold areas, including New England, the Upper Midwest, Scandinavia, Britain, and Iceland (Delabie and Bland 2002). This is a cryptic species whose wide range might not have been documented were it not for the large numbers of queens that emerge to disperse, stinging with relatively little provocation. In Florida, punctatissima is known from both the Peninsula and the Panhandle, but seems commonest in the south and central Peninsula.

Natural History: Hypoponera punctatissima occurs in small colonies in a variety of moist or well-drained habitats. In south Florida, it is often found in rotten logs and large grass tussocks, and especially in accumulations of organic matter, often open or disturbed areas. The key to the housefly-like success of *punctatissima* lies in its predilection for dung and various kinds of middens, which humans and their livestock supply in inexhaustible quantities. Delabie and Bland (2002) have compiled a whole series of breeding sites, ranging from earthworm cultures to heaps of chicken manure, and suggest that *punctatissima* emerged out of Asia following the domestication and extensive use of horses. While reluctant to challenge this attractive "Valley of the Horses" hypothesis, it seems to me that *punctatissima* only requires stables and dung heaps in the northern part of its range, and could have spread on its own through the Old World tropics and subtropics, although it would have undoubtedly benefited from the organic middens our species has generously provided throughout our history.

The tendency for *punctatissima* to produce disproportionately large numbers of dispersing females is appropriate for a species that colonizes resource-rich resources that are often ephemeral. This dispersal, in turn, leads to problems with stinging alate queens, first mentioned by Smith (1936), who reported on a Florida outbreak in which the ants in sugar cane fields were so numerous as to annoy field workers "like mosquitoes," also passing through screens to pester them by night. Many of the stings probably resulted from ants that got trapped in a film of sweat on the laborers' bodies. In May of 1991, the Archbold Biological Station was sent packets of alate *punctatissima* from pest control and agricultural extension agents. The complaints included the presence of "big, biting swarms" in the afternoon, with many stings on the back, neck, and arms; ants entering through screens at night; ants attacking and stinging racehorses (the horse connection again); heavy attacks in schoolyards and at afternoon sports events. There were no reports of serious reactions. My experience is that the sting, while somewhat painful, does not linger like that of *Wasmannia auropunctata* or cause a pustule like that of *Solenopsis invicta*.

Colonies of *punctatissima* can include normal workers, alate or dealate queens, intercastes without wings but with larger eyes than those of workers (these are probably flightless queens), and two types of flightless males. The range of morphs in *punctatissima* is like that of the rare Japanese species

Hypoponera bondroiti, studied by Yamauchi et al. (1996), and it is possible that many of the complex reproductive behaviors found in *bondroiti* also occur in the common and easily available *punctatissima*. There is already an observation by Hamilton (1979) that male *punctatissima* fight until one has possession of the nest chamber where females emerge from their cocoons.

Name Derivation: From "*punctatissima*" (Latin), meaning "extremely punctured," perhaps referring to the punctures on the head and body that are so numerous and tiny as to be almost invisible.

Genus *Leptogenys*

Beaked Hunter Ants

Members of the genus *Leptogenys* (pronounced **lep tō jen" ēz** or **lep toj" en ēz**) are long-legged, fast-moving ants. They usually lack teeth on the mandibles and the clypeus usually projects forward in a point, so that in frontal view it looks like the beak of a bird. Another unusual feature of the genus is the comb-like tarsal claws; the function of this apparent adaptation is unknown. Queens of most species are flightless and tend to look like workers. Approximately 236 species have been described (Bolton 1995; Lattke 2011), making *Leptogenys* one of the more speciose genera of ponerines. The genus occurs worldwide in the tropics, including species confined to isolated islands such as New Caledonia and Madagascar that have not been accessible for a very long time. There are also a few species whose distribution indicates that they have been transported by human commerce (Bolton 1975). One of these, *Leptogenys maxillosa*, apparently originating in Africa, has been found in Cuba (Bolton 1975) and might well survive in Florida, if it could get there. Southern Texas and northern Florida appear to be the northern limits of *Leptogenys* in the New World.

Neotropical species of *Leptogenys*, including 81 known species, tend to live in small colonies with only 20–30 workers (Lattke 2011). A few Old World species, such as *Leptogenys distinguenda*, have converged almost completely on an army ant way of life, with tens of thousands of workers (Gotwald 1995), but there are no known New World species with such habits (Lattke 2011). With the exception of the army ant–like species, *Leptogenys* seem to be specialized predators, usually feeding on terrestrial isopods (Lattke 2011).

Trager and Johnson (1988) suggest that the beak-like clypeus helps secure slippery, convex isopods as they are being carried back to the nest. One Old World species is a specialized predator of fungus-growing termites (Maschwitz and Schönegg 1983) and another specializes on earwigs (Steghaus-Kovak and Maschwitz 1993); it is always possible that species may be found with other strange prey preferences, as the habits of most Leptogenys species are not known.

Leptogenys species of the New World have recently been revised by Lattke (2011).

Name Derivation: From leptos (Greek), meaning "slender," and genys (Greek), meaning "jaw," referring to the long, slender mandibles.

Leptogenys manni Wheeler— Eastern Isopod Ant (Plate 12)

Taxonomy and Similar Species: Leptogenys manni looks slightly like similar-sized species of Aphaenogaster, but is even more slender, with an elongate gaster. The beak-like clypeus and elongate jaws are diagnostic. There is a closely related similar species west of the Mississippi, Leptogenys elongata. The distinctive males, which are orange with thick dark antennae and brown wings, can easily be picked out of light trap and Malaise trap samples; these help establish the range of the species.

Distribution: Known only from Florida, but undoubtedly occurs in southern Georgia, as it occurs at Tall Timbers Research Station (Leon County) just a mile or so from the Georgia border. There are scattered records through peninsular Florida, but as yet no Panhandle records west of the Ochlockonee and Apalachicola Rivers. This distribution suggests that the species was eliminated from much of its eastern range during cool periods of the Pleistocene, spreading back from southern Florida relatively recently but limited in its western expansion by north–south rivers.

Natural History: Leptogenys manni occurs on the Archbold Biological Station, as males have been collected at flight traps and at lights, and workers have occasionally been found in bowl traps filled with water and a little detergent. There are no observations, however, of this species out foraging, and it is probable that colonies are nocturnally active. This is a reasonable assumption, as isopods, the normal prey of L. manni, are also nocturnal (Hornung 2011). Elsewhere in Florida, L. manni is usually found in undisturbed mesic woodlands where limestone is at or near the surface (Trager and Johnson 1988). Terrestrial isopods have a relatively high requirement for calcium as the exoskeleton is calcified with calcite crystals and calcium carbonate (Hornung 2011). Nests of L. manni are often marked by a midden of isopod shells; examination of these middens suggests that manni prefers the native isopod Porcellionides virgatus rather than introduced species of Armadillidium (Trager and Johnson 1988). Colony reproduction is almost certainly some form of fission of mature colonies, but the process, not surprisingly, has not been observed. In the case of a species such as manni, there is no obvious reason why a new colony must begin with a large number of workers, so it is possible that a new colony is founded by a queen with a retinue of just a few workers. Males fly May through October (Trager and Johnson 1988). Continuously operating flight traps at the Archbold Biological Station captured males July–December 1983–1986, with the great majority (69 out of 77) in September–November, 40 in October alone.

Name Derivation: William Wheeler named this species for his former graduate student William Mann, who sent Wheeler the type specimens from Florida (Wheeler 1923). Wheeler and his graduate student did not operate on quite the same wavelength. Wheeler was a brilliant and versatile academician who enjoyed a well-organized field expedition. Mann, by his own admission, was forever yearning for adventurous collecting trips to exotic places. After several such trips, Wheeler insisted that Mann settle down and write his dissertation. Around that time, Mann's world was "completely demolished" by overhearing Wheeler turn down an offer to have Mann accompany an expedition to Borneo. He recovered from this shock in time to finish his doctorate and head off on an extended trip to the South Seas to collect ants and other animals, as related in his book, Ant Hill Odyssey (1948). After a series of productive ant-hunting safaris, eventually Mann displayed his own brand of versatility, becoming director of the National Zoo in Washington, DC.

Genus Odontomachus

Snapping Ants

The genus Odontomachus (usually pronounced **Ō' don tō mak" us**, but sometimes **Ō' don tōm"** a kus) occurs around the world in the tropics, including some remote islands in Melanesia. These are large stinging ants that are easily agitated, and would not seem preadapted as stowaways, but there is good evidence that two species are exotic in Florida, so the powers of independent dispersal may not be as great in some species as myrmecologists have assumed. There are approximately 60 described species (Bolton 1995), reduced from approximately 160 names. *Odontomachus* species are conspicuous ants, often with some intraspecific variation. In the early days of myrmecology, this resulted in a chaotic horde of specific taxa that were sorted and disciplined by Brown in his review of the genus (1976). Four species are known from Florida. North American species were reviewed by MacGown et al. (2014).

Members of this genus are distinguished from most other Florida ants by their elongate, apically bent jaws, and the absence of a conspicuous constriction between the first and second segments of the gaster. These features also occur in *Anochetus*, but the Florida species of *Anochetus* is only 4 mm long, much smaller than any species of *Odontomachus*, and has two points on the petiolar apex when viewed in front or at the back.

Odontomachus species are famous for the speed of their mandibular snap, which requires only 0.33– 1.00 milliseconds, the fastest animal movement known (Gronenberg et al. 1993). This is achieved by using stored tension, as in, for example, a crossbow. The jaws are cocked open and locked into place, and then tension is put on the massive jaw-closing muscles that run back into the rear lobes of the head. A trigger device releases the jaws, which snap closed with sufficient force to stun or crush prey, or drive off predators (Carlin and Gladstein 1989). A drawing of the jaw-snapping mechanism is provided by Barth (1960). The ends of the jaws are thickened, but the apical teeth often show wear or breakage from snapping shut. Despite these formidable jaws and an effective sting, Florida representatives of *Odontomachus* are not especially bold, and often retreat for a few minutes after stunning a prey, perhaps to allow dissipation of defensive chemicals (Brown 1976; Ehmer and Hölldobler 1995). This even allows at least one species, *Odontomachus bauri* (not a Florida species), to kill and carry off nasute termites, which squirt a glue of noxious chemicals (Ehmer and Hölldobler 1995). Although all species of *Odontomachus* are probably predaceous, at least one

species, *Odontomachus chelifer*, consumes arils on the seeds of a tropical tree and may have a role in seed dispersal (Piso and Oliveira 1998). The Florida species *Odontomachus brunneus*, which often occurs in open wet areas, could conceivably be a disperser of wetland milkworts (*Polygala* spp.). Several species are attracted to sweets (Oliveira and Hölldobler 1989), and this attraction can be seen, in a mild form, in Florida species.

There are behavioral studies of several tropical species, and these might serve as models for studies of Florida species, two of which are easy to find, and would make good research subjects. Oliveira and Hölldobler (1989) found that the forest species *O. bauri* uses forest canopy patterns in orientation, and foragers are able to recruit nest mates to bait by raising the level of foraging activity, but not by laying a trail to the bait. Medeiros et al. (1992) studied dominance hierarchies in queens of the polygynous species *O. chelifer*. Brandão (1983) studied division of labor in the Brazilian species *Odontomachus affinis*.

Name Derivation: From *odontos* (Greek), meaning "tooth," and *machetes* (Greek), meaning "fighter" or "warrior," referring to the toothed jaws and defensive behavior of *Odontomachus* species.

Odontomachus brunneus (Patton)— Southeastern Snapping Ant (Plate 8)

Taxonomy and Similar Species: The long, dense, appressed pubescence on the gaster distinguishes *brunneus* from other Florida species. Groups of these flexible hairs often cling together, a feature not seen in other Florida species. The pubescence of *haematodus* is also relatively dense, but the hairs are not flexible, and stand up from the surface of the gaster. The male is brownish yellow, with huge, protruding ocelli, each as wide as the space between the eye and the lateral ocelli. The males of Florida *haematodus* that I have seen are also brownish yellow but do not have extraordinarily large, protruding ocelli. Within the Southeast, there is some variation in workers, but apparently based on habitat rather than geographic variation. Specimens from dry habitats are often smaller and paler than specimens from wet habitats. Southeastern records of *insularis* (Smith 1979) actually refer to *brunneus*; *insularis* is a valid West Indian species that is not known from Florida. In the Neotropics, the entire *haematodus* complex designated by Brown (1976), which includes all of the five or six species found in the United States, needs some review, preferably with the inclusion of worker-associated males. In *Odontomachus*, male morphology has, at least in some cases, diverged more conspicuously than worker morphology (MacGown et al. 2014). The recognition of *ruginodis* as a species separate from *brunneus* (Deyrup et al. 1985) and *relictus* from the southwestern *clarus* (Deyrup and Cover 2004b) depended heavily on obvious differences among the males of these species.

Distribution: Southern Georgia, throughout Florida, although rare in the western Panhandle. There are records from Alabama (MacGown and Forster 2005), but the published records from Baldwin County (Deyrup and Cover 2004b) are from specimens of *haematodus*, which I identified through the sides of vials without taking them out to examine them more carefully. I have not seen specimens from the West Indies. Records from Central and South America, south into Paraguay and Bolivia (Brown 1976), are uncertain, as they date from a time when *brunneus* was combined with *ruginodis*, an apparently widespread and adaptable species. Biogeographically, therefore, *brunneus* allows highly disparate hypotheses. In North America, it could be a disjunct, possibly introduced, population of a widespread Neotropical species, or it could be an ancient, possibly Arctotertiary, endemic species. My gut feeling is that *brunneus* is restricted to southeastern North America. The name *brunneus* was first applied to a population from southeastern North America (Patton 1894), so this name should remain stable for the species in North America.

Natural History: This species occurs in both well-drained and poorly drained habitats, and nests may be in soil or in rotten wood. Van Pelt (1958) studied *brunneus* at the Welaka State Forest in Putnam County. He found many colonies in various habitats, including swamp forest, flatwoods, mesic forest, upland scrub, and sandhill. Nests were in various microhabitats: deep leaf litter, fallen logs, at the bases of trees, and open or sparsely vegetated sandy areas. At the Archbold Biological Station, Highlands County, *brunneus* is in moist areas and mesic hammocks, but is replaced by *relictus*, or occasionally *ruginodis*, in dry upland habitats.

Workers sometimes emerge to forage on cloudy days, but are generally nocturnal. The powerful jaws are not used as assertively as one might expect, and there are preliminary indications that *brunneus* is sensitive to chemical defenses. Foragers approach prey tentatively, and recoil immediately after striking (Brown 1976). There may be a delay before the prey is carried away; Brown (1976) suggests that this delay is a reaction to possible chemical defenses, which are allowed to dissipate before the prey is retrieved. Alex Wild (unpublished student notes in the files of the Archbold Biological Station) twice observed *brunneus* retreating hastily when confronted by aroused workers of *Dorymyrmex bureni*, a small species that can release strong-smelling defensive compounds. Van Pelt (1958) reported accumulations of head capsules in several nests of *Formica archboldi* and suggested the possibility that *brunneus* is a regular prey of *F. archboldi*. If predation is involved, it is more likely that *brunneus* is subdued by chemical means than by jaw-to-jaw combat.

Some aspects of the natural history of *brunneus* are known in great detail, thanks to Walter Tschinkel and his students and colleagues, who have studied ants in the Tallahassee area for several decades. Other research achievements of Tschinkel and his diverse team are discussed under *Solenopsis*, *Trachymyrmex*, *Pogonomyrmex*, *Pheidole*, *Dolichoderus*, and *Prenolepis*.

Colony organization was described by Powell and Tschinkel (1999). Colony structure is complex, but derived from a relatively simple mechanism of ritualized dominance contests between pairs of ants. Persistent losers end up at the periphery of the nest, whereas persistent winners are near the brood and near the queen, who is the most dominant individual in the colony. Persistent winners have larger ovaries and may be able to produce unfertilized (male) eggs. Persistent losers are probably important in foraging and in nest maintenance and defense. During a food shortage, contests intensify, resulting in more peripheral workers and more foraging. This class system is apparently based on dominance contests and stabilized by different work zones within the nest.

Nest architecture of *brunneus* varies with habitat. In wet areas, nests are often in rotten logs or branches on the ground; such nests have not been studied. In well-drained sandy areas, *brunneus* constructs elaborate nests consisting of a slender shaft with varying numbers of chambers along its length (Cerquera and Tschinkel 2010). Nests sometimes take advantage of cavities left by other animals or expand into hollow roots. Nests extend to a depth of 18 to 184 cm.

Hart and Tschinkel (2012) studied the seasonal cycle of *brunneus* in a well-drained sandy habitat in the Tallahassee area. Like most North American ants, *brunneus* produces larvae that will become males and queens once per year, in this case in spring. The effort of rearing sexual broods, added to the effort of raising worker larvae, depletes the energy reserves of the workers, as measured by the contents of the fat body (a storage organ found in most insects). In fall, after all larvae have matured, the fat content of workers quickly recovers. In winter, colonies are in a quiescent state.

The mating arrangements of *brunneus* might be unusual. Young queens still bearing their wings have been observed foraging away from the nest, rather than waiting in the nest for conditions appropriate for a mass emergence of males and queens (Hart and Tschinkel 2012). Males seem to fly around extensively. I find large numbers of males, but no females, in flight traps and at lights. When males and queens of other species of ants emerge en masse, as occurs in many species, there is no guarantee that couples will find each other, and these mass emergences usually attract predators. Young *brunneus* queens might incur a risk during foraging expeditions outside the nest (Hart and Tschinkel 2012), but might compensate by enjoying the comfort and security of the nest until the time they are out foraging and meet one or more roving males. An example of such a mating system is provided by the unrelated ant *Myrmecina graminicola* (Buschinger 2003). There is always the possibility that there are mass emergences of male and female *brunneus*, and normal mating aggregations; nobody knows what all those nocturnally active males are actually doing.

Name Derivation: From *brun* (Anglo-Saxon), Latinized to *brunneus*, meaning "brown."

Odontomachus haematodus (Linnaeus)— Two-Spined Snapping Ant (Plate 8)

Taxonomy and Similar Species: This widespread and variable tropical species is most clearly differentiated from a series of similar species by a pair of well-developed spines, usually of unequal lengths, located immediately in front of the hind coxae (Brown 1976). The discovery of these spines was a taxonomic triumph for aficionados of *Odontomachus*, but the spines can be difficult to see. The specimen must be viewed underneath from in front, and the hind coxae should be at right angles to the body (or removed). It helps raise the gaster so that the spines are not seen against the background of the gaster. If the specimen is astride a point between the second and third coxae, the spines are embedded in glue and covered by the point. One would not want to search for the spines on every specimen of *Odontomachus* found in Florida, so it is fortunate that there are alternative character states available for preliminary identification. The hairs on the gaster are densely arranged but do not lie flat on the gaster like those of *brunneus*, but stand up slightly off the surface, as can be seen by scanning the gaster from the side. There are no conspicuous ridges across the back of the petiole, as in *ruginodis*. The first gastral tergite of *relictus* is glassy smooth, with sparse hairs that are almost as far apart as they are long. The first gastral tergite of *haematodus* is covered with fine, wavy, transverse sculpture, and the hairs are close-set; the distance between them is less than a third of their length. Close-up photos of workers and males of *haematodus* have been published by MacGown et al. (2014).

Distribution: Guyana, west to the Andes; southeastern Brazil, the Amazon Basin, south into northwestern Argentina (Brown 1976). In southeastern North America, *haematodus* occurs as an exotic from southern Louisiana east into the western Florida Panhandle (MacGown et al. 2014). The debarkation point for this species is unknown, but likely to be Mobile or the Port of New Orleans. The earliest collection is from Alabama in 2001; the most recent is from Florida in 2006. I would not have predicted the establishment of this species in the Southeast, but it does fit a pattern of species that may have come from subtropical South America to ports on the Gulf of Mexico. Other species that may fit this pattern are *Solenopsis invicta*, *Brachymyrmex patagonicus*, *Cyphomyrmex rimosus*, and *Pheidole obscurithorax*. Two widespread species of *Odontomachus* that might be adapted to Florida, or at least the southern Peninsula, are *bauri* and *insularis*.

Natural History: This species is common in lowland tropical and subtropical forests (Brown 1976). In the Southeast, it has been collected in mesic disturbed areas (a zoological park, a residential area) and in salt marshes. It is probable that this species will prove poorly adapted to sandy uplands in Florida, but might spread widely through wet and mesic habitats, unless it is limited in its spread by competition with *brunneus*.

Name Derivation: From *haimatos* (Greek), meaning "blood," probably referring to the dark red head. This is one of only a few Florida ants that have a name given by Carl Linnaeus, in this case almost 250 years ago. Linnaeus named a relatively small number of ants, which he placed in only a few genera. Embarking on the first rigorously organized survey of biodiversity, Linnaeus greatly underestimated the subtlety of features that separate species of ants, to say nothing of the difficulty of seeing these features in such small organisms.

Odontomachus relictus Deyrup and Cover— Florida Scrub Snapping Ant (Plate 7)

Taxonomy and Similar Species: Workers of *relictus* can be differentiated from other Florida species by their shining gaster (unlike *haematodus*), sparse gastral pubescence (unlike *brunneus*), and smooth posterior face of the petiole (unlike *ruginodis*). For many years, I believed that this species was an isolated population of the southwestern *clarus*, and confidently cited it as an example of a western desert lineage that had moved east during dry periods of the Pleistocene, becoming isolated on the tops of dry, sandy ridges in Florida as the climate became less arid (Deyrup 1990). Workers of the two species are almost identical. Eventually, however, associated males were found for both species, and it became clear that the Florida form was a distinct species (Deyrup and Cover 2004b). Male *relictus* are distinguished by their enormous eyes and ocelli, and by their elongate propodeum. These features are shared with male *brunneus* (Deyrup and Cover 2004b), further undermining my hypothesis of a western origin for *relictus*. While it is satisfying to find such distinctive male features separating *clarus* and *relictus*, it does make one wonder how many other cryptic species of *Odontomachus* might be awaiting associations between males and workers.

Distribution: Apparently confined to a few high sand ridges in the interior of the Florida Peninsula. It occurs on the southern Brooksville Ridge (Citrus County), Lake Wales Ridge (Highlands and Polk counties), and the Orlando Ridge (Orange County).

Natural History: Nests are in dry sand, in either Florida scrub or sandhill habitats. There appear to be several nest entrances, and perhaps the entire nest is diffuse. I have never found a central chamber or discrete cluster of chambers, but nests might well be deep like those of *brunneus*. Nest entrances are marked by scattered pellets of sand. Foragers are active at night, and on cloudy days. I estimate that *relictus* now occurs on approximately 10% of its original habitat, the rest having been converted to citrus or housing. It still persists, however, in several large protected areas, and the species does not appear to be endangered at present.

Name Derivation: From *relictus* (Latin), meaning "left behind," referring to the distribution of the species in relict patches of Florida scrub and sandhill on a few ancient ridges in Florida.

Odontomachus ruginodis M. R. Smith— Rough Petiole Snapping Ant (Plate 8)

Taxonomy and Similar Species: The sparse gastral pubescence, combined with the conspicuous transverse ridges on the posterior side of the petiole, is diagnostic for *ruginodis*. These ridges are unlikely to serve in stridulating; they probably function in the same way (whatever that is) as the similar ridges on the mesosoma of *ruginodis* and other species of *Odontomachus*. This species was synonymized with *brunneus* in Brown's 1976 revision of the genus, reinstated as a species (Deyrup et al. 1985) on the basis of dramatic differences between the males of the two species. While the identity of *ruginodis* in Florida is clear, I do not know whether there are Neotropical species that might easily be confused with *ruginodis*.

Distribution: I have seen specimens from southern Florida and the West Indies. It is known from Costa Rica (Longino 2004), and I would expect it to be established from southern Mexico through Central America. In Florida, *ruginodis* occurs from the Keys north into Orange and Alachua counties. The earliest specimens from Florida were collected in 1931 (Deyrup et al. 2000), and it is possible that this widespread West Indian species is native to the Florida Keys, although I personally think it is exotic in Florida. My belief that this species is introduced is based on its preference for disturbed habitats and its apparently recent range extension north into the north-central Florida Peninsula. There is another ant species, the native *Dorymyrmex bureni*, which thrives in disturbed areas and has probably become much more abundant following the massive conversion of natural habitats in much of Florida. This species, however, has not changed its range, as far as I know, but has moved out from areas of scrub, sandhill, and other habitats that were kept open by natural disturbance in the form of fire. In a similar way, meadow larks in Florida probably moved out of fire-maintained native prairies and sandhills into cultivated fields. These fire-maintained open natural habitats do not usually contain *ruginodis*. The anomalous features of the distribution and ecology of *ruginodis* are best explained by the hypothesis that this species is introduced into Florida. The example of the recently arrived *haematodus* demonstrates that species of *Odontomachus* can be accidentally spread about by humans.

Natural History: This species occurs near beaches and in coastal hammocks in tropical Florida, but farther north in the state it usually occurs in disturbed habitats, including urban and suburban areas. In Puerto Rico, it differs from another sympatric species, perhaps *O. bauri* or *insularis*, in its preference for open, sunny areas, especially river bottoms (Smith 1936a). Nests may be excavated in soil, or may be under objects on the surface of the ground. I have seen several arboreal nests in bromeliads or in tree holes.

The defensive mandible snapping of *ruginodis* was studied by Carlin and Gladstein (1989). When a nest is attacked by other ants, the *ruginodis* workers rush out snapping their jaws, dismembering or knocking aside enemy ants. If the jaws strike a solid object, the defenders may be flung into the air for several centimeters. This does not seem to be an escape mechanism, as the worker, on landing, immediately charges back into the fray. Each nest entrance is usually guarded by a single worker, who stands with cocked jaws near or within the opening. If an intruder ventures within striking distance, the jaws snap shut, responding to signals from the antennae and long sensory hairs at the bases of the mandibles. The heavy apices of the jaws do not slice into the invader, but knock it away a distance of about 1–14 cm. Carlin and Gladstein (1989) call this the "bouncer defense." As might be expected, the jaws are often heavily eroded by this kind of usage. On one occasion, I imagined I had found a useful difference between two species of *Odontomachus* based on the size and shape of the mandibular teeth. Further inspection showed that I had been comparing recently matured specimens of one species with battle-worn specimens of another.

Name Derivation: From *rugis* (Latin), meaning "wrinkle," and *nodus* (Latin), meaning "knot or swelling," referring to the wrinkles on the posterior face of the petiole, which was sometimes called the "node" in early works of myrmecology. The history of this name is a good example of the intricacies of nomenclature. The name *ruginodis* was first applied to this ant by William Wheeler in 1905, but it was used as a "quadrinomial," or four-part name: *Odontomachus haematodus insularis ruginodis*. Such quadrinomials were not uncommon in early myrmecology and were sometimes called "varieties." They might be generously interpreted as a "subspecies of a subspecies," but sometimes they seem to have just been indicators of taxonomic confusion. In any case, the fourth name of a quadrinomial has no validity under the rules of zoological nomenclature. The first use of the name *ruginodis* as a trinomial, which would now be interpreted as a subspecies, was by Marion Smith in 1937: *Odontomachus haematodus ruginodis*. Marion Smith, therefore, becomes the author of this species, rather than William Wheeler, since Smith was the first person to use the name in an approved fashion (Bolton 1995). Apparently, the first person to recognize *ruginodis* as a distinct species rather than a geographic subspecies was Edward Wilson in 1964, but this does not affect the authorship of the name.

Genus *Pachycondyla*

Wolf Ants

The genus *Pachycondyla* (pronounced **pack' ē kon" di luh** or **pack' ē kon dī" luh**) is composed primarily of medium-sized or large ants that range singly and rapidly over the ground and through leaf litter in what might be imagined as a restless, wolf-like foraging pattern. A few species conduct large-scale raids on termite nests (Hölldobler et al. 1996), but North American species do not operate in this way.

The scope of this genus is unclear. A number of genera were informally synonymized with *Pachycondyla* by Brown (1973), perhaps anticipating revisions that are still in progress. It is probable that the genus will be split up as more phylogenetic evidence accumulates, but not necessarily along the lines of any of the former genera. Revising *Pachycondyla* is a daunting task, considering that it includes more than 200 described species, a big batch of subspecies, some of which are probably distinct species, and a backlog of apparently undescribed species. At the moment, nomenclature of this group is in a state of flux.

Pachycondyla is pantropical, with only one species, Pachycondyla stigma, established in Florida so far. Pachycondyla chinensis, one of the few species adapted to temperate climates, was imported from Asia and is well established as far south as South Carolina; this species might eventually reach Florida. Several tropical species appear to be adapted to disturbed conditions and might thrive in southern Florida if accidentally imported.

Name Derivation: From pachys (Greek), meaning "thick," and kondylos (Greek), meaning "knob" or "articular enlargement." This probably refers to the thick petiole of the type species, Pachycondyla crassinodis, which has an unusually thick, almost square petiole. The name crassinodis is derived from the Latin words for thick petiole, so that the name P. crassinodis is repetitive in two languages, much like the scientific name for domestic cattle, Bos taurus.

Pachycondyla stigma (Fabricius)—
Pantropical Wolf Ant (Plate 9)

Taxonomy and Similar Species: This species is relatively large (4–5 mm long), bigger than species of Hypoponera and Ponera. It has broad, triangular mandibles, unlike Stigmatomma pallipes, and lacks the strong striations of Gnamptogenys triangularis. There are similar species in the Neotropics, including Pachycondyla succedanea in Cuba and Puerto Rico.

Distribution: In North America, north of Mexico stigma is apparently restricted to Florida, but it is likely to occur in coastal Louisiana and Texas. In Florida, stigma is known from the Keys north into Orange and Volusia counties. It occurs through much of the tropics and subtropics of the New World and Asia (Smith 1979). It appears to have close relatives in both the New World and Asia (Brown 1963), so it is difficult to speculate on its origin on the basis of the homeland of its nearest relatives. The distribution of P. stigma in Melanesia suggests that it is still expanding its range (Wilson 1958). The mode of long-range dispersal of this species is not known. Colonies sometimes occur in the decomposed inner bark of dead trees, which might be carried about by commerce. Perhaps, alternatively, P. stigma found some hospitable microhabitat on wooden sailing ships that allowed it to be moved around the world.

Natural History: In Florida, P. stigma usually occurs in open pine stands, including wet or seasonally flooded flatwoods. It is usually found in habitats that are relatively undisturbed, except by natural disturbances such as floods or fires. It often lives around dead pines whose bark is intact. I have found little piles of dead or paralyzed termites in chambers of the nest. Smith (1937) reports that, in Puerto Rico, P. stigma forms colonies of a few hundred individuals in rotten logs and under stones. Haskins and Enzemann (1938) observed colony foundation by a dealate queen, who continued to forage as her larvae matured. In a study of behavior within a laboratory colony, Oliveira et al. (1998) found that the single inseminated queen in the colony produced almost all successfully developing eggs in the colony. She removed eggs of workers and unfertilized queens from the pile of eggs in the colony; workers also removed and destroyed eggs of individuals other than the queen. Oliveira et al. (1998) suggest that the queen's control over workers is mediated through chemicals produced by tibial glands in the front legs, which other ants rub with their antennae when they encounter the queen.

Name Derivation: From stigma (Greek), meaning "mark" or "spot." The application is not clear; perhaps it refers to a structural feature, such as the strongly developed basalar lobe.

Genus Platythyrea

Silvery Hunter Ants

The genus Platythyrea (usually pronounced **Plat' ē thir" ē uh**) is represented in Florida by a single species, a medium-sized (approximately 6 mm long) charcoal gray species with a distinctive silvery cast and conspicuous shallow punctures. The genus is pantropical with the majority of species in Africa or Australasia, but some in Southeast Asia, South and Central America, and the Caribbean (Bolton 1995). One Neotropical species ranges into the northern subtropics, including Florida and southern Texas.

Name Derivation: From platys (Greek), meaning broad or flat, and thyreos (Greek), a large, oblong, door-shaped shield (Brown 1975). This probably refers to the broad frontal area whose lobes shield the antennal bases.

Platythyrea punctata (F. Smith)—
Common Silvery Hunter Ant (Plate 12)

Taxonomy and Similar Species: Platythyrea punctata is distinguished from other Florida ants by its semirectangular petiole and conspicuous shallow punctures, as well as the frontal shield. Its silvery appearance, obvious under the microscope, is attributed to microscopic, appressed, shining, pale pubescence. These hairs are not deciduous, like the scales of Lepidoptera. This silvery coat resembles that of the otherwise dissimilar *Formica subsericea* and some other ants. The function of this dense covering of shining hair is unknown, at least to me. I have observed it primarily in diurnal ants that run about on the surface of the ground or on trees. It might serve to reduce heat absorption.

The male can be recognized by its robust and semirectangular petiole, as well as by its covering of fine hairs and conspicuous punctures, but the punctures of the male are far coarser and deeper than those of the female.

<u>Distribution:</u> *Platythyrea punctata* occurs from Texas into Central America and around the Caribbean, including the West Indies (Brown 1975). In Florida, it occurs from Brevard and Highlands counties south through the Florida Keys. The Florida population is isolated, but no more so than populations on various islands in the Caribbean. There is no reason to believe that the Florida population has been introduced. Heinze and Hölldobler (1995) studied a Florida population that had poorly developed wings, but some Florida queens have large and apparently functional wings.

<u>Natural History:</u> In Florida, nests are in logs and dead branches. In Puerto Rico, colonies are reported from rotten logs and stumps (Smith 1936a) or in hollow dead stems of *Psychotria berteriana* (Schilder et al. 1999). Workers forage singly, capturing adult and larval insects (Torres 1984a). The jaws are unique among Florida ants: they are shaped like concave triangles, completely lacking teeth on their sharp inner cutting edges. They appear to be adapted for carrying and chopping, or perhaps scraping. The inner basal margin has been reduced to a strong curved inner ridge fringed with long curved hairs. Its appearance suggests that *P. punctata* could transport a large droplet of liquid in its jaws. More detailed studies of the diet of this species might help explain its unusual jaws.

Relative to the conservative monarchies that hold sway among most ants, the social system of *P. punctata* is shockingly Bohemian and complex, as shown by Heinze and Hölldobler (1995) and Schilder et al. (1999). In many nests, all individuals give the appearance of workers. Unmated workers can produce female (diploid) offspring (this is called thelytokous parthenogenesis), as well as

occasional males. A single female usually lays most of the eggs in each colony. There are dominance hierarchies in these colonies, but the dominant female is not necessarily the individual who produces the most eggs. Some colonies produce a small number of males. Winged queens and individuals intermediate between queens and workers also occur. All females have a sac for storage of sperm (spermatheca), although it may be poorly developed, and a small number of females of all forms have sperm stored in the spermatheca. The flightless individuals with stored sperm may have mated with brothers in the nest, but this is not certain. It is also not clear whether this sperm is actually used in the production of young, at least in the Florida population (Hartmann et al. 2005). To make matters even more confusing, if possible, there is some indication that there are geographic differences in the reproductive system of this species. *Platythyrea punctata* may be an example of a pattern in which peripheral populations of a species exhibit various forms of parthenogenesis (Hartmann et al. 2005). There are some other ant species whose virgin workers can produce females, but the reproductive plasticity of *P. punctata* may be unique, unless it is shared with other members of its genus.

Male *P. punctata* appear in flight traps in Florida between August and December (11 specimens). In the case of most ants, one would expect that these males are seeking queens that are also flying about, but for all we know, these males may be mating with any female they see out foraging. It is possibly relevant that callow workers, recognized by their red color, are occasionally seen foraging.

<u>Name Derivation:</u> From *punctum* (Latin), meaning "a small hole," *punctata*, "covered with small holes"; this refers to the conspicuous punctures on the head and body.

Genus *Ponera*

Little Porthole Ants

Members of the genus *Ponera* (pronounced **Pō ner" uh** or **pō ner uh**) are small (2–3 mm long), subterranean or rotten wood–inhabiting species with triangular mandibles. They are distinguished from members of the very similar genus *Hypoponera* by the porthole-like translucent "window" in the keel-like process on the underside of the petiole. This important feature is often hidden by the hind coxae. It is easy to reposition specimens in alcohol, but dry specimens are a little more difficult. The hind coxae have a ball-and-socket basal joint, and it is possible to very carefully swing a hind leg forward to reveal the petiolar keel. The two Florida species of *Ponera* are much rarer than *Hypoponera* species, and *P. pennsylvanica*, rather similar to *H. opaciceps*, is confined to the northern part of the state. It may be easier to learn the identifying features of the species of *Ponera* and *Hypoponera* than to constantly check for the generic distinctions (see comments under *Hypoponera*).

The following information is summarized from Taylor (1967). Colonies are small, normally with approximately 30 workers, and typically occur in rotten wood, although *pennsylvanica* often occurs under stones. Most species live in mesic habitats. It is probable that all species are insectivorous, although food preferences are unknown. Larvae adhere to the walls or floor of the gallery by means of sticky, knob-shaped tubercles. Colonies are founded by individual queens, who continue to forage as their first brood develops.

Name Derivation: From *ponera* (Greek), meaning "toilsome," perhaps related to *ponikos* (Greek), meaning "hardworking." The general intent may be to describe an ant that labors away unseen in the earth or in rotten wood.

Ponera exotica M. R. Smith—
Southern Porthole Ant (Plate 11)

Taxonomy and Similar Species: Distinguished from other Florida species of *Ponera* and *Hypoponera* by its small size (length approximately 2 mm) and disproportionately large antennal club. The petiole is low and only weakly narrowed above, somewhat like that of *Hypoponera punctatissima*.

Distribution: North Carolina south into Florida, west into Oklahoma and southwestern Texas (MacKay and Anderson 1991). In Florida, it is known from scattered localities south into Highlands County. This species is rare or difficult to obtain. As its name implies, *exotica* was assumed to be introduced, primarily because of its apparent close relationship to Indo-Australian species (Smith 1962). The late date of its discovery (1959) and its appearance following a huge amount of poorly regulated (from the standpoint of introduction of species) air traffic in the latter part of World War II also seem to have influenced Smith's presumption that this species was exotic. Taylor, in his 1967 revision of *Ponera*, agreed that this species must be

exotic, and apparently discussed this with Smith before the species was described. *Ponera exotica* is so similar to the widely distributed (occasionally introduced) species *Ponera leae* Forel, not to be confused with *Hypoponera lea* (Santschi), that Taylor (1967) wondered whether *exotica* might be a variant of *leae*. Johnson (1987), after considering the wide distribution of *Ponera exotica* in undisturbed habitats, its absence from disturbed habitats, and the lack of any identical species from the Indo-Australian region, decided that *exotica* is probably native to North America. Its late discovery is not unusual for a small, cryptic ant of southeastern North America. I am provisionally accepting the native status of *exotica*. The question, which is more intriguing than urgent, might be settled with biochemical analysis if one could assemble specimens of *leae* representing the wide range of that species.

Natural History: *Ponera exotica* occurs in a wide variety of natural habitats, including mesic floodplain forests, dry, sandy upland forests, and prairies (Johnson 1987). Apparently, nobody has seen a nest, larvae, or males, and it is possible that *exotica* has never even been seen alive, only as extracted specimens.

Name Derivation: From *exotikos* (Greek), meaning "alien" or "foreign," referring to the supposed origin of the species.

Ponera pennsylvanica Buckley—
Common Porthole Ant (Plate 11)

Taxonomy and Similar Species: Distinguished from *exotica* by larger size (length approximately 3 mm) and relatively smaller antennal club. In general appearance, it resembles species of *Hypoponera*, especially *opaciceps*, but *Hypoponera* species lack a "porthole" in the keel-like process on the underside of the petiole. Unfortunately, in both genera, this area is usually hidden by the hind coxae, especially if the gaster is somewhat decurved. Even in the case of dry specimens, however, either the gaster or a hind coxa can usually be repositioned a little, thanks to their ball-and-socket joints, to reveal the ventral process of the petiole. The commonest species of *Hypoponera*, *opacior*, has its petiole conspicuously narrowed above (in lateral view), while the petiole of *P. pennsylvanica* workers (but not queens) is almost as wide above as below. By keeping this in mind, one can save oneself much manipulation of specimens when sorting large samples from areas, such as north Florida, where both species are common. Callow individuals are reddish, and give the impression of a separate species. *Ponera pennsylvanica* appears closely related to the Eurasian species *coarctica* and was usually considered a subspecies of *coarctica* until recognized as a distinct species by Taylor (1967). It would not be surprising if *coarctica* had been imported at some time into North America, where it could easily be misidentified as *pennsylvanica*. There are small differences, detailed by Taylor (1967), between the two species, but it is unlikely that anybody is looking for these.

<u>Distribution:</u> Distributed more or less uniformly from southern Nova Scotia south into northern Florida, west into eastern Minnesota, south into northern Florida (Taylor 1967). West of this area, there are some scattered records from North and South Dakota, Colorado, and New Mexico (Taylor 1967), with an outlier in Michoacán in south-central Mexico (MacKay and Anderson 1991). In Florida, *pennsylvanica* is known from northern Florida south into Marion County.

<u>Natural History:</u> Taylor (1967) reports that nests are usually found in broadleaf forests or in mixed conifer–broadleaf forests. In mesic forests, nests are usually in rotten wood or leaf litter, but in dryer or more open areas, nests are usually found under rocks. Colonies are founded by individual females, who continue to forage, at least until the first workers mature. Flights occur from mid-August to early October, according to collection data from various areas. There are no overwintering larvae or pupae. In some sites, this is a remarkably abundant species.

Wheeler (1900) studied colonies of *pennsylvanica* in the laboratory, making observations that display his usual curiosity and appetite for detail. He illustrated the larva, showing the sticky dorsal knobs that allow the larvae to adhere to the brood chamber so that the underside of the body becomes a kind of stable platform on which adults place bits of food. In northeastern North America, there is one generation per year. Males leave the nest and may be found at large in the field, but in artificial nests, they may also mate with young queens in the nest. Wheeler suspected that many colonies fission into satellite colonies, each with a young queen. Wheeler also experimented with combining two or more colonies. He found that adults went through a period of combat, usually not lethal, followed by integration of the colonies. Larvae and pupae of different colonies were grouped together immediately, "long before the ants have settled their various difficulties." It is possible that the original nests had social hierarchies that required reorganization when the nests were combined.

Wheeler (1900), in the vivid idiom of his day, characterized *pennsylvanica* as a "feeble little ant," and considered some of its behavior either primitive or degenerate. Many years later, however, Pratt et al. (1994) discovered that *pennsylvanica* has a complex behavioral repertoire well worth studying. Pratt et al. were able to maintain colonies in the laboratory, feeding them on brood of other ants, and distinguishing between individuals by gluing on combinations of minute colored flecks shaved from plastic Lego blocks. This meticulous project revealed 35 distinct behaviors. Foraging is done by individual workers, who do not cooperate or recruit, even when provided with a large prey, such as a *Camponotus* larva. Workers attack and kill small, live prey, such as isotomid Collembola, Psocoptera, and mites. Larvae are fed pieces of solid food. Workers assist in the emergence of young adults by cutting open the anterior end of the cocoon. There is a somewhat inconsistent division of labor between foragers and brood caretakers. I have often found accumulations of 5–20 pupae attended by only one or two workers.

<u>Name Derivation:</u> In his wholly inadequate description of this species, Buckley (1866) comments that it "Dwells beneath stones in the vicinity of Philadelphia." This piece of information is about as specific as the morphological features Buckley used to distinguish the species.

SUBFAMILY ECITONINAE

Genus *Neivamyrmex*

New World Army Ants

The term *army ants* does not actually designate a genus of ants, but rather a nomadic, predatory lifestyle that has evolved several times in different groups (Gotwald 1995). The orderly, large-scale rampages of army ants seem to reflect the persistent and mythic fantasy of a human warrior nation, glorified, for example, in the bloodthirsty and megalomaniac career of Alexander the Great. Human armies, however, tend to revert to communities, whose interests are not best served by the catastrophic disruptions of war, thus frustrating the military dictator's dream of endless conquest. In the army ants, we see the fulfillment of the mad dictator's fantasy, but it is accomplished without the benefit of generals or other military leaders. The orchestration of the complex and dramatic set of behaviors typical of army ants has attracted the attention of an illustrious group of behaviorists, including T. C. Schneirla, N. R. Franks, C. W. Rettenmeyer, W. H. Gotwald, Jr., H. R. Topoff, and E. O. Wilson. The large body of research on army ants has been compiled into a lively book by Gotwald (1995): *Army ants: the biology of Social Predation.*

The genus *Neivamyrmex* (pronounced **Nee vuh mur" meks**) includes the only Florida species of army ants. The New World army ants (subfamily Ecitoninae) include several genera, the largest of which is *Neivamyrmex*, with approximately 120 species. This is a genus of South and Central America and southwestern North America; only five species stray over into eastern North America. When one considers that queen *Neivamyrmex* are invariably flightless, such that the species must disperse on foot, it is surprising that there are any eastern species at all. It might seem impossible for army ants to cross the Mississippi, which is a very old river. It is conceivable, however, that changes in the course of the river, especially the cutoff of east-projecting oxbows, might have passively transferred a few species from one side to the other. The state of Mississippi has a series of former oxbows wrapped around land that was once on the west side of the river. Once western army ant species had made their way further east, they must have been completely isolated from their western relatives, so it is also a bit strange that no eastern populations have diverged into separate species.

Species of *Neivamyrmex* go through cycles of nomadic activity, as summarized by Gotwald (1995). During a "statary" (nonnomadic) phase, the queen becomes grossly distended with gigantic numbers of eggs. During this period, the larvae, which the workers had carried along during the nomadic phase, enter their pupal stage. Over a period of a few weeks or a month (in warm weather), the eggs hatch into hungry larvae and the pupae hatch into adults. The gaster of the queen contracts so that she can resume walking, and the new adults graduate from callows to mature workers. Soon, the whole colony hits the road, stopping in bivouacs while armies of workers go out to bring in arthropod prey and then move on as the foraging area becomes depleted.

When these larvae are mature and the queen begins to expand with eggs, a new statary phase begins.

A raid by one of the Florida species of army ants is an exciting spectacle for a myrmecologist, but would not be featured in the local paper. Most people find army ants far smaller than they expected, and the long, flowing trails of army ants seem hardly more dramatic than the defensive mass attacks of the fire ants nesting alongside a driveway. Raids by army ants are nevertheless remarkable, the more so because the participants are blind and independent of any strategic command center. Raiding behaviors of various species of army ants have been studied in detail and are discussed by Gotwald (1995). For simplicity's sake, one can imagine an army ant colony producing a narrow but dense stream of workers, which eventually begins to branch out like the delta of a river to form a swarm front that embraces and overwhelms any edible animal. Ants returning with food follow an odor trail back to the bivouac, and, as in many other kinds of ants, the return of successful foragers stimulates other individuals to follow the odor trail to the foraging area. Raiding depends on several factors, including weather, time of day, and the needs of the colony. Raiding patterns depend to some extent on terrain and are coordinated by odor trails of foragers, types and amount of prey brought back to the bivouac, defensive actions of prey (which may attract more ants), and size and portability of prey, which determine whether prey are cut up, brought back whole by a group of ants, or carried whole by individual ants.

Army ant colonies always include enormous numbers of workers. In the case of *Neivamyrmex nigrescens*, for example, there are 80,000 to 120,000 workers per colony (Schneirla 1958). This large colony size not only allows army ants to take on relatively large prey but also allows them to tackle well-defended nests of social insects, such as ants, termites, and social wasps. Some species of army ants seem to make their living largely by raiding nests of other types of ants. Army ant colonies must be large to be effective, and colonies reproduce by dividing in two, each with one or more queens, rather like fission of honeybee colonies. This is an expensive and slow method of colony founding, compared with sending out batches of queens, each of which can found her own colony. On the other hand, when a colony splits, each part has a good chance of success. Male army ants, which are large and bear absolutely no resemblance to workers of their species, are capable flyers, and presumably leave their colonies to mate with young queens from other colonies. The army ant system of colony reproduction should not lead to inbreeding, so long as there is a sufficient supply of local colonies.

Army ants can be difficult to find, even though their colonies are huge and engage in large-scale raiding behavior. There are no places I know of in Florida where I could be confident of finding army ants in a day or two of hunting, although I do know my chances would be better in north Florida than in south Florida. Southeastern army ants tend to be nocturnal, and much of their activity, even raiding forays, may be subterranean. The easiest way to confirm the presence of army ants in an area is to put out flight traps (Malaise traps), which eventually capture males.

The cryptic habits of Florida army ants prevent us from knowing whether any of our species are rare or endangered. It is reasonable to suspect that army ants are extinct in urban areas and are becoming rare in suburban and cultivated areas. They probably require a substantial acreage to support a population. Roads and ditches are possible death traps, and are inevitably significant barriers to dispersal.

Name Derivation: Named for the Brazilian entomologist Arturo Neiva (1880–1943), combined with *myrmex* (Greek), meaning "ant." The genus was erected by the Brazilian myrmecologist Thomas Borgmeier, whose early career was advanced by Neiva. Borgmeier went on to produce many papers on ants, especially army ants, of which he described numerous species. He also studied flies of the family Phoridae (hundreds of new species) and insect associates of army ants.

Neivamyrmex carolinensis (Emery)— Carolina Army Ant (Plate 13)

Taxonomy and Similar Species: *Neivamyrmex carolinensis* has a shining head, unlike those of *texanus* and *nigrescens*, whose heads (larger workers) are covered with granulate sculpture. The head of *opacithorax* is shining but has somewhat angulate occipital corners, while those of *carolinensis* are smoothly rounded. The petiole of *carolinensis*, in the dorsal view, is much less than twice as long as wide, while that of larger workers of *opacithorax* is almost twice as long as wide.

Distribution: North Carolina and western Tennessee south into Florida, west into Mississippi, scattered locations in Nebraska, Kansas, New Mexico, Arizona (Watkins 1985), and Illinois (DuBois 1988). In Florida, *carolinensis* is known from about 10 sites, mostly in north Florida, but there are records from Highlands County in south-central Florida where *carolinensis* was found in the stomach contents of narrow-mouthed toads, *Gastrophryne carolinensis* (Deyrup et al. 2013).

Natural History: This species is subterranean and rarely collected in Florida. In other states, it has been found under rocks (DuBois 1988; Rettenmeyer and Watkins 1978). One might justifiably argue that the absence of rocks in Florida has stunted our knowledge of the distribution of subterranean ants in general. Males are believed to fly during the day, based on the relatively small eyes and ocelli (Coody and Watkins 1986) and are seldom collected in flight traps. In Florida, *carolinensis* has been collected several times in sandhill habitats. A notable feature of *carolinensis* is that it regularly, perhaps always, has multiple queens in its colonies (Rettenmeyer and Watkins 1978). These are not colonies that are temporarily polygynous before splitting, because several or all queens may show the swollen abdomen characteristic of egg-laying queens. The rove beetle *Dinocoryna bisinuata* has been found with *carolinensis* (Frank and Thomas 1981).

Name Derivation: *Carolinensis* (Latin) means "an inhabitant of Carolina," referring to the collection site of the type series.

Neivamyrmex nigrescens (Cresson)— Blackish Army Ant (Plate 13)

Taxonomy and Similar Species: This species is extremely similar to *texanus*, and the two species were not separated until 1972 (Watkins 1972). Both species have nonshining heads, owing to heavy granulate sculpture, less developed in some smaller workers and in some California populations (Ward 1999). The declivity of the propodeum is not concave in the lateral view in *nigrescens* and slightly concave in *texanus*, with a weak angle where the declivity meets the dorsal surface of the propodeum. The gaster of male *nigrescens* is blackish, not reddish brown as in *texanus*.

Distribution: Known from scattered localities, sometimes with gaps of hundreds of miles, from West Virginia into Georgia, west into Nebraska and southern California (Watkins 1985). There appear to be no reliable records from Florida, but it occurs in both coastal Alabama and central Georgia. This species almost certainly occurs in Florida, unless there are ecological factors, such as a requirement for clay soil rather than sand that would make Florida unsuitable.

Natural History: In some places in the western part of its range, *nigrescens* is one of the more common and conspicuous army ants and has been the subject of several studies. The older studies are summarized and augmented by Rettenmeyer, whose 1963 paper gives the following information, most of it from areas where *nigrescens* is not likely to have been confused with *texanus*. In Kansas, *nigrescens* is usually found in moist, wooded areas. Raids occur along the surface of the ground; colonies of other ants are the primary prey. A captive colony refused various foods, such as nuts, raisins, dried coconut, sugar syrup, vegetable oils, bacon grease, and ground beef. Bivouacs are subterranean and may extend at least a meter below ground. Approximately 10,000 workers were collected from one colony.

In the laboratory, colonies in their nomadic phase are more likely to move the colony when underfed, and it seems likely that this is also the case in the field, as a probable adaptation to center the colony in productive sites (Topoff and Mirenda 1980). Male *nigrescens* fly during the day, and in the western part of its range, the flight period is from mid-August through most of November (Baldridge et al. 1980). Various insects and other organisms are reported as associates of *nigrescens*, but most of these are western, or may actually be associates of *texanus*.

Name Derivation: From *nigrescens* (Latin), meaning "blackish," referring to the blackish gaster of the male, from which the species was described.

Neivamyrmex opacithorax (Emery)— Shiny-Headed Army Ant (Plate 14)

Taxonomy and Similar Species: This species has a shiny head, distinguishing it from *texanus* and *nigrescens*. The petiole, in dorsal view, is almost twice as long as wide (in larger individuals), while that of *carolinensis* (which also has a shiny head) is much less than twice as long as wide.

Distribution: Southern Virginia south into southern Florida, west through Kansas into California (Watkins 1985), south into Mexico (Watkins 1982). In Florida, this appears to be the most abundant

army ant, or at least the most collected, with records scattered through the state, including, surprisingly, a record from the Florida Keys.

Natural History: Colonies have been found on several occasions in sandhill and Florida scrub habitats, once at the base of a tree at the edge of swamp forest. The following information is summarized from Rettenmeyer (1963). In Kansas, *opacithorax* is more common in open areas, where it seems to replace *nigrescens*, more typical of moist, wooded habitats. The diet is mostly ants and carabid beetles. Raids, which are usually nocturnal or crepuscular, concentrate on collecting subterranean and surface-dwelling arthropods. These army ants do not climb up short vegetation to capture arthropods that have sought refuge just above the heads of the raiders. Bivouacs are subterranean.

Males fly during the day in Florida, resembling local braconid wasps in the genera *Digonogastra*, *Doryctes*, and *Atanycolus*. *Digonogastra*, when handled, produces an acrid-smelling chemical. Males in Florida are most frequently collected in September and October. This species is associated with the staphylinid beetles *Ecitoxenidia alabamae*, *Microdonia occipitalis*, and *M. nitideventris* (Frank and Thomas 1981).

Name Derivation: From *opacus* (Latin), meaning "dark," or in this case "dull," and *thorax* (Greek), referring to the dull, nonshining mesosoma.

Neivamyrmex texanus Watkins— Texas Army Ant (Plate 13)

Taxonomy and Similar Species: Similar to *nigrescens*, which shares with *texanus* the nonshining, heavily granulate head. The propodeum of *texanus* is slightly concave in a lateral view, unlike that of *nigrescens*. The two species were not distinguished from each other until 1972 (Watkins 1972). There appear to be no valid records of *nigrescens* from Florida, but it occurs in adjacent states, so it is wise to check the identification of presumed *texanus* specimens.

Distribution: In eastern North America: Virginia south into south-central Florida, west through the Panhandle; in the West: Texas into California, south into southern Mexico (Ward 1999). Western populations appear to be separated from eastern populations by a large gap, but this might reflect difficulties in finding or identifying specimens.

Natural History: In Florida, *texanus* has been collected eight times from sandhill or Florida scrub habitats. Two colonies were seen raiding during the afternoon. Males are diurnal, resembling those of *opacithorax*, apparently belonging to the same mimetic complex. Florida flight records are from October and November. Elsewhere flights are from early September through mid-November (Baldridge et al. 1980).

Name Derivation: The type specimens are from Texas (Watkins 1972); hence, I assume the name *texanus*. At the time of its description, the extensive range of *texanus* was already known.

SUBFAMILY PSEUDOMYRMECINAE

Genus *Pseudomyrmex*

Slender Twig Ants

The genus *Pseudomyrmex* (pronounced **Sue' dō mur" meks**) is our only genus in the subfamily Pseudomyrmecinae. Every naturalist who sees these ants, some species of which are common, recognizes that these ants are peculiar, and I have often been asked whether these are ants, or some kind of wingless wasp. This must have been a problem early on, because the name of the genus translates as "false ant." These ants are morphologically unusual: they are very elongate, with huge eyes and short, thick legs (in the plates, the species are portrayed uncharacteristically "up on their toes" so that the body can be seen from the side). The feature, however, that is most "un-ant-like" about them is their habit of moving in short, rapid lunges, like a fishing lure that is being jerked forward along the surface of a pond. This pattern of locomotion does not seem to be associated with laying down a trail. Since *Pseudomyrmex* workers and queens are equipped with a powerful sting, perhaps this way of moving is a warning mechanism, appropriate to the exposed situation of a diurnal insect that moves about on leaves, twigs, branches, and tree trunks. It could just as easily be related to how vision works in these ants, or there could be pauses to pick up vibrations, or it might have some completely different significance.

Among the other peculiarities of adult *Pseudomyrmex* is their stridulating organ, located on the dorsal side of the base of the first tergite of the gaster and normally hidden by the apex of the postpetiole. Vibrations are produced by slight in-and-out or up-and-down movements of the first segment of the gaster. Such stridulating organs appear to be rare in arboreal ants, although they occur in several distantly related lineages of terrestrial species (Hölldobler and Wilson 1990). It is now known that the herbs and shrubs of a quiet meadow often resound with inaudible, substrate-borne communications of leafhoppers, treehoppers, and other insects; the technology for recording and playing back these messages has reached a high level (Hunt 1993). The application of this technology to the many species of *Pseudomyrmex* might reveal new aspects of ant communication.

Larvae of *Pseudomyrmex*, which are packed into hollow twigs and stems, are also strange. They subsist on dry pellets of partially ground up arthropod and vegetable fragments, tucked into a food pouch on the first abdominal sternite (Wheeler and Bailey 1920). Food may be softened or partially digested in this pouch, before being drawn into a special chamber in the mouth, where a series of fine ridges may serve to further mill and strain food particles (Wheeler and Bailey 1920). Wheeler and Wheeler (1956) present drawings of larvae of several species of *Pseudomyrmex*, including external, internal, and cross-section views of the remarkable feeding structures. The diet of Florida *Pseudomyrmex* is unclear. I have seen several species carrying small insects, and individuals feeding on nectar from flowers or extrafloral nectaries. Most individuals I have seen entering a nest are not carrying any visible resources.

All Florida species of *Pseudomyrmex* are equipped with a sting but are generally reluctant to use it, even on a myrmecologist who is breaking into a nest. This is unlike some highly defensive tropical species that protect acacia trees (Jansen 1966), or the Bahamian species *Pseudomyrmex subater*. Our species do sting readily if seized or slapped, and Floridians are often stung on the neck by one of these ants that has become dislodged from its home on a plant, and is climbing up the nearest tall object. The ascent of a human by *Pseudomyrmex* does not seem to be a random process and would make an entertaining undergraduate research project. There seems to be a tendency to avoid shaded areas or overhangs, which results in the ant climbing the back of the human, usually first hitting bare skin on the back of the neck. One can brush off the ant with a quick flick of the fingers, but a slapping reaction gives the ant a chance to sting, sometimes more than once. Most people seem to get a sensation of pain that starts slowly, builds over a few seconds to half a minute, and then begins to fade after a few minutes. A small number of people get persistent swelling, and I have heard of one case of anaphylactic shock. Considering that some species are very abundant, hundreds of Floridians must get stung by some species of *Pseudomyrmex* every month, without any lasting or traumatic effects. It is probably far safer to be outdoors with the stinging ants than indoors, glued to the TV, whose anxiety-raising stories cause all sorts of ulcers, heart attacks, stress-induced eating, and displacement activity in the form of domestic violence.

Pseudomyrmex is a large Neotropical genus, with an estimated 150 to 200 species, including many undescribed species (Ward 1992). It is fortunate for ant taxonomy that this difficult and complex genus has received the expert, meticulous attention of Philip Ward. Only eight species reach Florida, and their taxonomy is relatively well understood, thanks to the work of Ward (1985). Considering the number of tropical species and the ease of accidentally transporting an entire colony in a hollow stem, it would not be surprising if additional species eventually become established in Florida. Two likely immigrants are *subater* (common in the Bahamas) and *caeciliae*. The latter, somewhat problematical species (Ward 1989) could remain unrecognized in Florida for years because of its close resemblance to *cubaensis* and *elongatus*.

Name Derivation: From *pseudes* (Greek), meaning "false," and *myrmex* (Greek), meaning "ant," probably referring to an early confusion between these unconventional (to Europeans) ants and wasps.

Pseudomyrmex cubaensis (Forel)—
Cuban Slender Twig Ant (Plate 15)

Taxonomy and Similar Species: This species was originally described as *Pseudomyrma elongata* var. *cubaensis*, later synonymized with *elongatus*, and then revived from synonymy by Ward (1985). The species *cubaensis* and *elongatus* are similar in appearance, and I do not attempt to distinguish between them in the field. Under the microscope, the head of *cubaensis* behind the ocelli is weakly shining, with shallow punctures, while this area in *elongatus* has dense, close punctures and is not shining. The petiole of *cubaensis* in lateral view is relatively low, with more gently sloping sides than that of *elongatus*, but this difference is difficult to judge unless

one has specimens of both species. The similar species *caeciliae*, with a range extending into Texas (Ward 1989), is not known from Florida, but could probably survive here.

Distribution: Florida, Cuba, the Bahamas, Cayman Islands, Haiti, Jamaica; absent from Puerto Rico (Ward 1989). The mainland Neotropics have a "variable array of *cubaensis*-like populations" stretching from Mexico into Argentina (Ward 1989). In Florida, *cubaensis* occurs from Seminole County southward.

Natural History: In Florida, *cubaensis* is usually arboreal, in hollow twigs and vines, but in the Bahamas, it is also common in culms of sea oats (*Uniola*) and sawgrass (*Cladium*). A curious feature of this species is that males are notably large, even compared to queens. This suggests that there is an unusual degree of competition between males, or some other interesting feature of reproductive biology.

Name Derivation: The name *cubaensis* means an "inhabitant of Cuba."

Pseudomyrmex ejectus (F. Smith)—
Shining Dark Slender Twig Ant (Plate 15)

Taxonomy and Similar Species: This common and widespread Florida species went by the name of *P. brunneus* until Ward (1985) determined that *ejectus* and *brunneus* are separate species, and referred all Florida records to *ejectus*. It is the only dark, shining species of *Pseudomyrmex* in Florida. In extreme southwest Florida, around Copeland and Everglades City, there is a form of *ejectus* that is mostly yellow or orange, but retains enough dark markings to distinguish this species from the orange species that occur in Florida.

Distribution: East Coast of North America from Maryland through Florida, around the Gulf of Mexico as far south as Costa Rica; also Jamaica (Ward 1985). It is probably introduced into Jamaica, as it does not occur on other Caribbean islands. It occurs throughout Florida.

Natural History: In Florida, *ejectus* is usually in relatively open habitats and often occurs in disturbed sites. Nests may be in hollow weed stalks, grass culms, fine hollow stems of vines (especially grape), and small, hollow twigs on live trees and shrubs. It probably has a varied diet, although the only published prey record is a microlepidopteran leaf miner (*Eriocraniella* sp.) on oak trees (Faeth 1980). This common species is the host of the rare parasite *Pseudomyrmex leptosus*, discussed below. At the

Archbold Biological Station, it is also a host of a rare predaceous syrphid fly, *Rhopalosyrphus ramulorum* (Weems et al. 2003). *Rhopalosyrphus* is closely related to the more familiar *Microdon*, and its larvae, like those of *Microdon*, probably feed on larvae of ants, in this case *Pseudomyrmex*. The rarity of this fly may be connected to its affliction with a serious logistical problem: the *Pseudomyrmex*-sized exit hole is far too small for the adult fly (Weems et al. 2003). I have only found the pupa of *R. ramulorum* twice, and in both cases, the adult fly emerged the next day, suggesting that the pupa may be forced to wait until the twig is broken before emerging as an adult.

Name Derivation: Apparently from *ejectus* (Latin), meaning "thrown out" or "ejected." I have no idea why this name should have been applied, and there is no clue from Frederick Smith's terse description of the species. Frederick Smith, not to be confused with the more recent Marion Smith, was an early English myrmecologist (1805–1879) who spent many years at the British Museum. He described great numbers of ants, without much regard to what other myrmecologists were doing, and with such short and nondiagnostic descriptions that most of his species would be unidentifiable were it not for the preservation of types of many of his species (Creighton 1950). Subsequent, more capable myrmecologists were forced to spend so much of their time correcting Smith's mistakes that one might almost question whether he had a positive effect on myrmecology, despite his prolific descriptions of new species.

Pseudomyrmex elongatus (Mayr)—
Mangrove Slender Twig Ant (Plate 15)

Taxonomy and Similar Species: For some time, this species was combined with the similar species *cubaensis*; the two were separated by Ward (1985). The two species do not seem to intergrade in Florida, but they are less distinct in Jamaica (Ward 1985). I find these two ants indistinguishable in the field. Under the microscope, the head of *elongatus* behind the ocelli has deep, close punctures and is not weakly shining (as in *cubaensis*). The petiole in a lateral view is relatively high, with steeply sloping sides; it is best to have a specimen of *cubaensis* for comparison.

Distribution: Southern peninsular Florida, south coastal Texas to Panama; Jamaica (Ward 1985). In Florida, *elongatus* is restricted to the southern half of the Peninsula and is most common along the coasts

and in tree islands of the Everglades. The disjunct distribution of this species in south Florida, with no populations known from Cuba or the Bahamas, suggests that this species might be a long-established exotic in Florida that arrived before any ant surveys. An alternative is natural dispersal around the Gulf of Mexico during a period of warmer climate, although we have no other examples of tropical Florida species that are absent from the Caribbean. A third possibility is that *elongatus* does occur in the Caribbean but is relatively scarce and no specimens have come into the hands of myrmecologists. A final possibility is that the Texas species is not *elongatus*, but a species that looks just like it; I reckon this is a sufficiency of possibilities.

Natural History: This species is readily found in mangrove twigs along the coast, and in tree islands of the Everglades. Its distribution has been studied in red mangrove islets in the Florida Keys, assuming that these studies deal with *elongatus* and not *cubaensis*. Occurring on all but the smallest islands, *elongatus* also quickly recolonizes islands from which it has been eliminated (Simberloff and Wilson 1969). In a study of many small islands, Cole (1983a) found that *elongatus* could live on islands as small as 5.1 cubic meters, and on even smaller islands if other ants are not present. Nests are in the peripheral zone of the mangrove crown, which reduces interactions with the dominant species, *Crematogaster ashmeadi* and *Xenomyrmex floridanus*, nesting in the interior of the tree (Cole 1983b). The behavior of *elongatus* within its nest has been studied by Cole (1983a). Larvae are fed bits of insects or unidentified material from a storage pocket (infrabuccal pouch) in the mouth. Workers groom each other and exchange regurgitated food; they spend considerable time licking larvae and occasionally stridulate in an obvious way. In the Everglades, *elongatus* visits the extrafloral nectaries of various plants (Koptur 1992).

Name Derivation: From *elongatus* (Latin), meaning "elongate" or "drawn out," apparently referring to the slender form of this species. There is no requirement in nomenclature that a descriptive name should have any diagnostic value.

Pseudomyrmex gracilis (Fabricius)—
Big Bicolored Slender Twig Ant (Plate 16)

Taxonomy and Similar Species: This species, which is large (7–8 mm long), hairy, and bicolored, cannot be confused with any other species of *Pseudomyrmex* in Florida. As usual with this genus, the situation becomes much more complex in the mainland Neotropics. Not only are there several species that are similar to *gracilis*, but *gracilis* itself seems to vary in appearance and behavior (Ward 1993). This species was first reported in Florida as *Pseudomyrmex mexicanus* (Whitcomb et al. 1972b), a name that was synonymized by Ward in 1993.

Distribution: There is a disjunct, introduced population of *gracilis* in Florida; the native range is south Texas into Argentina, assuming this is not some complex of cryptic species (Ward 1993). Another exotic population occurs in Hawaii (Reimer 1994). It is possible that this disturbance-adapted species is in the early phase of a more widespread dispersion through the tropics and subtropics. In Florida, *gracilis* occurs from the Keys north into Alachua, Clay, and Leon counties. Its presence in northern, inland counties that have brief episodes of freezing temperatures every winter suggests that it could probably persist in the relatively moderate coastal region all the way around the Gulf of Mexico. The first Florida record is from 1960, and a decade later, it was well established (Whitcomb et al. 1972a).

Natural History: This species thrives in both natural and disturbed areas, wherever there are large shrubs or broadleaf trees; it is less common in frequently burned pine flatwoods. Nests are usually in large hollow twigs on living trees and shrubs but may also be in large dead stems of herbs, such as *Bidens alba*, or in beetle galleries in dead branches. If *gracilis* had become established in Florida 150 years ago, before the advent of local myrmecologists, we would have little reason to suspect that it was exotic, especially if the population had moved around the Gulf Coast and joined with populations in Texas and Mexico. There is a limited supply of good housing for ants and other insects that live in hollow twigs, so it is possible that *gracilis* is affecting the populations of some native insects. *Pseudomyrmex gracilis* would be unlikely to have an impact on other *Pseudomyrmex*, which are usually found in much smaller twigs, but it might usurp shelters suitable for various species of *Camponotus*, solitary vespids such as *Pachodynerus*, and megachilid bees. Small insects, especially caterpillars, are collected by workers, which also consume honeydew from sap-sucking insects (Whitcomb et al. 1972a).

This species is a common resident of oak trees in south and central Florida, where individuals fall out of oak trees with such frequency that *gracilis* is sometimes called the "oak ant." It is well known as a stinging species, with most stings occurring on the neck of the victim, as mentioned above in the discussion of the genus. When a nest is opened, however, they do not normally attempt to sting, but distinguish themselves by their jumping ability as they attempt to escape. Alates can be found from March through November.

Name Derivation: From *gracilis* (Latin), meaning "slender."

Pseudomyrmex leptosus Ward— Parasitic Slender Twig Ant

Taxonomy and Similar Species: There are no worker *leptosus*, only queens and males. Queens resemble those of *pallidus*, but are considerably smaller, and the front of the head lacks the conspicuous punctures and semimat look of *pallidus*. In practice, it is easy to identify the species because it is a rare species that is unlikely to be found except in the nests of its host *P. ejectus*, where it stands out because its bright yellow color contrasts strongly with the brown color of its host. It is possible that *leptosus* also parasitizes yellow species such as *pallidus* and *simplex*, and has been overlooked because of its resemblance to these species. This, however, is unlikely, as most (but not all) parasitic ants have a single host.

Distribution: Florida, collected in a few widely separated sites (Alachua, Highlands, and Monroe counties), presumably occurring in other sites in peninsular Florida, but rare. It has been collected only five times.

Natural History: This species is a workerless parasite that apparently kills its host queen (Ward 1985).

Name Derivation: From *leptos* (Greek), meaning "slender."

Pseudomyrmex pallidus (F. Smith)— Common Orange Slender Twig Ant (Plate 17)

Taxonomy and Similar Species: This is a difficult complex of similar yellow orange species, including four in Florida. The front of the head and the base of the gaster of *pallidus* are conspicuously punctate, unlike those of *simplex* and *leptosus*. Specimens of *simplex* and some specimens of southern populations of *pallidus* have a black spot at the base of the gaster; punctures, if they are present, show up well in this black spot when it is viewed with diffuse lighting.

The real problem is differentiating between *pallidus* and *seminole*. Specimens of *seminole* are larger than *pallidus*, especially the queens, which are more than 6 mm long. The clypeus of *seminole* has a small but definite angle at the center of its anterior margin, while the clypeus of *pallidus* is straight or slightly concave. Both species occur in culms of grasses and sedges and in dead stalks of herbs, but *pallidus* also lives in dead vines and in twigs of trees and shrubs, habitats where *seminole* seldom occurs.

Distribution: Atlantic Coastal Plain from New Jersey through Florida, around the Gulf of Mexico and south into Costa Rica; Cuba and the Bahamas (Ward 1985). The restricted distribution of *pallidus* in the Caribbean and its absence in amber deposits from the Dominican Republic (Ward 1992) suggest that *pallidus* reached the West Indies from the mainland relatively recently, perhaps in the Pleistocene.

Natural History: This is a common species found in hollow stems of grasses and herbs and in dead vines and twigs. It appears to me that *pallidus* is much more common than *ejectus* in grass stems, but somewhat rarer in shaded twigs. In marshes, it occurs together with *seminole* in dead culms of grasses and sedges. The habitat preferences of Florida *Pseudomyrmex* are still somewhat unclear, especially since most ecological studies predated Ward's 1985 review of the Florida species. Cole's 1982 study of sawgrass culms, for example, needs to be revisited, as some of the supposed *pallidus* may have been *seminole*. Cole's work could also be used as a model for more extensive studies of sawgrass culms in the Everglades, where there are probably many thousands of acres of sawgrass occupied by species of *Pseudomyrmex*. Members of this genus must be of considerable ecological importance in the Everglades system.

Name Derivation: From *pallidus* (Latin), meaning "pallid, pale."

Pseudomyrmex seminole Ward— Seminole Slender Twig Ant (Plate 17)

Taxonomy and Similar Species: The gaster of *seminole* is covered with fine punctures (best seen with diffuse lighting on the first gastral tergite), unlike that of *simplex*. It is most similar to *pallidus* (the differences between the two are listed above under *pallidus*), and the two were recognized as separate species relatively recently (Ward 1985). It is possible to learn to distinguish the two species in the field by size alone: *seminole* is larger and is usually

darker orange. In mixed collections in museum trays, these differences are harder to see; it is not unusual for ants (like the rest of us) to be more distinctive when alive than when dead.

Distribution: This species occurs in the Bahamas, Florida, and around the northern and western Gulf of Mexico into Mexico (Ward 1985). It probably also occurs in coastal Georgia. Since the species is apparently absent from Cuba, this is probably a mainland species that reached the Bahamas in the Pleistocene.

Natural History: This species is usually found in hollow culms of grasses and sedges, such as sea oats, sawgrass, and large species of broomsedge (*Andropogon* spp.), but it also occurs in dead stalks of herbs. These nesting preferences tend to restrict the species to marshes and beach dunes. Like other species of *Pseudomyrmex*, *seminole* is completely vulnerable to fire, and the frequent fires of earlier Florida landscapes would have restricted all *Pseudomyrmex* to fire-free habitats. *Pseudomyrmex seminole* and *pallidus* would have been in dunes and marshes, *ejectus* would have been in hollow twigs and vines in swamp forest and tropical hammocks, and *elongatus* and *cubaensis* would have been in mangroves and tropical hammocks. The recently reduced fire frequency over much of Florida has probably allowed these species to gradually expand out of their traditional habitats.

On two occasions, queen *seminole* have been found with workers of *pallidus*, suggesting that *seminole* might be a facultative, temporary nest parasite of *pallidus* (Ward 1985).

Name Derivation: Named for the Seminole Tribe of Florida, currently inhabitants of the northern Everglades.

Pseudomyrmex simplex (F. Smith)— Shining Orange Slender Twig Ant (Plate 17)

Taxonomy and Similar Species: This species, like *seminole*, was apparently confounded with *pallidus* until Ward (1985) reviewed the Florida species. Florida specimens always have a pair of black spots on the gaster, and when these are viewed with diffuse light, it can be seen that the gaster lacks the fine, dense punctures occurring in *pallidus*, *seminole*, and queen *leptosus*. The remainder of the first gastral tergite is also smooth, but this is more difficult to see.

Distribution: This is a widely distributed species, occurring throughout the Caribbean, south and central Florida, and southern Mexico south through Brazil (Ward 1985). In Florida, *simplex* occurs as far north as Brevard and Orange counties. It is a little odd that *simplex* is absent from Texas and most of Mexico, but it could have been overlooked. Arboreal species of *Pseudomyrmex* can be very difficult to find. I was in several sites in Dominica where I would have had no idea there was an apparently thriving population of *Pseudomyrmex* were it not for the fact that an occasional clumsy individual would fall out of the trees onto my head. There is no doubt that *simplex* is native to the Greater Antilles (and probably Florida as well), because there are subfossil specimens in copal from the Dominican Republic (Ward 1992).

Natural History: This is an arboreal species, usually found in hollow twigs or vines, and seldom in grass culms or terrestrial herbs. At the Archbold Biological Station, it often occurs in hollow twigs of scrub hickory in mature Florida scrub. I suspect that *simplex* spread to such habitats from swamp forests (where it also occurs) after scrub fires became less frequent.

Name Derivation: From *simplex* (Latin), meaning "simple" or "plain," perhaps referring to the lack of sculpture and conspicuous hairs, relative to most other species of *Pseudomyrmex*.

SUBFAMILY MYRMICINAE

Genus *Aphaenogaster*

Long-Legged Ants

The genus *Aphaenogaster* (pronounced **A fē" nō gas' ter**) is a large genus with almost 200 species, and probably a good number of species to be discovered. Members of this genus are slender bodied, with long legs and antennae. The head and mesosoma are usually finely or coarsely sculptured, and most species have well-developed propodeal spines (but not the southeastern species *A. floridana*). In the field, *Aphaenogaster* species have a strong resemblance to minors of some *Pheidole*, such as *dentata* and *obscurithorax*, but *Aphaenogaster* workers are considerably larger (usually 3.5 to 5 mm), and not, of course, accompanied by major workers.

Most North American species of *Aphaenogaster* are easily identified, but there are some difficult species in the *fulva–rudis–texana* complex, which was reviewed by Umphrey (1996). Several species, distressingly, can best be distinguished by genetic markers, leaving museum specimens unidentifiable. Two Florida species, *carolinensis* and *miamiana*, may not always be separable in parts of north Florida where the species are supposed to have overlapping ranges.

The species of *Aphaenogaster* have the look of archetypal ants and have long been considered primitive and generalized (Smith 1961). Several species have been described from Baltic Oligocene amber, along with species in the current myrmicine genera *Monomorium*, *Temnothorax*, *Myrmica*, *Oligomyrmex*, *Pheidole*, *Solenopsis*, *Stenamma*, and *Vollenhovia* (Bolton 1995). The distribution of living species seems to reflect both Madrotertiary and Arctotertiary patterns. There are many species in semiarid areas of Eurasia and Mediterranean North Africa, and a smaller number in Southwestern North America. A possible remnant of an Arctotertiary forest fauna extends into Central America, and through the Australasian Region into Australia. *Aphaenogaster* is not represented in southern Africa and South America.

North American *Aphaenogaster* occur in various habitats but are usually in natural or seminatural sites. There are no "weedy" species or invasive species. *Aphaenogaster miamiana* might be exotic on Gorda Key in the Bahamas, where it was found recently, but there is no good evidence of this. Otherwise, there are no indications of *Aphaenogaster* species being transported by humans. There are no pest species, with the possible exception of *pythia*, which is said to damage sugar cane in Australia (Smith 1961). Several North American species are known from only a few sites, but there is no good reason to believe that any species are endangered.

Florida species of *Aphaenogaster* have no pest status. It is true that four species from eastern North America appear in Smith's treatise on house-infesting ants (1965), but he included a number of species of ants that I suspect would only appear in houses as strays on rare occasions. Sometimes, I think that Smith wanted an excuse to present his great store of natural history information about as many species as possible; sometimes, I wonder whether he was taking advantage of the opportunity to have made a large number of high-quality pen-and-ink illustrations of ants, presumably paid for by the U.S. Department of Agriculture. Either motive seems valid to me. At any rate, none of the Florida species is a household pest, nor do they invade stored foods or even encourage sap-sucking Homoptera that attack useful plants. The sting apparatus is not a penetration device but has been modified to apply, or possibly spray, defensive chemicals (Kugler 1978, 1979). As predators, *Aphaenogaster* species might have a role in controlling populations of phytophagous insects, but this has not been documented. Certain species appear to be important as seed dispersers in various regions (Handel 1978; Handel and Beattie 1990; Higashi et al. 1989). Good candidates for Florida plants dispersed by *Aphaenogaster* are members of the genera *Polygala*, *Viola*, and *Trillium* and perhaps some sedges and rushes.

The natural history of Florida species of *Aphaenogaster* was described by John F. Carroll in a PhD thesis (1975) that, unfortunately, was never published, or even summarized in print. I have borrowed extensively and gratefully from this work in the species accounts that appear below. I hope that this will bring a little more recognition to Carroll's valuable contribution.

Males and queens of Florida *Aphaenogaster* have been described by Carroll (1975), except for *mariae* and *umphreyi*, whose males are unknown.

Name Derivation: From *aphaino* (Greek), meaning (in this case) "not shining," and *gaster* (Greek), meaning "belly," or, in the case of hymenopteran morphology, "gaster." This refers to the nonshining gaster of *Aphaenogaster sardoa*, the species for which the genus was established. Many North American species have a shining gaster.

Aphaenogaster ashmeadi (Emery)— Ashmead's Long-Legged Ant (Plate 18)

Taxonomy and Similar Species: This species resembles *treatae*; both species are large (often 7–8 mm), with a conspicuous expanded lobe (of unknown function, but possibly serving to protect the antennal base) at the base of the antennal scape. In *ashmeadi*, this lobe is broad in a dorsal view and thin and convex in a side view; in *treatae* it is thickened, such that there is a distinct lateral face in lateral view. In Peninsular Florida, *ashmeadi* is blackish brown, while *treatae* is reddish brown, but in the western Panhandle, both are reddish brown.

Distribution: North Carolina south into Florida, west into Missouri and Texas (Smith 1979). In Florida, *ashmeadi* is known from Highlands County northward and west through the Panhandle. There are some large gaps in this known distribution, but *ashmeadi* probably occurs in most of the upland natural habitats in central and north

Florida. Despite its large size, *ashmeadi* is often difficult to find, and workers have a tendency to freeze or hide under dead leaves when alarmed.

Natural History: The information in this paragraph is summarized from Carroll (1975). Nests are in soil in a variety of habitats, ranging from mesic hammocks to open xeric forest, and from undisturbed forest to shaded lawns. Nests usually have one or two entrances (up to six), which may be marked with a short turret of plant debris and insect remains. There are usually five to seven chambers, including a superficial chamber with pupae and prepupae, a refuse chamber, and a deeper chamber, often approximately 25 cm below the surface, containing the queen. Mature colonies (those producing alates) usually contain approximately 100 to 250 workers. The largest colony had 423 workers. Alates are in the nest by mid-April, and flights are in June and July. Workers forage both day and night in warm weather, usually in shaded or semishaded habitats. Almost all foraging is on the surface of leaf litter. The diet of *ashmeadi* is primarily live and dead arthropods. Foragers are able to subdue caterpillars, small spiders, Diptera, Orthoptera, and smaller ants. Certain mushrooms are cut up and brought back to the nest, the most frequently collected belonging to the genera *Russula* and *Marasmiellus*.

Van Pelt (1958), who observed *ashmeadi* in Putnam County, Florida, found it in well-drained habitats, especially in xeric forests and scrubby flatwoods, and more rarely in sandhills, mesic forests, and bayheads. One nest included 333 workers, as well as 250 pupae, eggs, and a queen. Workers did not forage during winter. In North Carolina, Carter (1962) found *ashmeadi* restricted to the Coastal Plain, where it usually occurred in open forests on sandy soil. On one occasion, I found the pupal skins of four mydas flies protruding from the ground adjacent to the nest hole of *ashmeadi* in low flatwoods, suggesting that the flies might have been inquilines as larvae.

Name Derivation: Carlo Emery presumably named this species (named as a subspecies of *treatae*) for the American hymenopterist William Harris Ashmead (1855–1908), although this is not mentioned in the description. The type is from Florida and Ashmead had worked in Florida for some years, as well as spending a year in Berlin, where Emery might have met him or encountered his specimen of *Aphaenogaster*. Ashmead was an important early hymenopterist, describing some

3100 new species and 607 new genera (Mallis 1971). This productivity may have been aided by the fact that Ashmead only slept three or four hours a night. From 1895 to 1908, Ashmead was a curator of insects at the U.S. National Museum.

Aphaenogaster carolinensis Wheeler— Carolina Long-Legged Ant (Plate 19)

<u>Taxonomy and Similar Species:</u> This species may be separated from most other Florida species of *Aphaenogaster* by the following combination of features: base of the antennal scape not expanded (eliminating *ashmeadi*, *flemingi*, and *treatae*); frontal lobes not notched (eliminating *lamellidens*); anterior edge of mesonotum only slightly raised (eliminating *fulva* and *umphreyi*); postpetiole without a ventral, forward-projecting protuberance (eliminating *tennesseensis* and *mariae*); propodeal spines well developed (eliminating *floridana*). There is, however, no simple way known to separate *carolinensis* from *miamiana* in Florida. This is an annoying problem for the field biologist, because *miamiana/carolinensis* is the most abundant Florida *Aphaenogaster*.

Both *carolinensis* and *miamiana* were described as subspecies of *texana*, but *miamiana* was raised to species level by Creighton in his manual of North American ants (1950). When John Carrol reviewed Florida *Aphaenogaster* (1975), he decided that neither *carolinensis* nor *miamiana* could be forms of *texana*, a conclusion that has been justified by subsequent work (Umphrey 1996). Since Carroll's work remained unpublished, *carolinensis* was not recognized as a full species in a taxonomic work for 20 years (Umphrey 1996). The remaining problem for Carroll was to deal with the *miamiana–carolinensis* complex. Carroll found that, in southern Florida, there is a large, long-spined species that is reddish brown with a darker gaster; this is the species Wheeler described as *miamiana*. In north Florida, however, there is a variety of forms, some larger, some smaller, varying in color and spine length. Carroll finally concluded that *miamiana* intergraded with *carolinensis* in the northern part of its range but did not go so far as to synonymize *miamiana*. When I began to work on this problem, I was confident that with more material from Florida and other southeastern states, I would be able to provide a definitive solution. All that I achieved was eye strain, and a reinforcement of Carroll's view that something was happening with *miamiana* in the northern part of its range. The frustration is exacerbated by the fact that *Aphaenogaster* species generally have plenty of features that look as if they would be perfect for species discrimination, especially the patterns of sculpture on the head and body. These, unfortunately, often seem to vary more within species than between species. A similar situation occurs in the genus *Myrmica* to the north of Florida.

Enter, in the mid-90s, Gary Umphrey, equipped with superior technology, intensity, and patience. He showed (1996) that the karyotype of *carolinensis* differs strikingly from that of *miamiana* and that both species, particularly *miamiana*, have greater geographic ranges than expected. Alas, there are still no convenient morphological traits that can be used to differentiate the species, although *carolinensis* seems to be lighter in color and have shorter propodeal spines than *miamiana* in north Florida. The number of colonies studied was limited by the problems of obtaining and preparing karyotypes, and making the precise measurements used in morphological analysis. Only one colony of *carolinensis* and eight colonies of *miamiana* were sampled from north Florida, so it is possible that the whole story is not known. For example, it is possible that *carolinensis* begins to converge in color with *miamiana* at the southern edge of its range, a phenomenon that is not so unusual in ants. North of Florida, the range of *carolinensis* overlaps with additional sibling species that do not yet have scientific names (Umphrey 1996). Eventually, these should be formally described. Two species with overlapping ranges are separate evolutionary (and probably ecological) units, even if they cannot be identified in any convenient way. Recently, some enthusiastic visionaries of DNA "barcoding" have suggested that someday we will be able to pop our specimens in a tiny blender and have their names appear on the screen of a compact genetic processor. This is probably the only way that it could become easy to identify some complexes of *Aphaenogaster* in some parts of eastern North America.

<u>Distribution:</u> According to Umphrey (1996), *carolinensis* occurs over the whole Coastal Plain, from New Jersey through the Florida Panhandle, and into the Piedmont of the Carolinas. In Florida, there is only one definitive record, a karyotyped series from Walton County, 7.5 km west of Bruce (Umphrey 1996).

<u>Natural History:</u> The natural history of *carolinensis* is unclear, as there are so few verified records. This species probably nests in rotten

wood, in deep accumulations of leaf litter, and under objects on the ground, much like *miamiana*. Carroll's work (1975) on the natural history of *miamiana/carolinensis* can probably be referred almost entirely to *miamiana*.

Name Derivation: The name *carolinensis* means "an inhabitant of Carolina."

Aphaenogaster flemingi M. R. Smith— Fleming's Long-Legged Ant (Plate 19)

Taxonomy and Similar Species: A highly distinctive species, *flemingi* has a shining pronotum, extremely long, slender propodeal spines, and the base of the antennal scape is expanded like the base of an arrowhead. Even in the field *flemingi* is easily identified as a large, reddish brown species with a shining pronotum.

Distribution: North Carolina south through Florida, west into Kentucky and Louisiana. In Florida, *flemingi* is known from scattered localities extending from the Keys through the Panhandle. It is probably more common than these records indicate: *flemingi* is usually difficult to find because the nest entrances are concealed and workers are often active at dusk or during the night.

Natural History: The following paragraph is summarized from Carroll (1975). Nests of *flemingi* are in soil in open pine flatwoods and stands of scrub oak, usually at the base of a plant or clump of grass. Nests have one or two entrances, one of which has a thatched turret of bits of vegetation. Nests are in either well-drained or poorly drained soil, and may be more than 25 cm deep. Brood are often in a superficial chamber under a clump of grass. Colonies contain up to approximately 300 workers. Foraging is on the ground in open or semishaded habitats. Workers bring arthropods to the nest and have been seen visiting mushrooms of the genus *Russula*. A colony kept in the laboratory fed on pieces of these mushrooms.

Van Pelt (1958) observed *flemingi* (which he called by the synonymous name *macrospina*) in Putnam County, Florida. He found it associated with pine-dominated habitats, including sandhills and flatwoods. He also found that *flemingi* was attracted to molasses traps. In North Carolina, Carter (1962) found *flemingi* restricted to the Coastal Plain, where it showed a preference for open, dry, grassy, sandy sites, including fields, sandhill, and coastal dunes, but two collections were made in wetter grassy areas. I have found *flemingi* in dry, grassy areas, wet flatwoods, and even salt marshes. On several occasions, I have seen columns of foragers emerging at dusk.

Name Derivation: In his description (Smith 1928) of *flemingi*, which he treated as a subspecies of *texana*, Marion Smith writes, "This new subspecies is named in memory of the late Mr. Andrew Fleming of Sibley, Mississippi, a man who made many important contributions to the knowledge of the ants of Mississippi."

Aphaenogaster floridana M. R. Smith— Florida Long-Legged Ant (Plate 21)

Taxonomy and Similar Species: This is a yellow species with an arrowhead-shaped base on the antennal scape, but its most distinctive feature is the absence of propodeal spines, which are well developed in all other Florida species of *Aphaenogaster*. Males and queens have small spines, suggesting that the loss of spines in the worker is a recent evolutionary event. The adaptive significance of this is unknown, but two other Florida ants of open, sandy areas, *Pogonomyrmex badius* and *Pheidole morrisi*, also seem to have recently dispensed with propodeal spines, although they occasionally appear in *P. badius*. The primary function of the propodeal spines is to protect the vulnerable petiolar joint, best achieved by raising the petiole so that it rests between the petiolar spines. This is probably most useful for species that grapple and sting, and would not be as useful for species that spray repellents. Three possible hypotheses explaining the loss of propodeal spines in *floridana* are as follows: this species uses another form of defense, such as powerful chemical deterrents; this species has a way of avoiding grappling encounters; this species lives in a habitat where most competitive, aggressive, or defensive encounters are with other ants that spray repellent chemicals.

Distribution: North Carolina south into Florida, west along the Gulf Coastal Plain into Alabama. In Florida, *floridana* is relatively common in uplands through north Florida, much rarer to the south, although it has been found on the Lake Wales Ridge in Highlands County.

Natural History: The following paragraph is summarized from Carroll (1975). Nests are in well-drained, sandy soil in open areas such as fields, sandhills, or Florida scrub habitats. There are usually one or two nest entrances, sometimes with a short turret of plant debris and arthropod remains, and often situated at the base of a clump of grass. The main entrance may be changed every few days or weeks. Nests are deeper than those of other Florida *Aphaenogaster*, sometimes 1.3 meters deep. Foraging is on the ground, crepuscular or nocturnal during warm weather. Foragers collect various arthropods, including *Pheidole morrisi*, and alate dispersing *Solenopsis invicta* and *S. geminata*. This species collects seeds of the hemiparasitic plant *Seymeria pectinata*, and pieces of *Russula* mushrooms. Alates are in the nest from late June and early July, and may be seen at nest entrances in late August and early September. The largest colony seen had approximately 200 workers.

In North Carolina, Carter (1962) found *floridana* on the Coastal Plain in grassy, open sandhill areas, and similar habitats, including grassy depressions between coastal dunes. In Florida, Van Pelt (1958) found *floridana* in open sandhill areas and open, sandy roadsides. He states that most foraging is nocturnal or on cloudy days.

Name Derivation: Named for the state of Florida, where the types were collected, near Gretna, in Gadsden County.

Aphaenogaster fulva Roger— Ridge-Backed Long-Legged Ant (Plate 20)

Taxonomy and Similar Species: Workers distinguished from the generally similar *miamiana* and *carolinensis* by the irregular, strongly raised ridge (sometimes with a saddle) running across the front border of the mesonotum. The base of the antennal scape is simple (unlike *flemingi*, *treatae*, and *ashmeadi*), the propodeal spines are well developed (unlike *floridana*), the postpetiole lacks a forward-projecting, ventral protuberance (unlike *tennesseensis* and *mariae*), and the frontal lobes are not notched (unlike *lamellidens*). The most similar species is *umphreyi*, a subterranean species with short propodeal spines and tiny eyes. The eye of *umphreyi* is about the same width as that of the last segment of the antenna. In Florida, *fulva*, which is not a subterranean species, is also darker than *umphreyi*, but elsewhere there are reddish forms of *fulva*. As with some other species of eastern *Aphaenogaster*, small individuals, which are often from young colonies, may not have some of the features of large individuals. In *fulva*, the head may be somewhat longer with respect to its width, the propodeal spines may be relatively short, and the sculpture may be less conspicuous.

Distribution: Vermont south into Florida, west into Colorado and Louisiana (Smith 1979). In Florida, *fulva* is known from the northern part of the Peninsula through the Panhandle. It is not known from the swamp forests of the southern Peninsula, although it occurs in such habitats farther north.

Natural History: The following paragraph is summarized from Carroll (1975). In Florida, this species usually lives in densely forested areas. It often occurs in seasonally flooded swamp forests, foraging in hummocks, logs, and trees when the water is high. Nests are in rotten wood or in cavities near the base of live trees, sometimes extending into the surrounding soil. Foraging is diurnal. Arthropods are the principal prey, but foragers in wet areas also capture mollusks and oligochaetes, and collect seeds and pieces of fallen flowers. In north Florida, alates are in the nest by early May, and flights occur from early June to mid-July. Colonies contain up to approximately 800 workers.

In North Carolina, Carter (1962) found *fulva* most commonly in mesic forests, especially in bottomlands and moist mountain forests. A few colonies were collected in xeric sites as well, including sandhills and xeric oak and pine stands. Colonies were in rotten logs and stumps, and in soil under stones. It is possible that some of these colonies were some other species, although the characters used to separate *fulva* in Creighton should work well in North Carolina, as far as is known. In Putnam County, Florida, Van Pelt (1958) found *fulva* common in floodplain swamps and rarer in scrub, flatwoods, wet hammocks, and marshes. He observed that *fulva* tends to replace *ashmeadi* in wet areas. Nests were found under logs, in deep litter, in dense mats of roots, and in rotten logs and stumps. Males were found from May through July. Foragers were attracted to peanut butter and oatmeal baits. Colonies were sometimes near colonies of the termite *Reticulitermes flavipes*, and foragers were seen carrying live termites. In Connecticut forests, Weseloh (1994) found *fulva* as one of the three commonest ants, the others being *Formica subsericea* and *F. neogagates*. All three of these species had superficial nests under leaf litter and were randomly distributed with respect to each other. In tests of the efficiency of finding gypsy moth larvae on the forest floor, *A. fulva* was not even mentioned, so this ant probably does not attack these larvae. In Maryland, Fellers and Fellers (1976) studied "tool use" in four species of *Aphaenogaster*, including *A. fulva*. The ants carried bits of debris to jelly bait, allowed the fragments to absorb and accumulate jelly, and then carried the jelly-coated material back to the nest. Work on the chemistry of the poison gland of *fulva* (Wheeler et al. 1981) may be pertinent to the tool-using study. Disks of paper treated with poison gland extract (anabaseine) produced by *fulva* were attractive to ants, who picked them up and carried them into the colony or around the foraging area. It is possible that this compound might be used to organize tool-using. On the other hand, when the normal food of the colony was laced with anabaseine, it was attractive to the ants, but prevented feeding. Mandibular glands of *fulva* produce a chemical, methyl anthranilate that appears to function as an alarm pheromone (Duffield et al. 1980).

Name Derivation: From *fulva* (Latin), meaning "reddish yellow," presumably referring to the color of the type specimen.

Aphaenogaster lamellidens Mayr—
Notched Long-Legged Ant (Plate 18)

Taxonomy and Similar Species: The frontal lobe of this species is elevated to form a thin ridge that is notched posteriorly so that it projects posteriorly as a tooth. The antennal scape can fit into this notch, perhaps protecting the base of the scape, or allowing it to fold back more compactly against the head. This diagnostic character is not as easy to see as it might seem, because the head of the ant must be positioned at just the right angle for this notch or tooth to be visible. In the field, workers are distinctively colored, a dark reddish brown with pale coxae that contrast with the dark femora and tibiae. The gaster is often paler than the rest of the body. In the Appalachians, workers of *fulva* and *tennesseensis* may be colored like *lamellidens*, perhaps forming a mimetic trio. Foraging *lamellidens* are usually seen walking slowly up or down the trunks of large hardwoods.

Distribution: New York south into Florida, west into Illinois and eastern Texas (Smith 1979). In Florida, *lamellidens* is known from Highlands County northward, including most of the Panhandle. Records are scattered because this species is often difficult to find without baiting a large number of trees.

Natural History: The following paragraph is summarized from Carroll (1975). In Florida, the nests of *lamellidens* are almost always in dead portions of living hardwood trees. There is some evidence that colonies may be founded in rotten logs on the ground, and these colonies either fail or relocate into standing live trees. In its range north of Florida, *lamellidens* colonies may be permanently in fallen logs, and may also be in standing pines, which do not seem to serve as nest sites in Florida. The shape and extensiveness of the nest depends on the preformed cavity in which it occurs. Old galleries of termites and other insects may be modified, and large holes or cracks to the outside may be sealed with wood chips, but *lamellidens* does not construct its own extensive galleries in sound wood. Foragers explore the trunk and large branches of the nest tree, and make expeditions to the surrounding ground, but do not forage on leaves and twigs. Workers collect arthropods, such as crickets, caterpillars, staphylinid beetles, small spiders, psocopterans, and chironomids, as well as flower petals and seeds. Alates were found in a colony in June. Mature colonies may contain more than 600 workers.

In a survey of the ants of Tennessee (Dennis 1938), *lamellidens* was the commonest *Aphaenogaster* found, except for *fulva* (the latter probably a combination of several species). It was collected in many forest types, up to a level of 2500 feet in the mountains, nesting in stumps and rotten logs. Dennis (1938) suggests that low temperatures restrict *lamellidens* to lower elevations. In North Carolina, Carter (1962) found *lamellidens* in hardwood, pine, and mixed forests, usually in mesic rather than xeric conditions. The species seemed to be more common in pine forests. Nests were in rotten wood, stumps, and standing dead trees. Large numbers of individuals could be found under loose bark. In Florida, I, like Carroll, have found *lamellidens* restricted to dead, often rotted out, portions of large, standing, live trees, always in mesic forests or edges of swamp forests. This regional difference in nest site may possibly reflect the history of Florida and adjacent areas of the Coastal Plain. During parts of the Pleistocene, the climate was much drier, and large trees and logs may have been confined to riverine corridors and bottomlands. Seasonal flooding could have precluded nesting in fallen logs, leaving the rotted out portions of standing trees as the only useable nest sites. As recently as two or three centuries ago, the uplands of Florida in their native state burned frequently, leaving open pine forest with little dead wood, as fallen dead trees were mostly consumed by fire. Swamp forests and floodplain forests would have been the primary refuges for *lamellidens*, with logs and stumps submerged or saturated during periods of high water.

Aphaenogaster lamellidens has a strong chemical repellent that is easily detected by the human nose and unpleasant to taste when specimens are aspirated from tree trunks.

Name Derivation: From the diminutive of *lamina*, *lamella* (Latin), meaning "a small thin plate," and *dens* (Latin), meaning "tooth," referring to the expanded frontal lobes that are notched, leaving a projecting tooth.

Aphaenogaster mariae Forel— Mary's Long-Legged Ant (Plate 19)

Taxonomy and Similar Species: This rare species has a conspicuous ventral protuberance on the postpetiole and sculpture of coarse, irregular ridges on the mesopleuron and both features are also found in *tennesseensis*. *Aphaenogaster mariae* is distinguished from *tennesseensis* and other species of *Aphaenogaster* by the fine ridges that radiate from the base of the first gastral tergite and cover about a quarter of the tergite. This feature appears here and there in other species of myrmicines, and is regularly present in the dacetines, but its significance is unknown. One possibility is that these areas serve as channels and evaporative areas for chemical repellents.

Distribution: New York south into Florida, west into Iowa and Kansas (Smith 1979). I have seen only two specimens from Florida, one from Liberty County and the other from Orange County. Both were taken in traps. The type specimen is also from Florida, with the locality unknown.

Natural History: Little is known about the biology of this rare species. Wheeler (1910a) suggested that *mariae* might be a temporary nest parasite of some other *Aphaenogaster*, such as *fulva*. He based this on the small size and long propodeal spines of the queen, both characteristics of the apparently related species *tennesseensis*, which is also thought to be a temporary nest parasite. In Ohio, however, Wesson and Wesson (1940) found *mariae* "frequently," in oak trees, the nests in branch stubs or in rotten holes in the trunk. The Wessons did not find any other arboreal *Aphaenogaster*, although they regularly found several other species of arboreal ants. It seems a bit unlikely that they would have missed a species of arboreal *Aphaenogaster* common enough to serve as the host for *mariae*. Of course,

the nesting ecology of a species of ant is not necessarily constant throughout its range; *lamellidens* seems to be an example of this. In Florida, one specimen was collected in a red maple swamp; the other was probably collected in a hardwood forest.

Name Derivation: August Forel named this species for Mrs. Mary Treat, who sent him a specimen from Florida. I am impressed that she found this species, which I have sought in vain for years. Mary Treat (1830–1923) was an important self-taught scientist who wrote several books on natural history, including one on butterflies. Her work was noted by Charles Darwin in his book on insectivorous plants, and the two exchanged letters on the subject. Her work was often published in popular magazines rather than scientific journals, and probably inspired a generation of naturalists. Many of her studies were done in her yard and garden in New Jersey, or in the nearby Jersey Pine Barrens. A patient, intelligent, original observer and excellent writer, she might have been an academic star if she had been born at a time or place when there was less disparagement of what was called "the female intellect."

Aphaenogaster miamiana Wheeler— Deep South Long-Legged Ant (Plate 19)

Taxonomy and Similar Species: This topic has already been covered above, under *carolinensis*.

Distribution: Colonies identified by karyotype are known from southern peninsular Florida (Highlands County) through north Florida, including the Panhandle, and from south-central Georgia (Toombs County) (Umphrey 1996). Umphrey (1996) suggests that the range of *miamiana* probably extends through the Coastal Plain from North Carolina into eastern Texas. In Florida, specimens I have provisionally identified as *miamiana* were collected throughout the state, from the Keys north. Zachary Prusak has collected *miamiana* on Gorda Key in the Bahamas. It is not known whether it is native or imported in the Bahamas. During the lower sea levels of the Pleistocene, the Little Bahama Bank was probably less than 100 km from Florida, so natural dispersal is not very improbable. On the other hand, it is easy to imagine importation of Florida cabbage palms, complete with ants, to landscape some resort.

Natural History: The following paragraph is summarized from Carroll (1975), who was probably observing *miamiana* rather than *carolinensis* most of the time, perhaps all the time. Nests are in a variety of habitats, including disturbed habitats, such as the edges of fields and lawns. Nests, which are almost always in shaded locations, may be in dead wood or pine cones buried in leaf litter, in rotten logs and stumps, in cavities at the bases of trees, in leaf bases of cabbage palms, or in soil. The nest entrance is not marked by a turret. When nests are in deep leaf litter and humus, there are one to three chambers, with the brood in an upper chamber. Foragers collect various arthropods, mollusks, annelids, mushrooms (especially species of *Russula*), seeds of sweet gum and dogwood, and flower parts of elderberry. Mature colonies may contain up to approximately 800 workers.

In south Florida, *miamiana* is usually in dense forests, including xeric hammocks, bayheads, and swamp forests. It can be found reliably in leaf bases of cabbage palms at the edges of swamps. The location of nests suggests that this species is not favored by burning, and the reduction in frequency of natural and control burns has probably led to an increase in abundance of *miamiana*.

Name Derivation: Named for Miami, where the type series was collected.

Aphaenogaster tennesseensis (Mayr)— Curved-Spined Long-Legged Ant (Plate 20)

Taxonomy and Similar Species: This species has a protuberance on the ventral side of the postpetiole, and coarse irregular ridges on the mesopleuron, both features shared by *mariae*. It lacks the long, fine ridges at the base of the first gastral tergite found in *mariae*. *Aphaenogaster tennesseensis* differs from all other Florida species in the lack of any erect hairs on the mesosoma, petiole, postpetiole, and gaster. It is also distinguished by its extraordinarily long propodeal spines, which are thick at the base and somewhat curved, tapering to a sharp point. In the field, *tennesseensis* might be mistaken for *lamellidens*.

Distribution: Quebec south into Florida, west into Minnesota and Oklahoma (Smith 1979). In Florida, *tennesseensis* is known from a few sites in the northern part of the state. It appears to be rare in Florida.

Natural History: This species usually occurs in mesic woodlands. It is believed to be a temporary nest parasite of other *Aphaenogaster*, on the basis of the small size and large spines of the queen, and the discovery of three small mixed colonies of *tennesseensis* and some species in the *fulva–rudis* complex

(Wheeler 1910a). These colonies were found under stones, rather than in rotten wood, where mature colonies of *tennesseensis* occur (Wheeler 1910a). Nests may be in rotting stumps or logs, in standing dead trees, and in dead portions of live trees (Smith 1965). Mature colonies have several hundred to several thousand individuals (Smith 1965). The latter estimate would be unusually high for a species of eastern *Aphaenogaster*. Foraging is usually on the ground, where the workers collect small arthropods (Carroll 1975). Alates have been found in the nest in August (Carroll 1975).

Name Derivation: Named for the state of Tennessee, where the type specimen was collected.

Aphaenogaster treatae Forel— Treat's Long-Legged Ant (Plate 18)

Taxonomy and Similar Species: This is a large species resembling *ashmeadi*. Both species have a conspicuous lobe at the base of the antennal scape; this might serve to protect the joint at the base of the antennae, but if this is the case, one might expect to see the structure in more species of ants. In *treatae*, this lobe is thickened, with a distinct lateral face, while in *ashmeadi* the lobe is thin, although it may look thicker from some angles because its dorsal surface is convex. In peninsular Florida, *treatae* is reddish brown, while *ashmeadi* is blackish, but in the Panhandle, both species may be reddish brown. A subspecies of *treatae*, *treatae pluteicornis*, has been described for the population occurring from Alabama into Texas; the two intergrade in Alabama (Creighton 1950), and there is no reason to believe that this subspecies is actually a separate species.

Distribution: The subspecies *treatae treatae* occurs from Ontario south into Florida, west into Michigan and Alabama; the subspecies *treatae pluteicornis* occurs from Alabama west into Oklahoma and Texas. In Florida, *treatae* occurs in sandy uplands from Collier County northward and west through the Panhandle, although there are many large gaps in this distribution. This species is often difficult to find even where it is known to occur.

Natural History: The following paragraph is summarized from Carroll (1975). Nests are in well-drained sandy soil, generally in open stands of pines and oaks. In Florida, the nest entrances are usually at the base of a clump of grasses or herbs, while in northern parts of its range, *treatae* is often associated with rocks or logs. Nests have one or two entrances, one of which is usually marked by a small turret of plant debris. Nests go down to approximately 45 cm, with pupae and prepupae in a superficial chamber and larvae in deeper chambers. Foraging is on the surface of the ground, always in open to partially shaded areas. Workers commonly collect arthropods, occasionally collect seeds and bits of *Russula* mushrooms. Alates are in nests from mid-May through mid-July. Florida colonies usually have approximately 200 workers.

In Michigan, *treatae* was studied in detail by Talbot (1954, 1966). Nests often had a chamber that was mostly above ground and roofed with a thatch of plant material and soil, presumably for warming pupae or prepupae when the soil temperature was cool. Talbot made a complete inventory of the individuals in 30 nests. Numbers of workers ranged from 65 to 1662, with an average of 682. The locations of nests were also plotted in a sandy, abandoned field. In a study plot 100 feet by 120 feet, there were 63 colonies, providing an average of 32 workers that might forage over every square yard. Alates fly under precisely controlled conditions, when the temperature is 78°F–88°F, the air is calm, the sun is partly or completely obscured by clouds, but the ambient light remains relatively high and there is no rain (Talbot 1966). Alates sometimes wait in the nest entrance and rush out as soon as a cloud obscures the sun. In North Carolina, Carter (1962) found *treatae* common throughout the state, living primarily in fields and in open forests, and only rarely in closed, mesic forests.

Name Derivation: August Forel named this species for Mary Treat, who sent him the type specimen. Forel named both *mariae* and *treatae* for Mary Treat consecutively in the same 1886 paper (Forel 1886). Some years earlier (1879), Mary Treat published an extensive account of a colony of a slave-making *Formica*, with detailed descriptions of raids and some preliminary experimental manipulations. Lacking scientific credentials, and perhaps primarily interested in stimulating general interest in the natural history of ants, Treat published her paper in *Harper's Monthly Magazine*. This article is currently available online, and can be found by entering "A chapter in the history of ants" as a Google query. Treat's observations are still useful and raise some interesting questions, assuming the account is accurate. There are plenty of startling anthropomorphisms, as one might expect for the time, considering that the first volumes of Fabre's *Souvenirs Entomologiques* (which mark the beginning of our modern understanding of insect behavior) were also published in 1879. This is not to say that insect behavior is widely understood even today. It is common to see a statement that the behavior of ants and other insects is "programmed," without much thought about what kind of programming would enable an insect to make a living in a highly complex, diverse, and inconstant environment.

Aphaenogaster umphreyi (Deyrup and Davis)— Umphrey's Long-Legged Ant (Plate 20)

Taxonomy and Similar Species: This species is distinguished from all other species of eastern North America by its tiny eyes, which are about the same width as the last segment of the antenna. This is a reddish brown, subterranean species that seems most closely related to *fulva*, but the resemblance could be superficial (Deyrup and Davis 1998).

Distribution: This species is known only from Florida and Georgia, but might have a wider range, as it is difficult to collect. In Florida, *umphreyi* is known from Highlands, Marion, Alachua, Putnam, Liberty, and Okaloosa counties. It has been collected in Emanual County in Georgia.

Natural History: This is a subterranean species that lives in sandy areas. It probably emerges only at night. On two occasions, it was collected below rotten stumps, and it is possible that this species preys on subterranean termites, which move freely through the soil of the Florida scrub and sandhill areas where *umphreyi* lives. There is no known way to collect *umphreyi* except by digging and sifting, and no nests have been discovered. One specimen was collected in a sunken bowl trap, so pitfall traps may be useful for collecting this species. It remains the least known species of *Aphaenogaster* in eastern North America.

Name Derivation: Deyrup and Davis (1998) explain the species epithet *umphreyi*: "This species is named in honor of Dr. Gary Umphrey, in recognition of his long labors working to elucidate the taxonomy and phylogeny of the intractable *A. rudis* group."

Genus *Cardiocondyla*

Sneaking Ants

The genus *Cardiocondyla* (pronounced **Kar' dee ō kon"di luh**) is characterized by small size, three-segmented antennal club, propodeal spines or angles (unlike *Monomorium*), postpetiole much broader than petiole, and an absence of more or less erect hairs on the dorsal surfaces of the head and body. This lack of standing hairs distinguishes *Cardiocondyla* species from some similar species of Florida *Temnothorax*, such as *Temnothorax torrei*. Field myrmecologists can learn to identify these ants by their small size; slender, elongate shape; and their characteristic way of moving discretely and singly among other ants, with the body almost touching the ground—hence the name "Sneaking Ants."

Cardiocondyla species are native to the Old World tropics and the Mediterranean region, but several species are "tramps" that have been widely distributed by commerce. The taxonomy of the group had slowly degenerated into chaos until 2003, when Bernhard Seifert brought order to a major portion of the genus, including the tramp species and the Palearctic species. This impressive revision synonymized 13 species, elevated 5 subspecies to species, and described 20 new species. Approximately 50 species of *Cardiocondyla* are now known, but the rich Old World tropical fauna has hardly been touched, and the actual number of species probably exceeds 100 (Seifert 2003).

These small ants frequently occur in disturbed habitats, and three species (*minutior, emeryi,* and *obscurior*) can often be found in cities. Colonies of most species are in soil, sometimes in cracks in pavement, but *obscurior* is usually arboreal, and *wroughtonii* may be in hollow twigs and weed stems on the ground. Workers usually forage independently, but "tandem running" (one individual following another) has been seen in one species (Wilson 1959). Foragers never retrieve large, solid food items, but scavenge fluids and minute bits of dead insects, or collect sweet substances such as nectar and honeydew (Creighton and Snelling 1974). Despite their small size and lack of any obvious defense, *Cardiocondyla* foragers are avoided by larger ants, such as workers of *Solenopsis* or *Pheidole* (Creighton and Snelling 1974). The basis of this avoidance is presumably chemical, as crushed gasters of *Cardiocondyla mauritanica* cause agitated and frenzied behavior in *Linepithema humile* (Gulmahamad 1997).

Members of this genus are often overlooked because they are small, move about singly, and are often in unattractive habitats. Longino, in his informative website on ants of Costa Rica (2004), cites additional reasons: "I have to admit that I have a very poor knowledge of *Cardiocondyla* in Costa Rica because studying these ants often involves putting your nose to the pavement in highly public places. It can expose you to considerable public scrutiny and perhaps even derision, and your introversion/extroversion ratio can determine whether this is a painful or pleasant experience. Studying *Cardiocondyla* also carries the risk of physical harm when the habitat under investigation is the side of a busy highway. Sniffing the gravel while being buffeted by the wind of passing trucks is decidedly disconcerting." I have accumulated many records of these ants in Florida, not because I am tough, but rather because the presence of an ant often makes me more or less oblivious of my immediate surroundings.

With six species introduced into Florida, *Cardiocondyla* has contributed more exotic species than any other ant genus, other than *Strumigenys* with 10. The runners-up are the much larger genera *Pheidole* (four species) and *Nylanderia* (five species). Features that may favor the emergence of tramp species in *Cardiocondyla* are as follows: small size, compact and cryptic colonies, optional or obligatory sibling mating within the nest, polygyny, the ability of colony fragments to reconstitute whole colonies by rearing queens and males from remaining larvae, and a general resistance to desiccation (Heinze et al. 2006; Seifert 2003). These natural history traits occur in all the species that have been studied, and one might wonder why the genus has not produced even more tramp species. Some species that have not achieved tramp status are already well adapted to urban habitats (Seifert 2003) and could be just waiting for the right boat or airplane.

Heinze et al. (2006) suggest that tramp *Cardiocondyla*, which they call "stealthy invaders," may affect populations of native ants. This is unlikely to occur regularly in Florida because most *Cardiocondyla* species are most common in highly disturbed habitats that are dominated by exotic animals and plants, and unsuitable for most native ants. The exception to this rule in Florida is *Cardiocondyla obscurior*, which may occur in large numbers in dead branches and under bark and moss on individual trees in natural habitats.

Myrmecologists who study sexual systems and kin selection are intrigued by *Cardiocondyla*. Many species, perhaps all species, produce ergatoid (worker-like) males that mate with young queens before the latter leave the nest. These males, unlike other male ants, continue to produce new sperm through their lives (Heinze and Hölldobler 1993). There is usually only one such male per nest, because the first individual to emerge kills rival males (Heinze and Hölldobler 1993; Stuart et al. 1987). These males, in a kind of parallel with steroidal male humans, are heavily built, with big jaws and hugely expanded shoulders. Some species also have winged males that do not fight; these mate with females within the nest, or they may leave the nest, possibly to mate with females elsewhere (Seifert 2003). Females may occur as fully winged or short-winged (flightless) forms, but no Florida species are known to have flightless females (Seifert 2003). Winged females, like winged males, might mate outside the nest as well as in their natal colony, but this outcrossing has not been observed, nor is it likely to be, considering the tiny size of the potential participants. Any attempt to overlay this multiplicity of reproductive options on the already Baroque theoretical structure of kin selection is enough to make the head spin.

In any case, it is probable that some *Cardiocondyla* lineages are exclusively sib-mating. This was realized as early as 1892 by August Forel in a paper whose title translates as, "The Male of *Cardiocondyla*, and Perpetual Consanguinous Reproduction." Such situations can lead to taxonomic conundrums. Endlessly diverging, inbred lineages do not fit within the normal biological species concept. Moreover, a phenotypically expressed mutation could quickly become fixed in a set of local lineages, confusing the interpretation of variation. There could be consistently different sympatric forms, giving the illusion of separate species. Differences between allopatric lineages could easily be an expression of founder effects (the morphological idiosyncrasies of the single colony or queen that founded the population). Given these potential problems in defining species, it is worrisome that Seifert's species definitions (2003) depend so heavily on complex morphometrics. At present, however, there is no evidence that any of the presumed species in Seifert's revision are morphs representing inbred lineages.

There are various ways to deal with sib-mating lineages. Habitual sib-maters may be rare in ants, but they are common in some groups of beetles in the subfamily Scolytinae in the family Curculionidae. The genus *Xyleborus* and some allied genera, and the genus *Hypothenemus* and a few allied genera, have a combined total of more than 1500 species. Sib-mating has not been observed in all these species; rather, it is inferred from strongly female-biased sex ratios, and flightless, small, weakly sclerotized males. When working with these species, I never saw a formal set of guidelines for defining species. It appeared that species analogs were determined by consistent morphological differences (including genitalic differences) analogous to those that define species in non–sib-mating, related genera, and by consistent ecological differences. Most species of Florida *Cardiocondyla* show consistent morphological differences, and several show ecological differences as well.

Name Derivation: From *kardia* (Greek), meaning "heart," and *kondylos* (Greek), meaning "knuckle" or the knob of a joint. This refers to the postpetiole of the type species, *Cardiocondyla elegans*, which is heart shaped in dorsal view.

Cardiocondyla emeryi Forel— Emery's Sneaking Ant (Plate 22)

Taxonomy and Similar Species: The black gaster of *emeryi* contrasts strongly with the brick red color of the rest of the body. The head may be brick red, or it may be darker, but is never black. *Cardiocondyla obscurior* has the same coloration and has often been confused with *emeryi*. In lateral view, *emeryi* has a more bulging postpetiole and a flatter head; it is relatively easy to learn to make these judgments, at least with Florida specimens. Nest series can be identified in the field, as *emeryi* nests in the ground, whereas *obscurior* nests in standing weed stems or in trees.

Distribution: The distribution of this species is somewhat unclear because it may have been confused with other species, primarily *obscurior*, but perhaps other species as well. Seifert (2003) provides the first completely reliable records of distribution. These records suggest that *emeryi* is primarily African, with scattered outposts in the Old World tropics and subtropics, including Hawaii, and the New World tropics and subtropics, including the entire Caribbean. Florida records extend as far north as Orange and Volusia counties. It might well occur farther north along the Atlantic Coast in urban areas. It is known from the southeast corner of Texas at La Feria (Creighton and Snelling 1974).

Natural History: This species nests in the ground in open areas, including lawns and trampled earth in urban areas. The nest crater is inconspicuous and may be concealed by vegetation. Workers forage for small bits of dead insects, and possibly for honeydew. Foragers can be baited with finely ground cookie crumbs or with granular sugar, but retreat when other ants appear in numbers. Creighton and Snelling (1974), observing a colony in the laboratory, found that foraging was haphazard, and even workers returning with food took a circuitous route and sometimes seemed to have trouble finding their nest. My few observations of this species in the field in Florida suggest that foragers returning to the nest are able to proceed directly and have little difficulty finding the nest. This species lives in open areas, and might be using sunlight or polarized light to assist in orientation, but there is no evidence at this point.

Cardiocondyla emeryi appears to have the multiple reproductive options found in some other species of *Cardiocondyla*. Large numbers of dealate queens may occur in the nest, so there is an opportunity for nest fragmentation or budding. I have taken a few winged males and queens in flight traps, so outbreeding and aerial dispersal are possible. There are also ergatoid males with heavy mandibles and expanded humeral angles; these are described and illustrated by Kugler (1983). In a series of 10 colonies collected in Barbados, there were winged males that mated with queens in the nest and later left the nest, as well as ergatoid males that mated with their sisters and attacked other ergatoid males (Heinze and Trenkle 1997).

Name Derivation: This species was named in 1881 by the great Swiss myrmecologist August Forel (1848–1931) in honor of the equally eminent Neapolitan myrmecologist Carlo Emery (1848–1925). In 1890, Forel carried this connection further by naming a new genus *Emeryia*, with the type species *Emeryia wroughtonii*. This genus, which was based on the weird, sickle-mandibled male of *Cardiocondyla wroughtonii*, was synonymized with *Cardiocondyla* in 1892 by Forel himself. Emery never named a *Cardiocondyla* for Forel, but did celebrate his colleague in 1888 with the genus *Forelius*. Forel achieved a lasting tribute to Emery in 1912 with the ferocious-jawed *Emeryopone*.

Cardiocondyla mauritanica Forel— Mediterranean Sneaking Ant (Plate 22)

Taxonomy and Similar Species: The short, triangular propodeal spines and the absence of heavy sculpture on the petiole, postpetiole, and sides of the pronotum distinguish this species from other Florida *Cardiocondyla*, except for *venustula*. There are two morphologically similar forms of *Cardiocondyla* in Florida. I use the name *mauritanica* for the smaller, lighter form that is more characteristic of dry sites. The mesosoma and petiole of mature individuals are reddish brown, much lighter than the gaster, and usually lighter than the head; total length of the worker does not exceed 2.3 mm. I tentatively use the name *venustula* for the larger, darker form that usually occurs in moist habitats. The mesosoma and petiole of mature individuals is brownish black, as dark as the head, which is also black; total length of the worker may reach 2.8 mm, although some individuals are smaller. Seifert uses a series of complicated morphometrics to distinguish between *mauritanica* and *venustula*, but there is no independent way to measure the variation within species. In normal situations, the presence of ant colonies representing two distinguishable forms in an area allows one to infer that these are two species that are not interbreeding. Inbreeding lineages are not so easy to interpret. I have fallen back on an ecological species concept, but this could also have its problems. Environmental conditions can directly affect the development of ants. In this case, the larger, darker individuals ("*venustula*") might really be *mauritanica* that have grown up in a wet habitat. Some dedicated taxonomist will probably straighten this out eventually.

The California species *Cardiocondyla ectopia* (Snelling 1974) has been synonymized with *mauritanica* (Seifert 2003).

Distribution: Throughout the southern Mediterranean Region, also southern Portugal and Spain, the Middle East, Zimbabwe, Afghanistan, and the Indian Subcontinent; in the New World in Puerto Rico, Arizona, California, and Florida (Seifert 2003). In the Southeast, I have found it as far north as the central Coastal Plain of South Carolina and west along the Gulf into the central Panhandle of Florida. It might well occur in scattered localities all around the Gulf of Mexico. The first North American specimens were collected in 1967 (Snelling 1974), and it is possible that this is a recently arrived species currently expanding its range. In Florida, it is known from a relatively small number of widely dispersed localities.

Natural History: This species generally occurs in warm, semiarid habitats throughout its range (Seifert 2003). Nests are in soil in open areas, usually where vegetation is sparse. A few Florida collections are from beaches. Workers sometimes forage on pavement. Foragers studied by Creighton and Snelling (1974) collected nectar from flowers of a low mat of the Chamaesyce serpens (Euphorbiaceae) and minute particles of organic matter. Foraging individuals went to food sources and returned to the nest in a generally unidirectional way, but with many divergences and some back-tracking. Tandem running (one individual following another to a food source) was frequently observed. Gulmahamad (1997) observed nectar-feeding on flowers of low-growing sweet alyssum (Lobularia maritima) and scavenging on soft-bodied arthropods.

The ergatoid male, which has broad mandibles and expanded humeral angles, is described and illustrated by Snelling (1974). An alate male has been observed mating within the nest (Creighton and Snelling 1974). It is not known whether alate queens leave the nest to start new colonies, but this seems likely. Creighton and Snelling (1974) suggested that this species might achieve outcrossing by adoption of virgin queens into established colonies, where queens would mate with an ergatoid male. This has yet to be demonstrated in the field; it might be explored by looking at genetic diversity between and within colonies.

Name Derivation: Apparently named for the African country Mauritania, although the types are from Tunisia, and the species is not recorded from Mauritania.

Cardiocondyla minutior Forel—
Little Black Sneaking Ant (Plate 23)

Taxonomy and Similar Species: Before Seifert's 2003 revision, North American populations of this species have usually gone under the name nuda, or nuda subspecies minutior. Seifert has presented evidence (2003) that nuda, which is confined to the Old World, is a separate species. Among Florida Cardiocondyla, this species is easily distinguished by the absence of a conspicuous groove or dip in the mesosomal profile (metanotal groove) and its (usually) dark color, in contrast to emeryi, wroughtonii, and obscurior. The other dark species, mauritanica and venustula, have a conspicuous metanotal groove and have the petiole and postpetiole shining and feebly sculptured. Some minutior specimens are pale reddish brown, but not bicolored like emeryi and obscurior, or yellow like wroughtonii.

Distribution: Southwestern Texas (Creighton and Snelling 1974), Costa Rica (Longino 2004), Florida, the Caribbean islands, India, and the islands of the Indian Ocean, Sri Lanka, Nepal, Japan, Polynesia, Indonesia, and New Guinea (Seifert 2003). This species is not known from the coastal cities of Central and South America, but is likely to occur there. In Florida, it occurs throughout the Peninsula and west into Leon County. There is a good chance that it occurs in parts of southern Georgia.

Natural History: Nests in Florida are usually in open areas, such as lawns, pastures, and partially paved areas. A few series are from wet areas, such as low pastures and grass tussocks in seasonally wet areas. This species has both ergatoid and alate males (Seifert 2003).

Name Derivation: The name minutior means "smaller" in Latin. It is smaller than some species, about the same size as others.

Cardiocondyla obscurior Wheeler—
Arboreal Sneaking Ant (Plate 22)

Taxonomy and Similar Species: It appears that obscurior has often been confused with emeryi, a superficially similar species (see the discussion of emeryi). Structurally, obscurior is more similar to wroughtonii and has sometimes been considered a color form of that species. The two differ not only in color but also in nesting preferences: wroughtonii is usually in hollow twigs and weed stems that are on the ground, while obscurior is usually arboreal.

Distribution: The known distribution seems remarkably scattered: Canary Islands, Germany, Israel, Kenya, India, Nepal, Taiwan, Hawaii, Mariana Islands, Brazil, Caribbean islands, Florida (Seifert 2003), and Costa Rica (Longino 2004). It seems unlikely that there is an outdoor population in Berlin, Germany. Some pre-Seifert records probably confused obscurior with emeryi.

Natural History: Unlike other Florida Cardiocondyla, this species is usually arboreal. Colonies are in dead twigs and branches, under clumps of moss and lichens on tree trunks, and in hollow sand pine cones in sand pine forests. Occasionally, colonies are in dead weed stems or in dead wood on the ground. Such nesting sites are unusual for the genus as a whole (Seifert 2003). The diet is unknown, but honeydew and nectar are probably important, as they are for most arboreal ants. Colonies usually have many queens and reproduce by splitting, but also produce alate queens and males that might found nests (Seifert 2003). On one occasion, I found a dealate queen in a hollow pine twig, apparently a case of solitary nest founding. Ergatoid males have long, sickle-shaped mandibles, shown in photographs in a paper by Stuart et al. (1987) (in this paper, *obscurior* was still considered a form of *wroughtonii*). These mandibles are used to kill sibling ergatoids, so there is usually only one ergatoid male per colony (Stuart et al. 1987).

Cardiocondyla obscurior probably arrived in Florida relatively recently, perhaps in the 1970s. The earliest Florida specimens that I know are from 1982. It might expand its range further in Florida and into the adjacent states. This species is sometimes abundant on individual trees and might have ecological effects, perhaps competing with other "ecologically subordinate" species (Heinze et al. 2006), such as the native *Solenopsis picta* or the exotic *Monomorium floricola*.

Name Derivation: The Latin name "*obscurior*," meaning "darker," was first applied as a subspecies name, *Cardiocondyla wroughtonii obscurior*, before Seifert (2003) took the time and effort to establish that *obscurior* was a distinct species.

Cardiocondyla venustula Wheeler— Larger Black Sneaking Ant (Plate 22)

Taxonomy and Similar Species: The problems of separating Florida specimens of *venustula* from *mauritanica* are discussed under *mauritanica*. Florida specimens that I call *venustula* have somewhat larger propodeal spines than illustrated by Seifert (2003), but Seifert states that for the genus as a whole, "a rather small contribution to species discrimination is given by spine length…." It is quite possible that the specimens that I consider *venustula* are actually unusually large *mauritanica*, but, if this is the case, I am confident that eventually some myrmecologist will kindly set me straight.

Distribution: Zimbabwe, Mozambique, Namibia, Puerto Rico, and Florida (Seifert 2003). This species has a modest distribution for a tropical and subtropical tramp, especially if some or all of the African records are from its native range. In Florida, it is known from a few sites in the southern half of the peninsula.

Natural History: Nests are in soil in open areas, including beaches. Food brought back to the nest includes small insects and spiders, which are probably picked up dead or moribund, since foragers shy away from live arthropods encountered in the field (Wilson 1959). There is no evidence of odor trails in this species (Wilson 1959). This is not surprising, because any large and concentrated resource would be likely to rapidly attract more dominant species, such as some of the Old World tropical *Pheidole*. Nonetheless, it would be advantageous to be able to recruit foragers to more scattered resources such as honeydew or the nectar of small flowers. This may be accomplished by tandem running, which has been studied in *venustula* by Wilson (1959). The leader is followed by a single ant, which is apparently stimulated by a short-range or contact chemical produced by the leader. The leader waits for the follower, only advancing when contacted from behind. The follower often seems to lose track of the leader when the leader moves, but the leader soon stops until contacted again. Although tandem running may seem a less efficient form of recruitment than following a trail, it is not necessarily simpler, or even more primitive (Wilson 1959): "Nevertheless, it will have to be remembered that in *Cardiocondyla*, at least, tandem running is a highly evolved behavioral pattern in its own right. It can be fairly said to include more complex individual behavior than trail-laying and trail-following." Tandem running may be an efficient form of recruitment to small or dispersed resources, and it does not result in a stream of ants that might attract a predator, such as a small toad, or a parasitoid, such as a phorid fly.

Name Derivation: Diminutive of *venustus* (Latin), meaning "beautiful" or "charming," hence, "beautiful little *Cardiocondyla*."

Cardiocondyla wroughtonii (Forel)— Yellow Sneaking Ant (Plate 23)

Taxonomy and Similar Species: This species was originally described as a separate genus, based on its peculiar ergatoid male. Workers are distinguished by their golden yellow color, usually with a pair of vague dark spots on the first gastral tergite. The propodeal spines are slender and well developed.

Distribution: India, Sri Lanka, Nepal, Taiwan, Japan, Hawaii, Thailand, Singapore, Malaysia, Brunei, Indonesia, Papua New Guinea, Australia, Tanzania, Florida, and Louisiana (New Orleans) (Seifert 2003). In Florida, it is known from scattered sites throughout the state, including the western Panhandle. It probably occurs in southern Georgia and Alabama.

Natural History: Nests are usually in hollow twigs, weed stems, grass stalks, and nuts, typically on the ground or close to the ground. This species is probably closely related to *obscurior*, which is similar in morphology (including males) and nest sites. Both species may occur in relatively undisturbed and unmodified habitats. *Cardiocondyla wroughtonii* is seldom very common, but it has turned up eventually in most Florida sites where ants have been intensively sampled. A community ecologist coming upon this ant, and unaware of the biogeography of the genus, might assume that *wroughtonii* was a native species. Kugler (1983) has published an illustration of the peculiar male, with its sickle-shaped mandibles, pronotal knobs, and one-segmented antennal club.

Name Derivation: August Forel named *wroughtonii* for R. C. Wroughton, the naturalist who collected the type specimen in Poona, India. Wroughton is best known for his studies of the mammals of the Indian subcontinent and had 13 species of mammals named for him by respectful and grateful colleagues to whom he sent specimens. Only one of these, a bat, has survived synonymy.

Genus *Cephalotes*

Turtle Ants

Cephalotes (pronounced **Sef al ō" tēs**) is a genus of 115 living species, plus some fossil species in amber. All are neotropical, except for one West Indian species occurring in tropical southern Florida and two southwestern species (one in southern Texas and the other in southern Arizona) (De Andrade and Baroni Urbani 1999). There is a beautiful monograph of the genus by De Andrade and Baroni Urbani (1999), including identification keys, species descriptions, photographs, drawings, and syntheses of phylogeny and natural history. This volume is enough to make the Nearctic myrmecologist yearn to race off to the tropics, where *Cephalotes* is strikingly diverse, with many common species.

The defenses of some tropical species have been studied by several myrmecologists. Some species appear to have powerful chemical deterrents, resulting in the evolution of warning coloration and a series of mimetic complexes based on different species in different areas (Hespenheide 1986). Some rain forest canopy species, when approached by a potential threat, drop and glide back the trunk of their tree (Yanoviak et al. 2005). They use visual cues to locate the tree trunk and perform a controlled glide backwards, with the gaster pointing down and the dorsal side facing the tree (Yanoviak et al. 2005).

Name Derivation: From *kephalotos* (Greek), meaning "with a head." Perhaps the suffix *-otes* in this context is an emphatic, as in "such a head!" At any rate, the name refers to the expanded, saucer-like head. Larger individuals may use the head to block the entrance to the nest, which is usually in a dead branch or twig. Away from the nest, defensive behavior includes hunkering down on the substrate and tucking away the appendages under the head and body, hence the name "turtle ants."

Cephalotes varians (F. Smith)— North Caribbean Turtle Ant (Plate 24)

Taxonomy and Similar Species: There are no similar Florida species. Several species of arboreal *Camponotus* have majors with enlarged heads adapted to block the nest entrance, but these have cork-shaped heads, unlike the manhole-cover head of *C. varians*. This species has bounced through a series of genera now synonymized with *Cephalotes*: *Cryptocerus*, *Cyathocephalus*, *Paracryptocerus*, and *Zacryptocerus* (De Andrade and Baroni Urbani 1999).

Distribution: Florida, the Bahamas, and Cuba. In Florida, this species is most common in the Keys, but has also been found in the Miami area, at Flamingo in Everglades National Park, and once in southern Brevard County. It appears to be restricted to frost-free sites.

Natural History: North American myrmecologists who wish to view this species in its natural habitat must visit southernmost Florida, but it is worth the trip. The black workers scuttling along flat against a branch or acting as gatekeepers in a hollow twig are unforgettable sights. It is a more common ant than it would appear from a visual survey, because most activity is at night.

This species inhabits mangrove zones, tropical hammocks, and sawgrass areas with scattered shrubs. Nests are in dead branches, hollow twigs, and hollow culms of sawgrass. These nests are usually in dead parts of living plants, not in dead trees or shrubs. Several researchers have maintained colonies in the laboratory; methods are briefly described in Wilson (1976a).

Workers are nocturnal, slow-moving scavengers, collecting nectar, honeydew, pollen, unidentified fluids, and minute fragments of dead insects (Cole 1980; De Andrade and Baroni Urbani 1999; Wilson 1976a). Tiny pellets of solid matter, carried back in the mouth cavity, are shared with adults and larvae, but most food is transported in liquid form in the crop (Wilson 1976a). Larvae are also fed miniature eggs produced by workers; these eggs are much smaller than the eggs laid by queens (Wilson 1976a). Pollen-gleaning from leaf surfaces, recorded in several species of *Cephalotes* (De Andrade and Baroni Urbani), is an unusual activity in ants. If pollen is an acceptable food for ants, it is strange that more species, including *Cephalotes*, do not gather pollen directly from flowers. Most foraging by *C. varians* is done by single individuals, but when there is a concentration of sweet liquid and the colony has been deprived of food, foragers are able to lay down a short-lived pheromone trail that recruits other workers (Wilson 1976a).

Defensive behavior is primarily passive. When molested outside the nest, these ants crouch down, retracting the appendages and firmly grasping the substrate. Majors have spines on the sides of the mesosoma, although much less developed than in some tropical species. The major uses its saucer-like head to block the nest entrance. The "saucer" may be filled with fine filaments, possibly extruded from pores (Wheeler and Hölldobler 1985). These might help camouflage the nest entrance, or function as a physical repellent, as fine waxy filaments can deter ants (Eisner et al. 2005), or hold repellent chemicals. There are no reports of strong repellent chemicals evident to the human nose, as in some tropical species. A putative alarm pheromone that, when presented to an ant, makes it back up as if in retreat has been discovered (Olubajo et al. 1980). There would seem to be a possibility that the function of this pheromone is to cause a major to back away from the nest entrance to allow another worker to go in and out. This hypothesis seems strengthened by the observation (Olubajo et al. 1980) that the pheromone causes ants to back up whether it is presented from the front or rear of the insect. When its nest is invaded by other ants under laboratory conditions, smaller ants are dragged out by the workers and larger ants are bulldozed out by the majors (Wilson 1976a).

Its various nest defenses may allow *C. varians* to live where there is severe competition for hollow twigs from spiders, eumenine vespids, crickets and other ant species. On very small mangrove islands, however, where resources are stringently limited and interactions between co-occurring species are likely to be frequent, *C. varians* is unable to coexist with *Xenomyrmex floridanus* and *Crematogaster ashmeadi* (Cole 1983a). In the absence of either of these species, *C. varians* persists on tiny islets; in the presence of these species, *C. varians* persists only on larger islands (Cole 1983a).

Genus *Crematogaster*

Acrobat Ants

There is no mistaking the genus *Crematogaster* (pronounced **Krē mat' ō gas" ter**). Members of this genus can swing the gaster up over their mesosoma and head, so that the rear precedes the head as the ant advances; this trick justifies the name "acrobat ant." The contortion is achieved by the combination of a flat petiole that is articulated with the mesosoma to allow unusual dorsal movement, and by attachment of the postpetiole to the dorsal surface of the gaster. A column of these ants moving along with their gasters raised and ready looks at first glance like a parade of half ants that have lost their gasters. The adaptive value of this posture is to allow the ant to present the spatulate stinger beaded with repellent chemicals in the face of an oncoming foe. The sharp jaws can also be brought to bear, and the combined weaponry helps protect the guidance system, otherwise known as antennae. The basal joint of the petiole and the posterior aspect of the ant are protected by the propodeal spines and by chemicals from a well-developed metapleural gland near the attachment of the hind coxae.

The advanced defense system of *Crematogaster* may help account for its success in terms of multiplicity of species, 425 of which are listed in Bolton's 1995 catalog of ants. This is a genuinely speciose genus, but the number of species that have been described is large, in part, because these ants often move about conspicuously in the open where they are likely to be seen by myrmecologists, in contrast to members of such genera as *Temnothorax* and *Strumigenys*. There is a taxonomic downside to catching the attention of every passing myrmecologist, as it tends to lead to large numbers of synonyms, many dating from the early days of myrmecology, before the current biological species concept was in use and before taxonomists had ready access to collections for comparing specimens. An ominous sign in the 1995 catalog is the huge number of named subspecies, only a small proportion of which are probably true geographic subspecies, the majority being either simple synonyms or distinct species. Obviously, *Crematogaster* is in great need of revision, but taking on those hundreds of species and subspecies, to say nothing of additional undescribed species, verges on a lifetime project.

The situation in Florida is not nearly so dire as in most tropical areas, or in southwestern North America including Mexico. We have only 10 species (Arizona probably has twice that number), but even our small fauna manages to display some of the problems with cryptic and ambiguous species that are typical of the genus. This will become apparent in the species accounts that follow. Southeastern myrmecologists owe debt of gratitude to Clifford Johnson, who provided a detailed guide to southeastern species (1988a). This built upon Buren's revision (1968), which provides no explanation for his treatment of eastern species; Buren died before he could finish his work on *Crematogaster*. Johnson's discussion of *pilosa* is particularly useful.

There is a strong tendency for *Crematogaster* species to be arboreal, but this is repressed in temperate areas, where the arboreal ant fauna is generally small, probably attributed in part to cold winter temperatures. In Florida and the remainder of the southeastern Coastal Plain, arboreal ants in the recent past faced the threat of frequent fires. In the case of *Crematogaster*, one species (*pinicola*) has a refuge in the tops of fire-resistant pines, three species (*vermiculata*, *atkinsoni*, and *pilosa*) are usually in swamp forests or marshes, three species (*cerasi*, *lineolata*, and *minutissima*) are in thick leaf litter, perhaps originally in mesic areas, one species (*missuriensis*) is subterranean, and one species (*ashmeadi*) is arboreal in mesic forests, also rapidly colonizing any unburned area of forest or brush.

A satisfying feature of Florida *Crematogaster* is that most species appear to occupy clearly distinct microhabitats. The importance of field natural history studies to understanding the taxonomy of *Crematogaster*, which includes both cryptic and variable species, was pointed out by Creighton (1950): "Because of the unusual plasticity of many of the species, any analysis based solely on cabinet specimens is apt to lose itself in a maze of incomprehensible details." Ecological niches, of course, are also subject to geographic variation.

Species of *Crematogaster* occur everywhere in Florida, but one species is known only from the Keys, and two others are primarily found in north Florida. Despite the important analysis of southeastern species by Johnson (1988a), it is possible that the last word has not been spoken on the nomenclature of eastern species in the *lineolata–cerasi* complex. In south Florida, there is a possible undescribed species in the Keys, and there is a chance that a widespread and abundant Caribbean species, *steinheili*, will colonize tropical Florida. An illustration of this species is included on one of the *Crematogaster* plates.

Name Derivation: From *kremastos* (Greek), meaning "suspended," and *gaster* (Greek), meaning "abdomen," probably referring to the habit among these ants of suspending the gaster over the rest of the body in the acrobatic defensive stance.

Crematogaster ashmeadi Mayr— Ashmead's Acrobat Ant (Plate 25)

Taxonomy and Similar Species: *Crematogaster ashmeadi* has short, thick propodeal spines, the sides of the pronotum are shining, and there are few erect hairs on the head or pronotum. These features distinguish *ashmeadi* from all Florida species except for *pinicola*. The latter species is bright reddish brown with a black gaster and lives in large, open-grown pines typical of sandhill or flatwoods habitat. The color of *ashmeadi* is dark brown to black, and colonies are in a great variety of situations, but seldom in large pines out in the open. I have not found structural characters to separate the two species, and it may not be possible to identify old and discolored specimens. There is also a rare, conspicuously large form of *ashmeadi* in the Keys that is possibly a distinct species, as normal *ashmeadi* also occurs in the Keys, but I have not found structural characters to distinguish between the two. In the field, it is usually possible to provisionally identify *ashmeadi* by its blackish color, arboreal habits, and relatively small size. Field misidentifications usually turn out to be *pilosa*. *Crematogaster vermiculata* is an even smaller black species that occurs in and at the edges of swamp forests in central and north Florida. The mesosoma of *vermiculata* is less shiny and smooth than that of *ashmeadi*, but it is usually necessary to use a hand lens to see this in the field.

Distribution: Coastal Plain and Piedmont from Virginia south through Florida, west into Texas

(Johnson 1988a). It is also reported from South Andros Island in the Bahamas (Wheeler 1905a), although this might be the large form or species that occurs in the Keys. In Florida, *ashmeadi* occurs throughout the state.

Natural History: This is our commonest arboreal ant in most habitats, but in open sandhills and flatwoods, it is replaced by *pinicola* in pine trees, but not in nearby oak trees. Extensive studies of the natural history of "*ashmeadi*" by Walter Tschinkel and others all refer to *pinicola* and will be covered under that species. A wide variety of habitats are suitable for *ashmeadi*, including mesic and xeric forests, forested roadsides, and isolated patches of hardwoods in pastures and landscaped areas. It is my impression that *ashmeadi* is more likely to occur along edges of heavily forested areas than in the forest interior, but it may be that it is more difficult to sample in mature forest because the ants are foraging in the canopy and do not descend to baits on the boles of trees. Individuals attracted to bait on trunks of large trees often seem to me to be larger and browner than individuals living out in the open, but this could easily be an effect of more mesic conditions in the nest. I prefer this hypothesis to the alternative, that there is another virtually indistinguishable species hanging about in the treetops. Nests are in rotten branches on live trees, and hollow branches, twigs, and vines. Foragers collect small insects, scavenge dead insects of any size, and feed on honeydew produced by sap-sucking insects. There are occasional examples of *ashmeadi* attacking nestling birds, but it is probable that these birds were already dead when attacked. When I have used shredded tuna fish as bait, I have seen several times an apparent group caching behavior, with the bait tucked under a flake of bark or in a tuft of lichen or ball moss a little way from the original bait site. Presumably, solid food is finely macerated by the workers, as larvae of *Crematogaster* feed exclusively on liquid food regurgitated by workers (Petralia and Vinson 1979). Workers are rapidly recruited to baits, primarily by means of a gland in the hind tibia that debouches into a fine duct running through the distal part of the leg to the fifth tarsal segment, where the trail substance is released (Leuthold 1968). The alarm pheromones of *ashmeadi*, produced by the mandibular gland, have been identified by Crewe et al. (1972).

Name Derivation: Gustav Mayr named this species for William Harris Ashmead, who is discussed under *Aphaenogaster ashmeadi*. Mayr described this species from specimens collected by Ashmead in Florida and from specimens collected by Theodore Pergande in Florida, Virginia, and Georgia. Most of Ashmead's numerous taxonomic contributions deal with parasitic Hymenoptera, but a few years before his death, he turned his attention to ants, publishing an arrangement of the higher taxa of ants (Ashmead 1905), which he stated was a "skeleton" version of a work of several hundred pages that was almost complete. This larger study was never published, but this may not have been a huge loss to myrmecology, as it probably would have been overshadowed almost immediately by the classifications of Carlo Emery and William Wheeler, who were far more experienced in myrmecology.

Crematogaster atkinsoni Wheeler— Atkinson's Acrobat Ant (Plate 26)

Taxonomy and Similar Species: *Crematogaster atkinsoni* has long, slender spines, and the sides of the pronotum are smooth and shining. Its smooth pronotum separates *atkinsoni* from *cerasi* and *lineolata*, two other species with relatively long spines. This species strongly resembles *pilosa*, another large, long-spined species that may occur in or near marshes, the usual habitat of *atkinsoni*. *Crematogaster atkinsoni*, however, has only a few erect hairs on the head (the others are closely appressed), and usually only two standing hairs on the pronotum, while *pilosa* has numerous standing hairs on the head (best seen in lateral view) and more than four standing hairs on the pronotum. These hairs are difficult to see on specimens in alcohol and, like all erect and semierect hairs, are best viewed against a dark background. Workers of most Florida colonies of *atkinsoni* are bicolored, reddish brown with black gaster, but some are entirely reddish brown or entirely dark brown. These color variants, distinctive though they may appear, seem to have no taxonomic significance and are not consistent within a region (Johnson 1988a).

Distribution: Coastal Plain (especially in coastal marshes) from North Carolina south through Florida, west into Mississippi (Smith 1979). In Florida, *atkinsoni* probably occurs in coastal and freshwater marshes throughout the state, but I have not seen specimens from the Panhandle.

Natural History: Studies of the natural history of *atkinsoni* precede its recognition as a species by several decades. In 1887, the conspicuous paper ("carton") nests of *atkinsoni* (identified as *lineolata*) were noted by G. F. Atkinson, who suggested that these nests were an adaptation for living in marshes (Wheeler 1919). This ant often tends aphids on semiaquatic shrubs, such as *Baccharis halimifolia*. The location of a nest must involve some interesting decisions, as the presence of both sap-sucking insects and a short canopy of contiguous shrubs and herbs suitable for foraging must be important. In addition, inland marshes are subject to burning, which would be disastrous to any colony not protected by standing water. It is probably fire that restricts *atkinsoni* to tidal salt marshes through much of its range. With their paper nests perched up in vegetation, *atkinsoni* are also vulnerable to predators, which have probably selected for their massive reaction to disturbance. This reaction is characterized by hundreds of ants rushing out, waving their gasters with sting extruded and fiercely biting the intruder. These defenses, however, do not protect the ants from black bears in the marshes and wet prairies of south Florida, where these ants are a favorite food (Maehr 1997) (ants misidentified as *pilosa*). The accessibility of colonies of *atkinsoni* should make them good subjects for natural history studies.

Name Derivation: This species, as mentioned above, was named by William Wheeler in memory of George Francis Atkinson (1854–1919), who had studied the species many years earlier. Atkinson was a multitalented biologist who taught entomology at the University of North Carolina, beginning in 1885; during this period, he discovered the paper nests of *atkinsoni*. Later, he changed his focus to botany and mycology, becoming a professor of botany, and later the department head, at Cornell University. He wrote several books on botany and mycology, but left his mycological work unfinished when he died of pneumonia contracted during a collecting trip in western Washington.

Crematogaster cerasi (Fitch)—
Cherry Tree Acrobat Ant (Plate 27)

Taxonomy and Similar Species: *Crematogaster cerasi* has fairly long, slender propodeal spines, much longer than those of *ashmeadi* and *pinicola*, but shorter than those of *atkinsoni* and *pilosa*. There is some fine sculpturing on the sides of the pronotum (visible in diffuse lighting), although the lower sides of the pronotum may be almost smooth. There may be some irregular ridges on the dorsal surface of the mesosoma, but this area is not a mass

of sinuate ridges, as in *vermiculata*. These features separate *cerasi* from other Florida species, except for *lineolata*. The latter species has a conspicuous row of flattened hairs across the front of the pronotum and is often more heavily sculptured, on both the sides and top of the pronotum. At various times, *cerasi* and *lineolata* have been combined, as in Creighton's ant manual (1950); the two species were separated by Buren (1968) and discussed by Johnson (1988a). In Florida, *cerasi* and *lineolata* are the only dark species that nest in leaf litter and rotten wood on the ground; thus, field identification can often be narrowed to those two species.

Distribution: Quebec south into Georgia, west into South Dakota and New Mexico (Smith 1979), also Florida (Johnson 1988a). In the southwestern part of its range, *cerasi* occurs together with a large number of confusing western species; records from that area should be considered with that in mind. In Florida, *cerasi* is known from scattered localities throughout the state, including a disjunct record from Miami-Dade County.

Natural History: To the north of Florida, *cerasi* nests in dead branches and logs on the ground, under rocks, and in the woodwork of buildings (Smith 1965). It is a common house-infesting species, probably moving into preformed cavities, but possibly gnawing new galleries in wood (Smith 1965). In my observations of northern *cerasi*, it seems to me that northern colonies are more active, less likely to nest in leaf litter, and composed of larger individuals than Florida colonies. In his revision of *Crematogaster* (1968), Buren seems to imply that *cerasi* is a northern species that is confined to upper elevations to the south, "reaching Georgia through the Appalachians, Arkansas in the Ozarks … and south to New Mexico in the Rockies." In his exhaustive survey of ants of North Carolina, Carter (1962) found that *cerasi* occurred in the Piedmont, but more commonly in the mountains; he did not find it in the coastal area or fall-line sandhills. In Florida, *cerasi* occurs in xeric uplands (Johnson 1988a), the equivalent of Coastal Plain sandy areas in North Carolina. At some point, therefore, *cerasi* comes down from the southern mountains and becomes a Coastal Plain species in Florida. It also becomes a species that nests in leaf litter in sandy uplands. The natural history of Florida *cerasi* seems to indicate that it is an ecologically distinct and possibly geographically isolated population. Genetic studies may be required to solve this riddle.

I have considered the possibility that *lineolata* in Florida is more variable than elsewhere, and that presumed Florida *cerasi* are specimens of *lineolata* that have fewer pronotal hairs and less pronotal sculpture than normal farther north. I have not, however, seen any mixed nest series. Both forms occur in Putnam, Marion, Okaloosa, Walton, and Leon counties, so there is not likely to be a geographic cline involved.

Name Derivation: From *cerasus* (Latin), meaning "cherry tree." The type series was found tending aphids on a cherry tree (Ellison et al. 2012).

Crematogaster lineolata (Say)— Lineolate Acrobat Ant (Plate 27)

Taxonomy and Similar Species: *Crematogaster lineolata* has fairly long, narrow propodeal spines, much longer than those of *vermiculata*, *ashmeadi*, and *pinicola*, shorter than those of *atkinsoni* and *pilosa*. The sculpturing on the sides of the pronotum is usually conspicuous. The most similar species is *cerasi*, but *cerasi* has only one to three long hairs on each shoulder of the pronotum, while *lineolata* has one or two rows of hairs across the front of the pronotum. As discussed under *cerasi*, the two species live in similar habitats and are the only two dark, ground-nesting species in Florida, although the nests I have seen are in leaf litter or dead branches on the ground, rather than subterranean. It appears to me that *cerasi* and *lineolata* are more similar to each other in Florida than they are farther north, and I have had an occasional crisis of confidence in the distinctness of two species that differ consistently only in the presence or absence of a few hairs on the pronotum. I have not, however, seen mixed colonies; the two forms do not occur in any geographic cline, and there are several counties where both forms occur together. It is unlikely that Florida specimens identified as *lineolata* are some other species, but I feel less sure of Florida *cerasi*.

Distribution: Quebec south into Florida, west into North Dakota and Texas (Smith 1979). In Florida, *lineolata* occurs in the northern part of the state, extending south into Sumter and Hernando counties.

Natural History: Some accounts of the natural history of *lineolata* published before Buren's 1968 revision are difficult to interpret because the name *lineolata* was applied to various species. An example of this is Van Pelt's 1958 study of the ants of Welaka (Putnam County), normally an excellent source of natural history. It is not clear how he combined species under the names *clara* and *lineolata*, names that probably include the species *pilosa*, *cerasi*, and *lineolata*. Smith (1965) seems to have had a modern understanding of *cerasi* and *lineolata*, but he uses information from earlier sources in an unspecified way, so it is not clear which authors shared our present understanding (such as it is) of *lineolata* and *cerasi*. My own limited experience in the northern range of *lineolata* supports Smith's report that this species nests under rocks, in stumps and in dead wood on the ground. I have always found *lineolata* in forested sites, including residential areas. In Florida, *lineolata* usually occurs in xeric upland forests with thick layers of leaf litter.

Name Derivation: From *linea* (Latin), meaning, originally, "linen thread," hence, "fine line," in this case "having fine lines," referring to the fine lines on the side of the head and mesosoma. This is one of a number of ants named by Thomas Say (1787–1834), although some of his names have been abandoned in cases in which there are no specimens to go with an insufficiently specific description. Say was a pioneering and enormously talented naturalist who described multitudes of species of insects (more than 1000 beetles), land snails, and other animals. His efforts were hampered by a deficient formal education, chronic poor health, repeated financial crises, distracting responsibilities (he was a leader in a doomed utopian venture called "New Harmony"), various kinds of bad luck, and an abbreviated life span.

Crematogaster minutissima Mayr— Forest Floor Yellow Acrobat Ant (Plate 28)

Taxonomy and Similar Species: *Crematogaster minutissima* is easily distinguished from most Florida species by its yellow color and small size. Almost nothing, however, is truly easy in *Crematogaster* taxonomy. There is another practically identical species, *missuriensis* (this spelling is correct). The two species are separated by the slightly granulate mesopleuron of *missuriensis*, compared to the smooth mesopleuron of *minutissima*. This difference can only be seen with good, diffuse lighting. *Crematogaster missuriensis* was long considered a subspecies of *minutissima* (Creighton 1939, 1950). Creighton was working with specimens from ranges that did not seriously overlap: Texas and Missouri for *missuriensis*, and the Southeast west into Texas for minutissima. It is now known that the two species occur together over a wide area in the Southeast. Even this would not be enough to generate much enthusiasm for a pair of species that can only be distinguished with the aid of unusually good lighting and an imagination stimulated by a good cup of coffee, were it not for the fact that the two species are ecologically distinct. In Florida, missuriensis seems to be restricted to the northern part of the state, while minutissima is more widespread. A third yellow species, steinheili, is so common in the West Indies that it could easily be imported into tropical Florida. The gaster of steinheili is densely hairy, as shown in the plate.

Distribution: North Carolina south into Florida, west into Texas (Smith 1979). In Florida, minutissima occurs throughout the state, including the Keys.

Natural History: This species usually occurs in densely forested sites, including mature sand pine scrub, as well as xeric hammocks, tropical and coastal hammocks, and mesic and wet hammocks. It nests in leaf litter, rotten wood, and hollow twigs and nuts on the ground. Workers may forage above ground at night, and I have collected stray workers in flight traps. The social system of minutissima is probably unusual, but has not been the subject of publications. This species is highly polygynous, sometimes to the degree that one wonders how the workforce could be sufficient. In such situations, one might expect colonies to be founded by queens that depart with a group of workers, but there is no evidence of this. All queens that I have seen in nests appear to have had wings formerly (dealate), and I have collected queens at light traps. I have made many collections of this species and have not seen males, even in colonies with alate females. On two occasions, I have collected "large workers" resembling those of Crematogaster smithi (Heinze et al. 1995), but these do not seem to be common. In colonies of smithi, these workers produce sterile eggs that are fed to the larvae and queen, or, more rarely, reared into adult males (Heinze et al. 1995). It would probably be relatively easy to do laboratory studies of minutissima.

Name Derivation: From minutissima (Latin), meaning "very small." This is one of the many species of North American ants that were ably described by Gustav Mayr. Not only did he describe numerous species, but he established many of the current ant genera, and was the first to write dichotomous keys to help entomologists identify ants. Previously, it was necessary to read all the descriptions and figure out how the species differed, a task impeded by the fact that different authors described ants in different ways. In this particular instance, however, Mayr's description (1870) does not seem adequate to distinguish between minutissima and missuriensis. Until somebody examines the type (from Texas), we have to take it on faith that the species is the same as the one that is called minutissima here.

Crematogaster missuriensis Emery— Subterranean Yellow Acrobat Ant (Plate 28)

Taxonomy and Similar Species: Crematogaster missuriensis is similar to minutissima, as discussed under that species. Crematogaster missuriensis is much less likely to be obtained from litter samples than minutissima and is absent from south Florida. In the field, missuriensis can be distinguished from minutissima by its subterranean nests. It might take genetic evidence to show whether the southwestern form smithi is the same as missuriensis or a separate species.

Distribution: This species is broadly distributed in the southern plains and prairies from north Florida west through Texas, and north through Missouri, Kansas, and Colorado. It may be considerably more widespread in pine savannahs and fields of the Southeast than current records indicate. It is a species that is easily overlooked, as it is probably active only at night and its nests are inconspicuous. In Florida, it is known from a few northern counties, primarily in the Panhandle.

Natural History: Nests are usually in open pine savannahs, fields, and prairies; I have not seen it in heavily disturbed open areas such as lawns. In Kansas, nests are in unshaded or shaded prairie, occasionally in deciduous forest (DuBois 1985). Nest entrances I have seen in Florida could easily be mistaken for those of some species of Nylanderia, but, unlike Nylanderia, there is only one nest entrance. At this point, it seems likely that missuriensis is monogynous, like smithi, but more nest excavations are needed to prove this. Colonies of missuriensis produce small numbers of workers that are conspicuously larger than normal workers. These workers probably produce unfertilized eggs that are fed to larvae and queens, as in smithi (Heinze et al. 1995).

Name Derivation: The name missuriensis means "an inhabitant of Missouri." American myrmecologists "corrected" the name to missouriensis until recently, although Carlo Emery (1895) used *missuriensis* in the description. The name of the state Missouri is spelled correctly elsewhere in the description, so *missuriensis* is either a misprint or, more probably, a Latinization of Missouri. The name designated in an original description is the official name, however it may be spelled. A similar story is associated with the name *Crematogaster*, which is "incorrectly" derived from the Greek *kremastos*. Some early myrmecologists used the name *Cremastogaster*, but it is not permissible to change a genus name on the basis of a faulty derivation.

Crematogaster obscurata Emery—
Tropical Granulate Acrobat Ant (Plate 26)

Taxonomy and Similar Species: *Crematogaster obscurata* is an introduced species restricted, so far, to the southern Florida Keys, where it might be confused with *ashmeadi*. Under the microscope, *obscurata* is easily identified by the granulate sides of the head and mesosoma, and by its erect, flattened hairs. In the field, it is notable for its tiny size, about two-thirds that of other dark species, such as *ashmeadi*. The Florida population was reported as *agnita* (Deyrup et al. 2000), a name that has since been synonymized with *obscurata* (Longino 2003).

Distribution: Mexico, Guatemala, Belize, Costa Rica, and Venezuela (Longino 2003). In Florida, *obscurata* is known from a few colonies on West Summerland Key, discovered in 1995 (Deyrup et al. 2000) and revisited in 2006; it was also found in 2010 by Lloyd Davis on Key West (Moreau et al. 2014).

Natural History: In Central America, *obscurata* is known from dry forest habitats and beach margins (Longino 2003). Florida colonies were in a tropical hardwood hammock and nearby coastal mangroves. Unlike most species of *Crematogaster* (except for *minutissima*), *obscurata* is highly polygynous. Nests are in hollow twigs and branches, and in galleries in larger dead branches. One nest was in the stub of a dead branch of *Piscidia piscipula*, whereas others were in *Rhizophora mangle*. Two of the red mangroves were isolated trees in the intertidal zone, so foraging may be confined to a single tree and exposed flats at low tide.

This species presents an unusual dilemma in conservation ecology. It is very likely that this species is not native to the Florida Keys, as it is not known from the West Indies, the usual source of ants that may have gotten to Florida by their own efforts before the arrival of Europeans. It might

have been introduced recently, or much earlier, for example, in the late 1800s, when there was a focused effort to introduce plants from Mexico and Central America (Deyrup et al. 2000). It is known from only a few sites, from which it could probably be eradicated relatively easily, using toxic baits placed on trees where colonies occur. This is not reported to be a pest species, and it does not seem to have dispersed widely since it was discovered. Nevertheless, it is always possible that *obscurata* could displace native mangrove ants, foster sap-sucking insects, or attack native tropical insects. Moreover, an introduced species may persist in an area for a long time and then achieve a critical population level or evolve a critical adaptation that permits invasive spread. Prudence suggests an attempt to eradicate *obscurata* from Florida. Before such an effort, there should be a more thorough investigation of the taxonomy of Florida *obscurata*. It would be unfortunate and embarrassing if the population was exterminated and then found to represent a rare endemic. There should also be a more thorough survey of West Summerland and Key West to make sure the species has not already established itself beyond the limits of practical eradication.

Name Derivation: From *obscura* (Latin), meaning "dark," referring to the dark color of the species.

Crematogaster pilosa Emery—
Hairy-Headed Acrobat Ant (Plate 26)

Taxonomy and Similar Species: *Crematogaster pilosa* has long propodeal spines and a smooth-sided pronotum, a combination of features that distinguish pilosa from *ashmeadi*, *pinicola*, *vermiculata*, *lineolata*, *cerasi*, and *obscurata*. It is similar to *atkinsoni*, but *pilosa* has numerous standing hairs on the head. This might seem like an easy difference, but it is difficult to see these hairs unless the specimen is clean and dry, and viewed in profile. As usual with ants, the hairs stand out better if a black background is placed below the specimen.

The nomenclature of *pilosa*, as worked out by Johnson (1988a), is not easy to follow. Apparently, less hairy variants of *pilosa* were once recognized as a separate species, *clara*, which is a synonym of a western species, *laeviuscula*. Buren's report of *clara* in the East (1968) refers to *pilosa* (Johnson 1988a). To summarize Johnson's 1988 discussion, historical records of *pilosa* really are *pilosa*, and eastern records of *clara* and *laeviuscula* are also *pilosa*.

Distribution: New Jersey south into Florida, west into Missouri and Texas (Johnson 1988a). In Florida, pilosa occurs throughout the state.

Natural History: In North Carolina, Carter (1962) collected pilosa primarily in open forest and at the edges of disturbed areas and marshes, but also has some records from forests with a closed canopy of tall trees. In Florida, I have usually found this species at the edges of swamps and marshes, as well as at the edges of open fields. Like most other arboreal Crematogaster in Florida, pilosa is highly vulnerable to fire, which determined the structure of Florida upland habitats until a little more than 100 years ago. A preference for marshes and their edges may have evolved to a greater degree in the Florida population of pilosa than in some populations to the north. Nests may be in grass tussocks, dead weed stems, small dead trees in marshes, and dead branches on live trees. In Alabama, I found a large, bicolored form of pilosa nesting in reeds in a coastal marsh, and a smaller black form inland in oak trees.

Wheeler (1933) described a supposed parasitic species, Crematogaster creightoni, from a colony of pilosa collected in Virginia by Creighton. This form was described from miniature females that occurred in a nest along with normal females. It was synonymized with pilosa by Smith (1958), along with kennedyi, found with cerasi. Creighton (1950) believed that these miniature females represented a rare dimorphism, especially since there were no conspicuous structural differences between the miniature and host species. This logic is possibly correct, but in ants with queen dimorphism, the alternative form is not usually especially rare. Parasitic ants, in contrast, often seem rare or local. I have not seen miniature queens of any species of Crematogaster, nor heard reports of such in Florida.

Name Derivation: From pilosa (Latin), meaning "hairy."

Crematogaster pinicola Deyrup and Cover— Pine Tree Acrobat Ant (Plate 25)

Taxonomy and Similar Species: This is an arboreal species with short, thick propodeal spines, no sculpture on the sides of the pronotum, and only a few standing hairs on the head and pronotum. This combination of features distinguishes pinicola from all other Florida species, except for ashmeadi, which appears to be structurally identical to pinicola. The color of pinicola in the field, and usually in museum specimens, is reddish brown with a black gaster; ashmeadi is completely black or blackish brown. Crematogaster pinicola lives in large, open-grown pine trees typical of sandhill and flatwoods habitats, while ashmeadi lives in a wide variety of situations, but seldom in open-grown pines. It would be useful to record the color and colony site of pinicola when it is collected, as the contrast between the red and black coloration is sometimes reduced in museum specimens. I sometimes see old specimens that I am unable to identify with certainty.

The logical basis for recognizing pinicola as a species distinct from ashmeadi is detailed in Deyrup and Cover (2007). In summary, it is difficult to explain why there should be a bicolored form in pine trees and a black form in nearby oak trees unless the two forms represent separate species. This correlation between color and microhabitat occurs over a large geographic area in Florida. Color is notoriously deceptive in ant taxonomy, and there are plenty of color-based "species" that have proven to be synonyms, including a batch in the genus Crematogaster. We should be grateful that a color difference works in this instance and reflect on the fact that if pinicola were uniformly black, we would not have the slightest idea that we were dealing with two species rather than one. Hand-labeled specimens collected by William Buren, an expert on Crematogaster, show that he recognized pinicola as a distinct species. Unfortunately, he died before he could complete his studies of the genus.

Natural History: Amazingly, the natural history of this recently described species is better known than that of any other North American Crematogaster, thanks in large part to Walter Tschinkel and his colleagues in north Florida. In the discussion accompanying the description of pinicola, Deyrup and Cover (2007) point out that this species appears adapted to life in the pine savannahs that once covered much of the Coastal Plain of southeastern North America. These landscapes include both sandhill habitat in dry sites and flatwoods habitats in wet sites. Both habitat types are maintained by frequent fires (Frost 1993), which tend to be lethal to colonies of arboreal ants. Pines in fire savannahs survive by fire adaptations that are exploited by pinicola. The outer bark of pines such as Pinus palustris (longleaf pine) and Pinus elliottii (slash pine) is thick, with an outer zone of many papery layers like the phyllo dough used in Greek cooking. This protects the living inner bark during the short-term, low-intensity fires typical of pine savannahs. *Crematogaster pinicola* lives in chambers in the thick bark, often using galleries created by bark-eating caterpillars of a moth (Tschinkel 2002). Another feature of these pines is that the twigs and young branches are relatively stout, making them more fire-resistant than fine twigs. A columnar growth form is typical of these pines, and the lower branches normally spontaneously die and fall off, lessening the chance that fire will be conducted up into the vulnerable crown of the tree. Various insects, especially longhorn beetles, hollow out the moribund lower twigs and branches, providing a succession of temporary residences for *pinicola*. In mature pines, larger dead branches in the crown and chambers in the outer bark can support large colonies of *pinicola* in relative safety.

There are several natural history studies of *ashmeadi* in the Apalachicola National Forest in northern Florida (Baldacci and Tschinkel 1999; Hahn and Tschinkel 1997; Hess and James 1998; Tschinkel 2002; Tschinkel and Hess 1999). These probably all refer to *pinicola*, based on conversations with some of the authors, descriptions of the biology of the species, and personal experience collecting *Crematogaster* on pines in the Apalachicola National Forest. Mating flights are in June and July (Tschinkel 2002). Nest-founding queens can be found in small dead branches on pine saplings (Baldacci and Tschinkel 1999; Hahn and Tschinkel 1997), where they might build up a colony that could later move to a larger tree that would accommodate a mature colony, which includes tens of thousands of ants (Tschinkel 2002). There is never more than one colony in a large pine, and few colonies occupy more than one tree (Tschinkel 2002). Surveys of pine stands showed that 55% to 90% of pine trees large enough for a mature colony were occupied by *pinicola* (Tschinkel 2002; Tschinkel and Hess 1999). There are no equally dominant arboreal ants in these pine forests, although several other species of ants, such as *Camponotus nearcticus* and *Solenopsis picta*, also occur in these large pines (Tschinkel and Hess 1999). There is no reason to suspect that pine-dwelling populations of these other ants are also cryptic species specialized for living in pine savannahs. The endangered red-cockaded woodpecker, another species adapted to pine savannahs, is dependent on *pinicola* for as much as 43% of its diet (by biomass) (Hess and James

1998). The woodpecker is directly threatened by the loss of large, contiguous tracts of old-growth, open pine savannahs, not by its remarkable dependence on *pinicola*. *Crematogaster pinicola* itself is not at all endangered as it is often abundant in pine forests that are too small or isolated to support red-cockaded woodpeckers.

Name Derivation: From *pinus* (Latin), meaning "pine tree," and *-cola* (Latin), a suffix meaning "dweller," referring to the microhabitat of the species.

Crematogaster vermiculata Emery— Cypress Acrobat Ant (Plate 27)

Taxonomy and Similar Species: The dorsal area of the pronotum of *vermiculata* is sculptured with distinctive, short, angulate ridges that are not aligned longitudinally, like the ridges found on *lineolata* and some other species. These make the pronotum nonshining, such that *vermiculata* can sometimes be identified in the field by a combination of this character and its swamp forest habitat. The most similar species in the field is *ashmeadi*, which may also occur in swamp forests. The propodeal spines of *vermiculata* are usually short, but are variable, sometimes relatively long and slender in large individuals.

Distribution: This species was described from Los Angeles, California, but it appears that the types were labeled incorrectly and were actually from some unidentified locality in the Southeast (Buren 1968). Such mistakes are uncommon but can occur when a collector has a collection box of superficially similar, unlabeled specimens from various places, the specimens separated only by a label at the beginning of each series. There are some species of ants, especially tropical species, first described from introduced populations, but this is extremely unlikely in the case of *vermiculata*. The real distribution of *vermiculata* is from North Carolina south into Florida, west into Arkansas and Louisiana (Johnson 1988a). In Florida, it has been collected in scattered sites from Hillsborough County north, and west through the Panhandle. I have not found *vermiculata* as frequently as I would expect in suitable habitat; perhaps it has a tendency to stay up in the canopy.

Natural History: This species appears to be confined to swamp forests, including both cypress swamps and hardwood swamps. This is another example of an arboreal ant that has found refuge in wet areas in the fire-prone Southeast. *Crematogaster vermiculata* often takes advantage of the railings of boardwalks to forage through cypress swamps in parks, as do other species of ants.

Name Derivation: From *vermiculata* (Latin), meaning "like little worms," referring to the short, irregular ridges on the dorsum of the pronotum.

Genus *Cyphomyrmex*

Little Fungus Garden Ants

Members of the genus *Cyphomyrmex* (pronounced **Sī' fō mur" mex**) are inconspicuous, knobby, non-shining little ants that can fold up into what looks like a crumb of dirt when alarmed. The corners of the head are extended up and outward, so in frontal view these ants appear to have ears. Species of *Cyphomyrmex* are similar to those of *Trachymyrmex*, but *Trachymyrmex* have irregular jagged spines on the head and body, and are larger, 5 mm or more in length, while *Cyphomyrmex* lack jagged spines and are smaller, 3.5 mm or less. *Cyphomyrmex* is a Neotropical genus of approximately 40 species, 2 of which live in Florida. Most Cyphomyrmex species live in mesic or wet areas, but a few, such as *wheeleri* in southwestern North America, live in arid or subarid habitats.

Cyphomyrmex belongs to the taxonomic tribe Attini, all of which cultivate and consume fungi. The attines include the famous leaf-cutter ants of the genus *Atta*, which excavate massive subterranean cities and are able to strip the leaves from whole trees in a matter of days. No species of *Atta* have yet invaded Florida; these ants are potentially so destructive that a plan to have an enclosed colony of *Atta* on exhibit at Disney's Animal Kingdom was nixed by the state. The attines are the only ants that can be considered true fungus gardeners and all are confined to the New World; it appears that fungus gardening evolved a single time in the ants (Chapela et al. 1994; Mueller et al. 2005), approximately 50 million years ago (Mikheyev et al. 2010). The complexity of fungus gardening, even in relatively primitive attines such as *Cyphomyrmex*, may help explain why this apparently efficient way of life did not evolve multiple times. It is not necessarily a problem of hooking up with the perfect type of fungi; apparently there have been multiple cases of fungal domestication, for example, in the genus *Cyphomyrmex* (Adams et al.

2000; Mikheyev et al. 2010). The real problems of fungus-gardening ants are agricultural: finding and maintaining a site with suitable temperature and humidity, collecting and processing the compost substrate for the garden, and, in particular, preventing weed fungi from overrunning the nutrient-rich garden bed. Just as certain weeds have evolved adaptations to efficiently infest agricultural fields, certain fungi have evolved to infest the fungus gardens of ants (Currie et al. 1999, 2006).

The evolutionary complexities of weed control in *Cyphomyrmex* gardens may help explain why nonattine ants have not repeatedly taken up the agricultural way of life. *Cyphomyrmex* recognize and pull out patches of weed fungi and have at least two forms of biological control of weed fungi. The garden fungi themselves produce chemicals that probably act as general fungicides (Wang et al. 1999). These chemicals are produced in small quantities in pure cultures of the fungus, but in large quantities in the fungal gardens (Wang et al. 1999), suggesting that the fungicides might be induced by the presence of other fungi, and hinting at other induced chemical defenses. In addition, *Cyphomyrmex* have specialized pockets and pits that hold antibiotic-producing bacteria that control specific weed fungi; these shelters are equipped with glands that probably nourish the bacterial cultures (Currie et al. 2006). *Cyphomyrmex* carries a large bacterial culture on the propleural plates that are just above the attachment of the front coxae and small cultures in pits distributed over the head and body (Currie et al. 2006). Each pit is marked by a silvery flat hair, so it is easy to identify their locations on a specimen of *Cyphomyrmex*. There may be multiple genetic strains or species of bacteria associated with one species of *Cyphomyrmex* (Currie et al. 2006). Ants can secrete their own antibiotics from their metapleural glands (Hölldobler and Wilson 1990), and perhaps *Cyphomyrmex* ants do so, but by outsourcing pesticide production to symbiotic fungi and bacteria they do something much better: they enlist the genetic systems of rapidly reproducing symbionts to combat any pesticide resistance evolving in the weed fungi (Mueller et al. 2005). Yet another level of complexity is suggested by genetic evidence that attines frequently recruit or pick up new strains of their fungicidal bacteria from soil, renewing their bacterial arsenal from its genetically diverse source (Ishak et al. 2011).

Dispersing queen *Cyphomyrmex* carry fungus starts with them and must forage for substrate after choosing a nest site. Although this foraging might expose the queen to danger, in terms of energetics, it is much more efficient than the more normal claustral form of colony initiation, which requires the queen to carry within her body all the resources she will need to raise the first set of workers in her sealed and secluded chamber (Seal and Tschinkel 2006b). Freed from the need to carry relatively massive resources, largely in the form of fat (Seal and Tschinkel 2006b), *Cyphomyrmex* queens can probably disperse efficiently relative to most queen ants and must be much less sought-after prey for aerial predators.

Name Derivation: From *cypho* (Greek), meaning "humpbacked," and *myrmex* (Greek), meaning "ant," referring to the conspicuous blunt dorsal tubercles on the mesosoma.

Cyphomyrmex minutus Mayr—
Least Fungus-Gardener Ant (Plate 29)

Taxonomy and Similar Species: *Cyphomyrmex minutus* is similar to *rimosus*, and both are somewhat variable, especially in different parts of their ranges. Older literature usually synonymized the two species or listed *minutus* as a subspecies of *rimosus*. In his manual of North American ants (1950), Creighton spends four pages on *rimosus/minutus*, finally deciding that future taxonomic revisions would have settled the matter, but ant taxonomy often proceeds slowly, in this case requiring 42 years. By the time Snelling and Longino (1992) had hacked their way through the nomenclatorial thicket that had grown up around *minutus* and *rimosus*, the two species had settled the matter among themselves by taking up residence in various sites without any evidence of interbreeding. *Cyphomyrmex minutus* is a little smaller, and paler brown than *rimosus*, and has relatively shorter legs, but it is difficult to conveniently quantify these differences, although they allow preliminary sorting of a box of specimens without resorting to a microscope. Under the microscope, the most easily seen difference is the much greater development of the dorsal thoracic tubercles in *rimosus*. If one views the mesosoma from the front and imagines that the ant is large enough to ride, *minutus* would be an acceptable mount, while *rimosus* would be too uncomfortable to even consider.

Distribution: *Cyphomyrmex minutus* occurs through the West Indies and Central America south into northern South America (Snelling and Longino 1992). In the United States, it occurs in southern Florida and southern Texas (Creighton 1950). Florida records extend north into Hillsborough, Orange, and Brevard counties. This species often lives in disturbed sites and might be able to live in the localized warmer areas of cities somewhat north of its general distribution. There is no reason to consider that *minutus* is not native in Florida, as it has managed to colonize even small islands in the Caribbean. Queens are hard bodied and do not have a swollen gaster, so they might be able to disperse long distances. This species could also be easily transported in containers of plants, and the Florida population may have absorbed more recent immigrants from the Neotropics.

Natural History: *Cyphomyrmex minutus* lives in both natural and disturbed habitats, although not in lawns or fields in my experience. It can easily withstand drought, as in the southern Florida Keys. Nests are not excavated in the soil but in cavities in rotten wood or under objects on the ground, in the debris at the base of trees, or in grass tussocks.

Weber (1955) found that *minutus* was sometimes associated with large nests of the stinging ant *Wasmannia auropunctata* at several Florida sites. *Wasmannia auropunctata* was introduced into Florida from the Neotropics, where it is also sometimes associated with species of *Cyphomyrmex* (Grangier et al. 2007). There is no evidence of any benefit accrued to either *Cyphomyrmex* or *Wasmannia* (Grangier et al. 2007). It is unlikely, however, that either species is forced to cohabit because of a lack of nesting sites. In the Neotropics, *minutus* is attacked by the diapriid wasps *Acanthopria* sp., which "sometimes inflict a heavy cost on *Cyphomyrmex* colonies" (Fernández-Marín et al. 2006). It is possible that *W. auropunctata* deters these larval parasitoids. There are no reports of larval parasitoids of *Cyphomyrmex* in Florida, but an undescribed species of *Acanthopria* is known from Florida (Masner and Garcia 2002); it is almost certainly an ant parasitoid, but might have a host other than *Cyphomyrmex*.

The fungal gardens of *minutus* are not fed pieces of leaves, but dead insects and insect excrement (Weber 1955). If the nest is severely disrupted, workers may use larvae and pupae from the colony to restart the garden (Fernández-Marín et al. 2006). In the gardens, the fungus grows in masses of single-celled yeasts, which can revert to a free-living filamentous state in the absence of *Cyphomyrmex* (Mikheyev et al. 2010). Larvae are reared separately from the fungus garden and fed on compact bits of fungus presented to them by workers (Weber 1955). Workers appear to be nourished by fungal liquid workers obtain by abrading fungal fragments and imbibing their juices (Weber 1955). The fungus associated with *rimosus* produces antifungal compounds that probably inhibit weed fungi or pathogenic fungi (Wang et al. 1999).

Alates fly May into October in Florida (17 specimens); flights seem to be in early morning, judging from the time of day males and queens appear at lights.

Name Derivation: From *minutus* (Latin), meaning "small," referring to the relatively small size of the species.

Cyphomyrmex rimosus (Spinola)—Immigrant Little Fungus-Gardener Ant (Plate 29)

Taxonomy and Similar Species: *Cyphomyrmex rimosus* is similar to *minutus*; the taxonomy and distinguishing features of these species are discussed above under *minutus*.

Distribution: Argentina, Brazil, Panama, the Guianas, and Venezuela (Snelling and Longino 1992); I have collected *rimosus* in the botanical garden in Santo Domingo, Dominican Republic, where it is probably introduced. *Cyphomyrmex rimosus* was probably introduced in the Mobile, Alabama area, and has spread south through Florida. The first known Florida collection was in 1957 (Deyrup et al. 2000). It is common in northern and central Florida, but there are only a few specimens from the southern Peninsula. Time will tell whether *rimosus* will become as abundant in tropical and subtropical Florida as it is in northern Florida. Outside of Florida, it occurs in some tropical regions, but the Florida population might be derived from a population living in temperate parts of South America.

Natural History: *Cyphomyrmex rimosus* lives in both natural and disturbed areas. My impression is that *rimosus* is less resistant to dry conditions than *minutus*. Nests are usually under objects on the ground, although I have found nests in grass tussocks. I have repeatedly found nests in potted plants, which helps explain the rapid and long-range dispersal of the species, but makes its relative scarcity in tropical Florida more puzzling. Nests of *rimosus* resemble those of *minutus*, with fungus gardens growing on bits of dead insects and insect feces, and the larvae in a chamber separate from the fungal garden. There is no reason to think that *rimosus* and *minutus* interact in any way, but this has not been studied. In Panama, *rimosus* larvae are attacked by diapriid wasps, including four species of *Acanthopria* and one species of *Mimopriella* (Fernández-Marín et al. 2006). It is not known whether any of these immigrated with *rimosus*, or whether *rimosus* picked up any larval parasitoids from ants native to Florida. In Florida, alates fly before dawn from May to December (44 specimens) and often appear at lights.

Cyphomyrmex rimosus in Florida is an excellent example of an invasive species of ant with no probable deleterious effects. It is a scavenging species that is unlikely to compete with native ants or other insects either for the substrate that it uses in its gardens or for nesting sites. Most other ants that have become common in Florida are predatory or have an association with sap-sucking insects. It is easy to imagine that these might have an impact on prey populations, or on the welfare of plants, or as competitors of native species, although there is seldom any evidence of this as yet.

Name Derivation: From *rimosus* (Latin), meaning "with clefts," perhaps referring to the deep incisions in the frontal carinae at eye level. Spinola provisionally assigned this species to the genus *Cephalotes*, species of which have the frontal carinae greatly expanded, somewhat as in *Cyphomyrmex*, but without the deep incisions.

Genus *Eurhopalothrix*

Cryptic Club-Haired Ants

The genus *Eurhopalothrix* (pronounced Ū' rō pal" ō thrix, or sometimes Ū' rōp al oth" rix) is a small genus distributed through the New World and Old World tropics. The single species found in Florida represents an unusual northern range for the genus. Species of *Eurhopalothrix* and related genera have a mandibular snapping mechanism somewhat like that of the tribe Dacetini (Bolton 1999) but are not specialized predators of Collembola, like most dacetines that have been studied. The single species whose behavior has been studied, *Eurhopalothrix heliscata*, forages singly but has an interesting adaptation that increases the efficiency of "central place foraging" (Wilson and Brown 1984). Groups of workers may congregate away from the nest, moving out individually from the group. This might represent a form of staged recruitment that allows organized exploitation of dispersed small prey by slow-moving predators. Wilson and Brown (1984) found that foragers of E. *heliscata* show a tendency to thrust their heads into small openings and tunnels, where they might find prey, such as termites and beetle larvae, missed by other ants. Foragers only become aware of potential prey a short distance away, approximately 3–4 mm. With jaws gaping, the ant slowly approaches its prey, seizing it with a sudden snap and then stinging it. While the prey is struggling, the ant's antennae are clamped tightly against the head, the concave antennal scapes protecting the rest of the antenna. If the prey is large, the ant may be flung about until the venom takes effect. If several termites are presented simultaneously, there is an increase in the number of ants in the foraging arena, suggesting some form of recruitment. *Eurhopalothrix* and other members of its tribe, the Basicerotini, have peculiar, enlarged, brush-like hairs on the head and upper surface of the body. These may serve to sweep up minute particles of dirt that are held in a thin film on the surface of the head and body by a covering of appressed plumose hairs (Hölldobler and Wilson 1986b). This, combined with a tendency to freeze when alarmed, allows these ants to vanish against the substrate when confronted by a small predator such as a frog or lizard (Hölldobler and Wilson 1986b).

Name Derivation: From *eu* (Greek), meaning "good," or used as an emphatic, *ropalon* (Greek), meaning "club," and *trichos* (Greek), meaning hair. This probably refers to the strange, club-shaped hairs on species of both *Rhopalothrix* and *Eurhopalothrix*. The complete name *Eurhopalothrix* might be loosely translated as "ant with amazing club-shaped hairs."

Eurhopalothrix floridana Brown and Kempf— Florida Cryptic Club-Haired Ant (Plate 24)

Taxonomy and Similar Species: This species is superficially somewhat similar to some dacetines, such as *Strumigenys membranifera*, but can be easily distinguished by examination of either end of the ant. The upper surface of the gaster is covered with deep, closely arranged punctures, a feature never found in dacetines (Bolton 2003). The eye is perched on the upper rim of the deep antennal scrobe (antennal groove), while the eyes of Florida dacetines are near the lower edge of the antennal scrobe. There are also a few club-shaped hairs

that stick up from the surface of the body. The pair that is on the postpetiole of the worker and queen also occurs on the male, allowing identification of males found in flight traps. In the field, *Eurhopalothrix floridana* somewhat resembles a small *Cyphomyrmex*, but is never seen walking about in the open.

Distribution: Widely distributed in peninsular Florida, extending west into Gulf and Jackson counties. It is probable that this species also occurs along the southern border of Georgia, and it is possible that it is expanding its range westward. Brown and Kempf (1960) suggest that E. *floridana* might be introduced into Florida from the Neotropics, although it had not been found elsewhere. Deyrup et al. (1997) presented some circumstantial evidence for this: there is an unconfirmed record from Mexico; the species was not collected in the Gainesville area when Van Pelt was sampling for dacetines (1958), although it is common there now. It is, however, known from Cuba (Baroni Urbani and De Andrade 2007) and might be a West Indian species that was introduced to Florida or arrived on its own. The first Florida specimen was collected in Key West in 1887 (this headless specimen was spotted by David Smith at the Smithsonian), at a time when that city was a center of trade in Mexico, Central and South America, and the West Indies. A recent accidental experiment is testimony to the durability of E. *floridana*. A 2-L plastic bag of leaf litter and soil from Gainesville was misplaced, and finally run through the Berlese funnel 38 days later. A small colony of E. *floridana* was still alive, along with colonies of the exotic *Strumigenys eggersi* and *Pheidole moerens*, and the native *Nylanderia faisonensis*.

Natural History: There are 39 Florida collections with habitat records: 17 from mesic hardwood or hardwood and pine forest, 9 from coastal tropical hammock, 8 from xeric forest, and 5 from wet forest or swamp forest. Some sites were forested vacant lots in urban areas. *Eurhopalothrix floridana* appears to be the only member of its genus that can be easily found. Almost all the biological information known about *Eurhopalothrix* is derived from observations of a single nest of *heliscata* (Wilson and Brown 1984), so *floridana* provides unusual opportunities for original natural history studies by a patient naturalist. The male was described by Deyrup et al. (1997).

Name Derivation: Named for the state where the type series was found.

Genus *Monomorium*

Trailing Ants

Members of the genus *Monomorium* (pronounced **Mon' ō mō" ree um**) are slender, elongate, and small (generally approximately 2 mm long). All but one of the Florida species are shiny. Species of *Monomorium* closely resemble small species of *Solenopsis*: they are small, elongate, and slender, lack propodeal spines, and are usually shiny. Under the microscope, it is easy to see that *Solenopsis* have two very large segments in the antennal club, while *Monomorium* have three more moderate segments. In the field, however, it is easy to confuse the identities of species such as *M. floricola* and *S. picta*, both of which inhabit hollow twigs. *Xenomyrmex floridana* is another small, dark species that could be mistaken for an arboreal *Monomorium*, but under the microscope, *X. floridana* can be seen to have a low petiole (elevated as a blunt cone in *Monomorium*), and strongly thickened hind coxae.

Monomorium species are well known for their trailing behavior, with narrow files of ants closely following a winding scent trail. It is often possible to diagnose over the telephone an infestation pharaoh's ants (*Monomorium pharaonis*) by asking whether the ants are very tiny and reddish, and follow each other along narrow trails that wander up the walls and over the counters in the kitchen. *Monomorium pharaonis* and *Monomorium destructor* often invade buildings in large numbers and are difficult to control with spray insecticides because colonies are polygynous and dispersed in a series of separate nests. They can usually be safely and easily controlled with baits made from sugar water and boric acid.

The genus *Monomorium* includes almost 300 species, the great majority of which are native to the Old World tropics, the Mediterranean Region, and Australia (Bolton 1995). North America has a small fauna of native species, all of which are small and black, perhaps the product of a minor radiation of a lineage that crossed the Bering Land Bridge (DuBois 1986). When Creighton wrote his manual of North American Ants (1950), he was able to claim that there were only two or three species of native *Monomorium*, outnumbered by the four exotic species. DuBois' groundbreaking revision of native species (1986) produced nine species native to North America north of Mexico. This study straightened out a great number of misconceptions, misidentifications, and mistaken geographic ranges, but it may have been somewhat hampered by the fact that there are fewer good collections of these ordinary-seeming ants than are needed, especially collections including queens, which are often much more distinctive than workers. It seems likely that there are additional species in the Southwest. The new species discovered by DuBois fall into two categories: parasitic species that are so localized and difficult to find that they escaped detection, and species whose workers are so similar to each other that they could not be reliably separated until they were associated with queens and males. In Florida, there are, as far as is known, only two native species, and five exotics, one of which is from the Caribbean, the others from the Old World tropics and subtropics. It is not very probable that additional species will be found in Florida, unless there is an undiscovered parasite of the abundant *viridum*, or additional Old World exotics appear in south Florida. There is a slight possibility that *minimum* will be found in the Red Hills area of Leon and Jefferson counties, where the soil has more clay. *Monomorium minimum* is a soil-dwelling species resembling *viridum*, but without the hard-to-see bluish reflections and with a queen that is black rather than reddish and bluish black.

Name Derivation: From *monos* (Greek), meaning "single," and *morion* (Greek), meaning, in this case, "segment," referring to the single-segmented maxillary palps.

Monomorium destructor (Jerdon)— Destructive Trailing Ant (Plate 30)

Taxonomy and Similar Species: The brownish yellow color of *destructor* distinguishes it from other Florida *Monomorium*, except for *pharaonis*. *Monomorium destructor* is shiny (*pharaonis* is not) and is variable in size within colonies (all *pharaonis* are the same size), and the first segment of the antennal club is about the same size as the second, a rather unusual feature for ants, most of which, including *pharaonis*, have the second segment conspicuously larger than the first. In Florida, *destructor* is also much more localized, possibly restricted to Key West, where it was once common. It appears that *pharaonis* may have occasionally been mistaken for *destructor* in published literature.

Distribution: This is a pantropical species that originated in the Old World tropics, probably in the Indian subcontinent, where it would have had thousands of years to adapt to permanent human settlements. In North America, *destructor* is known from Tennessee and from "a number of localities" in Florida (Smith 1965). It has not spread widely in the Southeast, and it is probable that the population in Tennessee no longer exists. In Florida, *destructor* was not found in the earliest ant surveys (Smith 1930a; Wheeler 1932), even though these surveys emphasized tropical parts of Florida. It was first reported in 1933 (Smith 1933), from specimens collected in St. Petersburg and Bradenton. Wilson (1964) found *destructor* on Key Largo and Plantation Key. In a subsequent, more intensive survey of the ants of the Florida Keys (Deyrup et al. 1988), *destructor* was found only on Key West, where it was easy to find; it was not found in a more recent survey of the Keys (Moreau et al. 2014). My conclusion is that this species was never fully established in Florida outside of the Keys, and even there it is now rare.

Natural History: My observations of *destructor* on Key West agrees with Smith's statement (1965) that this species lives in large polygynous colonies that consume many kinds of human food that are composed of protein or fats or sugar. Outside of buildings, *destructor* feeds primarily on insects and honeydew. Smith (1965) reports that *destructor* gnaws holes in fabric and may damage insulation of electric wires. It is also said to sting "fiercely" (Smith 1965); this is plausible, but it would be prudent to make sure that any Florida records do not refer to *Wasmannia auropunctata*. There is also the implication that *destructor* might spread bubonic plague by feeding on infected rats (Smith 1965). This is completely contrary to modern understanding of the transmission of this disease, which is mediated through fleas, and probably cannot even be spread through contact with crushed, infected fleas (Kettle 1984). Needless to say, the connection between the bubonic plague and *destructor* appears on a number of websites, bringing new life to old misinformation. It appears that we are overdue for a new study of the natural history of this common pest ant. From an ecological standpoint, *destructor* does not appear to be a problem in Florida.

Name Derivation: The name *destructor* (Latin) means "destroyer." This species was described (very briefly) by Thomas Claverhill Jerdon (1811–1872), a British medical doctor and naturalist working in India. Jerdon worked with both plants and animals, including insects, but was especially interested in birds. These ants were evidently a problem, as documented in his description (Jerdon 1851). "They live in holes in the ground, or in walls, etc.,

and are very numerous in individuals. They prefer animal to vegetable substances, destroying dead insects, bird skins, etc., but also feed greedily on sugar. They are common in all parts of India, and often prove very troublesome and destructive to the Naturalist."

Monomorium ebeninum Forel— Caribbean Trailing Ant (Plate 31)

Taxonomy and Similar Species: This shiny black *Monomorium* resembles two other Florida species, *trageri* and *viridum*, whose known ranges end just north of that of *ebeninum*. The petiole of *ebeninum* is flattened across the top when viewed from behind (not in lateral view), while the petiole of *trageri* and *viride* is narrowly rounded above. This may seem like a small difference, but by the standards of native North American *Monomorium*, it is huge, and one can only wish that the other species were separated so easily.

Distribution: Caribbean islands, including Trinidad; the east coast of Mexico and Central America, with scattered records from the Pacific Coast of Mexico and Central America (DuBois 1986). Creighton, who was well aware of the difference between *ebeninum* and other species, reports "scattered records" from southern Florida and the Brownsville area of Texas (1950). Presumably, there are specimens somewhere, although *ebeninum* does not appear in the lists of ants in the Hymenoptera catalogs (Smith, D. R. 1979; Smith, M. R. 1951). The records of "*minimum*" from Lower Matacumbe Key (Wheeler 1932) probably refer to this species. Wilson (1964) did not find *ebeninum* in the Keys, but it was reported later from Upper Matacumbe and Big Pine Keys (Moreau et al. 2014). At the moment, *ebeninum* appears to be a precariously established exotic confined to the Florida Keys (Deyrup et al. 2000).

Natural History: In the Bahamas, where *ebeninum* is common, nests are in the ground, or in dead branches on living trees, or in the dead basal leaves of bromeliads. Smith (1937) found it in similar situations in Puerto Rico and comments on the unusual adaptability in nesting sites relative to other species of *Monomorium*. Like other species of *Monomorium*, *ebeninum* is predaceous and also feeds on honeydew (Smith 1937). Colonies are polygynous and usually composed of hundreds of workers. I have not observed this species invading buildings in the West Indies, and it is of negligible importance as a pest. The feeding behavior, habitat preference, and temperature tolerance of *ebeninum* in Puerto Rico were quantified by Torres (1984a,b). He found that *ebeninum* is an open-site species, and colonies brought into the forest generally die out or disappear, although supplemental feeding can possibly prevent or postpone this fate.

This species has flightless queens, which, combined with its generally coastal distribution, suggests that its dispersal to numerous Caribbean islands must have been achieved by stowaways in shipments of products. This, however, is not necessarily the case. A laboratory colony of *trageri*, a related polygynous species with flightless queens, produced a few winged queens (DuBois 1986). It is possible that *ebeninum*, as well, has not completely abandoned its aerial options.

Name Derivation: From *ebenus* (Latin), from the ebony tree, which has black wood, meaning "black."

Monomorium floricola (Jerdon)— Bicolored Trailing Ant (Plate 30)

Taxonomy and Similar Species: Unlike all other shiny, elongate little ants with which it could be confused, *floricola* is bicolored: black with a conspicuously orange red mesosoma, petiole and postpetiole; it can be described as orange in the middle, black at both ends. Despite the tiny size of *floricola*, this coloration can be seen in the field. Somewhat similar species that are not bicolored in this way are *Solenopsis picta* and *Xenomyrmex floridana*.

Distribution: This is a pantropical tramp species that probably originated in tropical Asia. In North America, outdoor populations are known from Florida. There is a record from an unspecified site in Alabama (Smith 1965), but a recent survey of the ants of Alabama (MacGown and Forster 2005) did not produce any specimens, and the apparent absence of *floricola* from the Gulf Coast of Florida suggests that the northern Gulf Coast may be too chilly for this ant. It is probable, however, that there are outdoor colonies in the Brownsville area of Texas. Elsewhere, greenhouses and indoor displays in zoos might be good places to look for *floricola* (Wetterer 2009c). In Florida, *floricola* is common in the southern part of the Peninsula and occurs in scattered localities, usually near water, through the north-central Peninsula.

Natural History: This polygynous species with flightless queens forms large colonies in dead branches on living trees and smaller colonies in hollow twigs, vines, and weed stems. Files of workers may invade buildings in Florida, entering through window screens or small cracks. It is not unusual for an outdoors colony to fission and one part take up residence in a building, for example, near the coffee pot in the dining room of the Archbold Biological Station, where these ants take their milk and sugar without coffee. In the field, *floricola* captures or scavenges small insects, feeds on honeydew, and takes nectar from extrafloral nectaries, such as those of *Passiflora* species. Foragers also creep into flowers to collect nectar; the morphology and behavior of *floricola* preclude any role as a pollinator. *Monomorium floricola* seems well positioned to prey on beetle larvae found in dead wood and plant stems, but there are no reports of this. The feeding, temperature, and habitat preferences of *floricola* in Puerto Rico were quantified by Torres (1984a). *Monomorium floricola* occurred in grasslands and agricultural areas but was not found in the forest. I have found it in forested areas elsewhere, but in dead twigs in the tops of low trees. Foragers collected insects and insect parts, as well as unidentified liquids.

Name Derivation: From *floris* (Latin), meaning "of a flower," and the suffix -*cola*, meaning "inhabitant," probably referring to the nectar-feeding habit of the species.

Monomorium pharaonis (Linnaeus)— Pharaoh's Ant (Plate 30)

Taxonomy and Similar Species: Easily distinguished from *destructor*, the other brownish yellow *Monomorium*, by its nonshiny head and body (*destructor* is shiny), and by the absence of obvious variation in worker size (groups of *destructor* usually consist of workers that obviously vary in size). The first segment of the antennal club of *pharaonis* is smaller than the second, rather than similar in size, as in *destructor*. *Monomorium pharaonis* in Florida is also much commoner and more widespread than *destructor*, which appears to have disappeared from the sites where it used to be common in the Florida Keys (Moreau et al. 2014).

Distribution: This species originated in the Old World tropics or subtropics and has been spread throughout the world by commerce. It is not uniformly distributed in North America and is probably commonest in relatively warm areas, where colonies in siding, foundations, and elsewhere on the periphery of buildings can recolonize the interior when insecticides have eradicated colonies from the prime locations within. In Florida, I collect specimens of *pharaonis* whenever I see them in a new site. This has produced only a sparse pattern of distribution, whereas a survey of pest control records would probably blanket the state with distribution points. Smith's statement (1965) that "The Pharaoh ant probably occurs in every town or city of commercial importance in the United States" undoubtedly still applies to Florida, but this is not a species that I can confidently expect to find every time I go to a cheap hotel or restaurant.

Natural History: This may be the only true domestic species of ant, that is, the only species well adapted to breed inside buildings without any link to the outdoors. The group of arthropods that can live in buildings for an indefinite number of generations is relatively tiny, including some species of mites, spiders, centipedes, cockroaches, bedbugs, Psocoptera, clothes moths, fleas, stored products insects (primarily beetles and moths), and a few parasitic and predatory insects. This group is distinct from the huge diversity of species that regularly move into buildings from outdoors, or move freely in and out of permeable buildings such as log cabins and bamboo-slat homes. All the true domestic arthropods must share a few characteristics: resistance to a permanently dry atmosphere, resilience in the face of frequent disturbance, behavioral and genetic acceptance of inbreeding among the offspring of a single foundress, adoption of a very limited set of microhabitats that will both provide a living and promote dispersal as stowaways, and the absence of any requirement for autonomous long-distance dispersal. Given this demanding list of prerequisites, it is easy to understand why only one species of ant has become truly domesticated and pioneered a parasitic existence with human hosts. It is also relevant to consider how recent are the changes in design of human habitations that provide a modern inquiline of humans an ecological niche from which peridomestic species are excluded. If the pest control industry were to cease its efforts, one can imagine that over a few thousand years, other ants preadapted to a domestic existence might evolve a few final adaptations and move into this poorly occupied niche.

The same ecological tolerances that allow *pharaonis* to become a domestic parasite of humans make it suitable for intensive studies in the laboratory. Much of this work appears in Hölldobler and Wilson's book, *The Ants* (1990), from which the following account is compiled. *Monomorium pharaonis*, like several other species in its genus, is polygynous with flightless queens. Colonies reproduce by budding, one or more queens departing from the nest with a large group of workers. Mating occurs within the colony. Reflecting the capricious nature of their host, colonies of *pharaonis* are quick to react to disturbance by locating a new nest site and organizing an exodus. Colonies can reunite, and in some regions, colonies from presumably long-separated lineages can be brought together without aggression. While the colonies of *pharaonis* are, theoretically, immortal, queens have the shortest life span of any known queen ant, approximately 200 days. During this period, after mating in the nest, a queen goes through three reproductive phases: an initial period in which all her eggs become workers, a middle period in which her eggs can become queens if provided with extra food, and a final period in which her eggs again produce only workers. This means that by the time a queen dies, she has provided a mechanism for a large group of replacement queens (depending on food availability) and two major cohorts of workers. The foraging behavior of *pharaonis* is an obvious focus of interest. There are two sets of compounds involved, faranol from the Dufour's gland, and monomorine I and III from the poison gland. Initial or exploratory foraging is not random: individuals, probably following a chemical remnant, start out following the path that was previously productive and then branch out to wander about looking for resources. This explains how *pharaonis* can so quickly zero in on each meal's crumbs that appear on the kitchen counter, or the dead insects that occasionally appear on the windowsill.

Pharaoh's ants can form gigantic and destructive infestations. An 1879 report described an infestation "whose myriads were past comprehension and in some places it distinctly colored the white wall with its hosts" (Donisthorpe 1915). The resource-finding ability of *pharaonis* is legendary, causing Donisthorpe (1915) to credit these ants with "a very remarkable sense of smell, or some other faculty which enables them to discover the whereabouts of any comestible suitable to them." Entomologists know of the speedy devastation that can be visited on a box of insect specimens left out overnight in an apparently ant-free room. A more familiar example is the desecration of the secret, emergency candy bar kept in a drawer of a desk where no ants had been observed. There is no reason to believe, however, that any process is at work other than the persistence of great numbers of wandering foragers exploring a (to them) vast territory. These ants can also cause problems by their mere presence, as reported by Eichler (1990). "Once I entered an administrative building overrun by ants. Their base was the restaurant kitchen in the building's basement. For weeks, the employees discussed the ants or made shallow experiments with them. This cut their work efficiency and total output significantly." One could probably predict, however, that eradication of the ants would not transform these particular employees into models of industriousness.

In hospitals, pharaonis can be a serious problem and health risk, as described in some detail by Eichler (1990), whose discussion is based on personal experience. Although the sting of this species is weak and of little consequence, the jaws are sharp, and can cause lesions and raw places where the skin is week or recovering from damage. Large numbers of these ants can work their way under casts and bandages, causing intolerable itching by feeding around wounds. Certain ointments can attract enormous numbers to lesions and areas of burned skin. Pharaoh's ants can gnaw on the skin of newborn babies and sedated or comatose patients. Various alarming pathogenic bacteria can be obtained by allowing the ants to traverse sterile agar plates, and it is likely that this species can spread diseases in hospitals. The ants make their way into supposedly sterile cases of surgical instruments and tunnel into glucose bottles and intravenous feeding tubes. Insecticidal sprays usually have little lasting effect; Eichler calls spraying in hospitals "a form of malpractice." Several kinds of baits are convenient and efficacious.

Name Derivation: Carl Linnaeus described the appearance of pharaonis using only six words. This brevity is typical of his insect descriptions, and most of his species would be impossible to recognize if he had not attached his names to actual specimens, many of which still exist. Evidently, his specimens of pharaonis came from Egypt ("Habitat in Aegypto"), hence the name pharaonis, meaning "of the pharaoh."

Monomorium subcoecum Emery—
Golden Trailing Ant (Plate 23)

Taxonomy and Similar Species: This is an easily identified species: it is golden yellow, has tiny eyes, and has a small but distinct angle on the propodeum in lateral view. It is more likely to be mistaken for a small yellow Solenopsis such as Solenopsis abdita, but the three-segmented antennal club and the propodeal angle readily distinguish this species from any species of Solenopsis.

Distribution: Monomorium subcoecum was described from Saint Thomas (U.S. Virgin Islands) and I have collected specimens from Dominica in the Lesser Antilles. It also occurs in Puerto Rico (Torres and Snelling 1997); these authors list M. subcoecum as an exotic species introduced from the Old World; they cite no evidence of this, but it seems likely. Monomorium has undergone most of its evolutionary radiation in the Old World and appears to be represented in the New World by a compact lineage that is sometimes called the M. minimum complex. Monomorium subcoecum obviously belongs to a different lineage within Monomorium. In Florida, it is known from Matheson Hammock in South Miami. This species is listed as occurring occasionally among the prey of the tropical blind snake Typhlops platycephalus in Puerto Rico, a species that feeds extensively on ants (Torres et al. 2000b).

Natural History: In Dominica, specimens were found by digging in an open, dry, scrub area. In Puerto Rico, Monomorium subcoecum occurs in subterranean nests with multiple queens in open areas, according to an unpublished manuscript by Roy Snelling and Juan Torres on the ants of Puerto Rico (manuscript date, 1993; both authors are deceased, and there seem to be no plans to publish this valuable work). In Florida, specimens were collected from soil in dense tropical hammock.

Name Derivation: Probably derived (with a minor typo) from caecus (Latin), meaning "blind," combined with sub- (Latin), meaning in this case "almost," referring to the minute eyes. Emery's description (1894) begins (my translation), "A species notable for the extreme smallness of its eyes, which are reduced to a single facet...."

Monomorium trageri DuBois—
Trager's Trailing Ant (Plate 31)

Taxonomy and Similar Species: This is a black, shiny species, similar to ebeninum, viride, and minimum; the latter species is apparently absent from Florida. The flattened crest of the petiole of ebeninum (viewed from behind) separates that species from trageri. Monomorium trageri nests in trees and shrubs, occasionally in rotten logs, while viridum nests in the ground. Queens of trageri are black and flightless, while those of viridum are dark red with a blue black gaster. Workers of *viridum* have slightly bluish reflections, but this is difficult to see, much more difficult than the bluish sheen of *Pheidole metallescens*.

Distribution: At the time of its description, *trageri* was known from only four sites, all in northern peninsular Florida (DuBois 1986). It is now known from numerous Florida records, and I have taken it in coastal North and South Carolina, so it may be generally distributed on the southeastern Coastal Plain. There are unpublished records from southern New York (S. Cover 2007, personal communication). This gives some hope that populations will be found within the range of *minimum*, eliminating any possibility that *trageri* is a coastal form of *minimum* that has largely or completely shifted to the production of flightless queens. Although *trageri* is arboreal while *minimum* usually is not, there are examples of ants, such as *Aphaenogaster lamellidens* and *Dolichoderus pustulatus*, which are arboreal in Florida but not elsewhere. In the future, it should be practical to investigate whether *trageri* is genetically distinct from nearby populations of *minimum*.

Natural History: This species usually occurs in areas that are wet for part of the year, although not necessarily flooded. Nests may be in dead branches of shrubs, such as wax myrtles, and may also be in the thick bark of standing pines. One nest was in the hollow stem of an herbaceous plant near the edge of a swamp. Like other *Monomorium* species, *trageri* responds eagerly to bait. If the number of foragers coming to a bait is large, queens may be drawn out as well, as in *pharaonis* and *floricola*. In the laboratory, James Trager observed males and queens mating within the colony and also reared a small number of winged queens (DuBois 1986).

Name Derivation: Named by Mark DuBois for James Trager, a contemporary myrmecologist and general naturalist, who collected or raised many of the specimens that were used in the description of the species. James Trager is a pioneer in the taxonomy of Florida ants, providing the first revisions of *Nylanderia* (1984), *Dorymyrmex* (1988), the confusing *Solenopsis geminata* complex of fire ants (1991), *Polyergus* (2013), and, with coauthors J. A. MacGown and M. D. Trager, the difficult *Formica pallidefulva* group (2007).

Monomorium viridum Brown—
Metallic Trailing Ant (Plate 31)

Taxonomy and Similar Species: A shiny black species similar to *ebeninum*, *trageri*, and *minimum*; the latter species is not known from Florida. The crest of the petiole of *ebeninum* in posterior view is flattened, narrowly rounded in *viridum*. Colonies of *viridum* are terrestrial and in well-drained sandy areas, while those of *trageri* are in dead wood or under pine bark. Queens of Florida *viridum* are dark red with a metallic blue black gaster; those of *trageri* and *minimum* are black; queens of *trageri* are flightless, whereas those of *viridum* have functional wings. *Solenopsis nickersoni* is a shiny black, ground-nesting species, but has contrasting yellow legs and a two-segmented antennal club.

Distribution: New Jersey (pine barrens and sandy coastal areas), North Carolina (one site), and Florida. There is a tentative (queenless) record from Alabama (MacGown and Forster 2005), where the species is certain to occur in sandy coastal areas. *Monomorium viridum* may be expected in sandy sites throughout the Coastal Plain of eastern North America north to southern New England, but records should be confirmed by obtaining queens. The similar species *minimum* can also be in sandy areas (DuBois 1986). In my experience, it is relatively easy to find queens, as the nests are relatively shallow and there are usually several queens in each colony. Records of the distribution of *viridum* and its synonym *peninsulatum* are suspect before DuBois' 1986 revision, except for those from New Jersey and Florida. In Florida, *viridum* occurs throughout the state in open sandy areas.

Natural History: This is a polygynous species with winged queens. I have taken a queen in a flight trap, so there is some dispersal by air, but it is possible that colonies bud as well. The distribution of colonies at the Archbold Biological Station is suspiciously patchy. The habitat is always open, well-drained sandy areas (DuBois 1986). Nests are conspicuous by their pile of excavated sand, frequently with large teams of workers bringing up sand a few grains at a time. It is common to see large-scale movements of workers going from one nest mound to another a short distance away. The purpose of these expeditions, whether colony relocation or budding, is unknown. I have occasionally seen small larvae of some species of *Microdon* (Syrphidae) in nests.

Name Derivation: From *viridum* (Latin), meaning "green," referring to "the metallescence of the gaster, which surpasses in intensity any other green that I have yet seen in the *Formicidae*" (Brown 1943). Florida queens appear to me more blue than green. The description of this handsome ant (Brown 1943) was the first publication by William L. Brown, Jr. (1922–1997). It was the harbinger of 54 years of outstanding publications in myrmecology, evolution, systematics, and biogeography. Wherever he went, Brown found new species and new ideas.

Genus *Myrmecina*

Mite-Eating Ants

The low, squarish, ridged petiole distinguishes members of the genus *Myrmecina* (pronounced **Mur' me sī" nuh**) from other Florida ants. Most of the 50 or so species of *Myrmecina* (AntWeb 2015) live in Southeast Asia or Australia (Shattuck 2009), but there are several northern species that might represent a limited evolutionary radiation of species associated with temperate Arctotertiary forests. The number of North American species is small but uncertain, despite analyses by Smith (1948), Creighton (1950), Brown (1949d, 1951, 1967), and Snelling (1965). Brown (1967) recognized two species, one from Mexico, and a widespread species in North America north of Mexico, but there is evidence that there may be at least four species north of Mexico. In *Myrmecina*, the chief impediment to taxonomic clarity is intraspecific variability. This does not mean that there could not be distinct species hidden within this variation: variable lineages provide the fodder for natural selection leading to speciation.

The diet of two species of *Myrmecina* has been studied by Masuko (1994) in Japan. These species are general predators of soil microinvertebrates and have additional specializations for attacking a particular prey, hard-bodied soil mites of the family Oribatidae. Masuko found *Myrmecina* nests with middens of oribatid mites showing characteristic trauma to their ventral regions. As reported and illustrated by Masuko, workers open mites while standing on both rear legs and one middle leg, bending the gaster forward for additional stability. The ant grasps the mite with one middle leg and both forelegs and then peels open the mite, using its serrated and somewhat scoop-shaped jaws. The head of larval *Myrmecina* is peculiarly long and narrow, suited for consuming oribatids on the half-shell, as observed and photographed by Masuko. Specialized predation on oribatid mites is a useful talent, as these mites are among the most numerous of soil invertebrates: a square meter of temperate forest soil typically supports 50,000 to 500,000 oribatids (Coleman et al. 2004). With such a resource available, one might expect *Myrmecina* to be more speciose and its species to be more abundant than they are in most habitats.

The reproductive biology of *Myrmecina* species has gotten considerable attention. Several species, including *Myrmecina americana*, show queen polymorphism, consisting of a winged form that sheds its wings like a normal ant, and a smaller "intermorphic" form without wings, but larger than a worker, with somewhat larger eyes (Murakami et al. 2002). In *nipponica*, removal of the queen from a colony stimulates the production of intermorphs, but not alate queens (Murikami et al. 2002). These intermorph queens apparently mate with males that are attracted from other nests. In Japan, intermorph queens occur only in colder regions and possibly represent an adaptation for producing smaller but functional queens under conditions of reduced resources or a shorter season suitable for development (Murakami et al. 2002). In the European species *graminicola*, colonies with normal queens always have a single queen, but colonies with intermorph queens sometimes have multiple queens (Buschinger et al. 2003). Normal queens can produce either normal or intermorph queens, but not both; intermorph queens can produce both kinds of queens (Buschinger et al. 2003). A further study (Steiner et al. 2006) suggests that queen polymorphism may be the ancestral condition in temperate-climate *Myrmecina*. The mating behavior of *graminicola*, studied by Buschinger (2003), may reflect the necessity of attracting males to flightless queens. The queen leaves the nest, climbs up a convenient vertical object, and dabs a spot of pheromone, produced by the poison gland, on the substrate. Males are attracted to this pheromone and encounter the nearby queen.

Name Derivation: From *myrmex* (Greek), meaning "ant," and the suffix -*ina* (Latin), in this case probably a diminutive, meaning "little ant."

Myrmecina americana Emery— American Mite-Eating Ant (Plate 32)

Taxonomy and Similar Species: Distinguished from species of other genera of Florida species by the low, squarish petiole. Almost all specimens of Florida *Myrmecina* are *M. americana*. There is a single Florida record of an undescribed parasitic species: unlike *americana*, this species is reddish and lacks heavy sculpturing with ridges and reticulations.

Myrmevina cooperi is a tiny, rare species (under 2 mm in length) that has a strongly projecting keel on the underside of the postpetiole. The postpetiole of *americana* is conspicuously concave below in lateral view, with a thorn-like anterior tooth.

Myrmecina americana has been the subject of superabundant taxonomic discussion, fueled by its morphological variability. The various eastern subspecies and forms of *americana* appear to have been successfully synonymized by Brown (1949d, 1951, 1967). In the West, however, there is the problematic species *californica*, which has been called a subspecies (Snelling 1965) or synonym (Brown 1967) of *americana*. It is possible, however, that *californica* is a valid species, as its distinguishing character states (flattened antennal bases, a median clypeal ridge ending in a tooth) are typical of the Old World species such as *graminicola*, not *americana*.

Distribution: Quebec south through Florida, west into Iowa and Texas; in the Southwest unevenly distributed and subject in part to taxonomic dispute. In Florida, this species is relatively common in the northern third of the state, but discontinuously distributed to the south. The most unexpected record is from Elliot Key in Miami-Dade County.

Natural History: *Myrmecina americana* is found in mesic forest habitats, in Florida usually restricted to coastal forests and edges of seasonally flooded areas. It is seldom found in xeric hammocks, even where the leaf litter is deep. It sometimes occurs in moist, shady, urban gardens, especially in the Mid-Atlantic States, occasionally in north Florida. I have twice collected colonies with middens of oribatid mites as described by Masuko (1994) and assume that *americana* shares the feeding habits of its Japanese relatives. Queen polymorphism has been found in colonies from Ohio, Indiana, and North Carolina (Steiner et al. 2006), but I have not seen Florida intermorphs.

Name Derivation: Long considered a North American subspecies of a European species: *Myrmecina graminicola americana*. It was elevated to a separate species more than 50 years after its description (Brown 1949d). There is no evidence that any species of ant ever migrated naturally from Europe to eastern North America.

Myrmecina cooperi Deyrup— Cooper's Mite-Eating Ant (Plate 32)

Taxonomy and Similar Species: This is a rare and geographically restricted *Myrmecina* found in northwestern Florida and adjacent Alabama. *Myrmecina cooperi* is characterized by small size (length under 2 mm, usually approximately 1.8 mm), minutely shagreened (nonshining) gastral tergites, short and triangular propodeal spines, and a projecting, keel-like ridge, somewhat like that of "species A," on the underside of the postpetiole. For years, I convinced myself that these were undernourished and undersized *americana*, like the miniature specimens reared by Brown (1949d) from starving colonies of *americana*, used as evidence for synonymy of a small, short-spined form that Emery had described as the subspecies *brevispinosus*. Some Florida Panhandle specimens of *americana* have shagreened tergites, as do some Texas specimens that Wheeler (1908) placed in the subspecies *texana*, later synonymized by Brown (1949d, 1951). Discounting these kinds of character states, the major remaining diagnostic feature of the miniature Florida form is development of the postpetiolar keel, a feature not found in any *americana*, large or small, among the hundreds of specimens I have examined. This keel, combined with the general structural uniformity of a series of specimens that represent seven collections from the Florida Panhandle and an adjacent site in Alabama, indicates that this is a rare, localized species that has escaped the attention of myrmecologists. In the field, this species may be identified by its small size (under 2 mm); among the 597 specimens of *americana*, collected in 12 states, there are no examples of specimens of *americana* that are 2 mm long or smaller.

Distribution: Alabama: Houston County (one specimen), Florida: Okaloosa County (two specimens, one collection), Liberty County (three specimens, three collections), and Walton County (five specimens, two collections).

Natural History: The small number of specimens that I have seen was collected by litter extraction or litter sifting in broadleaf forests. Several collections were from the upper slopes of "steepheads," which are naturally occurring hollows in sandy uplands in the Panhandle. One series was from a suppressed pine and magnolia stand near a Gulf Coast beach. Like *americana*, this species remains curled up in the sifting tray for a minute or more, so it is easily overlooked.

Name Derivation: Named in honor of the Robert J. Cooper family of Palm Beach, Florida, in recognition of strong support for the biodiversity program of the Archbold Biological Station (Deyrup 2015).

Myrmecina Species A—
Parasitic Mite-Eating Ant (Plate 32)

<u>Taxonomy and Similar Species:</u> This rare, undescribed species lacks conspicuous sculpture and is reddish in color. The postpetiole has a conspicuous ventral keel in lateral view.

<u>Distribution:</u> Found through most of the range of its host, *M. americana* (Fisher and Cover 2007). In Florida, it has been collected once, in Bradford County, by Lloyd Davis.

<u>Natural History:</u> This species appears to be a social parasite of *M. americana* (Fisher and Cover 2007). It occurs together with its host, and queens of both species can be found in the same nest. The biology of this species has not been studied.

Genus *Myrmica*

Northern Furrowed Ants

The genus *Myrmica* (pronounced **Mur mī" kuh**) includes approximately 112 species, although the species-level taxonomy has been in a state of flux for decades, so this number is constantly changing. Basic variation within the genus is small, with most species distinguished by features involving the shapes of the antennal scapes and propodeal spines, and by variations in sculpture. The genus is informally recognized by its uniform size (usually 4–5 mm) and heavy sculpture, usually with prominent grooves or furrows on the sides of the mesosoma. A few species of North American *Tetramorium* might be mistaken for *Myrmica*, but these *Tetramorium* have either conspicuous antennal scrobes or, in the case of the exotic *caespitum*, very even, parallel ridges on the head. A few large *Stenamma* look a bit like small *Myrmica*, but have a long petiolar peduncle. Members of the genus *Myrmecina* are also slightly similar to *Myrmica*, but have a distinctive rectangular petiole and are also smaller. All this is academic with respect to the Florida fauna, as our single *Myrmica* species can be distinguished by its long, slender propodeal spines and obvious seta-bearing punctures on the gaster.

Myrmica is a cold-adapted Holarctic genus that moves into southern areas of the Northern Hemisphere primarily in mountainous areas and is completely absent from the Southern Hemisphere. Species can accept harsh winters in a variety of habitats, but do not tolerate tropical, subtropical, or true desert conditions. Florida reflects the southward decline in *Myrmica* diversity, with only one species, confined to the northern border of the state. This highly restricted diversity has its convenience, as more ink has been spilled over species taxonomy of *Myrmica* than for any other ant genus, and this has not even resulted as yet in a satisfactory arrangement of North American species. After one of my annual trips to New England, where confusing *Myrmica* swarm over the ground, it is something of a relief to get back to Florida.

Within its stomping grounds, *Myrmica* is an interesting and ecologically important genus, foraging with almost the same fervor as *Pheidole* to the south. Species of *Myrmica*, like many other ants, tend to be dietary generalists: opportunistic predators, scavengers, and honeydew feeders. Ecological separations between species, as far as is known, seem to be based primarily on habitat and substrate. Although many species prefer some kind of open habitat, some occur in moist forested areas. The Florida species belongs to this latter group.

<u>Name Derivation:</u> From *myrmex* (Greek), meaning "ant."

Myrmica punctiventris Roger—
Punctate Northern Furrowed Ant (Plate 21)

<u>Taxonomy and Similar Species:</u> In the Florida fauna, *punctiventris* is distinguished from all other ants by its very long, slender, curved propodeal spines, the furrowed sides of the mesosoma, and hair-bearing punctures on the gaster. A little to the north, in Alabama and Georgia, there is a very similar species, *pinetorum*. I have included an illustration of *pinetorum*, although it may not occur in Florida. The propodeal spines of *pinetorum* are usually straight and its frontal lobes are more elevated; the whole insect, in my experience, is slightly smaller, more slender and paler than *punctiventris*.

<u>Distribution:</u> Massachusetts south into Georgia, west into Iowa, Nebraska, and Arkansas (Smith 1979); I have also collected it in Maine. In Florida, *punctiventris* is known from a few sites in Walton, Santa Rosa, and Escambia counties.

<u>Natural History:</u> This species lives in hardwood forests in Florida and elsewhere. The nests I have seen have been in thick leaf litter, but may also be in rotten wood (Dubois 1985), in soil, or under stones (Carter 1962). The colonies I have seen are relatively small, with no more than approximately 50 workers, but it is possible that these have been colony fragments.

Name Derivation: From *punctum* (Latin), meaning "a small hole," and *venter* (Latin), meaning "belly," referring to the hair-bearing punctures of the gaster. It is reasonable to suspect that these punctures found in *punctiventris* and *pinetorum* have some adaptive function, such as protecting the bases of the gastral setae. Any function one can think of, however, would seem to be equally useful for other species that lack these punctures. The functional significance, if any, of most species-level differences between ants remains mysterious.

Genus *Pheidole*

Big-Headed Ants

In Florida, and in other regions of the world, members of the genus *Pheidole* (pronounced **FĪ dō" lē**) are almost always instantly recognizable for their trademark dimorphism: every colony is composed of larger workers with disproportionately large heads, and much smaller workers without enlarged heads. *Pheidole* is a gigantic genus, with approximately 900 described species, 624 of which live in the New World, according to Edward Wilson's amazing and monumental revision of the New World *Pheidole* (2003), a hefty volume that I would not want to drop on my toe. As big-headed primates, it is difficult for us to resist the speculation that the key to the success of *Pheidole* lies in its big-headed majors. Wilson (2003) points out that the extreme dimorphism in *Pheidole* allows an equally extreme division of labor, in which the majors can be specialized for defense, heavy lifting, seed-milling, or other tasks where large size is an advantage. Meanwhile, the much smaller minors carry out the other functions of the colony, such as caring for young and fanning out to find resources, tasks in which the important factor is numbers of individuals rather than size. For many species, Wilson suggests, defense is a key role of majors and involves not only large size and bigger jaws but also investment in relatively large volumes of defensive chemicals and the ability to move faster on longer legs.

I find this kind of reasoning highly persuasive, but it does not explain why species in many other genera of ants have not gone down this same evolutionary road. The production of variable-sized workers, often with disproportionately large heads, has evolved independently in lots of ant genera. These ants, one might think, could have attained all the advantages of the super-successful *Pheidole* simply by losing the intermediate-sized group between the largest majors and smallest minors. Groups of ants with variable-sized workers do not necessarily lack the evolutionary vigor of *Pheidole*. Among these is the gigantic genus *Camponotus*, another genus that Wilson (2003) labels "hyperdiverse." Some individual species with variable-sized workers are excessively successful generalists, such as *Solenopsis invicta* and *S. geminata*. Others, such as some species of *Atta* and *Eciton*, are dominant within their ecological niche. All of these variable ants tend to be considerably larger sized than the vast majority of *Pheidole* species; thus, size might be a factor in these ecological considerations.

The simplest way to think about this is probably to fall back on a variant of a dynamic familiar to all ecologists. Specialized species benefit from greater efficiency, but have a smaller resource base, while generalist species are less efficient in using any one resource, but have a larger resource base. (Most species are specialized in some ways but less so in others, so one may need to specify the resources involved.) Applying this to division of labor, the workforce of variable-sized *Camponotus* species is more flexible, whereas that of *Pheidole* is more dedicated to specific tasks. Focusing for the moment on defense, simplistic though it may seem, size may provide a tipping point in this trade-off. *Camponotus* is like a farming village whose burlier citizens can come up with defensive alternative uses for scythes and pruning hooks. For smaller ants, however, a somewhat larger individual is still a small ant; Hobbits do not need larger Hobbits, they need NFL defensive linemen. Put another way, the smaller the ant, the greater the importance of absolute size. This may help explain why *Pheidole* species are so dominant among species of small ants in warmer parts of the New World: their majors are much bigger than any monomorphic ants that are the same size as the *Pheidole* minors.

Explaining how something works is not the same as explaining how it got that way. It is as if the genus *Pheidole* was benefiting from what the myrmecologist Bill Brown used to call "an evolutionary breakthrough." I would not be surprised if he had actually applied that term to *Pheidole*. What barrier was overcome by the early *Pheidole* lineage? Perhaps the answer might be approached by thinking of *Pheidole* majors as representing a more radical innovation, not a subcaste, but a completely new caste, analogous to queen or male. From that standpoint, the challenge faced by the proto-*Pheidole* would have been primarily developmental: what would be the mechanisms triggering the production of the right number of majors at the appropriate time; how would a network of divergent behavioral patterns be segregated out and genetically fixed; how would the morphological divergences coexist within an already complex set of developmental options. These might not be simple incremental steps that could easily occur in any lineage. No matter how much bigger, stronger, faster, tougher, more aerodynamic you make a car, you would never segue into an airplane, unless, of course, you happen to be Batman. The evolutionary pressure needed to overcome these difficulties might have resulted from a rare ecological situation, which would help explain why there are a small number of genera of dimorphic ants. Imagine a species of ant faced with the unusual situation of having a single, major, uniform-sized threat or resource that was present for a short time each year. For a variable-sized species, the evolutionary pressure would be toward dimorphism. A uniform-sized resource or threat is not hard to envisage; seeds of a single species of plant, or very large numbers of dealate queens of another species of ant might be enough. The factor that would be ecologically rare would be a scarcity of additional resources or threats within a wide range of sizes.

In the last few paragraphs, I have indulged in the pleasures of virtually fact-free speculation. Aside from the observation that unusual hypotheses may sometimes reside within thickets of rank speculation like a unicorn in the forbidden forest, anybody who is attempting to tackle *Pheidole* taxonomy deserves a little enjoyment. It is also pleasant to imagine all the myrmecologists who will themselves derive pleasure from demolishing the above speculations and setting up their own. A solid foundation from which to do this has been provided by Moreau (2008), who subjected approximately 140 species of *Pheidole* to genetic analysis. Among the findings of this study are the New World origins of *Pheidole* with only one or a few introductions into the Old World, and the somewhat discouraging discovery that the earlier attempts to come to grips with the huge numbers of species by organizing them into species groups

are belied by phylogenetic analysis. Seed-milling is widespread within the genus, but it is not clear that this rather than defense or some other function was the seminal adaptive advance in the group.

The Florida *Pheidole* fauna includes 17 species, an absurdly small number compared with most Neotropical areas, or even southwestern states such as Arizona. Most Florida species are morphologically distinctive, and several have restricted distributions. Although one might expect that there would be several species confined to tropical Florida, considering the huge tropical fauna to the south, there is only one such species (known so far), while there are four northern species whose ranges a short way into northern Florida. Four species have been introduced into Florida.

Name Derivation: From *pheidos* (Greek), meaning "thrifty." Westwood named this genus in 1839 to accommodate a species previously known as *Atta providens*. The name *providens* was bestowed by Sykes (1835) on a species that he observed bringing seeds up out of the ground after a rainy period and spreading the seeds out in the sun to dry before storing them again underground. Sykes describes the difficulty faced by the ants dragging seeds up the vertical shaft of the nest, sometimes tumbling back down, but trying again "with a perseverance affording a useful lesson to humanity." Westwood obviously agreed with Sykes when he decided on the genus name *Pheidole* for an ant that is both thrifty and provident.

Pheidole adrianoi Naves—
Florida Rosemary Big-Headed Ant (Plate 33)

Taxonomy and Similar Species: *Pheidole adrianoi* is a small (minor, under 2 mm long) dark species; it is somewhat similar to *metallescens* and *littoralis*. The head of the major *adrianoi* is dark red to black, strongly shining with variable scattered hair-bearing punctures on the occipital lobes. The occipital lobes of the head of major *metallescens* and *littoralis* are strongly sculptured; both the major and minor of *metallescens* have strong bluish iridescence. The minor *adrianoi* is black and shining; it is similar to that of *littoralis* but lacks the granulate sculpture found to a varying degree on the head of *littoralis*. The pronotum of the *adrianoi* major is usually smooth and shining, but I have seen specimens with varying numbers of transverse carinae on the pronotum. *Pheidole adrianoi* is similar to a larger and more northern sand-dwelling species,

82   ANTS OF FLORIDA   *Deyrup*

davisi, and it is possible that *adrianoi* is a Florida isolate of the *davisi* lineage. The known ranges of the two species come within approximately 60 miles of each other: *adrianoi* occurs in Emanuel County, Georgia, and *davisi* occurs in Aiken County, South Carolina (Van Pelt and Gentry 1985). It is unlikely that anybody has sought these two species in sandy barrens up the Atlantic Coast from the Florida border into South Carolina to see if the two species ever intergrade or occur together.

Distribution: *Pheidole adrianoi* is known to occur in scattered sites throughout Florida, from Collier County into Nassau County, and west into Santa Rosa County. I have collected this species in the Ohoopee Dunes habitat in Emanuel County, Georgia, and it almost certainly occurs in remnant coastal scrub habitat in Alabama.

Natural History: In Florida, *adrianoi* can almost always be found in Florida rosemary (*Ceratiola ericoides*) barrens, and this is the most typical and possibly original habitat of the species. It can also be found along open sandy roadsides and in open patches of sandhill vegetation. It may be found in the same sites as *metallescens* and *littoralis*, but *metallescens* is usually in a somewhat shaded microsite with some leaf litter, while *adrianoi* and *littoralis* are in bare expanses of sand with small plants such as *Stipulicida setacea* or *Paronychia* species. Majors of *adrianoi* can easily be lured out of the nest entrance with cookie crumbs. A colony excavated by Naves (1985) had approximately 60 majors and more than 240 minors. The diet includes both seeds and small insects (Naves 1985). Queens have been captured in flight traps in June into August (three specimens).

Name Derivation: Naves (1985) named this species for his son, Adriano de Resende Naves.

Pheidole bicarinata Mayr—
Variable Big-Headed Ant (Plate 34)

Taxonomy and Similar Species: *Pheidole bicarinata* is a wide-ranging species that varies from place to place in color and sculpture and the development of propodeal spines (Wilson 2003). This has led to several subspecific names, one of which, *P. bicarinata vinlandica* Forel, was raised to species level by Naves (1985), subsequently merged as an unnamed variant into *bicarinata* by Wilson (2003) along with other subspecies and forms of *bicarinata*. In his study of Kansas ants, DuBois (1985) rejected two "subspecies" of *bicarinata* because he found them in adjacent nests and even intergrading within nests. *Pheidole bicarinata*

is a climatically tolerant generalist species of fields, prairies, and open sandy riverine forests; it is likely that with increasing land clearing in the 19th and 20th centuries, variants found in divergent isolated populations began to move about and even merge.

In the Florida fauna, *bicarinata* is distinctive and easily identified. The major has parallel carinae on the lower half or two-thirds of the head, and the upper part of the head is shining with sparse punctures. The major can be mistaken for that of *moerens*, but the postpetiole of *bicarinata* is widened, its pronotal "shoulders" are accentuated, and *bicarinata* lacks the pair of stout hooked teeth (hypostomal teeth) found on the underside of the head at the edge of the mouth opening in *moerens*. The minor of *bicarinata* is dark brown, the head and pronotum are shining, and the eye is conspicuously longer than the distance between the eye and the mandibles.

Distribution: *Pheidole bicarinata* has a wider range in temperate North America than any other *Pheidole*, with the possible exception of *Pheidole pilifera*. Its range extends from New Jersey south into northern Florida, across southern North Dakota south into Texas, and from Kansas west into Colorado, Utah, and Nevada (Wilson 2003). Its range may be even wider than reported, as it can be found not only in many types of natural habitats but also in disturbed sites, lawns, and golf courses. Lawns and golf courses are seldom sampled by myrmecologists, who are more susceptible than one might suppose to the romantic allure of pristine habitats filled with aboriginal ants.

Natural History: *Pheidole bicarinata* lives in a variety of open habitats, including open forests, rocky hillsides, meadows, pastures, cultivated fields, open roadsides, and lawns (Carter 1962; Wilson 2003). In areas with surface rocks, colonies are often found under stones, which warm up in the sun faster than surrounding areas of soil. Under such stones may chambers with ants and brood be found, as well as stores of seeds of grasses and other plants (Cole 1940). Nests have also been reported in rotten logs (Gregg 1963), although not in the eastern part of the range of *bicarinata*. In Florida, nests are in open sand or near clumps of grass in sandy uplands. Although apparently suitable habitat extends far south into peninsular Florida, this species is restricted to the northern border of the state. Colonies are usually small, with 200 or fewer workers (Cole 1940).

Wheeler and Nijhout (1981) used laboratory colonies of *bicarinata* as convenient subjects for a study of the mechanism of production of majors in *Pheidole*. These researchers discovered that applying juvenile hormone to third-stage larvae delayed the pupal stage, and during this delay, the larvae continued eating and growing. The attainment of larger size apparently switched on genetic pathways that resulted in morphologically divergent adult workers. The majors that Wheeler and Nijhout induced were identical morphologically to *bicarinata* majors produced spontaneously. Wheeler and Nijhout suggest that there is also a level of nutritional resources available to the colony below which majors will not be produced. While this study nicely shows the hormonal mechanism that produces soldiers, the mechanism must work with surprising precision or *Pheidole* majors would not be so uniform in size; there must be a second mechanism that stops the larvae destined to be majors from growing irrespective of the nutrition available or the time since juvenile hormone levels dropped off. Meanwhile, the balance between major and minor workers must be determined by colony-level mechanisms as hypothesized in the case of fire ants (Tschinkel 2006).

Name Derivation: The name of this species means bicarinate. According to Mayr's description (1870) of the species, this seems to refer to a pair of notable carinae that project forward and backward on the propodeum as a continuation of the propodeal spines. ("Metanotum spinis 2 haud longis, antice et postice in carinam longitudinalem continuatis, inter carinam oblique descendens, transverse concavum et transverse striatum.") These unusual carinae do not occur in specimens I have seen from Florida, or from anywhere in eastern or western North America, but are conspicuous in specimens from Iowa. Mayr's types are from Illinois (Mayr 1870).

Pheidole carrolli Naves—
Carroll's Big-Headed Ant (Plate 35)

Taxonomy and Similar Species: This species is closely related to *pilifera*, a widely distributed variable species that includes several geographic forms (Wilson 2003). The differences that distinguish between these variants seem roughly equivalent to those that are used to characterize *carrolli*, and it is logical that *carrolli* should be considered no more than a distinctive isolated population of *pilifera*. It is also possible, however, that *carrolli* is a sand-loving lineage that has inhabited the sandy uplands of Florida for a long

time like, for example, *Odontomachus relictus*. Within the Florida *Pheidole* fauna, *carrolli* is easily recognized. It is an orange brown species whose majors have the massive bilobed head associated with heavy dependence on seed-grinding in *Pheidole*. The heads and mesosomas of both majors and minors are densely sculptured and not shiny. The occipital lobes of the major are described as smooth and shining (Naves 1985), but those of specimens I have seen are at least finely sculptured and are feebly shining or not shining. The petiolar spiracles of the major are on the underside of conspicuous sharp tubercles. The eye of the minor is relatively large, clearly longer than the distance between the eye and the mandible.

Distribution: Northern Florida, west to the Apalachicola River in the Panhandle. It probably also occurs in southern Georgia, as it has been collected in Leon County near the Georgia border (Naves 1985).

Natural History: *Pheidole carrolli* appears to be a sandhill species, dependent on open forest with a sparse or dense ground cover of grasses and herbs that provide seeds. The apparently small and weak colonies that Naves (1985) studied were in shaded forest that probably had gone unburned for a long time. The first Florida specimens of *carrolli* were collected and identified as *pilifera* by Van Pelt (1958) on the Welaka Reserve in Putnam County. Van Pelt found specimens in lawns and orange groves, but not in natural areas. The nest entrance of this species, in my experience, is inconspicuous, but foraging minors accept cookie crumbs and can be followed back. The minors can be recognized in the field because they are considerably larger than other local brownish, nonshining species such as *floridana*, *dentigula*, and *moerens*. As Naves (1985) notes, minors freeze and remain motionless when alarmed; this behavior is not unique to *carrolli*.

Name Derivation: Naves named this species in honor of John F. Carroll, a University of Florida myrmecology graduate student contemporary of Naves. Carroll made the earliest collection of the species examined by Naves. Carroll produced an excellent dissertation on the biology and ecology of Florida *Aphaenogaster* (1975); unfortunately, this work was never published.

Pheidole crassicornis Emery—
Thick-Scape Big-Headed Ant (Plate 36)

Taxonomy and Similar Species: *Pheidole crassicornis* is one of approximately 13 apparently closely related species whose majors have conspicuously thickened antennal scapes. This group is centered in southwestern North America, with two species having ranges extending into southeastern North America (Wilson 2003). The adaptive significance (if any) of the thickened antennal scapes is unknown; this trait has appeared independently in a few unrelated species of *Pheidole* (Wilson 2003). In addition to its thickened antennal scapes, a trait shared with a similar Florida species (*diversipilosa*), the major has unusual hairs on its gaster: short, shining, evenly spaced hairs, strongly appressed to the gaster; there are few long, semirecumbent hairs. Minor workers lack these features. The queen of *crassicornis* is undescribed, but may well have the unusual hairs found on the major. In the field, *crassicornis* might be confused with the ubiquitous *dentata*, but seems more compact and slower. In the field, *crassicornis* and *diversipilosa* are not distinguishable without microscopic examination of the hairs on the gaster of the major.

Distribution: North Carolina south into northern Florida, west into western Texas (Wilson 2003).

Natural History: *Pheidole crassicornis* is a relatively uncommon species whose natural history is poorly known. All specimens that I have seen were from wooded areas, including mature forests of deciduous trees and overgrown sandhill. In North Carolina, most colonies were found in scrub oak forests, but colonies were also found in a hillside pasture (Carter 1962). Nests are in soil. Workers have been seen carrying small arthropods (Naves 1985), and this species probably also gathers seeds.

Name Derivation: From *cornu* (Latin), meaning "horn," and *crassus* (Latin), meaning "thickened," referring to the thickened antennal scapes.

Pheidole dentata Mayr—
Versatile Big-Headed Ant (Plate 37)

Taxonomy and Similar Species: *Pheidole dentata* is versatile in its choice of habitat and nesting sites and this ecological plasticity is correlated with confusing differences in color and, to a much lesser extent, differences in morphology and size. In a few places, there are two sympatric color forms, giving an impression of two closely related species. As a rule, colonies that are almost black live in mesic woodland, while brown or yellowish brown colonies are in more open sites, but in some places, all colonies are reddish brown in any habitat. Colony structure is also variable from place to place (Wilson 2003). Naves (1985) states that this variation has no taxonomic significance, but Cover, quoted in Wilson (2003), notes that consistent differences between colonies in different habitats and in different areas suggest a complex of sibling species. It is not necessary to choose between these hypotheses, as variation within a species may provide the raw material for natural selection leading to speciation. What we call *P. dentata* might include both a variable species and a species complex. Several Florida ant species show local variation in color, for example, *Pseudomyrmex ejectus*, *Temnothorax pergandei*, and *Camponotus impressus*. *Pheidole dentata*, however, is uniquely variable, as if a suite of character states were more genetically malleable or pleiotropic than in other Florida ants.

Pheidole dentata, despite its variation in color, is usually easy to distinguish from most other Florida *Pheidole*. It is a large species, with the upper half of the head and pronotum of majors and minors shining, with little or no sculpture; the head of the major is not as disproportionately large as in species that are primarily seed-eaters; the propodeal spines are small but sharp and directed upward; the color is usually blackish brown to reddish brown. In the field, *dentata* could be confused with *crassicornis* or *diversipilosa*, but those species, even in the field, can be seen to have proportionately larger heads (majors); under the microscope, those species can be seen to have broad, flattened antennal scapes. In the field, *dentata* also resembles the introduced species *megacephala*. The latter species is usually found in dense populations in urban areas, with swarms of minors running about, but few majors. Under the microscope, in lateral view, *megacephala* has no declivity between the pronotum and mesonotum: the pronotal/mesonotal profile is smoothly convex. The head of the major *megacephala* in a frontal view is narrowed below; the head + mandibles are heart shaped. Occasionally, it is difficult to distinguish between *morrisi*, which is always yellow, and yellowish forms of *dentata*. *Pheidole morrisi* has poorly developed propodeal spines that do not project upward. On two occasions in the Florida Panhandle, I encountered slightly darker *morrisi* in a habitat intermediate between an open area with *morrisi* and a nearby more shaded site with yellowish brown *dentata*. Fortunately, one such colony had the distinctively large queen of *morrisi*; workers of *morrisi* also have relatively longer and more slender appendages. According to a study of the phylogenetics of *Pheidole* (Moreau 2008), *dentata* and *morrisi* are closely related, and both are in the same area of the phylogenetic tree as *hyatti*, a variable western species similar to *dentata* in ecology. In this tree, however, the closest relative to *dentata* is *violacea*, a Central American species that does not resemble *dentata* and lives in strange "ant gardens" that it constructs on shrubs in tropical clearings (Wilson 2003). This pairing seems very unlikely, but phylogenetic reconstruction should be given a little latitude; it does not, as yet, pretend to be an exact science.

Distribution: Assuming all the populations thought to be *dentata* really are that species, it occurs from Maryland south through Florida, west into Illinois, and south through Kansas into Texas into northern Mexico (Wilson 2003).

Natural History: In Florida, *dentata* is the usual large *Pheidole* of forested areas, a habitat it shares with smaller species such as *dentigula* and *moerens*. In open sites with patches of bare sand, it is usually replaced by *morrisi*. In densely vegetated sites, *dentata* sometimes occurs in grass tussocks or in rotten wood, more rarely in soil. In Deland (Volusia County) in the 1990s, I observed a population thriving in a semiurban situation, with large colonies foraging over pavement. These ants were more defensive of their colonies than in other populations of *dentata* that I have observed and appeared to be holding their own against *Solenopsis invicta*.

The complex defensive behavior of *dentata* against *Solenopsis geminata* is the subject of a classic study by Wilson (1976b). When *dentata* minors encounter one or a few *geminata*, some minors attack the *geminata*, whereas others race back to the nest, leaving a pheromone trail. This pheromone, combined with odors from *geminata* adhering to the minors, mobilizes both majors and minors, which charge out to attack the *geminata*. The majors do not follow the trail closely, but range out in looping paths that increase the likelihood of encounters with scouting *geminata*. A small number of *geminata* can be rapidly chopped up by the *dentata* majors, which do not immediately return to the nest but remain patrolling the area. If a large number of *geminata* are approaching the nest, the minors return to the nest without leaving trails. The alarm pheromone alerts the majors, which form a defensive blockade around the nest entrances. If this second defensive phase is not successful, the whole colony becomes increasingly agitated and eventually absconds

from the nest carrying brood and accompanied by the queen. This complex defense behavior appears to be specifically elicited by *Solenopsis geminata* and *invicta* (Johnson and Wilson 1985; Wilson 1975). Defense against *Solenopsis* depends on the number of majors that can be mobilized. Johnson and Wilson (1985) hypothesized that *dentata* colonies that are constantly exposed to *Solenopsis* might respond by producing more majors, but this does not seem to be the case. Ecological conditions in general do not seem to influence the ratio of majors to minors in *dentata* colonies (Calabi and Traniello 1989). Studies of *dentata* and other ants with specialized worker morphs suggest that these specialized workers may be more flexible in their behavior than has been assumed, allowing individual colonies to adapt quickly to changes in labor requirements (Calabi 1988; Calabi and Traniello 1989).

This abundant and widespread species is exploited by several natural enemies. The wasp *Orasema robertsoni* (Eucharitidae) attacks larvae and pupae in the nest (Van Pelt 1950). The fly *Apocephalus tenuipes* (Phoridae) oviposits in adult majors, attacking them outside the nest (Burges 1979; Hill and Brown 2006). The presence of *Apocephalus* causes majors to go into hiding, interfering with defense against *Solenopsis* as well as normal foraging, possibly changing the competitive balance between *dentata* and species of *Solenopsis* (Feener 1981b). The single known specimen of the ant *Solenopsis phoretica*, presumably a nest parasite, was found attached by specialized mandibles to the petiole of a queen *Pheidole dentata* (Davis and Deyrup 2006).

Name Derivation: From *dentata* (Latin), meaning "toothed," referring to the propodeal teeth. Mayr described this species as a variant of *P. morrisi* distinguished by having sharp propodeal teeth rather than blunt propodeal angles as in *morrisi*. The two species are similar in many ways, so Mayr's determination was reasonable at the time.

Pheidole dentigula M. R. Smith— Woodland Big-Headed Ant (Plate 38)

Taxonomy and Similar Species: *Pheidole dentigula* is one of five small Florida species with yellowish or orange brown minors and similarly colored majors, with the exception of the variably colored major of *moerens*. These five species have been assigned to the large and taxonomically difficult "*flavens* group," which is the largest group of

New World *Pheidole*; its members are also the most abundant *Pheidole* in tropical and subtropical habitats (Wilson 2003). In the Florida fauna, *dentigula* is easily identified by the dense reticulate sculpture on the upper part of the head of the major and minor. *Pheidole metallescens* has somewhat similar sculpture in the major, but the body of *metallescens* is black with metallic reflections. The postpetiole of the major and minor is expanded, conspicuously wider than the petiole in dorsal view.

Distribution: North Carolina west into Tennessee, south through Florida, and west along the Gulf Coastal Plain into Texas (Wilson 2003).

Natural History: In Florida, this common species is associated with mesic habitats, including hammocks, moist flatwoods, moist shrub lands, mesic microhabitats in scrub and sandhill habitats, and occasionally in large grass tussocks. Nests are usually in rotten wood, including logs, stumps, and sticks buried in leaf litter. This species may also nest in thick layers of bark at the base of slash pine or longleaf pine. Majors in nests often have the gaster strongly distended with reddish fluid, as if they were functioning as repletes. The similar species *floridana* is found in these same habitats, but also occurs in dryer sites and microhabitats. Recently, the introduced species *moerens* has become the dominant member of the *flavens* group in many Florida sites and seems to be especially well adapted to all the habitats favored by *dentigula*. I now find *moerens* in most sites where I would have expected to find primarily *dentigula*. It is possible that *moerens* is undergoing the kind of temporary population explosion associated with the invasive phase of many introduced insects, but it is also possible that *dentigula* will be permanently replaced and will slowly become a rare species. *Pheidole dentigula*, unlike *moerens*, is not usually associated with highly disturbed habitats, so habitat destruction and conversion in Florida favor *moerens* over *dentigula*.

In Florida, queens have been captured in flight traps in June (five specimens) and July (one specimen).

Name Derivation: From *dens* (Latin), meaning "tooth," and *gula* (Latin), meaning "throat," referring to the tooth on each side of the oral cavity of the major. This pair of teeth (some species have two pairs) occurs in most species of *Pheidole*. It is easy to imagine, without any real evidence, that they help clamp an object into place while the jaws are slicing or grinding. These teeth are so widely distributed among apparently distantly related lineages of *Pheidole* that it seems likely that they are an original or very early feature of the group; they even have a special name: hypostomal teeth. Some species that apparently depend heavily on seeds, such as *carrolli*, *metallescens*, *adrianoi*, and *bicarinata*, have hypostomal teeth that are slender and sharp, projecting straight forward from the edge of the oral cavity so that they are not visible in lateral view. If we knew more details about the natural history of species of *Pheidole*, we might have some idea why the hypostomal teeth of some species, such as *dentigula*, are unusually large and conspicuous.

Pheidole diversipilosa Wheeler— Diverse-Haired Big-Headed Ant (Plate 36)

Taxonomy and Similar Species: This species is similar to *crassicornis*, with the distinctive thickened antennal scapes found in the *crassicornis* group. Under the microscope, *diversipilosa* may be seen to be lacking the short, appressed hairs found on the gaster of *crassicornis*, and there are abundant longer gastral hairs. These two species are closely associated in Moreau's phylogenetic analysis of *Pheidole* (2008). In Creighton's manual (1950), *diversipilosa* is presented as an intergrade between the eastern and western forms of *crassicornis*, the latter called *crassicornis tetra*, but Naves (1985) recognized *crassicornis* and *diversipilosa* as separate species, as they occurred together at Tall Timbers Research Station. The form *crassicornis tetra* is now also recognized as a distinct species (Wilson 2003). The taxonomy of the eastern and western populations of the *crassicornis* group might benefit from further study.

Distribution: Florida (Naves 1985), southwestern Texas, and southern and central Arizona (Wilson 2003). It is not reported from North Carolina, although *tetra* occurs there (Guénard et al. 2012). Apparently, both species are rare east of the Mississippi.

Natural History: This species is common in some places in the Southwest, where it occurs in large colonies, each with a single queen; some colonies were found with seed caches (Cover, referenced in Wilson 2003). Habitats in Florida are not known.

Name Derivation: From *diversus* (Latin), meaning "different," and *pilosus* (Latin), meaning "hairy," referring to the gaster of the major, which has both long and short hairs.

Pheidole flavens Roger—
Golden Tramp Big-Headed Ant (Plate 39)

Taxonomy and Similar Species: This is a small yellow or pale reddish species similar (in the Florida ant fauna) to *moerens*, to which it is closely related (Moreau 2008). The majors of the Florida population of *flavens* have fine granulate sculptures in the antennal scrobes and the upper part of the head, except for the tips of the occipital lobes. The upper part of the head of the major of *moerens* is strongly shining, a conspicuous distinguishing characteristic. The antennal scrobes of the major of *moerens* are not finely and evenly granulate, and the inner margins of the antennal scrobes are set off by a conspicuous ridge, so the scrobes are clearly concave. The antennal scrobes of *flavens* are not set off by a ridge, and the scrobes themselves are flattened rather than obviously concave. *Pheidole flavens* is also much rarer than *moerens*, and almost completely confined to tropical parts of the Peninsula, whereas *moerens* is an abundant, invasive species found throughout Florida. *Pheidole flavens* is also similar to *floridana* but easily distinguished by its much narrower postpetiole. The postpetiole of the *floridana* major is more than twice as wide as the petiole in dorsal view, with the sides developed into conspicuous points (among aficionados of *Pheidole* these are known as "connules"); the postpetiole of *flavens* is less than twice as wide as the petiole, without protruding lateral angles. The minor of *flavens* also has a relatively narrow postpetiole that is not twice as wide as the petiole, as in the case of *floridana*. *Pheidole dentigula*, another small Florida species, has a postpetiole like that of *floridana*, or even wider, and also has conspicuous reticulations on the upper part of the head of the major and minor.

Distribution: In the United States, *flavens* is known only from the southern tip of the Florida Peninsula, although it was probably also introduced into the Brownsville area of Texas in the early days of plant importation. In the Neotropics, *flavens* is a tramp species that is readily moved about in pots of plants; it occurs through the West Indies and is ubiquitous in disturbed tropical and subtropical habitats in Central and South America. Wilson (2003) writes, "*Pheidole flavens* rivals *P. jelskii* as the most widespread and abundant species of the genus in the New World." Shuffling through the various published lists of Florida ants does not reveal when *flavens* arrived on our shores, as taxonomic confusion has made published records unreliable right up to Wilson's 2003 revision of the genus. As recently as 1985, *flavens* was given a new name, *greggi* (Naves 1985), synonymized by Wilson in 2003. It would be possible to look for old Florida specimens in various museums, but this would not be an especially valuable contribution to science: it may be assumed that the actual arrival of the species in Florida happened when it was being broadcast throughout the West Indies. This must have occurred early on, as the Roger-described *flavens* in 1863 from specimens collected in Cuba.

Natural History: In contrast to its prevalence in most of the Neotropics, *flavens* is a relatively rare species in Florida, occasionally found in tropical hammocks. This habitat is also occupied by three other small *Pheidole*: *dentigula*, *floridana*, and *moerens*. It is possible that *flavens* is being replaced by its close relative *moerens*, which appeared in Florida in the 1970s (Wojcik et al. 1975) and rapidly became the dominant small *Pheidole* in disturbed habitats throughout Florida. In the tropics, nests are often in rotten wood on the ground, but may be under rocks, or arboreal in epiphytes and rotten portions of trees (Wilson 2003). In the tropics, nests may have thousands of individuals (Wilson 2003), but I have not seen large nests in Florida. A captive colony fed on small arthropods, including oribatid mites (Wilson 2003).

Name Derivation: From *flavens* (Latin), an adjective meaning "yellow" or "golden."

Pheidole floridana Emery—
Florida Big-Headed Ant (Plate 38)

Taxonomy and Similar Species: Among the other small, reddish yellow *Pheidole* in Florida, *floridana* is easily identified by the combination of a broad postpetiole, at least twice as wide as the petiole, and fine, even, granulate sculpture on the head and mesosoma of both major and minor workers. Most Florida populations have an opaque opalescent sheen on the base of the gaster of both the major and minor, as well as a granulate postpetiole, but I have seen specimens from the southern peninsula that lack this opalescent area on the gaster and have a shining postpetiole. Naves (1985) believed that the more shining form found in the Miami area was the genuine *floridana*, and the form with the nonshining basal area of the gaster was a distinct species named *anastasii*, described long ago by Emery; the type locality was Costa Rica. There is considerable variation in the degree of shininess of the gaster and postpetiole in south Florida specimens, and it appears that, in Florida, the finely granulate *Pheidole* with a broad postpetiole is a single species.

This does not solve taxonomic questions about *floridana*. There is a widespread neotropical species called *bilimeki* that might easily be the species that I am calling *floridana*. The two species are closely related (Moreau 2008) and difficult to distinguish (Wilson 2003). Wilson (2003) retained *floridana* as a distinct species, but synonymized *anastasii* with *bilimeki*. It seems unlikely that *bilimeki* was never brought to Florida, considering that it is widespread in natural and disturbed habitats in Central and South America, as well as the West Indies. If it is ever discovered that the population in the southern Peninsula is a genetic mingling of neotropical and northern Florida lineages, that would be good evidence that *floridana* is synonymous with *bilimeki*. *Pheidole floridana* does not give the impression of being a recent importation, as it is at home in a variety of native and disturbed habitats, both open and forested. Its range includes the southeastern Coastal Plain from Georgia through Florida and west into Texas and Mexico. In summary, in the context of the North American fauna, *floridana* appears to be a native species analogous to *dentigula*, but less likely to be found in moist forest habitats. In the context of New World *Pheidole*, *floridana* looks like a northern isolate of a *bilimeki* lineage, possibly arriving in Florida from the Southwest. Some genetic sleuthing might straighten this out, but such an effort could be complicated if Neotropical *bilimeki* is itself a species complex, as suspected by Wilson and others (Wilson 2003).

<u>Distribution:</u> Southern Coastal Plain from South Carolina north through Florida, west into Texas and Mexico (Wilson 2003).

<u>Natural History:</u> In Florida, *floridana* is most often found in dry forest or in open areas. In open wet habitats, nests are in rotten wood or in grass tussocks. Nests are generally associated with dead wood, but may also be in deep leaf litter or in layers of bark at the base of pine trees. Cover, as referenced by Wilson (2003), reports that nests may contain 1000 or more ants and have a single queen. Naves (1985) found colonies easy to keep in the laboratory, and usually willing to accept workers from other colonies. Foragers scavenge dead arthropods and are strongly attracted to baits, including both sweets and various types of protein. I have never found seeds in nests of this species. Queens have been captured in flight traps

in May (one specimen), June (five specimens), and July (five specimens).

<u>Name Derivation:</u> From *floridana* (Latin), meaning "pertaining to Florida."

Pheidole lamia Wheeler— Monster Big-Headed Ant (Plate 40)

<u>Taxonomy and Similar Species:</u> The head of the major of *P. lamia* is unlike that of any other Florida species. The head is unusually elongated, with parallel carinae; its most unusual feature is the modification of the lower part of the face and the expanded jaws into a circular flattened area somewhat resembling that of soldier stopper ants (subgenus *Colobopsis* of the genus *Camponotus*). Minor *lamia* workers are yellow, shining, and almost devoid of sculpture, and have the propodeal spines reduced to angulate projections.

<u>Distribution:</u> Northern Florida into Texas (Naves 1985). In Florida, *lamia* is known from the Red Hills region near Tallahassee.

<u>Natural History:</u> In Florida, *P. lamia* occurs in open fields and open pine stands (Buren et al. 1977) in the Red Hills region, an area with more clay in the soil than in other parts of Florida. In Texas, nests are in woodland, often hidden by leaf litter (Feener 1981a). Nest openings are small but may sometimes be indicated early in the morning by small honeycombed hillocks of clay excavated from the nest (Naves 1985). Colonies may be large, an estimated 1000 minors and 200 soldiers found in a single colony (Buren et al. 1977). Minors forage in the morning but avoid sunlight; Naves (1985) suggests that this species is nocturnal. In Texas, Feener (1981a) found large numbers of minor workers foraging both day and night, with most movement of foragers occurring under the shelter of surface leaf litter. Feener (1981a) never found majors foraging. The peculiar modification of the head of the major into a cork-like structure (phragmosis) also occurs in the unrelated Florida species *Cephalotes varians* and in species of *Camponotus*, which use their head to block the nest entrance against intruders. Several authors made the reasonable assumption that the head of the major of *lamia* served a similar phragmotic function (Gregg 1956). Challenging the obvious, Buren et al. (1977) checked the nests of *lamia* and found that the entrances were not plugged by the head of a major. This inspired Buren and his coworkers to consider another hypothesis: the majors might be defending against subterranean enemies,

especially the ubiquitous *Solenopsis* thief ants. Using laboratory colonies of *lamia*, Buren et al. (1977) set up confrontations between thief ants and *lamia*. When thief ants were introduced into the colony, they pursued minor *lamia* workers, which rushed back to a central chamber containing majors. The latter immediately became agitated as if receiving a chemical stimulus. While some minors and soldiers remained in the nest chamber, groups of soldiers and minors then set off along several passages, not just the ones used by retreating minors. On meeting a *Solenopsis*, the *lamia* soldiers delivered a bite to the mesosoma or petiole, causing immediate immobility and apparent lethal paralysis (Buren et al. 1977). The speed and permanence of the *Solenopsis* reaction caused Buren et al. to suggest that a toxin might be involved. Buren et al. interpret the flat and heavily armored jaws and clypeal area as a type of shield projecting the soldier from the sting of thief ants.

Name Derivation: From *lamia* (Greek), referring to a monster said to drink blood and eat the flesh of humans. This seemingly inappropriate name derives from the circumstances in which Wheeler (1901) discovered the species. The type series was associated with a nest of *Camponotus*, and Wheeler thought, possibly correctly, that the *lamia* had been feeding in a refuse pile of dead *Camponotus* within the nest. Wheeler interpreted this as the normal behavior of the species, and the weird head of the soldier, combined with the supposed habit of scavenging on the bodies of dead "host" ants, led him to name the species after a sinister monster.

Pheidole littoralis Cole—
Seaside Big-Headed Ant (Plate 34)

Taxonomy and Similar Species: *Pheidole littoralis* is another small black species like *adrianoi* and *metallescens*, with a highly distinctive major. The head of the major is brick red to very dark red, elongate, and so large in proportion to the body that walking might be difficult. The sculpture of the head of the major is more variable than has been previously recognized. In some specimens, the head is striate except for the occipital lobes, which are shining and strongly punctate; in others, the head is striate and strongly reticulate, and the occipital lobes are punctate-reticulate and not shining; other specimens show an intermediate condition. Cole (1952) somewhat hesitantly described *littoralis*

as a subspecies of *sitarches* (the latter synonymized with *soritis* by Wilson in 2003) but considered that *littoralis* might be a distinct species, as did Gregg (1958). *Pheidole littoralis* was eventually elevated to species rank by Naves (1985).

Distribution: *Pheidole littoralis* is not confined to central Florida as stated by Naves (1985) but is known from southern Florida (Collier County) north to the Georgia border and into Escambia County in the westernmost Panhandle. Records are scarce and widely scattered. This species almost certainly occurs in the small area of Alabama coastal scrub and beach habitat west of Florida. The distribution of *littoralis* is limited to areas with extensive open patches of deep, well-drained sand. These occur on inland ridges in Florida and along the beaches. *Pheidole littoralis* may not occur everywhere where the habitat seems appropriate. One reason why records of this species are scarce is that majors seldom, if ever, venture out of their nest, while minors are tiny and easily confused with those of *metallescens* and *adrianoi*. Majors are most easily obtained by baiting with cookie crumbs, which eventually usually brings one or more majors up close to the surface.

Natural History: *Pheidole littoralis* inhabits open sandy areas, where it gathers small seeds, including those of *Lechea deckertii*. The jaws of majors, which are thickened and lack sharp teeth, are adapted for milling seeds. Naves (1985) reports finding chambers of small grass seeds approximately 10–20 cm deep. Queens have been captured in flight traps in July (3 specimens).

Name Derivation: From *littoralis*, more properly *litoralis* (Latin), meaning "of the beach."

Pheidole megacephala (Fabricius)—
Pantropical Big-Headed Ant (Plate 41)

Taxonomy and Similar Species: *Pheidole megacephala* is a dark brown species about the size of *dentata*, *crassicornis*, and *diversipilosa*. In a lateral view, *megacephala* lacks a declivity between the pronotum and mesonotum, so the pronotal/mesonotal profile is evenly convex. The head of the major is narrowed below; the head + mandibles are heart shaped in frontal view. In the field, *megacephala* usually occurs in dense aggregations in urban areas, with great numbers of minors running about and very few majors.

Distribution: *Pheidole megacephala* originated in Africa, where it has close relatives and was apparently being distributed through the tropics before any ant surveys, as the earliest ant surveys in any tropical region usually included this species (Wetterer 2012a). In Florida, it was found in one of the first ant surveys in 1932 (Smith 1933). According to Wetterer's map (2012) of the records of *megacephala*, it is especially likely to occur in coastal areas and on islands. It sometimes is found in warm-temperate climates, but northern records, especially those from the Mediterranean region, may frequently be based on misidentifications, the ant in question being *Pheidole pallidula* (Wetterer 2012a). In Florida, *megacephala* occurs as far north as Gainesville (Alachua County) and might occur farther north in urban areas, but Florida records are scarce. There are no records from the Gulf States other than Florida, which would suggest that *megacephala* is sensitive to cool weather.

Natural History: I have usually found Florida nests of *megacephala* under rocks or trash, or in rotten wood. It is easy to imagine that it would survive well in the damp and verminous holds of wooden sailing ships in the early days of European colonization. Since its nests are diffuse and polygynous, *megacephala* would be difficult to eradicate and easy to accidentally offload with supplies. *Pheidole megacephala* has a history of causing major ecological destruction, especially in Hawaii, where it directly threatens many endemic insect species, disrupts some specialized pollination systems, and probably indirectly affects various endemic birds by reducing or eliminating their normal insect prey (Reimer 1994). Hawaii is a special case, as there are no native ants; the endemic fauna and flora lack specialized defenses against ants (Reimer 1994), while the fauna of other tropical areas has always been exposed to a wide variety of ants. In most tropical areas where *megacephala* has been studied, it is likely to have moderate effects on endemic fauna because it tends to be found in open, disturbed areas, and more rarely found in relatively undisturbed natural habitats (Wetterer 2012a). In Florida, *megacephala* has not become an ecological threat; I have found it occasionally in large but very local populations. There are no records that might suggest that *megacephala* was ever a dominant species in southern Florida. It is possible that it has been repressed by occasional cold weather even in the southern Peninsula. Once eliminated from a site by weather or other factors, *megacephala* might have difficulty recolonizing because colonies apparently spread by budding rather than by long-range dispersal of solitary queens (Hölldobler and Wilson 1990). Hölldobler and Wilson (1990) point out that extreme polygynous species have traded dispersal ability for potential colony immortality. When such a species finds itself in an area where suitable habitat is discontinuous and entire colonies are sometimes eradicated, the advantages of a "supercolony" are probably greatly reduced.

Name Derivation: From *mega* (Greek), meaning "big," and *cephalo* (Greek), meaning "head," referring to the large head of the major. The big head of this species might have seemed distinctive back in 1793, when nobody would have imagined there were hundreds of similarly big-headed species. This species was described by Johann Christian Fabricius (1745–1808), a Danish professor of economic science who also managed a career in natural science, becoming the foremost entomologist of his time, revising the classification of insects as well as describing almost 10,000 new species. His description of *Pheidole megacephala* (under the name of *Formica megacephala*), like many early descriptions of insect species, is not diagnostic enough to allow one to identify specimens. Fortunately, Fabricius' collections are preserved in various museums, so the names he bestowed are not in doubt.

Pheidole metallescens Emery— Blue Big-Headed Ant (Plate 33)

Taxonomy and Similar Species: *Pheidole metallescens* is shining black with distinct blue, sometimes slightly purple reflections, the only Florida *Pheidole*, perhaps the only North American *Pheidole*, with this coloration. Such reflections or iridescence also occurs to a lesser degree in *Monomorium viridum*. While this kind of black coloration with metallic reflections may be rare in ants, it is common in some other groups of insects, such as many flea beetles (Chrysomelidae), some beetles of the genus *Lebia* (Carabidae), and several genera of bees, including *Ceratina* (Apidae), *Osmia* (Megachilidae), and some *Lasioglossum* (Halictidae). It is unlikely that such a widespread phenomenon is neither adaptive itself nor associated with an adaptive trait. It is possible that metallic coloration facilitates signaling within a species, or it might be warning coloration associated with a defense, such as chemical repellents. In *metallescens*, the latter hypothesis seems more useful. The head of the major is dark red, sometimes almost black, in frontal view almost completely sculptured, with fine parallel carinae centrally and reticulations posteriorly and laterally.

Distribution: This is a widely distributed southern species occurring from coastal North Carolina (Carter 1962) through the Gulf States including Texas and north into Oklahoma (Wilson 2003).

Natural History: In Florida, metallescens is an upland species usually associated with open forest, such as sandhill habitat, open scrub habitat, along the open edges of sand roads, and sandy pastures. It is also found in the dryer areas of flatwoods and along sandy beaches of lakes. This species harvests seeds (Naves 1985) and appears to require a relatively well-drained sandy habitat with some herbaceous ground vegetation, with or without a forest overstory. Nests are usually in the soil, the entrance sometimes hidden by light leaf litter; occasionally, nests are in fallen logs (Van Pelt 1958). In North Carolina, Carter (1962) found metallescens in open woodlands. Cover (quoted in Wilson 2006) found metallescens nesting in both soil and rotten wood in Texas. Van Pelt (1958) counted individuals in a nest in a rotten log, finding 505 workers, 29 soldiers, and 1 queen. This species recruits readily to cookie crumbs. Naves (1985) states that metallescens collects small grass seeds and scavenges for dead insects. I have seen workers colleting bits of crushed acorns in a parking lot, but this opportunity may not often occur in nature. Queens have been collected in flight traps in July (10 out of 11 specimens).

Name Derivation: From metallum (Latin), meaning "metallic," and -escens (Latin), meaning in this case "almost," referring to the subdued metallic coloration.

Pheidole moerens Wheeler—
Little Caribbean Big-Headed Ant (Plate 39)

Taxonomy and Similar Species: Pheidole moerens resembles flavens, to which it is closely related (Moreau 2008). The occipital lobes of major moerens are strongly shining, the antennal scrobes have obvious carinae and are not granulate, and the inner side of the antennal scrobes has a strong ridge. In flavens, the occipital lobes are mostly granulate and the antennal scrobes are partly granulate and weakly bordered on the inner side. Pheidole moerens occurs throughout Florida while flavens is restricted to tropical areas and frost-free microsites in southernmost Florida. Three other small

reddish yellow or brownish Florida species, dentigula, bicarinata, and floridana, may be distinguished by the wide postpetiole, among other characteristics.

Distribution: Pheidole moerens is possibly native to the West Indies and has been imported to the cities of Mobile, Alabama, and Houston, Texas (Wilson 2003), as well as California (Garrison 1995) and Hawaii (Starr and Starr 2013). I have found this species common in potted plants from Florida nurseries, and it is possible that it will eventually have a worldwide range. Many records of P. floridana from potted plants imported into California appear to be misidentifications of moerens (Garrison 1995).

Natural History: Pheidole moerens is most common in disturbed or natural mesic habitats, where nests are commonly found in rotten wood or hollow twigs, under objects on the ground, or beneath loose bark on dead trees. It also occurs in rotten wood or grass tussocks in open wet areas. It seldom occurs in dry open sandhill or Florida scrub habitats. It is monogynous; although several queens may cooperate in nest founding, eventually one queen kills the others (Naves 1985). Queens fly primarily in July (Naves). I have seen enormous numbers of dealate queens walking about the pumps of gas stations, having been attracted by lights the previous night. Considering its abundance, Deyrup et al. (2000) suggested that moerens might be suppressing populations of native Pheidole with similar habits, specifically floridana and dentigula. King and Porter (2007), in a study of exotic ants in both undisturbed and disturbed upland habitats, found moerens to be the commonest and most widespread exotic species. Contrary to the speculation by Deyrup et al. (2000), moerens had not displaced dentigula or floridana.

Name Derivation: Possibly from moerus or murus (Latin), meaning "wall." Pheidole moerens often occurs around structures, and I have seen many instances of colonies emerging from cracks in pavement, so it would not be surprising if William Wheeler first observed this species emerging from a crevice in a masonry wall.

Pheidole morrisi Forel—
Morris Big-Headed Ant (Plate 37)

Taxonomy and Similar Species: Pheidole morrisi resembles dentata in size and structure but has reduced propodeal spines, slightly more elongate appendages, and is yellow or pale yellowish orange. The minor of *morrisi* has propodeal spines reduced to blunt angles. *Pheidole dentata* is sometimes orange, and the two species are occasionally difficult to distinguish. In Florida, *morrisi* is usually easily recognized as a very active yellow species common in open sandy sites.

Distribution: New York (sandy coastal areas) south to southern Florida, local in sandy areas west into Illinois, Missouri, Oklahoma, and Texas (Wilson 2003).

Natural History: *Pheidole morrisi* is found in open areas with sandy soil. This species often lives along sandy roadsides in Florida and is dependably found in natural habitats such as open scrub and sandhill. Nests may be in open sand or at the base of a grass tussock.

Pheidole morrisi is a widespread and conspicuous species whose aboveground behavior can be observed relatively easily, thanks to the open habitat where nests are located. Showing remarkable persistence, Johnson (1988b) kept a five-year diary of three clusters of nests near his Florida home. He discovered that each cluster represented a single colony that can have up to 16 nests. Foragers encountering large baits or large dead insects can be seen carrying bits of food to various nest entrances. In the late afternoon or at night, columns of minors and majors emerge and go from one nest to another. Johnson suggests that this traffic helps maintain the integrity of the colony. Subsequent studies by other authors (Brown and Traniello 1998; Patel 1990; Yang 2006; Yang et al. 2004) have showed that *morrisi* majors have an unusually broad behavioral repertoire, including colony defense, recruiting to carry off larger food items, acting as fat-storing repletes in winter, and a limited amount of brood care. The intracolony movements reported by Johnson might be related to balancing the number of majors and minors in the various nests. Since Florida colonies of *morrisi* appear to have a single queen (Johnson 1988b), at some point, brood or eggs must also be moved from nest to nest.

Pheidole morrisi, like *dentata*, often occurs in an area where it may be exposed to the aggressive, nest-raiding native fire ant, *Solenopsis geminata*. Unlike *dentata*, which has a modulated response to this fire ant (Wilson 1976b), *morrisi* reacts by immediately retreating, followed by mobilization of majors just inside the nest entrance (Feener 1987). Since *morrisi* and *S. geminata* both prefer open habitats, Feener (1987) suggests that *morrisi* encounters

geminata too frequently to allow the evolution of an active defense that often results in the sacrifice of a number of majors. It is possible that the multiple nests of *morrisi* in Florida are an adaptive form of insurance against raids by fire ants or army ants. Johnson (1988b) discovered that *morrisi* has a preemptive defense against raiding or competing ants. Johnson repeatedly observed narrow columns of rapidly moving ants, apparently responding to a specific recruiting pheromone, heading out of the nesting area to dig up and kill founding queens of other species, including *S. geminata* (13 of 21 raids) and *Dorymyrmex bureni* (8 raids). Individuals from several nests may join in one of these hunting parties, which Johnson calls "regal raids."

While colonies of *morrisi* in Florida appear to have only one queen, in the northern part of its range, *morrisi* is polygynous. This appears to be the result of nest founding by several queens, which remain together permanently, each contributing eggs to the colony (observations by S. Cover, reported in Wilson 2003). While cooperative nest founding by unrelated queens is known for many ant species, in almost all cases, the nest eventually ends up with only one queen, the others having been killed by workers or by the surviving queen (Hölldobler and Wilson 1990). In the northern part of its range, *morrisi* appears to be especially dependent on fat storage by replete majors and minors during the relatively long cool northern spring (Yang 2006). It is possible that several queens are more likely to be able to build up the colony fast enough to store adequate food for the winter. Perhaps, on the other hand, the unusual long-term polygyny with unrelated queens in *morrisi* is not a settled adaptation. The temporary cooperative nest founding seen in many other species probably had an initial phase in which there was selection for cooperative behavior, followed by a second phase in which there was selection for the ability to eliminate rival queens at the appropriate time. *Pheidole morrisi* might still be in phase 1.

In Highlands County, I have seen several large flights of alates following the first rainy days in May.

In Florida, *morrisi* nests sometimes include an undescribed species of *Terapus* (Histeridae). This beetle, which can be obtained by sifting, appears to be highly modified for living with ants, with a pair of brushes, probably secreting some attractant or pheromone, in saucer-like depressions on the pronotum.

Name Derivation: Forel evidently named this after Reverend G. K. Morris, who collected the type specimens. The types are from Vineland, New Jersey, where Mary Treat also collected ants for Forel.

Pheidole obscurithorax Naves—
Large Imported Big-Headed Ant (Plate 41)

Taxonomy and Similar Species: *Pheidole obscurithorax* is the largest species of *Pheidole* in Florida, about twice the size of *dentata* or *morrisi*. Minors are black, and the majors are blackish brown with a dark reddish brown, nonshining head. This species is easy to find and identify in the field, where the black minors can be seen busying themselves around a large nest mound built in an open area. Majors are easily lured out of the nest. The majors are the only Florida *Pheidole* with a strong repellent odor easily detected by humans. *Pheidole jelskii* is a similar species that is common in the West Indies but has not yet appeared in Florida. *Pheidole jelskii* is closely related to *obscurithorax* (Moreau 2008) and shares its smell, but its major has a shinier head.

Distribution: The native range of *obscurithorax* is the relatively flat region of northern Argentina and central and southern Paraguay (Wilson 2003). In North America, as of 2007, *obscurithorax* occurred in Florida, Georgia, Alabama, and Mississippi, with an isolated population in Texas (King and Tschinkel 2007). In Florida, populations occur at least as far south as Highlands County.

Natural History: *Pheidole obscurithorax* in Florida thrives in open disturbed sites, including fields, pastures, and roadsides, producing large, conspicuous nests with up to approximately 10,000 workers (Storz and Tschinkel 2004). The few nests I have attempted to excavate were deep, and I have not seen nests in wet areas, but in its native habitat, it occurs in floodplains and near rivers (Storz and Tschinkel 2004). It can live in urban areas, at least in sections that are not densely built up, such as parts of Tallahassee, Gainesville, Orlando, and Avon Park. It is frequently found near nests of *Solenopsis invicta*, and it appears that the two species are, to some extent, compatible (King and Tschinkel 2007). One might suspect that *obscurithorax* might displace fire ants, considering the phenomenal speed with which *obscurithorax* recruits to baits. Batches of fire ant corpses often occur around the nest entrance of *obscurithorax*. Despite these observations, the significance of which could be accentuated by wishful

thinking, there is no evidence so far that *obscurithorax* is able to kick out *Solenopsis invicta* (King and Tschinkel 2006; Storz and Tschinkel 2004). There is also no evidence that *obscurithorax* is perturbing Florida's native ant fauna, as this species is generally confined to disturbed areas that have few native ant species (Storz and Tschinkel 2004). Despite its recent increase in abundance, there is no reason to consider *opacithorax* a pest, as it does not sting or invade houses (King and Tschinkel 2007). On the contrary, this would be an excellent ant for high school science fair projects, as it is so conspicuous and dynamic. One could test, for example, what is the smallest bait that would send a worker back for assistance, or how far from the nest foragers can be induced to go for various baits, or what happens if several foragers attempt to recruit to separate baits at the same time.

Pheidole obscurithorax was first collected in North America by Wilson in 1950 (Wilson 2003). The colony was found about a kilometer from a ship docking area in Mobile, Alabama, and could easily have arrived there from South America (Wilson 2003). In 1992, I found *obscurithorax* in the western Panhandle (Pensacola), so it took *obscurithorax* more than 40 years to travel approximately 35 miles eastward. Six years later, it was in Tallahassee (Storz and Tschinkel 2004), approximately 195 miles east of Pensacola. By 2007, there were dense populations in the Tallahassee area (King and Tschinkel 2007); it appeared in Gainesville around 2006, in western Hillsborough County (Tampa) around 2004, in the east edge of Hillsborough County around 2006, and in northern Highlands County in 2011. *Pheidole obscurithorax* is a large ant with conspicuous nests, so I could not have missed it for long while doing surveys of Florida ants, beginning in the early 1980s. Moreover, throughout the 1990s, there were extensive surveys of roadside fire ant populations that should have encountered *obscurithorax* if it were present. One explanation for this sudden range expansion is that *obscurithorax* evolved some unknown trait that made it better adapted to southeastern North America. Another possibility is that there was a second introduction from a population better adapted to North America. A third possibility is that North American populations were unable to expand rapidly until they reached a certain density. This last possibility is reasonable for a species with colonies established by single queens that must mate with males encountered in swarms found well away from the natal nest. Colonies of *obscurithorax* have a single queen (King and Tschinkel 2007), but we still lack detailed information on nest founding in *obscurithorax*. Two other Florida ants that have had a long-delayed, rapid range expansion are *Camponotus planatus* and *Odontomachus ruginodis*.

Name Derivation: From *obscurus* (Latin), meaning "dark," and *thorax* (Greek), in this case meaning "chest," referring to the blackish mesosoma, which contrasts somewhat with the dark red head.

Pheidole tysoni Forel—
Tyson Big-Headed Ant (Plate 35)

Taxonomy and Similar Species: *Pheidole tysoni* is a shining yellow species with little conspicuous sculpture except for strong parallel carinae on the lower sides of the head and, in the case of the major, above the frontal lobes. There are a few carinae on the pronotum of the major and a few small granulate patches on the mesosoma of both major and minor. The minor is somewhat similar to that of *lamia*, but that species lacks propodeal spines and has unusually small eyes.

Distribution: New York south and west into northern Florida and eastern Missouri; there are scattered records from upper elevations in Arizona and Texas (Wilson 2003). *Pheidole tysoni* could be more widespread than it seems as it is apparently at home in lawns (Wilson 2003), the most ubiquitous habitat in North America. In Florida, *tysoni* is known from only three sites in northern Florida, but it might be more widespread, especially in lawns and roadsides.

Natural History: The natural habitat of *tysoni* appears to be open rocky hillsides and open woodland of south-temperate North America (Carter 1962), with relict southwestern populations that retreated into midlevel biotic zones after the Pleistocene. The natural habitats of *tysoni* have been greatly expanded by pastures, roadsides, and lawns (Wilson 2003). Unlike *metallescens*, *adrianoi*, and *littoralis*, this species does not occur in hot, dry, sandy habitats in Florida or in fire-maintained pine flatwoods. Leon County records are from Tall Timbers, which has long-maintained open fields on soil with some clay. Alachua County records are queens from a flight trap that was sampling a residential area with lawns; Madison County records are queens from a light at a gas station. *Pheidole tysoni* collects seeds and also tends aphids and collects nectar from low herbaceous plants (S. Cover referenced in Wilson 2003). In Florida,

queens have been collected in flight traps in June, July, and October (five specimens).

Name Derivation: Forel collected the types from a site above 1000 meters at the base of Mt. Mitchell in North Carolina. Mt. Mitchell is the highest peak in eastern North America, perhaps reminding Forel of the lower reaches of the Alps of his native Switzerland. The specimens were collected near the Tyson Farm, and the name is apparently a locational reference, not in honor of the owner of the farm.

Genus *Pogonomyrmex*

Harvester Ants

The name *Pogonomyrmex* is usually pronounced **Pō' gō nō mur" mex**, or **Pō gon' ō mur" mex**, but in casual conversation, this is contracted to "Pogos," as there are no other ants whose generic name begins in a similar way. This is a New World genus of approximately 60 species, most of which live in open habitats in southwestern North America or regions of Mediterranean climate in southern South America, with a few tropical species in South and Central America and on the island of Hispaniola. *Pogonomyrmex* is a lineage dating from the Miocene or earlier, as shown by fossils from the Florissant beds in Colorado (Carpenter 1930), and must have been strongly affected by the periods of dry climate in the Pliocene through the mid-Pleistocene. The *Pogonomyrmex* lineage ancestral to Florida's *P. badius* probably made its way across the Mississippi River during one of these dry periods when savannah stretched across southern North America. The most likely time for immigration of *Pogonomyrmex* into the Southeast was in the late Pliocene through early Pleistocene, when a wave of western animal lineages moved into the longleaf pine habitat of Florida (Webb 1990). The majority of these animals, such as jackrabbits and pronghorn antelopes, disappeared later in the Pleistocene, and wet floodplains along the Mississippi River isolated the remaining Florida fauna associated with a semiarid climate (Webb 1990). The Florida harvester ant is the most obvious example of this relict semiarid fauna among the ants.

Members of this genus are easily identified by their conspicuous sculpture on the head and body, and "beard" of curving hairs on the underside of the head (absent in a few species);

many species are also relatively large, 7–9 mm in length. The curving hairs on the underside of the head are usually called a "psammophore," meaning "sand-carrying organ." Experimental studies of the psammophore show that this basket of hairs greatly improves the ability to carry dry soil, although psammophore removal has little effect on the ability to carry moist soil (Porter and Jorgensen 1990). Large seeds are carried in the mandibles, but a small seed may be carried in the psammophore. If the ants are offered piles of small seeds, they may tuck several into the psammophore, but if small seeds are scattered, the ants return to the nest with the first one they find; they do not keep searching to fill the psammophore, even though it would appear more efficient to do so (Porter and Jorgensen 1990).

Seeds brought back by the harvesters are hulled and stored in the nest, and then milled by workers. The ground-up seeds are placed on the underside of the larva, where an inward-pointing set of spinules and barbed hairs holds the food in place (Petralia and Vinson 1979). The anterior part of the body forms a slender "neck" that allows the head to be brought down to feed on the ground seeds held by abdominal segments 3–5 (Petralia and Vinson 1979).

Name Derivation: From *pogon* (Greek), meaning "beard," and *myrmex* (Greek), meaning "ant," combined as "bearded ant," referring to the conspicuous psammophore.

Pogonomyrmex badius (Latreille)— Florida Harvester Ant (Plate 40)

Taxonomy and Similar Species: *Pogonomyrmex badius* is easily distinguished from all other Florida ants by its relatively large size (7–9 mm) combined with its conspicuous sculpture of fine parallel or concentric ridges on the head and mesosoma, and the array of long, curving hairs on the underside of the head. Unlike similar western species, *badius* usually lacks propodeal spines, and varies in size, with some large, big-headed workers. No western species have been brought to Florida, although a few of them could probably survive in the state. Occasional specimens from the western Florida Panhandle have spines on the propodeum. The population on the Lake Wales Ridge has males that are completely reddish brown, while the males from elsewhere are bicolored, the head and mesosoma are blackish, and the gaster is red. The Lake Wales Ridge, which runs down the center of the south-central Florida Peninsula, has many isolated populations and endemic species of sand-loving arthropods (Deyrup 1990).

Distribution: North Carolina south into Florida, west along the Gulf Coastal Plain into eastern Louisiana (Cole 1968). Within Florida, *badius* is missing from areas that do not have deep sandy soil.

Natural History: *Pogonomyrmex badius* requires open, well-drained sandy areas with patches of bare sand. With the exception of beach dunes, such habitats are maintained in the Southeast by fires that temporarily remove a canopy of trees and shrubs, and foster a ground cover of herbs and grasses that provide the seeds needed by *badius*. The natural habitat of *badius* is primarily Florida scrub and sandhill, the latter also called "high pine." These habitats are maintained by different fire frequencies: every few years for sandhill and one to several decades for most types of Florida scrub habitat (Myers 1990). This means that in the absence of fire, *badius* could be quickly eliminated from sandhill habitat but could persist longer in Florida scrub. Rosemary balds are a form of Florida scrub dominated by shrubby Florida rosemary (*Ceratiola ericoides*) and characterized by persistent open patches between the rosemary plants. I have never seen a rosemary bald that lacked *badius*. Florida rosemary also provides seeds that are harvested by *badius*, the workers often climbing the plants to clip off the seeds. During drier climatic conditions that ended 7000 to 10,000 years ago, Florida rosemary was a widespread and dominant plant (Delcourt and Delcourt 1981; Myers 1990). This dry period may have been ideal for *badius*.

Nests of *badius* are conspicuous, often as much as a meter in diameter, the perimeter marked with bits of twigs, plant husks, droppings of scrub millipedes, and charcoal. The charcoal has attracted the attention of myrmecologists, as it is often gathered from a little distance away as if it were a valuable resource. Gordon (1984) experimentally removed the charcoal from around the nest, finding that it would be replaced in approximately seven days. Since charcoal is known for absorbing certain chemicals, Gordon tested the hypothesis that territorial chemicals adhering to charcoal fragments deterred other ants from entering the nest area. There seemed to be evidence that removing charcoal resulted in more ants of other species entering the mound area, requiring extra defensive behavior by the *badius* workers. Smith and Tschinkel (2007) tested the deterrent charcoal hypothesis directly with charcoal from *badius* nests ringing pitfall traps, but found no evidence of any deterrence. Ants of various species, in my experience, seem attracted to areas with recent surface disturbance, so the incursions of ants on mounds with charcoal removed may have been associated with this disturbance effect. Smith and Tschinkel (2007) also tested hypotheses that the charcoal changed soil temperature or hydrology, but found no convincing effects. Western *Pogonomyrmex* species also gather small objects, such as pebbles, to put around the nest, so the collection habit seems established in the group. Smith and Tschinkel (2007) suggest that large mounds of pebbles may protect western *Pogonomyrmex* from sheet flow during heavy rain, but bits of charcoal on deep sandy soil would have no such effect. As a kind of null hypothesis, Smith and Tschinkel (2007) speculate that foraging for small, inedible objects may be built into the overall ancestral foraging patterns of *badius* while it looks for seeds and dead insects, and there may not be a simple genetic mechanism to delete this particular aspect of foraging.

The nest architecture of *badius* has been studied in depth (so to speak) by Walter Tschinkel, a preeminent behavioral ecologist of the ant world. The nests of *badius* are complex and elegant, as can be seen in Tschinkel's paper (2004), featuring photographs of nests that have been painstakingly pieced together from plaster casts. The larger chambers in the top third or quarter of the housing are for foragers and other mature workers (Tschinkel 1999). The large upper chambers often seem underutilized (Tschinkel 1999); perhaps they have an important but temporary role as a sheltered staging ground when the colony is about to relocate or has just relocated, or when alates are waiting for the appropriate moment to emerge. One to several tunnels spiral down from the upper chambers, with shallow, horizontal, pancake-like chambers extending from sides of tunnels at various levels (Tschinkel 2004). Seeds are stored in chambers 40 to 100 cm deep (Tschinkel 1999). Brood and callow adults are in the deepest chambers, some of which may be 2 meters or more below the surface, depending on the size of the colony and the depth of the water table (Tschinkel 1999). The adaptive value of such deep bunkers is not known, but it does shelter the most vulnerable colony members in a zone of reduced fluctuations in temperature and humidity, and far from the upper, biologically active strata, permeable to burrowing predators such as thief ants (*Solenopsis* spp.) and army ants (*Neivamyrmex* spp.).

The nest of a mature colony of *badius* is a large and complex structure in which the colony members and their stored seeds have been organized in specific areas. One would not expect a colony to excavate a new nest and accept the hassle of moving the brood, the various stages of workers, and the granary to their designated chambers except under relatively extreme duress, such as a major change in the environment around the nest, or perhaps a rival colony moving into the neighborhood. Surprisingly, *badius* colonies relocate an average of about once per year, with the relocation rate ranging from two years between moves to four moves per year (Tschinkel 2014). Even more surprisingly, meticulous observations by the indefatigable Walter Tschinkel have not shown any compelling reason for relocation (Tschinkel 2014). Colonies usually move a short distance, averaging 4 meters, into a similar-appearing home site, and there is no correlation between relocation and distance to other colonies (Tschinkel 2014). Once a move has begun, it is completed in only four to six days, including excavating the complex subterranean structure, stocking it with seeds and colony members in their appropriate chambers, and retrieving charcoal and other objects to arrange around the nest mound (Tschinkel 2014). There is evidence that nest relocation exacts a toll on colony growth (Tschinkel 2014). In simplistic terms, *badius* colonies abandon their fancy and well-organized homes and endure the disruption of constructing and moving to an identical mansion just across the street. As the owner of an older home, I am often reminded of problems accumulating with prolonged occupancy; if I could, *badius*-like, construct and move into a similar brand-new home in four to six days, I would not hesitate. There is no evidence, however, of ecological problems accumulating with residency in a *badius* nest.

The reproductive cycle of *badius* has been investigated by Smith and Tschinkel (2006). Colonies begin producing queens and males when there are approximately 700 workers in the nest. The eggs destined to become alates are produced in spring and reared to adults before the colony begins to rear workers. Males from larger colonies are conspicuously larger than males from smaller colonies, but queen size does not increase with colony size. The ratio of males to queens also increases as colonies increase in size. It is difficult to interpret the fitness of this pattern of male to queen investment, in part because theories governing male and queen investment sometimes contradict each other. Female *badius* mate multiple times; genetic studies of worker heterozygosity suggest that all queens mate several times, and some mate at least 10 times (Rheindt et al. 2004). Perhaps higher production of males has a fitness linked to polyandry. Mating occurs in the afternoon around the nest entrance in June (Harmon 1993; Van Pelt 1953). I have observed many *badius* mating sequences at the Archbold Biological Station (Highlands County), but plenty of questions remain. During a typical episode, males, alate queens, and workers emerge from the nest and mill about, with workers often seizing queens or males and dragging them around. Queens mate with males emerging from their own nests, as well as with males that fly in from elsewhere. The latter males usually approach flying low over the sand, landing a short distance from the nest and running with buzzing wings to join the group. It is not clear that all males emerging from the nest are mating with siblings, because if the mating aggregation is interrupted by rain, both males and females usually return to the nest. Both females and males may mate more than once. It is easier to count mating sequences in females; the largest number recorded was seven. Males attempt to mount queens, workers, and each other, but only persist when mounting queens. When a male has mounted a queen, the pair is usually immediately mobbed by other males. After mating for approximately 15 to 20 seconds, the queen curls under to pinch the gaster of the male, who disengages. The newly mating queen runs off, usually pursued by males, until she is alone, when she cleans off the tip of the gaster, to which sand can be seen adhering. She then returns to the group or, if she has already mated several times, she flies away. One aggregation included only three males and approximately 15 females. Queens were seen following males, and on two occasions, a queen seized a male and dragged him out of the group; one of these episodes resulted in a successful mating. On the Lake Wales Ridge, where these observations were made, *badius* colonies are often sparsely distributed; this may have provided evolutionary pressure for flexible mating behavior. One can imagine a shifting balance between more dependable sib-mating and riskier outcrossing, probably affecting queens and males differently.

Pogonomyrmex badius workers are often seen running about with the gaster tucked under their body. Their long legs, which permit them to carry long seeds slung partially under the body, facilitate this gaster-tucking behavior. I have wondered whether this reduced exposure to the heat of the sun or readied the ant for defensive action, but the behavior seems too inconsistent for those interpretations. A recent study of gaster-tucking in a species of *Pseudomyrmex* seems to show that this behavior may allow ants to run faster (Amador-Vargas et al. 2011). With the availability of inexpensive systems for filming small animals, it would be easy to set up a grid over a pogo nest and use movies of worker behavior to see whether the tucked-gaster individuals are running faster. If this is the case, however, there would still be the question of why some workers are in more of a hurry than others.

Anecdotes about the sting of *badius* claim that it is extremely painful, with the pain sometimes lasting for hours and, in the case of multiple stings, causing muscle soreness that lasts several days (Wray 1938). Symptoms of stings of other species of *Pogonomyrmex* species include tenderness in the nearest lymph nodes and hairs standing up around stings on the arms and legs (Creighton 1950; Schmidt and Blum 1978). I have never availed myself of innumerable opportunities to accumulate my own anecdotes, but one accidental sting did indeed cause the hair on my arm to stand up and discomfort of the armpit lymph node. A test of *badius* venom on mice (Schmidt and Blum 1978) showed a greater toxicity for mammals than any other insect venom known at the time. The venom is not especially toxic to insects (Schmidt and Blum 1978). If *badius* were as defensive around the nest as the fire ant *Solenopsis invicta*, it would be a significant hazard in Florida's scrub and sandhill areas, but workers seldom attempt to sting unless they are picked up and squeezed. Workers mill about when the nest mound is disturbed but do not swarm up the legs and hands of the disturber. I have plenty of experience of this, as I often dig in the mounds to look for larvae of the antlion *Brachynemurus nebulosus* (Bahls and Deyrup 1988) or the tiny scarab *Geopsammodius relictillus*. Unlike *Solenopsis invicta*, *P. badius* has no brood or recently emerged adults anywhere near the surface, so superficial disturbance of the nest mound causes little damage to the colony.

Pogonomyrmex badius foragers use their sting in personal defense, and perhaps to defend their seed stores from rodents, as the granaries, 40–100 cm underground (Tschinkel 1999), are well within the burrowing range of the Florida mouse (*Podomys floridanus*) or the oldfield mouse (*Peromyscus polionotus*). A dose of 0.42 mg/kg of *badius* venom is lethal to 50% of mice tested in the laboratory (Schmidt and Blum 1978); the authors of this venom study found this toxicity impressive, although I see no easy way to convert this into ant stings per mouse. While there is a study on colony-level response to predation on foraging *badius* (Gentry 1974), the predator was Gentry himself removing foragers from colonies. With persistent predation, colonies eventually cease aboveground activity; Gentry (1974) concluded that a failure to recruit workers engaged in subterranean tasks to replace lost foragers benefited colonies by preventing the complete destruction of the workforce. There is no evidence, however, that there is ever any large-scale predation on *badius* foragers under natural conditions.

Pogonomyrmex badius is unique among North American species of *Pogonomyrmex* in having dramatic differences in worker size, including large workers with disproportionately enlarged heads. A community-level study of seed-gathering *Pogonomyrmex* and other ants in the Southwest showed that ants of different sizes preferred seeds of different sizes, with preferred seed size related to ant size (Davidson 1977). This discovery inspired a hypothesis that *badius*, as the only *Pogonomyrmex* species in the Southeast, had evolved polymorphic workers that were able to deal efficiently with a wide range of seed sizes in the absence of competition from species with specific size preferences. This hypothesis pointed toward a phenomenon called "ecological release," meaning diversification within a lineage that has escaped from close competitors (Traniello and Beshers 1991). So beguiling was this notion of Florida as the land of opportunity for a pioneering *Pogonomyrmex* that for years I happily repeated it to visiting naturalists, without worrying whether there was any evidence for this correlation. Fortunately, there are myrmecologists willing to do the hard work of testing ecological speculations. Traniello and Beshers (1991) found little evidence that big foragers preferentially gather big seeds and small foragers gather small seeds. Statistical studies of *badius* head morphology strongly suggest that the big-headed individuals are specialized seed grinders, analogous to the big-headed seed-grinding specialists of *Solenopsis geminata* and some species of *Pheidole* (Ferster et al. 2006). By analogy with the omnivorous *Solenopsis geminata*, it seems likely that *badius* has a broader diet than most southwestern *Pogonomyrmex*, so that seed-milling, rather than being a task for all workers, can be more efficiently performed by a specialized caste of relatively few workers (Ferster et al. 2006). A more varied diet might be correlated with Florida's warm, moist climate that provides varied resources for about eight months of the year.

Name Derivation: The name *badius*, although it sounds as if it should be the name of a comic book villain, is from the Latin *badius*, meaning "chestnut brown," referring to the color of this *Pogonomyrmex*.

Genus *Solenopsis*

Fire Ants and Thief Ants

The genus *Solenopsis* (pronounced **Sō' len op" sis**) includes, as far as is known, 17 Florida species. A couple of these, *S. geminata* and *S. invicta*, are relatively large species that live in colonies with many thousands of workers. The workers effectively defend their nests by vigorously biting and stinging, so that a slipper-clad homeowner sleepily crossing the lawn to pick up the morning paper, or a rancher proudly surveying his herd of sleek Angus, may abruptly commence an energetic foot-slapping dance, while the surrounding air turns blue with expletives. These ants are, of course, the fire ants. These days, almost all fire ant interactions with humans in Florida may be attributed to *S. invicta*, which has largely replaced the similar *S. geminata*. Plants and animals that thrive in the kinds of drastically disturbed habitats provided by humans often become widespread "weedy" species, but such species may have originally been more narrowly restricted to natural habitats that were chronically disturbed. In such habitats, populations of these species build up rapidly and produce large numbers of propagules that disperse to other temporarily underoccupied areas. Some fire ants are probably examples of this, although it would be hard to prove at this point. *Solenopsis invicta* might have been adapted to take advantage of low inland areas subject to variable flooding. *Solenopsis geminata*, at least in the Southeast and eastern Texas, might have responded to occupy areas cleared by natural fires or those set by indigenous humans; elsewhere, it might have responded to disturbances caused by coastal storms and hurricanes.

Not many naturalists were out documenting fire ant populations at the time when fire ants may have been confined primarily to naturally disturbed areas. The exception was Henry Bates (1825–1892) who traveled up a tributary of the Amazon, the Tapejós, in 1852 when that river was still sparsely settled. The Tapejós rises and falls dramatically in response to rains upstream, and wherever there were wide exposed shores of sandy soil, fire ants (*Solenopsis saevissima*) were so abundant that "… it was impossible to walk about, on account of the swarms of the terrible fire ant…." (Bates 1864). In two places, Bates mentions enormous windrows of drowned winged fire ants along the river, where flights had been knocked into the river by a sudden squall and then washed ashore. I have seen a similar windrow of *S. invicta* along Florida's East Coast where offshore winds carried flights out to sea, the waves bringing them back. Bates called the small village (14–15 houses) where he stayed "the head-quarters" of the fire ants, which had moved back from the river into open sandy areas of the village. His complaint sounds familiar: "They seem to attack persons out of sheer malice: if we stood for a few moments in the street, even at a distance from their nests, we were sure to be overrun and severely punished, for the moment an ant touched the flesh, he secured himself with his jaws, doubled in his tail, and stung with all his might." The assumption that any insect that could mount such a ferocious attack must be a male did not disappear in the 1800s; it is common today.

As the victim of a fire ant attack stomps away, smarting and grumbling, he is almost certainly passing over the subterranean nests of other species of *Solenopsis*, the tiny thief ants. The 12 or so species of Florida thief ants, whose name refers to their frequent occurrence in or around the nests of other ants, do not challenge our dominance of the land, but they definitely challenge our taxonomic abilities. Although a fire ant nest may have 150,000 workers (Tschinkel 2006) and thief ants often appear to live in relatively small nests of maybe a few hundred or a few thousand, I am guessing that numbers of individual thief ants far outnumber fire ants in almost any Florida habitat and are the most abundant of all ants throughout the state.

Florida species of *Solenopsis* are shiny, lack propodeal spines, and have little surface sculpture other than the hair-bearing punctures. Some species of thief ants could easily be mistaken for *Monomorium*, but the antennal club has only two segments.

The genus *Solenopsis* occurs in both the Old and New Worlds, with the great majority of species confined to tropical or warm temperate regions (Bolton 1995). In the Old World, which has approximately 70 species, *Solenopsis* is strongly associated with arid and semiarid climatic zones, primarily in the Mediterranean and Middle East, as well as in Africa and Australia. Humid tropical zones have few species; it is possible that in the Old World, *Monomorium* occupies the ecological roles that are occupied by *Solenopsis* in the humid neotropics. The New World fauna of approximately 155 described species is about twice as large as the Old World Fauna and includes a large number of species that inhabit humid regions and damp habitats. In the New World, there are 19 species of fire ants and close relatives (Trager 1991), 130 described species of thief ants (Pacheco and Mackay 2013), and a few odd parasitic species. There are certainly many undescribed species of thief ants, whose numbers are likely to rise to more than 200. Even in Florida, it would not be surprising if there were a couple of native species and a couple of introduced species that have not yet been found.

Thief ants are among the most difficult ants to study, with large complexes of similar species and many species that are difficult to find. Fortunately for myrmecologists, José Pacheco and William Mackay had the courage to tackle this formidable group, providing the first revision of New World species (Pacheco 2007; Pacheco and Mackay 2013). While additional species are sure to be found, such as *Solenopsis* species A and B discussed below, we are permanently indebted to the pioneering work of Pacheco and Mackay. Florida thief ants are easily collected in numbers from Berlese funnel samples. Workers of some of the yellow species are difficult to identify, although their queens and males are usually distinctive. Unfortunately, queens and males associated with workers are often difficult to obtain. Two species, *pergandei* and *tonsa*, are seldom collected in litter samples because they live in pure sand. Baits buried in perforated vials often produce subterranean *Solenopsis* and other ants, although *tonsa* remains rare in collections. Workers of species that forage on the surface can be collected by baiting leaf litter with fine crumbs of cookie and then returning approximately 15 minutes later and sifting the litter. These, of course, are not accompanied by queens and usually belong to one of the yellow species that are difficult to identify. Two species, *picta* and *zeteki*, usually live in hollow twigs and are collected by breaking small twigs on live trees.

Thief ants get their name from their horror-film habit of excavating their tiny galleries within the walls of larger ant species, emerging from these galleries to prey on the larvae and pupae of the larger ants. This was documented in the 1800s for the European species *Solenopsis fugax* (Wheeler 1901). After finding the North American species *Solenopsis molesta* nesting adjacent to the nests of a variety of larger ants, Wheeler was convinced that *molesta* had habits similar to those of *fugax* (Wheeler 1901). Hayes (1920) lists many more associations between *molesta* and other ants. As mentioned below, Schneirla (1944) describes finding large nests of *pergandei* completely surrounding a colony of *Formica archboldi*. Anyone who has persistently looked at thief ants in the field, however, knows that colonies are often not clearly associated with other nearby ants. Wheeler (1901) suggests that there could be long galleries connecting the apparently isolated nests with distant nests of other ants, but he accepted that both *fugax* and *molesta* could occur both independently and in association with other ants. This kind of thieving (or, more accurately, murderous) relationship with another ant is called "lestobiosis" (Hölldobler and Wilson 1990). If one wanted to flaunt myrmecological jargon, thief ants such as *molesta* and *pergandei* could be called "apparently opportunistically lestobiotic *Solenopsis*."

The sandy soil found in much of Florida is highly permeable to ants, as is easily demonstrated by burying baits in perforated vials. Thompson (1980) experimented baiting with larval insects and with queen fire ants. Many of these bait insects were promptly killed and consumed by thief ants. Although these experiments were done on a small scale, they suggest that thief ants may be important subterranean predators, especially in sandy soil.

Several species of Florida thief ants can also be found in rotten wood, especially the small-queen species in the *carolinensis* complex, *terricola*, *tennesseensis*, *abdita*, and Species B. In rotten wood, it is common to find large aggregations of larvae and pupae packed into many small chambers in the wood, accompanied by workers, but the nest queen or alate queens and males are somewhere else. This is a reminder of one of the more obscure benefits of sociality accruing to some ants: very

large numbers of small larvae can be warehoused in relative safety, ready to allow efficient exploitation of a pulse of resources. When a colony of tiny *Solenopsis*, *Pheidole*, or *Monomorium* obtains a large insect prey, nothing needs to be wasted because there are plenty of larvae that have been waiting for such an event.

Name Derivation: The name *Solenopsis* is supposedly derived from *solon* (Greek), meaning "pipe or channel," and *opsis* (Greek), meaning "appearance or sight" (Tschinkel 2006). At first glance, this name makes no sense. The genus *Solenopsis* was invented by Westwood in 1840 to accommodate a species he named *Solenopsis mandibularis* (Westwood 1840), which later turned out to be a synonym of *S. geminata*, at that time still in the genus *Atta*. Although *mandibularis* was later synonymized, the new genus name, *Solenopsis*, remained valid. Westwood's description (back in 1840) clearly refers to a major worker of *geminata*, as it mentions the toothless mandibles and the enlarged head divided into two lobes. Westwood named the genus *ob faciem caniculatum*, "on account of (its) grooved face," a conspicuous feature of major workers of *geminata*. It turns out that *opsis* can also mean face, so *Solenopsis* does not indicate a pipe-like or channel-like ant, but rather an ant with a groove up its face. Not many fire ant species have an outsized head whose bulging lobes are divided by a groove, but while the name *Solenopsis* is not always appropriately descriptive, it does make sense.

Solenopsis abdita Thompson— Overlooked Thief Ant (Plate 42)

Taxonomy and Similar Species: As Thompson explains in her description of *S. abdita*, the translation of the name of the species refers to the close resemblance of *abdita* workers to those of *carolinensis* (Thompson 1989). Fortunately, queens and males of *abdita* are distinctive. The workers of *abdita*, *carolinensis*, and *texana* are, as Pacheco and Mackay remark (2013), almost impossible to separate. The workers of the three species are yellow and small (1.1–1.3 mm long); the eyes have three to five ommatidia. The upper surface of the head has inconspicuous punctures and semidecumbent hairs that, in lateral view, appear to make a moderately dense layer of short hairs, although it is obvious that these are not short, straight hairs that are sticking straight up. Hairs on the pronotum are sparse and curved, noticeably longer than those on the head. Workers of *abdita* average a little smaller than those of *carolinensis* and *texana*, but this is not useful in identification without some sophisticated measuring. Published keys for identifying worker *abdita* are not very helpful. Those of Thompson (1989) and Thompson and Johnson (1989) lead with a difference in eye size of five thousandths of a millimeter and go on to a subjective difference in the mesosomal profile and a distinction that I have never been able to see in the shape of the vaguely visible promesopleural suture. Pacheco and Mackay (2013) separate *abdita* out by region (Florida), while following couplets of supposedly non-Florida species include *carolinensis* and *texana*, both of which appear to occur in Florida.

Queens and males of *abdita* are easily identified: they are coffee brown to dark brown with milky white wings, and females have large punctures on the face. In Pacheco and Mackay (2013), the color of queens is given as "golden brown," but I have not seen any that I would describe this way. There is no doubt that *abdita* is a valid species, and its range overlaps with that of the *carolinensis/texana* complex with their yellow queens.

Distribution: Throughout Florida, extending into southern Georgia, and undoubtedly southern Alabama, as it occurs close to the northern borders of several Florida Panhandle counties, as well as Mississippi (MacGown et al. 2009a).

Natural History: *Solenopsis abdita* occurs in a wide variety of habitats in Florida, ranging from sandhill to swamp forest, but the great majority of specimens that I have seen are from wet habitats to mesic, including grass tussocks in seasonal ponds, the bases of pines in wet flatwoods, wet hammocks, and mesic forest (only collections with queens recorded here). It can be found reliably in open low flatwoods, and the vast expanses of wet flatwoods in Florida and surrounding states are probably the primary habitat of this species. This would help explain why this abundant Florida species has not become common in upland areas of the Atlantic Coastal Plain to the north of Florida. Nests appear to be superficial, not deep in the ground. *Solenopsis abdita* is polygynous and Thompson (1989) kept large numbers of queens in a laboratory colony for more than two years. Queens fly primarily in June and July (22 of 29 specimens collected in flight traps).

Name Derivation: From *abdita* (Latin), meaning "hidden," referring to the fact that this common species was long overlooked because of its resemblance to local species with similar small yellow workers (Thompson 1989).

Solenopsis carolinensis Forel, *Solenopsis texana* Emery—Carolina Thief Ant (Plate 42)

Taxonomy and Similar Species: I am treating these two named species as an unresolved species complex, in the sense that there is some uncertainty about which name pertains to which ants, and also in the sense that there is a good probability that these names have been applied to more than two species. In her 1980 and 1989 reviews of Florida thief ants, Thompson states that *carolinensis* is common in Florida, but records of *texana* are doubtful. Pacheco and Mackay (2013) state that *texana* is common in Florida, but *carolinensis* is absent from Florida and from the southern portions of all the southeastern states. On the other hand, in one place in this same 2013 revision, it is stated that *texana* apparently does not occur in Florida. In the key to species in this review, both species are excluded from Florida. It is also worrisome that Pacheco and Mackay attribute an enormous range to *texana* stretching from Virginia into northern South America, within which *texana* occupies almost every available habitat, with nesting habits ranging from subterranean to dead twigs, the latter in cohabitation with *Pseudomyrmex*.

I am generally eager to accept the assumption that thief ants with similar workers have provided us with distinctive queens and males that allow us to tell the species apart, but these ants are under no obligation to be so considerate. Creighton (1950) complains that workers of many thief ant species show "convergence," but from an evolutionary point of view, in most cases, it is much more likely that the alates of many species have diverged. Detailed evolutionary convergence among workers is possible, but less likely. This being the case, there are probably species whose alates who have not diverged in any way that has been noticed by the small number of myrmecologists who have turned their attention to the thief ants.

Whatever names one might apply to them, there are probably at least two Florida species in the *texana–carolinensis* complex. There is a widespread form that Thompson calls *carolinensis*, with relatively small queens (approximately 3.2–3.5 mm long) that vary slightly in color and size and occur in a great variety of habitats, ranging from open sand to rotten logs in dense forest.

This is the species that Pacheco and Mackay interpret as *texana*, according to Florida specimens that they identified and sent to the Archbold Biological Station collection. On the other hand, the queens identified by Mackay and Pacheco are golden yellow, while they describe the queen of *texana* as medium to light brown or, elsewhere in the revision, "black to light brown." There are also larger yellow queens, primarily in northern Florida (approximately 4.0–4.5 mm long). Small and large queens occur in the same area. The large-queen form is somewhat similar to a queen sent from Texas, identified as *carolinensis*. Creighton (1950) states that the eye of queen *carolinensis* is proportionately larger than that of *texana*, covering more than half the side of the head. I have not seen any queens that show this character state. The queen sent by Mackay as *carolinensis* has proportionately smaller eyes, not larger. In the Archbold Biological Station collection, there are many queens of the *texana–carolinensis* complex, most of which were extracted from soil or leaf litter. These vary in size, and it might be possible to set up a series progressing gradually from small to large, although I continue to believe that there is one species that generally has small queens and another that generally has large queens. The anterior ocellus of the largest queens is much larger than the lateral ocelli and wider than the distance between the anterior and lateral ocelli, but among flight trap specimens, there is much variation in the size of the ocelli. One set of specimens not only has unusually large queens, but the petiole of associated workers is (in posterior view) unusually wide, contrasting with the narrowly rounded petiole seen in other series of workers.

Working out this taxonomic tangle is not a lost cause. Morphological analysis, however, may have reached its limits of usefulness in this complex. More extensive collecting should help solve the problem by providing additional comparative material, but this collecting may need to be combined with genetic analysis. Once species limits have been established, it might be possible to apply the currently available names, as well as describe any new species.

Solenopsis geminata (Fabricius)— Big-Headed Native Fire Ant (Plate 43)

Taxonomy and Similar Species: *Solenopsis geminata* is easily distinguished from other fire ants found in eastern North America by the disproportionately large head of larger workers. This can be seen on a less exaggerated scale even on midsized specimens and makes identification easy even in the field. The largest workers are seed-millers, like the largest workers of *Pogonomyrmex badius*, the enlarged head accommodating the mandibular muscles. The jaws of the largest workers have been described as lacking teeth (Creighton 1950), but these workers emerge with teeth that are soon worn away by grinding seeds (Trager 1991).

Distribution: *Solenopsis geminata* occurs along the eastern Atlantic Coastal Plain from southern North Carolina (Guénard et al. 2012) through the Florida Peninsula and into Walton County in the western Florida Panhandle. Trager (1991) reported it rare or absent in Alabama, Mississippi, and Louisiana, but occurring from Texas through Mexico and Central America, and most of the length of South America. Creighton (1950) did not report a gap in the distribution along the northern Gulf, probably because no such gap existed before the introduction of *richteri* and *invicta*. It was certainly present in Alabama and Mississippi in the 1940s (Murphree 1947). This species has been introduced into Africa, the islands of the Caribbean, and Asia from Taiwan through India, and through the Malay Archipelago, Polynesia into Australia (Gotzek et al. 2015; Trager 1991; Wetterer 2011a). While it has been assumed that *geminata* must have started globetrotting long ago, a remarkable piece of genetic detective work (Gotzek et al. 2015) presents convincing evidence that *geminata* was transported in the 1500s by Spanish galleons from Acapulco to Manila and thence to China and beyond. Spanish galleons used soil and rock for ballast and also carried crop plants such as sweet potatoes that might have been infested with *geminata* (Gotzek et al. 2015). The spread of *geminata* must have been greatly facilitated by its ready acceptance of beaches as nesting areas. As an extreme example of the dispersal of *geminata*, in the Archbold collection, there are specimens from Johnston Atoll, a tiny, remote island occupied by albatrosses and a U.S. air base.

The remarkable colonization ability of *geminata*, combined with the supposed gap between southeastern and western populations in North America, has raised the question of whether the southeastern population might have been introduced in the early days of maritime commerce (Trager 1991; Wetterer 2011a). My suspicion is that this species is both native and introduced in Florida. *Solenopsis geminata* might well have come east during dry periods of the late Pliocene through the mid-Pleistocene, along with the lineages that produced *Pogonomyrmex badius*, the sand roach *Arenivaga floridensis*, the tortoise *Gopherus polyphemus*, and other animals with isolated southeastern populations. It is highly likely that *geminata* was also introduced by commerce into southern Florida at the same time that it was being transported to all the islands of the Caribbean, and it is possible that the population in tropical and subtropical Florida is Neotropical in origin or has a mixed Neotropical and western heritage.

Natural History: The habitats where *S. geminata* presently occurs in Florida may be determined largely by its relationship with *S. invicta*. According to Smith (1965), "Until the introduction of the imported fire ant into Florida, *geminata* was not only the most common but the worst fire ant pest in that State." More recently, in an area of mixed natural habitats in northern Florida, *geminata* occurred in well-drained upland habitats while *invicta* was in habitats with a high water table (Tschinkel 1988). Disturbance can allow *invicta* to invade well-drained habitats where one would expect *geminata* (Tschinkel 1988). In peninsular Florida, *geminata* is common in the Keys and in some beach areas but sporadic in both upland and wet areas inland. In much of interior peninsular Florida, *geminata* might be in the late stages of total replacement by *invicta*, or it might have achieved a persistent status as a relatively rare species. In addition to its probable replacement by *invicta* in most Florida habitats, *geminata* may have been affected by the drastic reduction of high pine (= Florida sandhill) habitat, of which only 10% remains, mostly in a degraded state (Myers 1990). This habitat, with its well-drained soil and dense carpet of grasses and other plants, would have been ideal for a seed-harvesting *Solenopsis*. Assuming that *geminata* is native in southeastern North America, it would have benefited from the open fields, sometimes maintained by fire, where indigenous tribes cultivated crops (Swanton 1946). Indeed, one of the best arguments for widespread southeastern distribution of *geminata* in pre-Columbian times is the extensive distribution of apparently suitable habitat around the Gulf of Mexico during the early Holocene.

In at least some tropical areas, *geminata* is an ecologically important predator of ground-dwelling arthropods (Risch and Carroll 1982a,b). This was shown experimentally in plots of squash and corn with and without *geminata*: there were far fewer individuals of both phytophagous and predaceous species in plots without *geminata* (Risch and Carroll 1982b). In tropical agricultural systems, *geminata* can have negative effects by eating crop seeds before they germinate, stinging humans and farm animals, and encouraging sap-sucking insects (Risch and Carroll 1982a). In balance, however, this species appears to be beneficial in biocontrol of pest arthropods (Risch and Carroll 1982a,b). Programs of using insecticides to eliminate *geminata* are doubly misguided: they are ineffective, because *geminata* in the absence of other ants soon recolonizes to become even more dominant, and meanwhile crop pests can multiply until *geminata* returns (Risch and Carroll 1982a).

A special feature of *geminata* natural history is its proficient use of seeds. Major workers of *geminata* have massive jaws powered by strong muscles originating in the bulging occipital lobes of the head. These jaws are for milling seeds and are not generally used for defense of the nest, foraging, carrying objects, excavating, or caring for young; in a laboratory study, majors of *geminata* were found doing only two kinds of work, grinding seeds and self-grooming (Wilson 1978). In the field, majors occasionally appear at baits or in confrontations with other ants (Carroll and Risch 1984). *Solenopsis geminata* harvests and consumes certain seeds so efficiently that it can alter the abundance of preferred host plants, especially certain grasses (Carroll and Risch 1984; Risch and Carroll 1986). This might even be beneficial to humans when preferred seeds are weeds in crops (Carroll and Risch 1984). *Solenopsis geminata* stores seeds, after first preventing sprouting by destroying the seed embryo, but storage may be short term rather than sustaining the colony during long periods between harvests (Risch and Carroll 1986). The incompatibility of *geminata* and *invicta* in most habitats deprives us ecologists of the kind of plausible explanations that we so enjoy. If the two species frequently occurred on the same site, we would hypothesize that the importance of seed harvesting in *geminata* placed that species in a different ecological niche than *invicta*. If *geminata* were the dominant species, we would hypothesize that a larger and more diverse resource base consisting of both insects and seeds gave *geminata* an advantage over *invicta*. As it is, the mechanisms by which *invicta* replaced *geminata* are unknown, and we cannot even be sure this replacement is temporary.

The sting of *geminata* is painful, like that of *invicta*, but the sting does not result in a white pustule like that produced by the sting of *invicta*. *Solenopsis geminata* may be responsible for the plagues of stinging ants that disrupted life in the colonial West Indies, perhaps even briefly impeding the European effort to convert most of the arable land in the islands into sugar plantations managed under a brutal system of slave labor. Sifting through accounts of the Hispaniola ant plagues of 1517–1519, Wilson (2005) determined that the probable culprit was *geminata*, based on reports of the ferocity of the ants and an examination of the fauna occurring on the island today. *Solenopsis geminata* is found today throughout the Caribbean, but there are no recent reports of outbreaks. As usual, in the case of animals whose exploding populations collapse suddenly and permanently, there is no obvious reason for the decline, aside from the impossibility of sustained exponential population growth. On the Caribbean islands that I have visited, *geminata* is a common but minor member of the insect fauna. It is reassuring to know that population outbreaks can be naturally and permanently quelled (although worrisome when referring to our own species), but during the outbreak phase of *geminata*, there might have been many unremarked extinctions of island vertebrates and insects. Cowan (1865) refers to deaths of livestock during the Caribbean ant plagues, and terrestrial vertebrates and invertebrates must also have been affected.

Name Derivation: From *geminata* (Latin), meaning "double," probably referring to the double lobes of the head of the major. This species was named by the Danish entomologist Johann Christian Fabricius (his name is often abbreviated to "F." in the names of insects) whose life and career (1745–1808) overlapped with those of Linnaeus (1707–1778).

Solenopsis globularia (F. Smith)—
Globular Thief Ant (Plate 42)

Taxonomy and Similar Species: *Solenopsis globularia* is easily identified by its enormously swollen, shining postpetiole, which is more or less globular. This feature is also found in queens and males of *S. globularia*. The function of this postpetiolar enlargement is apparently unknown; it also occurs in some other ants, such as species of *Cardiocondyla*

and *Temnothorax*. An enlarged postpetiole might accommodate a hypertrophied postpetiolar gland (the function of this gland also appears to be unknown), or it might simply provide a more robust attachment to the gaster or support larger muscles for the postpetiolar joints. Florida specimens are usually orange yellow with a darker gaster, but specimens from some Caribbean islands may be dark brown or yellow. *Solenopsis globularia* is said to be polymorphic, with workers of differing sizes within a nest (Pacheco 2007). In Florida, within-nest polymorphism seems to be negligible, although individuals from different colonies may vary conspicuously in size.

Distribution: *Solenopsis globularia* seems to thrive on beaches and in other open coastal habitats, and this predilection probably explains its wide distribution along the coasts of North America from North Carolina (Guénard et al. 2012) south through Florida and west into Alabama, through much of the Caribbean and northern South America and coastal and inland Brazil (Pacheco 2007). This species almost certainly came to Florida naturally by island-hopping from beach to beach, but it is likely that colonies have also been occasionally brought through commerce to Florida from Caribbean islands.

Natural History: Nests are in open habitats, including both coastal and inland areas, usually but not always near water. I have found several large colonies in clay or under clumps of grass on rocky coastal flats in the Florida Keys. These colonies must have been occasionally if not regularly inundated by sea water. On Florida beaches, workers may be found foraging under dry wrack at the upper tidal level. A nest found in a rotten pine log at the Archbold Biological Station had a large cache of tiny seeds, possibly seeds of *Paronychia* sp. (Caryophyllaceae). Queens have been taken in flight traps in December and June (two specimens). Many alate queens were found in a nest in August.

Name Derivation: From *globulus* (Latin), meaning "small spherical object or bead," referring to the globular postpetiole.

Solenopsis invicta Buren—
Red Imported Fire Ant (Plate 43)

Taxonomy and Similar Species: One might say that this ant needs no introduction, as it has introduced itself to almost everybody in Florida. It is quite similar, however, to *S. geminata*, from which it may be easily distinguished by the proportional size of the head of major workers, in great contrast to the majors of *geminata*, whose heads are so outsized that one wonders how they can walk without tripping over their own mandibles. If no majors are seen out foraging, they can easily be observed by stirring up the nest and jumping back a bit, a perennial Florida pastime. I did not need to invent an English name for such an important species of ant, but if I had, I probably would not have called it "red," as this species is usually so dark that it looks blackish. This, however, causes no problems in communication, as everybody in Florida, with the exception of a few entomologists, calls *invicta* "The Fire Ant," not "The Red Imported Fire Ant." *Solenopsis invicta* has largely replaced *geminata* in Florida, so any nonentomological mention of fire ants almost always refers to *invicta*. While discussing the common name of *invicta*, I would like to mention that it does not include the word "your," as in, "Your Fire Ants attacked me this morning," a usage that every Florida myrmecologist has encountered innumerable times. We did not create this species.

Amazingly, around 1970, when Buren was attempting to straighten out the taxonomy of the fire ants imported into the United States, he discovered that the commonest species, at that time the target of a faltering eradication campaign, seemed to lack any scientific name. Buren clearly had his assessment of the outcome of that eradication attempt in mind when he named the species *S. invicta*, "the unconquered fire ant." Later, when Trager (1991) completed the difficult task of revising the New World fire ants, a confusing group of variable species, it turned out that *invicta* had a forgotten name, *wagneri*, bestowed in 1916 (Bolton 1995). After a period of years in which taxonomic correctness struggled in vain against an onrushing flood of publications that continued to use the name *invicta*, a petition was presented to the arbiters of zoological nomenclature to preserve the name *invicta*, and permanently suppress the name *wagneri* (Shattuck et al. 1999). At the time the name *invicta* was reinstated as the permanent species name, there were more than 1800 papers already published using the name *invicta* (Shattuck et al. 1999).

Distribution: In South America, the original home of *invicta*, this species is widely distributed in an area that includes parts of Brazil and Bolivia, all to Paraguay and Uruguay, and part of Argentina (Tschinkel 2006). We do not know how much of this large territory was occupied by *invicta* before large-scale ecological disturbance that may have provided a larger area of preferred habitat. In North America, *invicta* occurs through the Southeast, west through much of Texas, sporadically in the Southwest; its distribution is limited by persistent freezing winter temperatures, by extreme dry conditions, and probably by cool summer temperatures (Tschinkel 2006). A general warming trend across central North America will not necessarily allow northern expansion of *invicta* because it is extreme rather than average cold temperature that determines distribution. The fantastic abundance of this species in the southeastern United States and Texas has guaranteed worldwide transportation, and there are many published notes of populations discovered here and there in warm climatic zones.

Natural History: At this point, I refer the reader to Tschinkel's highly readable, even entertaining, book on the biology of *invicta*. This book also includes an eye-witness cautionary account of the depressing, absurd, and counterproductive attempts at eradication and large-scale control. So great is my gratitude that I myself do not need to sift through the aforementioned 1800+ papers on *invicta* that I will keep discussion here extremely general and urge any Florida myrmecologist or general ecologist to run out and buy Tschinkel's book, *The Fire Ants* (2006).

The most basic fact that any Floridian biologist should know about *invicta* is that it is a weedy species. Like other weeds, it rapidly invades recently disturbed open habitats and quickly builds large populations that produce vast numbers of propagules that disperse to other recently disturbed habitats. Like many weeds, established populations of *invicta* can persist for generations as long as the habitat remains open and somewhat disturbed. When *invicta* began to increase exponentially through the millions of acres of disturbed habitat, mostly agricultural land, in the Southeast, there was a disastrous attempt to control and even eradicate the species. This effort, beginning in the late 1950s and petering out in the early 1980s, resulted in extreme ecological damage accompanied by increasing dominance of *invicta*. In the giant apparatus of the U.S. Department of Agriculture, and in all the state and local agricultural organizations, there was apparently almost nobody capable or willing to recognize a weed when they saw one. Enjoying a hyped-up scenario of an all-out assault on a deadly invader, the southern states blanketed hundreds of counties with wide-spectrum insecticides, not even bothering to carefully test the toxicity to vertebrates. Among all the agriculturalists, with or without advanced academic degrees, involved in this campaign, few were willing to state what every farmer knows: if you clear the land, weeds come in first. The story of *invicta* in the Southeast, stripped to its essence as I have done, seems like an inexplicable folly, but it is easier to understand in Tschinkel's more detailed telling, which includes the involvement of politics, propaganda, the blindness of denial, and the momentum of a government-sanctioned program blanketing counties and states with money as well as insecticides. Tschinkel also gives credit to researchers, such as Buren, who recognized the counterproductive nature of the campaign to eradicate or control *invicta*. Moreover, careful studies of *invicta* biology, which began belatedly, have continued to date. *Solenopsis invicta* is now the best known ant in North America, which means that it is currently possible to ask interesting and sophisticated questions about the species.

To act like a weed, a species of ant must not only efficiently colonize disturbed areas, which winged, fertile queens may do easily, but also rapidly occupy territory. This is a challenge for the many species of ants whose founding queen hides in a relatively secure chamber while raising a small number of small workers fed with excess eggs or with secretions produced by the queen herself. This means that the colony of a founding queen may not be able to occupy much territory until there have been at least three generations of workers. Demonstrating an evolutionary creativity that helps explain Tschinkel's respect for the species, *invicta* has met this challenge in three ways. When the number of colonizing queens is high, two or more often end up founding a colony together. This is not a long-term cooperation; one queen ultimately inherits the nest. Multiple founding on average increases the number and size of the first generation and the weight of the successful queen, and Tschinkel shows how the gamble of multiple colony founding can work out for the successful queen. From the population point of view, multiple queen colony founding under conditions of high density of potential founders might increase the number of successful nest founding attempts. The next stage of territory occupation is brought about by nest raids in which the first-born

miniature workers steal brood from a nearby nest and add them to their own nest. Nests with more workers are more successful raiders, and the final product is to consolidate many tiny colonies into a few large ones. Sometimes queens that have been abandoned by their workers can enter a colony of winning nest raiders and take over queenship. Putting aside the evolutionary forces that operate on individual ants to select for these complex behavioral patterns, there are consequences at the population level. Instead of many small, weak colonies, there are a few stronger colonies. The time it takes to build colony size has been greatly reduced, so colonies can go directly into production of large numbers of larger ants. Theoretically, this should also reduce the time before the first set of alate reproductives. A third way in which *invicta* changes its ability to occupy territory is through polygyny, a process in which animosity between queens disappears and workers likewise cooperate. *Solenopsis invicta* now occurs in a mosaic of single-queen and multiple-queen colonies. The mechanisms and consequences of polygyny fill 94 pages of Tschinkel's book. From an ecological perspective, this allows nests to fission, like those of army ants, so that there are no long and risky delays while colonies grow from tiny to large, and the polygyne form can move in a single front to occupy territory. Moreover, polygyny is accompanied by a shift from producing large numbers of alates to producing large numbers of workers, so polygyne colonies grow much faster. Tschinkel remarks that, "Polygyne colonies might be said to have carried weediness one step further than monogyne colonies by emphasizing 'vegetative' rather than sexual reproduction." One disadvantage encountered by the polygynous phase of *invicta* is that queens remain in the colony rather than flying, so the extended colony spreads on foot.

Solenopsis invicta is well adapted to seasonally flooded habitats, such as the flooded grasslands of the Pantanal in central South America. A generously watered suburban Florida lawn is like a welcome mat to *invicta*, a bit of the old country that would elicit tearful gratitude, if *invicta* were capable of emotions. As every Floridian eventually learns, as the water table begins to reach the surface of the ground, *invicta* builds up a loose mound of soil, often around a shrub or clump of grass. Such a mound often seems to be built of ants as much as of soil; larvae, pupae, young alates, and the queen are all up in this mound. This easily explains the apparent ferocity of invicta: the entire colony is now accessible to any animal or bird that eats ants, and there are many such. Vigilance and immediate, vigorous defense are survival skills of invicta.

Some native ants are also well adapted to seasonally flooded areas. These usually live in cavities among the roots of large bunch grasses. Around the seasonal ponds at the Archbold Biological Station, there is often a bare-sand area between the drier edges with shrubs and palmettos and the bunch grasses in the ponds that have grasses. These bare sand areas constitute a naturally disturbed area that is one of the few natural areas on the Station where invicta can be found. One of the best Florida habitats for invicta is wet pasture, a highly productive open area with an abundance of insect prey, where the bunch grasses that harbor native ants have been replaced with expanses of Bahia grass or other grasses with a turf growth habit. As a special treat, the cow pats produced by cattle are excellent hunting areas for larval beetles and flies; invicta can usually be found under every pat. Grazing and occasional fires keep the pasture in perfect condition for fire ants. Fires are usually during or at the end of the dry season, when the old grass burns easily and the fire ants are mostly underground in the dry soil. If we had set about designing a habitat specifically for invicta, we could not have done better than the wet pasture found on hundreds of thousands of Florida acres.

When we accidentally introduce a species into a new area where it spreads explosively, we like to think that we can atone by going back to its place of origin where benign nature has kept its populations in check with natural enemies. All we have to do is import some of these. From the beginning, prospects of biocontrol of invicta never seemed very probable, as this species is a common weedy species even where it is native. If it is less abundant than in parts of Florida, that might be caused by the much greater variety of potentially competing ants in its native South America. There are South American insect enemies of invicta, but we can hardly expect them to be more effective here than in their native environment, unless we subscribe to the theory that these enemies might themselves have enemies that we could leave back in South America. Nonetheless, considerable effort was made to import several species of tiny parasitic flies that attack invicta. If I had been asked to review the grant proposals for studying and importing these flies (fortunately for everyone, I was not), I would have written that I did not know that the project had good chances of success, but it was so interesting that it should definitely receive funding. These flies, several species of Pseudacteon in the family Phoridae, hover over foraging fire ants, occasionally darting down to insert an egg into the body of an ant. The microscopic larva emerging from the egg feeds for a time in the body of the ant and then moves into the head, which it hollows out completely. At this point, the head falls off the ant with the larva inside. The larva completes its development in the loose ant head and eventually emerges as an adult fly. There are several species of these flies that attack invicta, and a batch of others that are native to Florida, attacking other ants, including geminata and species of Pheidole. It was not expected that the specialized invicta decapitators would directly suppress populations of their host. The hope was that they would significantly disrupt foraging workers, as the presence of the flies hovering over them prompts the ants to go into hiding. This might, it was theorized, tip some kind of ecological balance, perhaps making invicta a less effective competitor with other ants. One or more of these species of flies can be found almost anywhere in Florida. They can be observed by stirring up a nest of invicta and waiting a few minutes, when the flies usually appear. Remember, one of these flies can fit as a pupa inside an ant head, so expect them to be extremely small. Florida invicta populations as yet show no signs of collapse, but if they are declining very slowly, it might not be noticeable. If geminata was actually the plague ant of Caribbean islands, there must be factors that can bring fire ants under control. These can be categorized as "biological or ecological resistance," a phrase that is conceptually impressive, specifically meaningless, and possibly true.

The long-term and short-term ecological effects of invicta have been the subject of numerous studies, those up to around 2005 reviewed by Tschinkel (2006) in his book. Many of these studies, especially those dealing with vertebrates, have flaws in either design or analysis, but some seem to stand up to the Tschinkel Test. Several studies of field crops showed that the presence of invicta was correlated with a decrease in the abundance of major crop pests, to the extent that use of insecticide for these pests was superfluous. At the least, it would be unnecessary to attempt to control fire ants in such a situation. This might be described as "set a weed to catch a weed." Studies showed that *invicta* can strongly reduce the flies (including the horn fly, a major irritant of cattle) breeding in cow dung. Even since 2005, however, changes in the drugs fed to cattle may make this study less applicable. One would expect *invicta* to have major effects on other ants, but these effects seem to be concentrated on *geminata* and *xyloni*, while diversity of other ants is unaffected or rebounds. This was tested experimentally by King and Tschinkel (2006), who removed *invicta* from pasture plots by flooding mounds with boiling water, a method that has little effect on other ants. The plots without *invicta* showed no increase in populations or diversity of other ants. The take-home message from most of Tschinkel's analyses of studies on ecological effects is that habitats change, namely, increased disturbance of habitats and greater divergence from their natural state, dwarf and, to some extent, mask the effects of *invicta*.

It might be useful to put the work summarized by Tschinkel and his own work into an evolutionary perspective. Everybody agrees that before *invicta* came to Florida, there was *geminata*. If it was imported, it must have been around for several centuries; if it was native, it must have been around for several millennia at least. Studies of *geminata* discussed above show that it is a weedy generalist that readily stings (some *Solenopsis* researchers prefer the word *envenomates*) vertebrates. Therefore, even though it is possible that *invicta* is more abundant than *geminata* used to be, native Florida species have been exposed to some level of fire ants for centuries or millennia. This may be one reason why it is difficult to show ecological effects of *invicta* independent of major increases in the amount and types of habitat disturbance. Moreover, adaptations for dealing with *geminata* might just need to be revved up a bit (in an evolutionary sense) to deal with *invicta*. In a more general context, Florida is an extraordinarily anty place. If it were possible to design a pervasiveness test for ants, Florida would probably win over other states, thanks to its warm climate and highly permeable soil. Every species of terrestrial animal in Florida must be able to co-occur with ants, lots of them. *Solenopsis invicta* is one among many.

Likewise, *invicta* itself has an evolutionary history. In the introduction to the fire ants (above), I speculated that some species of fire ants are adapted to take advantage of natural disturbances,

and this has serendipitously prepared them to take advantage of man-made habitat disturbance. Alternatively, these fire ant species might have evolved with humans to begin with, but this involves a more complicated scenario, as we already know that there are various types of frequently disturbed natural habitats. King and Tschinkel (2006, 2008) have shown that in most situations, *invicta* is not an autonomous invasive species. It is merely a member of the human entourage like many plants, appearing in the wake of man-made habitat disturbance. The situation is a little more complicated, however, because the natural disturbances where *invicta* first evolved may also occur in its new range. *Solenopsis invicta* occurs around the edges of seasonal ponds in the Apalachicola National Forest (Tschinkel 1987) and also at the Archbold Biological Station in Highlands County. There may be more extensive natural habitats, such as seasonally wet prairies, where *invicta* is also able to invade in the absence of man-made disturbance.

Solenopsis nickersoni Thompson— Elegant Thief Ant (Plate 44)

<u>Taxonomy and Similar Species:</u> This handsome, although tiny (approximately 1.3 mm long) species is easily recognized by the contrast between its glossy, deep brown body and its yellowish white legs and antennae. It is somewhat surprising that such a distinctive species remained unknown until 1982 (Thompson 1982, 1989), but its limited range, subterranean habits, and upland habitat make it somewhat difficult to collect. There are still no queens or males clearly associated with this species, although there are two specimens in the Archbold collection that I believe are queens of *nickersoni*. This lack of associated queens is unfortunate because, as Pacheco (2007) remarks, associated queens might definitively separate *nickersoni* from the similar Neotropical species *castor*. I have looked at specimens of *castor*, kindly supplied by William Mackay, and find some small differences between workers of *nickersoni* and *castor*: lateral clypeal teeth, found in *castor*, are lacking or very small in *nickersoni*; the postpetiole of *nickersoni* has a pair of elongate spatulate projections ventrally, these are longer than wide; the specimens of *castor* that I have seen are approximately one-eighth larger than any of the *nickersoni* I have seen from numerous collections; *castor* has more or less conspicuous irregular striae and sculpturing on

the lower third of the propodeum; the pronotum of *castor* is more convex than that of *nickersoni*; the clypeal profile of *nickersoni* is rounded and declivitous with a distinct acuminate apex, while that of *castor* is much less convex without a distinct angle before the apex. The antennal scape of *nickersoni* has long bristling hairs that stick up at a level of approximately 45°; these are also found in the possible queens of *nickersoni*. These may seem small differences, but they are as diagnostic as many of the character states that are used in Pacheco's 2007 revision to separate the species in the difficult *S. molesta* group to which *nickersoni* belongs. There are no Florida queens that I have seen in flight samples that look like queens of *castor*, but there are two queens from litter samples that are probably *nickersoni*. A supposed queen of *nickersoni* was described by Thompson (1989), but this queen was not associated with workers, and is almost certainly the queen of *S. tennesseensis*, judging by the dark wings and by Thompson's determination of a queen in the Archbold Biological Station collection.

Distribution: *Solenopsis nickersoni* is known only from Florida, although its occurrence in Leon County suggests that it probably occurs in sandy areas over the border in Georgia. It is known from sandy uplands as far south as Collier County, and into the northern Peninsula, but not, so far, from the western Panhandle, despite expeditions to appropriate habitats in Walton and Escambia counties.

Natural History: *Solenopsis nickersoni* is strongly associated with open sandy natural habitats, such as Florida scrub, sandhill, and flatwoods. Individuals may be seen foraging on the surface and can often be found in good numbers by sifting the roots of grass tussocks found in open areas. The best way to find this and other small ants associated with grass tussocks is to dig up the grass clump, bring it back to the laboratory, pull apart the clump, and sift out the sand with a 1/8-inch mesh. Ants can be extracted from the sand by use of a Tullgren funnel; the sand must be on a double thickness of cheesecloth to prevent the sand from falling through while allowing the tiny ants to burrow through the cheesecloth. While specimens are most easily found around grass roots, they can also be extracted from thin leaf litter in dry areas. *Solenopsis nickersoni* is apparently adapted to habitats that are maintained in an open condition by periodic fires and was probably one of the most common Florida ants when much of the Peninsula was open grassy natural habitats.

Name Derivation: "This species is named in honor of J. C. E. Nickerson, in recognition of his contributions to myrmecology" (Thompson 1982).

Solenopsis pergandei Forel—
Pergande's Thief Ant (Plate 45)

Taxonomy and Similar Species: *Solenopsis pergandei* could be confused with *Solenopsis tonsa*; both are pale yellow with heavy punctures and abundant hairs, and have minute eyes of only one or a few ommatidia, nonflattened heads that are convex below in lateral view, and both are larger (length, more than 1.5 mm) than other pale yellow Florida *Solenopsis*. *Solenopsis pergandei* is somewhat larger (length, approximately 2 mm) than *S. tonsa*, and the hairs on the dorsum of the pronotum are of obviously varying lengths, fine, and usually curved when seen in lateral view, whereas those of *S. tonsa* are short, even, thick, and straight. *Solenopsis pergandei* is much more frequently collected than *S. tonsa*, thanks to the conspicuous mounds made by *S. pergandei* when alates fly. Queens and males of *S. pergandei* are yellow with clear or slightly yellowish wings. Queens, which may vary considerably in size, have large eyes and ocelli; the distance between the lower edge of the eye and the ventral surface of the head is less than the length of the median ocellus.

Distribution: Virginia south through Florida, west into Louisiana (Smith 1979), Texas, and New Mexico (Pacheco 2007). The New Mexico record, from the Chihuahuan Desert, seems a little surprising, but no more so than the well-documented occurrence of *Temnothorax pergandei* in the southwestern deserts. Still, it would be useful confirmation to have alates from the Southwest.

Natural History: In Florida, I think of *S. pergandei* as a species of open sandy areas, but it is also reported nesting in normal soil and in rotting stumps (Smith 1931b). I seldom extract workers from leaf litter, and the occurrence of this species is revealed by the conspicuous mounds, honeycombed with exit holes, that are constructed in preparation for flights of alates. These flights occur before dawn in the morning the day after a rain. Workers and usually a few alates may be obtained from these mounds even after the sun has been up for an hour or two. Forel discovered *S. pergandei* during a collecting trip to Faison, North Carolina, in July of 1899, having been alerted to its presence by the elaborate mounds *S. pergandei* creates in preparation for mating flights (Forel 1901).

Mating flights and the construction of flight mounds were studied by Thompson (1980), who showed that flights occurred in the Gainesville area from June into September following rains, and flights from a single spot, presumably staged by a single colony, may occur several times during the summer. Thompson (1980) found that *S. pergandei* killed and consumed insect larvae in buried traps and made the reasonable assumption that this species is a predator of subterranean larvae. During his careful excavation of a colony of *Formica archboldi*, Schneirla found five colonies of *S. pergandei* surrounding the *Formica* colony near the surface (Schneirla 1944). The *Solenopsis* swarmed over and attacked the *Formica* when the two species came into contact. Queens and males of *S. pergandei* have been found parasitized by a large mermithid nematode (McInnes and Tschinkel 1996).

Name Derivation: Forel (1901) declared, "I dedicate this beautiful species to my tireless colleague, Monsieur Pergande" (my translation). Theodore Pergande, who emigrated from Germany as a young man, fought for four years in the Civil War, and subsequently spent many years working at the U.S. Department of Agriculture as an entomologist. He was a remarkable observer of insect natural history and an eager collector, famous for his finely prepared specimens. He sent many ants to Carlo Emery, and even named some species himself.

Solenopsis phoretica Davis and Deyrup— Florida Piggy-Back Fire Ant (Plate 43)

Taxonomy and Similar Species: This strange species, which has been collected a single time, may be recognized by its elongate mandibles with vestigial teeth except for an enlarged basal angle and the apical point (Davis and Deyrup 2006). The clypeus is concave and smooth. *Solenopsis phoretica* is similar to *Solenopsis enigmatica*, known from two queens and three workers collected on the island of Dominica in the West Indies (Deyrup and Prusak 2008). The two species differ in the shape of the mandibles: those of *enigmatica* are shorter and have a well-developed median tooth that creates a deep emargination between itself and the basal angle. These two species appear to form a distinctive lineage within *Solenopsis*.

Distribution: Known only from Gilchrist County, Florida.

Natural History: This ant is known from only a single queen, collected by Lloyd Davis in 1992. For many years, we waited for more specimens to show up, but finally published the description in 2006, hoping that this description might get others to keep an eye out for more specimens. This queen was attached to the petiole of a nest queen of *Pheidole dentata*. The jaws and clypeus of *phoretica* seem modified from the usual *Solenopsis* configuration to clamp effectively onto the petiole of the host, and the gaster was unusually flat ventrally, fitting snugly onto the gaster of the *Pheidole*. There seems to be little chance of spotting another *S. phoretica* riding about a the nest queen of *P. dentata* among the teeming workers of a disturbed nest, but at some point in the life cycle of the parasite, there should be a conspicuous batch of alates, or perhaps workers, which would be easily distinguished from the host. By a strange coincidence, in 2006, I, along with Zachary Prusak, found another similar *Solenopsis* associated with two colonies of *Pheidole antillana* in rotten wood on the island of Dominica in the Lesser Antilles. A few workers were associated with the queens. The queens were not attached to the queen *Pheidole*. It is possible that these two species of *Solenopsis* are temporary nest parasites of *Pheidole*, but this is not at all certain. It seems almost certain that the distribution of this peculiar group of *Solenopsis* is much more extensive than Dominica and northern Florida. Parasitic ants, however, are usually rare compared with their hosts, and thus difficult to find. In a study of another parasitic *Solenopsis*, *Solenopsis daguerrei*, only approximately one colony in 160 was parasitized (Calcaterra et al. 2000). While it is possible that species in the *phoretica*–*enigmatica* lineage of *Solenopsis* are widespread through the New World range of *Pheidole*, it is also possible that this group of inquilines is most likely successful in communities in which *Pheidole* is represented by a few relatively abundant species, such as on islands or near the edges of the range of *Pheidole* (Deyrup and Prusak 2008).

Name Derivation: From *phoretos* (Greek), meaning "carried," referring to the phoretic relationship between the holotype found by Lloyd Davis and a nest queen of *Pheidole dentata*.

Solenopsis picta Emery— Northern Twig-Nesting Thief Ant (Plate 44)

Taxonomy and Similar Species: This is a yellowish brown to dark brown, shining, tiny (length, approximately 1.5 mm) thief ant. *Solenopsis nickersoni* (a ground-nesting species) is somewhat similar but has contrasting pale appendages. *Solenopsis terricola* (a tropical species nesting in the ground or in fallen rotten wood or twigs) is similar in size and coloration. The two species are compared below morphologically under *S. terricola*, but can usually be separated by range and habitat. There are at least three arboreal species with sparse, long, fine hairs that arise from inconspicuous punctures on the head: *picta*, *corticalis* (apparently not in Florida, see *zeteki* below), and *zeteki*. *Solenopsis picta* may be recognized by its darker coloration. Queens, which may be collected in flight traps, are dark brown with fine punctures on the head. They are similar to queens of *terricola*, but have smaller eyes: the eye of *picta* in lateral view is hardly more than half the length of the distance between the margin of the eye and the vertex of the head, whereas the eye of *terricola* is about two-thirds the length of the distance between the eye and the vertex.

Distribution: North Carolina (Guénard et al. 2012) south through Florida, west around the Gulf of Mexico into Texas (Smith 1979). In adjacent tropical regions, *S. picta* appears to be replaced in dead twigs and branches by *Solenopsis corticalis* and *Solenopsis zeteki*.

Natural History: *Solenopsis picta* inhabits dead twigs on living trees. In northern Florida, it dependably occurs in dead twigs and under the outer bark of *Pinus palustris* and *P. elliottii* (Tschinkel and Hess 1999). In one study, *S. picta* was approximately 12% of the biomass in the diet of the endangered red-cockaded woodpecker (Hess and James 1998). If one were on a mission to find this species, the best places to look would be in small dead twigs part way up the trunks of open-grown young slash or longleaf pines. In general, however, *S. picta* is not particular about the host plant species, and I have found colonies in twigs of *Carya*, *Fraxinus*, *Zanthoxylon*, *Gleditsia*, *Vitis*, and even in dead stems of herbs. The main requirement is that the twigs be small, dry, at least partially exposed, and accessible for foraging on leaves and branches. Colonies are polygynous (Thompson 1980).

Name Derivation: From *picta* (Latin), meaning "colored," probably referring to the dark color of this species, in contrast to the light coloration of small European *Solenopsis* species, such as *S. fugax*, with which Carlo Emery was familiar.

Solenopsis tennesseensis M. R. Smith—
Least Southeastern Thief Ant (Plate 45)

Taxonomy and Similar Species: *Solenopsis tennesseensis* is Florida's smallest native ant; its workers are approximately 1.2 mm long. Specimens extracted from leaf litter into alcohol look even smaller, as they curl up into a tiny yellowish white ball. The head is flattened with conspicuous punctures, and the eyes are reduced to a single ommatidium. The dorsum of the pronotum, seen from the side, appears to have a dense layer of short, erect hairs, with a few longer ones along the side. *Solenopsis tennesseensis* is most similar to *Solenopsis* species A, discussed below, but that species has fine punctures on the head and the dorsum of the pronotum in lateral view shows a sparse scattering of short and longer hairs. Workers of species A do not curl up in alcohol. There has been some confusion about the queen of *S. tennesseensis*: the description of the queen of *S. nickersoni* in Thompson's 1989 review of Florida species apparently refers to *S. tennesseensis*. It is difficult to get associations between workers and alates of most subterranean species of *Solenopsis* (*S. pergandei* is a welcome exception). I have associated workers and alates of this species only twice, even though this is a very common ant. The queens are surprisingly large (approximately 3.5 mm) to be associated with such tiny workers and are blackish brown with strongly tinted brown wings. The crest of the petiole in a posterior view is broadened and slightly bilobed. Two unassociated queens in the Archbold Biological Station collection identified by Thompson as *S. nickersoni* look like queens associated with *S. tennesseensis*. The queen of species A is unusually small, approximately 2 mm long and brownish yellow.

There appear to be a number of similar, minute species with flattened heads and greatly reduced eyes (Pacheco and Mackay 2013); not all of these have associated queens, and there are probably additional species to be found in this group. One lookalike species is *Solenopsis subterranea*, which occurs west of the Mississippi from Louisiana and Texas south through Texas, Central America, and northern South America (Pacheco and Mackay 2013). I have examined a few paratypes of this species and am unable to distinguish them from *S. tennesseensis*, even with the help of the Pacheco's and Mackay's key. I mention this not to cast doubt on the validity of *S. subterranea*, but as a confession that *S. subterranea* could be in Florida and I wouldn't know it. It is possible, however, that alates of *S. subterranea* (assuming this is a valid species) are distinctive and should appear in Florida flight traps or litter as an unknown species. I have not seen queen thief ants that do not correspond to known workers, except for confusing situations like the *texana/carolinensis* tangle.

Distribution: North Carolina (Guénard et al. 2012) and Tennessee, south through Florida and west into Texas and south into Mexico (Pacheco and Mackay 2013). *Solenopsis tennesseensis* occurs throughout Florida.

Natural History: *Solenopsis tennesseensis* can be found in an extraordinary range of habitats, from tropical hammocks in the Florida Keys to north Florida forests and fields, from swamp forests and low flatwoods to Florida scrub and sandhills. I have even collected this species in urban and suburban sites. Unlike the other subterranean species *S. tonsa*, *S. tennesseensis* forages readily in surface leaf litter and is frequently extracted from litter samples, usually along with one or more other species of small *Solenopsis*. Alates fly from early May into September (51 records) and are most likely to be found in July (30 records).

While the toxic chemicals known as pyrrolizidine alkaloids are known mainly from plants, one such substance is produced by *S. tennesseensis* (Jones et al. 1980). Pyrrolizidine alkaloids are known to be repellent to ants (Hare and Eisner 1993) as well as to other arthropods, including spiders (Eisner 2003). This may help explain how the minute *S. tennesseensis* holds its own in a great variety of habitats swarming with larger ants.

Name Derivation: Marion Smith named this species from specimens collected in Hamilton County, Tennessee (Smith 1943b). The original name was *longiceps*, changed to *tennesseensis* because *longiceps* had already been bestowed by Forel on a species from Tunisia (Bolton 1995).

Solenopsis terricola Menozzi—
Dark Dead-Wood Thief Ant (Plate 44)

Taxonomy and Similar Species: This is a dark brown, shining, tiny (length, approximately 1.5 mm or less) thief ant. *Solenopsis nickersoni* is similar, but has contrasting pale appendages. The most similar Florida species is *S. picta*. Workers are distinguished as follows: *terricola*: eyes with four to five ommatidia in a round or somewhat square arrangement; hairs on front of head arising from obvious punctures; faint to conspicuous irregular sculpturing on the lower mesopleuron (area above the middle coxae); petiole in profile rounded-truncate above, not narrowly rounded; *picta*: eye conspicuously longer than high, almond-shaped,

with three to four ommatidia; hairs on front of head arising from almost invisible punctures; no sculpturing on lower mesopleuron; petiole in profile narrowly rounded above. Queens: *terricola*: ocelli large, median ocellus almost as wide as the distance separating it from the lateral ocelli; coarse striae extending from the frontal area adjacent to the antennal insertions about halfway to the level of the median ocellus; *picta*: ocelli small, median ocellus little more than half as wide as the distance separating it from the lateral ocelli; very fine striae extending only a short distance posteriorly from the level of the antennal insertions. It is unlikely that there will be many instances of difficulty in distinguishing between these two species, as *terricola* is found on the forest floor and appears to be tropical, whereas *picta* is arboreal, living in exposed twigs and branches, and has a more northern range; it has not so far been found in tropical areas of Florida.

The identification of the Florida specimens is based on comparisons of Florida specimens with Costa Rican specimens compared with type material by William Mackay, compared to photos and the description in the online version of *Ants of Costa Rica* (Longino 2005), and by use of the key and description in Pacheco's doctoral dissertation (2007). Large numbers of specimens, including queens, were collected from rotten wood in a dense tropical hammock adjacent to the Matheson Hammock parking lot on Old Cutler Road, South Miami. This is the first report of this species from North America.

Distribution: Central America and the Lesser Antilles (Pacheco and Mackay 2013), introduced into Florida.

Natural History: In the Matheson Hammock, *terricola* is a common species in rotten wood and in dead twigs on the ground in dense tropical forest. It can also be extracted from leaf litter samples. In rotten wood, colonies appear to be concentrated in cavities left by beetle larvae. This species is probably not polygynous; in rotten wood, there are many pockets of workers and larvae with no queens, and I have never seen two queens in one chamber. Considering its abundance in the Matheson Hammock, it is surprising that this species has not been found elsewhere in tropical Florida. It might be a recent import, or it might be a species that spreads slowly. I have collected winged queens, so it should be able to disperse relatively easily. It appears to be absent from the Florida Keys, where there have been several ant surveys (Moreau et al. 2014), either because it has yet to arrive or because the Keys have a prolonged dry season. In Dominica, I found *terricola* abundant in a wet forest, but did not find it in a dry forest in the rain shadow portion of the island. In a study of limiting factors for ants at a site in Panama, *S. terricola* did not increase in abundance in response to supplements of sugar, protein, or both (Kaspari et al. 2012).

Name Derivation: From *terra* (Latin), meaning "earth," and *-cola* (Latin), meaning "dweller." Presumably, the type specimens were found either in soil or on the ground, although this is not reported in the description of the species.

Solenopsis tonsa Thompson—
Thompson's Subterranean Thief Ant (Plate 45)

Taxonomy and Similar Species: *Solenopsis tonsa* is one of four subterranean Florida species with tiny eyes often reduced to a single facet. Two of these species, *S. tonsa* and *S. pergandei*, are slightly larger species (length, more than 1.5 mm) that do not have flattened heads: the head is notably convex below in lateral view. The other two, *S. tennesseensis* and *S.* species A, are 1.5 mm long or less and the head appears flattened in lateral view. *Solenopsis tonsa* is, as Thompson (1989) notes, remarkable for its dense, short, stiff-looking hairs, with those on the dorsal surface of the mesosoma looking in a lateral view as if they had been evenly mown off. The mesosomal hairs of *S. pergandei*, in contrast, are sparse and of varying lengths; *S. pergandei* is also larger, approximately 2 mm long, while *S. tonsa* is approximately 1.5 mm long. Queens of *S. tonsa* are not likely to be confused with those of other Florida *Solenopsis*: they are large, dark brown (*S. pergandei* has yellow queens), with slightly milky wings; the petiole is widened and slightly bilobed above in posterior view, with a strong thin keel on the underside of the petiole.

Distribution: This species is known from a few sites in northern and central Florida: Alachua, Leon, Orange, and Levy counties. It must also occur in southern Georgia, as one of the sites in Leon County is only about 3 km south of the Georgia border. Workers from eastern Texas (Rusk County) were also identified as *S. tonsa* (Pacheco 2007). I have never collected queens in flight traps in south-central Florida.

Natural History: *Solenopsis tonsa* workers appear to be completely subterranean in their habits, and if Catherine Thompson had not been using buried perforated vials to bait for this species (Thompson 1980), we still might not have seen any workers. Most specimens were collected in open, sandy habitat with some grassy vegetation. I once found queens and males associated with a few workers by sifting the roots of wiregrass (*Aristida stricta*). Sandhill appears to be the natural habitat of this species. I also found colonies at the edge of a lawn at Tall Timbers Research Station. Queens and males have been found in July.

Name Derivation: "The word 'tonsa' is Latin for 'shaven' and is dubbed upon this species because of the distinctive hairless vertical stripe down the center of the worker's head" (Thompson 1989). There are other Florida species of *Solenopsis* with a hairless stripe, but it does not always show up as well because the hair-bearing punctures to the side of this stripe are not as large and dense.

Solenopsis validiuscula Emery

Taxonomy and Similar Species: *Solenopsis validiuscula* is a relatively large (2–2.5 mm long) brown species that has long been known as *Solenopsis truncorum* (Pacheco and Mackay 2013). The latter name, however, was bestowed by Forel (1901) as a subspecies name on a form of *S. carolinensis* described as pale yellow and small (approximately 1.5 mm). The confusion apparently arose because the paratype selected for *truncorum* by Wheeler was a different species, presumably accidentally included in Forel's type series that he collected in North Carolina. It would not be so strange to base a species concept on a larger, darker individual when dealing with a genus such as *Solenopsis*, which sometimes has majors that are especially distinctive. The *carolinensis/texana* complex, however, has no such majors. Forel himself did not apparently choose a type specimen, but it is clear from the description that he was dealing with a small yellow species, and it seems reasonable to make *truncorum* a synonym of *carolinensis* (Pacheco and Mackay 2013).

All this, however, does not really mean that the largish dark specimens identified as *truncorum* in North Carolina (Forel 1901; Guénard et al. 2012) and Florida (Thompson and Johnson 1989) are the same as some largish dark thief ants found in western North America, identified as *validiuscula*. According to the revision (Pacheco and Mackay 2013), there is a huge gap, including all the Gulf States and the entire Midwest, between western populations of *validiuscula* and the specimens reported from the Southeast. Southeastern records are not included in the revision of thief ants (Pacheco and Mackay 2013). It is quite possible that the southeastern population that has gone under the name of *truncorum* is undescribed. This ant, however, is apparently rare, even in North Carolina (Guénard et al. 2012), and, to make matters worse for the Florida ant survey, the Florida specimens reported by Thompson and Johnson (1989) have disappeared. Southeastern myrmecologists should be on the lookout for this distinctive species, easily distinguished from similar *Solenopsis* by its larger size and dark color. In the field, it might be confused with the smallest individuals of *S. geminata* and *invicta*, but in the laboratory, the large eyes of *geminata* and *invicta* will give them away. I have not illustrated this species because I have never seen any specimens.

Distribution: North Carolina and Florida, widespread in the Southwest if this is the same species as *validiuscula*. Unless this species was imported from western North America, which seems unlikely, it should occur in several southeastern states.

Natural History: The Florida specimens are from pine and hardwood litter along the edge of a field in Alachua County, two to three miles southwest of Archer (Thompson and Johnson 1989). Western *validiuscula* is found in a variety of habitats that do not occur in Florida, but a number of records suggest associations with other ants, usually *Lasius* or *Camponotus* (Pacheco and Mackay 2013). It is possible that this species can be overlooked because it is found with larger and much more active species of ants.

Name Derivation: From *validus* (Latin), meaning "strong" or "robust," and the diminutive *iuscula*. This species was described as a more robust and darker form of *molesta*, so maybe the meaning is "bigger little *molesta*."

Solenopsis xyloni McCook— Small-Headed Native Fire Ant

Taxonomy and Similar Species: *Solenopsis xyloni* strongly resembles *S. invicta* in general appearance, although some populations of *xyloni* are much lighter in color than North American populations of *invicta*. Under the microscope, *xyloni* workers lack the median tooth on the lower margin of the clypeus found in *invicta*, and on the underside of the petiole of *xyloni*, there is a prominent flange, often with an anterior tooth or expansion; this is lacking in *invicta*. The head of larger workers of *xyloni* and *invicta* is not disproportionately large, unlike that of *geminata*. *Solenopsis xyloni*, which was reported from extreme northwestern Florida (Smith 1965), was almost certainly extirpated in Florida by *invicta* (Trager 1991), but persists in the Southwest and also in some parts of the Southeast just north of the range of *invicta* (Trager 1991). There is probably no need for Florida biologists to concern themselves with *xyloni* or to check specimens of *invicta* in case they are really *xyloni*, considering the apparent rare and local distribution of *xyloni* in Florida even before the advent of *invicta*. Smith (1965) provided *xyloni* with the English name of "southern fire ant," but as all eastern fire ants are southern, this name is not very useful.

Distribution: The current distribution of *S. xyloni* apparently does not include Florida, but does include parts of North and South Carolina and Georgia, west through lowland Tennessee all the way to the Pacific Coast in California, and south into Mexico (Trager 1991). It appears, however, that the eastern distribution is sporadic, perhaps based largely on historical records predating the conquest of the Southeast by *S. invicta*. On the other hand, maybe it still persists, overlooked in the throngs of *invicta*. *Solenopsis geminata* seems uncommon through much of its range in Florida, and if it were as difficult to distinguish in the field as *xyloni*, we might well conclude that it had been displaced completely by *invicta*. I have not illustrated this species because it probably does not occur in Florida.

Natural History: The natural history of *S. xyloni* was studied by Smith, mostly in the 1930s, before *invicta* was widespread in the Southeast (1965). This species usually nests in the ground, either in the open or under objects, but may also nest in dead wood. The objectionable features of *xyloni* are mostly like those of *invicta*: rushing out to attack interfering humans, entering homes, making mounds in lawns, eating recently sown seeds, girdling nursery stock, protecting sap-sucking insects, attacking hatchling birds, and moving into telephone equipment. Smith's account of *xyloni* gives the impression that it is even more generalized than *invicta*, moving into houses to obtain all kinds of human food, especially foods with high protein content, attacking fruits and vegetables in the field, and biting holes "in various fabrics such as woolens, silks, linen, and nylon." It is possible, however, that some of these activities are based on a few anecdotes and are not characteristic.

Buren (1972) made the interesting suggestion that xyloni might be a relatively recent invader in the Southeast, arriving from the west to colonize heavily disturbed fields that replaced more natural habitats, only to be displaced by invicta. Buren explains that xyloni is "a much less vigorous species" than invicta, but it might be more logical to speculate that if xyloni had arrived recently from the west, it was probably adapted to a dry climate, whereas invicta might be better adapted to the wetter southeastern climate.

Name Derivation: From xylon (Greek), meaning "wood." One might assume from this that McCook's original specimens were from dead wood, but he specifically states that "Solenopsis is a mining ant, and lives in nests made in the ground" (McCook 1879). McCook described S. xyloni in the course of his studies of the natural enemies of the cotton leafworm, Alabama argillacea, whose larvae are sometimes attacked by xyloni. This must be the only time that a species of fire ant was described so that it might be given proper credit for beneficial behavior.

Solenopsis zeteki Wheeler— Zetek's Thief Ant (Plate 46)

Taxonomy and Similar Species: Solenopsis zeteki is a yellow arboreal species. The long, fine hairs on the head emerge from almost invisible punctures, in contrast to species such as S. abdita, which has many short hairs (as well as some long ones) that emerge from readily visible punctures. The eyes of the worker are usually slightly elongated, like those of S. picta. The best way to identify this species, however, is by the unusually large eyes and ocelli of the queen. The eyes are oval, and extend down toward the ventral side of the head, so that the area between the eye and the ventral surface of the head is narrower than the width of one of the ocelli.

For many years, I have mistakenly listed Florida specimens of this species as S. corticalis, for example, in my list of Florida ants published in 2003. The first mention of the occurrence of S. corticalis in Florida is in Catherine Thompson's PhD dissertation on the thief ants of Florida (1980). This work and the 1982 publication derived from it do not include a description of the queen; it was not until Snelling described S. torresi from Puerto Rico (2001) that it became apparent that there were two similar yellow arboreal species of Solenopsis in the Caribbean area. Pacheco and Mackay's monograph on New World thief ants (2013) clearly differentiates between the queens of corticalis and those of zeteki, and lists torresi as a synonym of the latter. Both zeteki and corticalis could have been imported into Florida, but I have not seen specimens of corticalis. The workers of the two species are apparently so similar that it would be best to assume all specimens are zeteki unless they are accompanied by a queen that is clearly some other species, such as corticalis. In the Archbold Biological Station collection, there are several series from the Bahamas that might be the true corticalis, on the basis of the relatively small, round eyes of the queen. Fortunately, this species is often highly polygynous, so queens are relatively easy to find.

Distribution: Southern Mexico, through Central America into Colombia, Puerto Rico (Pacheco 2007); Florida. In Florida, this species is known from Miami-Dade and Monroe counties.

Natural History: Nests are in twigs and plant cavities, including cavities in "ant plants," plants that grow chambers for harboring ants. The type series was in such a cavity in Costa Rica (Wheeler 1942). In Florida, almost all nests that have been found were in red mangrove twigs. In Costa Rica, colonies are found not only in hollow twigs and plant cavities, but also under mats of epiphytes (Longino 2005). In Costa Rica, this species occurs in both wet and dry forests, and even in city parks (Longino 2005). In Costa Rica, S. zeteki has been found in dry forest (Snelling 2001). In general, S. zeteki seems to be a widespread, generalist, even weedy species. It is always possible, however, that S. zeteki is really a complex of two or more similar species with differing habits.

Name Derivation: This species was named for Dr. James Zetek, the founding director of the Barro Colorado Island Laboratory. Zetek assisted Wheeler in his studies of ants living in ant plants, including harvesting the ant domatia from the field, putting them in jars where the ants were asphyxiated with chloroform, and dissecting out and photographing the nests (Wheeler 1942). The description of S. zeteki was published in 1942, after Wheeler's death in 1937, but Zetek lived until 1959, researching various insect pests in Panama.

Solenopsis Species A (Plate 46)

Taxonomy and Similar Species: This unidentified species is apparently a member of the S. pygmaea species complex, in which Pacheco and Mackay (2013) place 13 species. Most of these, like the Florida Species A, are unusually small (approximately 1.2 mm long, except *tonsa*) and pale yellow, with tiny eyes that may lack pigment, and have elongated heads (except *tonsa*) and well-developed clypeal teeth, with the postpetiole almost circular viewed from above. Members of the *pygmaea* group are also characterized by having coarse punctures on the upper surface of the head. Species A has fine and inconspicuous punctures, distinguishing it from other members of the *pygmaea* group, including the Florida species *tennesseensis* and the Neotropical species *minutissima*. In Berlese funnel samples collected into isopropyl alcohol, *tennesseensis* always curls up into a ball, usually retracting the antennae as well; Species A does not curl up, and its antennal scapes usually stick out from the sides of the head. This is not a species that has been overlooked because it looks like *tennesseensis* or any other Florida *Solenopsis*. Queens of Species A are remarkably small, approximately 2 mm long, brownish yellow, with a heavily punctate head. Actually, the tiny size of the queen of Species A, while notable in the Florida fauna, might not be unusual in the *pygmaea* group as a whole, as queens are known for only 4 of the 13 species. One of these, *minutissima*, has a tiny queen, although its size is not recorded (Pacheco and Mackay 2013). Aside from its tiny size, the queen of Species A is remarkable (in my experience) in having the back of the head concave, with concavity beginning immediately behind the lateral ocelli, such that, in a frontal view, the head ends just beyond the level of the lateral ocelli.

Distribution: This species is known from Matheson Hammock County Park in south Miami. It has not been found during intensive collecting trips to the Florida Keys and is possibly confined to south Miami. There is no reason to believe that this species is a recent import that is likely to spread through Florida.

The native range of this *Solenopsis* is unknown at this point, but this species was almost certainly brought to Florida from some locality in the Neotropics. The Matheson Hammock is adjacent to Fairchild Tropical Gardens, in an area that received shipments of plants from all over the world, beginning more than a century ago (Fairchild 1939). Although a local plant quarantine station was established before 1910 (Fairchild 1939), it is probable that there was more concern about insect pests on foliage and stems than about soil arthropods. Moreover, Fairchild complains that the quarantine greenhouses were small and overcrowded, and it would not surprise me if many commercial growers of tropical plants decided to quietly omit the quarantine process.

Natural History: *Solenopsis* Species A has been found in thin leaf litter in dense tropical hammock. It is probably polygynous, as several queens were found in each of two small samples. It occurs together with *terricola*, *tennesseensis*, and *Solenopsis* Species B.

Solenopsis Species B (Plate 46)

Taxonomy and Similar Species: This species is a member of Pacheco and Mackay's *molesta* group. I have collected this species on several Caribbean islands and now in Florida. It is surprising, therefore, that it does not appear to be in the 2013 revision, but not many people collect this group of ants. The tiny, dark queen, 2.0 to 2.2 mm long, is distinctive and separates this species from a long list of species with larger queens. *Solenopsis* Species B is polygynous and queens are easily obtained. The worker is approximately 1.5 mm long, orange yellow, and translucent, with submedian and lateral clypeal teeth and with hairs on the top of the head curved and emerging from inconspicuous punctures. The worker is similar to those of *texana*, *carolinensis*, and *abdita*, but more orange, and with a thicker petiolar node; in lateral view, the petiolar node is wide and the anterior and posterior slopes are similar, without the steeper posterior slope found in the other three species.

Distribution: In Florida, this species is known from Matheson Hammock in south Miami, from Brooksville (Hernando County), and from Gainesville (Alachua County). I collected the Brooksville specimens in 1990, without associated workers, and did not see this species again until 2014. The Brooksville specimens were from the now abandoned Chinsegut agricultural experiment station, a project of David Fairchild and the U.S. Department of Agriculture. By 1911, large numbers of plants from the Miami experiment station had been established at Chinsegut (Fairchild 1939). *Solenopsis* species B may have been brought to Brooksville around this time. The Gainesville population is in a bamboo thicket at the Kanapaha Botanical Garden. The Kanapaha bamboo collection was founded many years ago from specimens from the Fairchild Tropical Gardens. The connection between Fairchild and this species of *Solenopsis* seems reasonably clear. In any case, *Solenopsis*

Species B is probably not a recently arrived species but one that was in Florida long before its first collection in 1990. I have collected specimens that appear to be the same species (with queens) in Dominica, the Dominican Republic, and the Bahamas. This is almost certainly an exotic species in Florida, originating somewhere in the Neotropics.

Natural History: *Solenopsis* Species B nests on the ground in small beetle borings in rotten wood and in small hollow twigs. Most Florida specimens were found in dense tropical hammocks, but one collection was from under a sea grape (*Coccoloba uvifera*) above the mangrove zone in the Matheson Hammock Marina. The Brooksville specimens were from dense plantings of cultivated trees. Caribbean specimens believed to be the same species were from forested areas except for a Bahamian series from a grass tussock. This species is polygynous; two to five queens were found in the same chamber.

Genus *Stenamma*

Retiring Ground Ants

The genus *Stenamma* (pronounced variously **Sten" am muh** and **Sten am" muh**) includes 45 described species (DuBois 1998). *Stenamma* is best represented in North America, temperate Eurasia, and North Africa, but there are a few tropical species. There are probably some undiscovered species, as these ants are easily overlooked. They are retiring, nondominant ants, usually seen as solitary foragers. In the field, they often resemble some other commoner ant, such as species of *Tetramorium* or *Solenopsis*. Many species appear to be confined to relatively small geographic ranges, and almost all require relatively undisturbed, usually forested habitats (DuBois 1998). DuBois (1998) warns that many species in the genus may be endangered, and some might even be extinct.

Species of *Stenamma* appear to be predators of small invertebrates found on the ground or in leaf litter and do not appear to tend honeydew-producing insects (Smith 1957). I have found northern species accepting cookie crumbs, but a little less readily than crumb-seizing *Myrmica* species found in the same habitat. They do not, in my experience, recruit to large crumbs. Nest entrances are usually extremely inconspicuous

(Smith 1957). Some species are restricted to cool, moist sites, and species in more open sites may forage primarily in cool weather (Smith 1957).

There is a review of Nearctic species by Smith (1957) and of western North American species by Snelling (1973), but even these highly competent myrmecologists were unable to make identification easy. Part of the problem is the frustrating variation within some species. Only one species is known from Florida, but it would not amaze me if an additional species, such as *meridionale*, turned up slowly traversing the leaf litter of some north Florida ravine on a cool December day.

Name Derivation: From *stenos* (Greek), meaning "narrow," and *hamma* (Greek), meaning "knot, "probably referring to the narrow, elongated petiole when viewed from above. This is a useful generic field mark when one of these ants is seen foraging or walking about the sifting tray.

Stenamma foveolocephalum M. R. Smith— Sand-Loving Retiring Ground Ant (Plate 21)

Taxonomy and Similar Species: Easily distinguished by the combination of short propodeal teeth, foveolae (conspicuous pits) on the sides of the head, and lack of antennal scrobes (grooves on the head that receive the antennae). In the field, *foveolocephalum* resembles a midsized, somewhat pale *Solenopsis invicta*. The species *carolinensis* was synonymized with *foveolocephalum* by DuBois and Davis (1998).

Distribution: Known from North Carolina, Florida, Alabama, and Mississippi (DuBois and Davis 1998). The only Florida records are from Walton County.

Natural History: All collections of *foveolocephalum* have been from sparsely wooded or open sandy areas (DuBois and Davis 1998). Foraging may be restricted to winter, as all collections have been in January or February, and *foveolocephalum* has not been found in some of the same sites during the summer (DuBois and Davis 1998). It is possible that *foveolocephalum* is more widespread in the Southeast, but overlooked because of its winter seasonality and resemblance to *Solenopsis invicta*, but it is also possible that this is a genuinely rare ant. It has been found less than 10 times (DuBois and Davis 1998).

Name Derivation: From *foveola* (Latin), meaning "little pit," and *cephalo* (Greek), meaning "head," referring to the pits on the sides of the head.

Genus *Strumigenys*

Mustache Ants and Pygmy Snapping Ants

I am, of course, too scientific and objective to choose a favorite genus of ants, but were I less virtuous, I would unhesitatingly choose the genus *Strumigenys* (pronounced **Strew mij" en is**). Amazingly ornate and strikingly diverse, these tiny ants are as remarkable as they are mysterious. They are especially notable for their variety of mandibular shapes and teeth, and many species sport peculiarly modified hairs on the head, especially on the clypeus (hence the name "mustache ants" applied to most of these species). In the species *pulchella*, for example, the clypeus is flat, wide, semicircular, and scalloped along the edge; along each side there are four large, equally spaced, spoon-shaped, silvery hairs that curve gracefully forward, while two apical enlarged hairs curve backward. All individuals in this widespread Eastern and Midwestern species show these precise clypeal features. *Strumigenys ornata*, in contrast, has an elongate, pointed clypeus, turned up at the tip, from which spring eight long, slender hairs with expanded ends. Many species have a large, triangular tooth at the base of each jaw, followed by a toothless gap, and then a series of sharp, slender teeth, sometimes with long and short teeth alternating. Some species lack the toothless gap, and some species have elongate mandibles with teeth near the end. All these kinds of elaborations must reflect the various ways in which the species make a living, but we have little idea of their precise function. The species with long mandibles are here labeled "pygmy snapping ants."

Ants of the genus *Strumigenys* belong to the tribe Dacetini, small ants that usually live in leaf litter, capturing their prey with an abrupt snap of their relatively large mandibles. Over a long period, the definitions of dacetine genera, which were often based on specialized features of the mandibles, gradually became blurred and phylogenetically suspect. Dacetine genera were redefined by Bolton in 1999, resulting in numerous genus-level changes in the Florida fauna. A somewhat more drastic revision of the Dacetini (Baroni Urbani and De Andrade 2007) reassigned all Florida dacetines to *Strumigenys*. Thanks to these revisions, *Strumigenys* has swallowed up a couple of dozen genera and includes approximately 350 species (Baroni Urbani and De Andrade 2007). This must be one of the largest simultaneous formal conglomerations of genera in zoological history. A session flipping through portraits of these ants suggests that this is the most morphologically diverse of all ant genera. One can only sympathize with the decades-long effort to assign an ever-increasing array of forms to various genera, sometimes erecting new genera for especially bizarre species or groups of species. Eventually, this nomenclatorial extravaganza began to collapse, partly because species that linked various genera were found. More importantly, it became obvious that there has been such dramatic radiation within numerous lineages that any attempt to diagram relationships in a phylogenetic tree would probably result in a phylogenetic thicket. In the long run, genetic analyses may be able to disentangle certain ancient divergences within the evolutionary brushwood, and *Strumigenys* will be redivided into several genera.

An extravagantly heterogeneous genus is a bit of an oxymoron but delivers the myrmecologist from the task of memorizing the character states of dozens of genera. It is still possible to discuss *Strumigenys* with some specificity because the genus has been informally broken into batches of "species groups," such as the "*pulchella* species group." This allows a kind of fluid hypothesizing about lineages, and spares us a bunch of formal subgenera that would probably succumb to the same problems that afflicted the subsumed genera. In any case, species are much more interesting than genera, because genera are artificial compilations, while species are real, in the sense that they represent actual populations.

With 36 species, including 10 introduced species, *Strumigenys* is, by a good margin, the largest genus of Florida ants. The native Florida *Strumigenys* are part of a larger fauna of at least 35 species that are largely confined to mesic forests of eastern and Midwestern North America. The fauna of southwestern North America is depauperate, with only four species, two of which are known from single collections from isolated moist areas (Ward 1988). Brown (1953) suggested that the southeastern fauna of *Strumigenys* is most closely related to that of warm-temperate eastern Asia, rather than to the well-developed Neotropical fauna. This is possibly apparent in Asian members of the "*rostrata* complex" that seem to share character states with the Palearctic species *Strumigenys rostrataeformis* and *Strumigenys incerta* from Japan and

Strumigenys emeswangi from China. Japanese species of the *circothrix* group, *Strumigenys circothrix* and *Strumigenys hiroshimensis*, show similarities to the North American *pergandei* group (Bolton 2000) and have at least a superficial resemblance to some Nearctic members of the *pulchella* group, such as *S. abdita*. Photographs of the Japanese *Strumigenys masukoi* resemble some species in the Nearctic *clypeata* group. Images of these Japanese ants may currently be viewed on the Internet on the Japanese Ant Image Database organized by Hirotami Imae (Imae 2003). The distribution of native southeastern *Strumigenys* is similar to that of warm-temperate Arctotertiary flora discussed by Raven and Axelrod (1978). The southeastern *Strumigenys* species represent the only extensive radiation of woodland ants in North America, and the only obvious assemblage of Arctotertiary ants (Deyrup and Cover 2009). Future analysis may show that there are other groups of ants that fit this pattern, perhaps in the genus *Proceratium* and in species groups of *Aphaenogaster* and *Camponotus*. There is, however, no such group that begins to match the diversity of *Strumigenys*. In *Strumigenys*, we get a last glimpse of what must have been a fantastically rich ant fauna on the floor of the Arctotertiary forest (Deyrup and Cover 2009).

The remnant diversity we see today probably survived because these ants usually occur as small colonies of slow-moving individuals feeding on ubiquitous micro-arthropods in the shelter of moist leaf litter. These traits should have allowed populations of *Strumigenys* species to survive in small refuges, such as ravines, valleys, and swamp forests, sites buffered from extremes, especially drought, during the repeated glacial episodes of the Pleistocene. These refuges were probably scattered through the southern Appalachians and the southern Mississippi Basin, where mixed mesic forests persisted in ravines and valleys (Delcourt and Delcourt 1984). The north–south orientation of both the Appalachians and the river valleys to the west would have allowed vegetation to move north and south with climatic fluctuations, as opposed to the situation in Europe and parts of Asia, where east–west mountain ranges blocked such movements (Delcourt and Delcourt 1984). Isolation of populations and the evolutionary pressure exerted by changing environmental conditions may have caused radiations within remaining temperate lineages of *Strumigenys*, as is documented in the groups of oaks (Axelrod 1983).

There are 10 introduced species of *Strumigenys* known from Florida: three from the Old World Tropics, six from the New World Tropics, and one from Japan. These ants desiccate easily and seem unlikely immigrants, but there is a long history of importation of containerized plants under conditions that might suit dacetine hitchhikers (Deyrup and Cover 2009). Now that there are restrictions on casual importation of plants in soil, it is less likely that additional species of *Strumigenys* will be brought to Florida. It is possible, however, that additional species are already here. These are inconspicuous insects, and an enclave of an exotic species could go undetected for decades before representatives happened to fall into the hands of a myrmecologist.

Species of *Strumigenys* hunt small invertebrates that live in leaf litter. Springtails (Collembola), which often occur in enormous numbers, seem to be important in the diet of most of the relatively few species of *Strumigenys* that have been studied. The springtail diet of *Strumigenys* was first described by Laurence Wesson in 1936. Many springtails, especially those that are eaten by ants, have a tail-like appendage that bends under the abdomen and is held in place by a type of hook. When a springtail needs to jump, it releases the tail, which strikes the substrate with enough stored energy to hurl the springtail into the air. This process can occur with tremendous speed, but the jaws of ants of the tribe Dacetini can snap shut in a small fraction of a second.

Despite the minuscule size of all *Strumigenys*, the mechanics of their jaws have been carefully studied. The jaws are equipped with basal projections that catch on the lateral edges of the labrum. The jaws are cocked by spreading them until the bases are engaged with the labrum and then applying pressure with the massive muscles that close the jaws. This pair of muscles can take up about two-thirds of the total volume of the head (Gronenberg 1996), bringing new meaning to the epithet "muscle head." A pair of trigger hairs project forward between the open jaws. Some *Strumigenys* species stalk their prey, while others seem to wait in ambush. In either case, when the trigger hairs touch the prey, the labrum is retracted, releasing the jaws, which close with a convulsive snap on their victim (Brown and Wilson 1959; Wilson 1950a). A simplified diagram of this mechanism is provided by Brown and Wilson (1959), and a more detailed treatment can be found in Gronenberg

(1996). The muscles that close the mandibles do not show the morphological features of "fast" muscles, as they are contracted before the strike is initiated, but the muscles that retract the labrum have features associated with rapid retraction (Gronenberg 1996). The strike begins within 5 milliseconds of contact by the trigger hairs, and the strike itself takes less than 2.5 milliseconds, considerably faster than the maximum speed of muscle contraction (Gronenberg 1996). The jaws of some species of Strumigenys species are narrow, so their ends achieve great velocity. They may strike with enough force to instantly kill or disable the prey, and it is not always necessary to sting the prey after it is caught (Bolton 1999). The principal teeth are near the tips of the jaws, where the strike force is greatest, usually impaling the prey (Bolton 1999). The majority of Florida Strumigenys have short jaws with series of sharp teeth along the inner margins. In this group, which used to comprise the genus Pyramica, the jaws only open to an angle of 60° to 90°, and the snap serves to seize the prey rather than stun and immobilize it. The teeth firmly grasp the prey, assisted by a mechanism whereby a process on each mandibular base slides between the labrum and the clypeus to form a lock that prevents the jaws from being twisted open by a struggling prey (Bolton 1999). Prey are approached stealthily, usually grasped by an appendage rather than the body, and subdued by stinging (Bolton 1999).

A functional analysis of the jaws of Strumigenys explains their general structure, but does not explain in any satisfying way the variation between species. For example, the jaws of the Florida species emmae, rogeri, membranifera, and talpa may share a similar mechanism, but they are very different in form. There are even more extreme differences between some tropical species. Does this reflect differences in hunting techniques or different prey? It is also possible that mandibular structures in Strumigenys are complex and interdependent in ways that make it unlikely that there will be strong morphological convergence even when species converge ecologically.

Meanwhile, the ecological significance of the diversity in Strumigenys is a separate, and more intriguing, question. For the most part, we are ignorant of the meaning of the radical differences between the jaws of various species, to say nothing of such lesser characters myrmecologists use for identification of species, such as the presence or absence of a long, crimped hair on the basal segment of the hind tarsi, or whether there are curved hairs on the antennal scape, and whether these hairs curve toward the base or the apex of the scape. The diversity of modifications of the clypeal hairs is even more impressive. It has been suggested (Brown and Wilson 1959) that these hairs, in such proximity to the trap-jaws, might be tactile lures or tactile "camouflage." It is also possible that these hairs dispense chemical attractants. In line with this last hypothesis, Dejean (1985) has shown that when species of Strumigenys are held in chambers under a fine screen, there is some aggregation of entomobryid Collembola on the screen above the ants.

I have run thousands of Berlese samples in an effort to survey the cryptic ants of Florida, especially Strumigenys, and this task has been assisted by other litter samplers, including Walter Suter, Lloyd Davis, Clifford Johnson, and Zachary Prusak. All this activity, however, does not fill me with confidence that the Strumigenys fauna of Florida is completely known. There is no known way to attract or concentrate species of Strumigenys. Each litter sample is taken in ignorance of what ants it contains, and only so many samples are possible. It is easy to become almost overwhelmed with a combination of anticipation and frustration when standing in some mesic hardwood and pine forest, looking out over a mass of microhabitats, such as rotten logs and stumps, tree buttresses, leaf-filled abandoned mammal burrows, small depressions and elevations, accumulations of hollow hickory nuts, beds of pine needles, and who knows how many additional microhabitats that species of Strumigenys recognize, but we do not. Several species of Florida Strumigenys have been found only once or only a few times, which suggests that there may be other species that no myrmecologist has been lucky enough to encounter.

Identification of Florida Strumigenys is challenging because there are so many species, because there are several sets of similar species, and because all of them are so tiny. If your microscope is not good enough to show the arrangement of hairs on the end of the clypeus of an ant 2 mm long, forget it. It is always important to have good, diffuse lighting to pick out the details of ants, but this is absolutely crucial when working with dacetines. In some instances, it is useful to spread the jaws and look at the teeth, and it is almost always necessary to have clean specimens unencumbered with grains of soil and not immersed in the drop of glue that holds them to the point. Specimens are often encrusted with what appears to be a dried, grayish film of some unidentified substance, probably applied by the ants themselves (Masuko 1985), but this is not a great impediment to identification.

No species of *Strumigenys* have any known economic significance. There are three imported species that are abundant in both natural and disturbed habitats; these should be considered invasive exotic species, but their ecological impacts are unknown. These species are *Strumigenys rogeri*, *Strumigenys emmae*, and *Strumigenys eggersi*.

Name Derivation: From *struma* (Latin), meaning "a glandular swelling," and *genys* (Greek), meaning "jaw." The name refers to the type species, *Strumigenys mandibularis*, which has heavy, bowed, swollen jaws. Many other *Strumigenys* have relatively long, slender jaws, or short triangular jaws, or almost any other mandibular conformation, and the genus name does not fit them quite so well.

Strumigenys abdita Wesson and Wesson— Poodle-Faced Mustache Ant (Plate 47)

Taxonomy and Similar Species: The head of *abdita* is covered with neatly arranged, wide, curled, shining hairs, hence the name "Poodle-Faced." Other, less exclusive features are the rectangular anterior portion of the head in a dorsal view (some *Strumigenys* species have a more wedge-shaped head), a toothless section (diastema) of the jaws that can be easily seen from above when the jaws are closed (some *Strumigenys* lack this extensive gap), and the more or less widened hairs on the antennal scapes and on the legs.

Distribution: Pennsylvania south into Florida, west into Iowa and Oklahoma (Bolton 2000), but not, as yet, Alabama, Louisiana, or Texas. In Florida, *abdita* is apparently known from only five specimens, with localities in Marion, Alachua, Dixie, and Walton counties.

Natural History: This species is either rare or difficult to collect in Florida, where it has been found in forested areas, including one site in low scrub oak forest. In more northern areas, specimens are usually found under stones, and Brown (1953) suggests that the rarity of *abdita* in collections may be attributed to its subterranean habits. Many specimens have a portion of their enlarged hairs removed by abrasion (Bolton 2000), another clue to a subterranean lifestyle. In Florida, our understanding of the subterranean fauna is greatly impeded by the lack of rocks on the ground. The male of *abdita* is apparently unknown.

Name Derivation: From *abdita* (Latin), meaning "hidden" or "concealed," presumably referring to the discovery of the type specimens concealed under rocks in a backyard in Jackson, Ohio (Wesson and Wesson 1939). This same backyard is also the type locality of *P. reflexa*. We should all have such a backyard, and the observational skills of the Wessons.

Strumigenys angulata M. R. Smith— Square-Faced Mustache Ant (Plate 48)

Taxonomy and Similar Species: The squared-off clypeal margin, long slender jaws, and strongly angulate antennal scape are distinctive. It could be confused with *pergandei*, but that species does not have a strongly angulate antennal scape and differs in the arrangement of mandibular teeth; *pergandei* is not known from Florida, but I have found *pergandei* and *angulata* together at a site in Emanuel County, Georgia.

Distribution: South Carolina into north Florida, west into Illinois and Oklahoma (Bolton 2000). In Florida, *angulata* appears to be rare or difficult to collect; it is known from a few sites in the northern part of the Peninsula and a few sites in the Panhandle. It may be more abundant in the Mississippi valley region.

Natural History: This species occurs in a variety of forested habitats, from xeric sand pine scrub to edges of swamp forests. The type series (Louisville, Mississippi) was from a crevice in a rotten log in a swamp thicket (Smith 1931a). Colonies have also been found under the bark of rotten pine stumps (Brown 1953), and I found a colony in a rotten pine log on a seepage slope in Georgia. From this limited information, it appears that nests are in rotten wood, although foragers may be extracted from leaf litter. The male is unknown.

Name Derivation: Marion Smith (1931a) named this species *angulata*, meaning angulate, both for its rectangular clypeus and for its angulate antennal scapes.

Strumigenys apalachicolensis (Deyrup and Lubertazzi)— Apalachicola Mustache Ant (Plate 49)

Taxonomy and Similar Species: This is one of several Florida *Strumigenys* that have a wedge-shaped head in dorsal view. The jaws of *apalachicolensis* have a conspicuous toothless gap (diastema) that can be seen even when the jaws are closed; this separates this species from some other wedge-headed species, such as *laevinasis*. The long, erect or posterior-pointing hairs on the clypeus of *apalachicolensis* distinguish this species from other wedge-headed species with a diastema, such as *archboldi* and *filitalpa* (the latter not known from Florida).

Distribution: Described from a small area in the Apalachicola National Forest, in Leon County (Deyrup and Lubertazzi 2001). Douglas Booher recently collected this species in Aiken County, South Carolina, so it probably occurs in various places between Florida and central South Carolina.

Natural History: The type series is from a large nest (332 individuals) that was in buried rotten wood in open, low pine flatwoods subject to frequent floods and fires (Deyrup and Lubertazzi 2001). In the laboratory, workers stalked and killed entomobryid Collembola. If colonies of *apalachicolensis* normally occur in soil in pine forests, this may explain why the species escaped notice until recently, as this is not a situation in which one would expect to find dacetines. The South Carolina specimens were from deep pine litter in an upland open pine forest. The male of *apalachicolensis* is unknown.

Name Derivation: Named for the Apalachicola National Forest of Florida, the only known place where this species was discovered by David Lubertazzi.

Strumigenys archboldi (Deyrup and Cover)— Archbold Mustache Ant (Plate 50)

Taxonomy and Similar Species: This is one of several species with a wedge-shaped head, seen in a frontal view. It differs from some other Florida wedge-headed *Strumigenys*, such as *laevinasis*, in having a conspicuous gap (diastema) in the teeth on the inner margins of the jaws. This is easily visible when the jaws are closed. *Strumigenys archboldi* differs from other Florida species in having a long, low, triangular, translucent blade filling much of the diastema. I need hardly say that the function of this blade is unknown; it also occurs in various forms in many other species. The hairs on the head are somewhat similar to those of *talpa*, and specimens have occasionally been confused in collections, which might seem unlikely, until one considers the minute size of both species.

Distribution: This species is known from about 12 sites, mostly in northern peninsular Florida, with the westernmost site in Jefferson County, and from one site in Georgia (St. George, Charlton County) near the Florida border (Deyrup and Cover 1998). Many of the collections were made by Clifford Johnson or Lloyd Davis in north-central Florida. This species is generally obtained from leaf litter extractions, and there is no reason to believe that it is a widespread species that has escaped detection through much of its range.

Natural History: Most specimens of *archboldi* have been collected near edges of lakes or streams, and it is possible that its habitat is restricted to forested waterside habitats in sandy uplands where flooding is not prolonged (Deyrup and Cover 1998).

Name Derivation: Named for Richard Archbold, who founded the Archbold Biological Station, which has supported this survey of Florida ants for more than 15 years.

Strumigenys boltoni (Deyrup)— Bolton Mustache Ant (Plate 51)

Taxonomy and Similar Species: This species is deceptively similar to *dietrichi*, and I had confidently placed specimens of *boltoni* in the trays of *dietrichi* for years before I noticed the differences between the two species, requiring the description of *boltoni* as a new species (Deyrup 2006). Both species have an elongated, pointed clypeus that has a set of long hairs sticking up from its tip, rather like the whiskers on a fox, and a pair of even longer hairs rising from the median area of the clypeus. In *boltoni*, there are four of the subapical whisker-like hairs (rather than the six in *dietrichi*); the jaws of *boltoni* do not protrude as far beyond the clypeus, and the sides of the pronotum are mostly shining, rather than finely granulate. I believe that these minor differences denote separate species rather than variation within *dietrichi* because the two forms never occur together in a colony, and both forms occur together over a large area of Florida.

Distribution: This species is known only from Florida, where it occurs from Martin and Highlands counties in the south, north into St. John's County, and west into Jackson County in the Panhandle (Deyrup 2006). It is to be expected in southern Georgia, but appears to be far less widespread than *dietrichi*.

Natural History: Specimens are known only from leaf litter samples. Most samples are from mesic and xeric forest sites, and *boltoni* seems less tolerant of wet habitats than *dietrichi* (Deyrup 2006). The male is unknown.

Name Derivation: *Strumigenys boltoni* is named in honor of Barry Bolton, a myrmecologist of incredible erudition and productivity. One of his many achievements is the complete reorganization of the dacetine ants (1999), in the course of which he described hundreds of new species and dealt with the intractable and outdated system of generic names. Barry Bolton's 1028-page revision of the dacetines (2000) is the modern foundation of dacetine taxonomy.

Strumigenys bunki (Brown)—
Bunk Johnson Mustache Ant (Plate 52)

Taxonomy and Similar Species: This species is distressingly similar to *creightoni*, although the two species might not be closely related. There are consistent differences in the large subapical mandibular teeth, five in *bunki* and four in *creightoni*. These differences are not visible unless the jaws have been spread. There are a few other minor differences that are also consistent, the most obvious of which is the size of the toothless gap (diastema) when viewed from above with the mandibles closed. This gap is almost invisible in *bunki* but obvious in *creightoni*. In Florida (this may also be true elsewhere), the shining, spoon-shaped hairs on the head of *bunki* are widely spaced over most of the head, contrasting sharply with the much more closely spaced hairs on the clypeus. In *creightoni*, the hairs on the clypeus and those on the rest of the head are distributed with almost equal density. Another similar species is *carolinensis*, which has hair patterns on the head and clypeus similar to those of *bunki*. The jaws of *carolinensis* (in frontal view) are heavier than those of *bunki*; it is difficult to quantify this, but look in collections of *bunki* for thick-jawed individuals, which may be the even more rarely collected *carolinensis*. The jaws of *carolinensis* have no gap between the large, triangular, basal tooth and the apical series of teeth. A conspicuous ridge (in diffuse lighting) runs the length of the mesosoma of *carolinensis*; this ridge is lacking or vague in *bunki*.

Distribution: Florida, Georgia, Mississippi, and Louisiana (Bolton 2000); it is also likely to occur in Alabama, but has not been reported from there. In Florida, *bunki* is known from a relatively small number of sites, mostly in the northern half of the state, but with one record from Highlands County.

Natural History: This is a woodland species that is normally represented in collections by stray individuals extracted from leaf litter. A colony was found in a hickory nut on the ground, and another in a small cavity in soil (Brown 1953). Some specimens are from wet sites, such as hardwood swamp forests, cypress swamps, low oak hammocks, lake edges, and I have even found a few specimens in sphagnum moss. On the other hand, Clifford Johnson of the University of Florida extracted specimens from xeric scrub forest on several occasions. The male of *bunki* is unknown.

Name Derivation: Named to commemorate Willie Gary "Bunk" Johnson of New Iberia, Louisiana (Brown 1950). Bunk Johnson was a major early jazz musician, specializing in the trumpet, which he both played and taught; he died shortly before Brown described *P. bunki*.

Strumigenys carolinensis (Brown)—
Carolina Mustache Ant (Plate 52)

Taxonomy and Similar Species: This species is similar to *bunki* and *creightoni* (see discussion under *bunki*). Despite these similarities, there is no reason to believe that they are especially closely related.

Distribution: North and South Carolina, and Florida (Bolton 2000). In Florida, there are a few records from north Florida. Five of the six known sites are clustered in adjacent areas of Putnam and Alachua counties, and one is in Liberty County. Amazingly, these meager records seem to be the majority of collections of this species, which is otherwise known from only two sites.

Natural History: Habitats where specimens have been found include a mature stand of live oak and loblolly pine (Brown 1964), and a xeric, south-facing stand of post and blackjack oaks (Carter 1962; this author evidently knew the name of *carolinensis* before the species description). Florida habitats are dry, turkey oak sandhill, an oak–hickory hammock on the edge of a lake, sand pine scrub in scrub rosemary (*Ceratiola*) litter, and open pine-oak forest. There are only seven Florida specimens in all, four of them isolated queens that may not have been in ideal habitat for producing a colony. All this suggests that, at least in Florida, this species might be a primarily subterranean ant, often in dry sites, that is unlikely to be collected, even with intensive litter extraction. The male of *carolinensis* is unknown.

Name Derivation: The name *carolinensis* means "of," or "living in," Carolina.

Strumigenys cloydi (Pfitzer)— Cloyd Mustache Ant (Plate 53)

Taxonomy and Similar Species: This species has an extended, slightly concave clypeus, with a subapical set of enlarged hairs that stick up from the surface. In these features, *cloydi* has some resemblance to *boltoni*, *dietrichi*, and *ornata*, but *cloydi* has much shorter subapical hairs, lacks the long, curving, medial pair of clypeal hairs usually found in those species, and has three long, backward-pointing, marginal hairs on each side of the clypeus.

Distribution: The type series is from Knoxville, Tennessee (Pfitzer 1951). The species was not rediscovered for several decades, when a few specimens were extracted from leaf litter in Florida, one specimen from Lake County, and two from St. John's County. This is the kind of discovery that makes one wonder how many ants are still missing from the Florida inventory. It seems likely that *cloydi* occurs here and there in the region between Tennessee and peninsular Florida, but it is evidently rare, or difficult to collect, or both.

Natural History: The type series consists of about 50 workers found in a chamber under the bark on the underside of a rotten oak log (Pfitzer 1951). The habitat was a small oak woodlot with good drainage in a residential section of Knoxville (Pfitzer 1951). One Florida specimen is from a tree buttress in a seasonally flooded swamp forest, whereas the others are from xeric coastal scrub. Apparently, habitat does not offer any clues for myrmecologists seeking this species. The male of *cloydi* is unknown.

Name Derivation: Named for John Will Cloyd, who helped collect the type series (Pfitzer 1951). Donald W. Pfitzer, a professional wildlife biologist who collected a batch of *Strumigenys* specimens on the side, did not publish additional descriptions of ants, but did become a semiprofessional outdoorsman who wrote a guide to the hiking trails of Georgia.

Strumigenys clypeata Roger— Shining-Snouted Mustache Ant (Plate 49)

Taxonomy and Similar Species: This species has a wedge-shaped head and no visible toothless gap on the jaws when the jaws are closed and viewed from above. There are several Florida species that share these features, but *clypeata* is the only one with a smooth clypeus that is neatly covered with elongate, scale-like hairs that pave the surface of the clypeus. The similar species *pilinasis* has hairs that stand up from the surface and are bent over at the tips; *ohioensis* has a tuberculate clypeus without scale-like hairs; *laevinasis* (very rare in Florida) has long, fine hairs that stick out to the sides of the clypeus.

Distribution: New York south into Florida, west into Illinois, and south into eastern Texas (Bolton 2000). In Florida, *clypeata* is known from scattered localities throughout the northern part of the state and south into Martin and Sarasota counties.

Natural History: In a survey of ants of North Carolina (Carter 1962), *clypeata* was collected 13 times, always from mesic forest, both pine and hardwood forests. In Florida, *clypeata* is one of the more frequently encountered native species of *Strumigenys*. Of 37 Florida collections with habitat information, 17 were from mesic hardwood or hardwood and pine forest, 14 were from xeric forest, and 6 were from swamp forest or low flatwoods. These collections were not part of an organized sampling program with equal effort devoted to each habitat type. Enough work was done in wet areas, however, to strongly suggest that this species is less likely to be found in wet habitats. The male of *clypeata* was described by Brown (1953).

Name Derivation: Derived from *clypeus* and the suffix *-atus*, meaning "having" or "in possession of." In this case, the suffix appears to be used as a kind of emphatic, meaning "having a notable clypeus," referring to its pavement of shining scale-like hairs.

Strumigenys creightoni M. R. Smith— Creighton Mustache Ant (Plate 52)

Taxonomy and Similar Species: This species is most similar to *bunki* (see discussion of that species), but differs in dentition, most conspicuously in the more pronounced toothless gap on the mandibles. The enlarged clypeal hairs of *creightoni* are more widely spaced than those of *bunki* or *carolinensis*.

Distribution: Virginia south into Florida, west into Tennessee, and south into Mississippi (Bolton 2000). In Florida, *creightoni* is known from about eight sites, most of which are widely separated, from Highlands County north into Putnam County and west into Calhoun County.

Natural History: In North Carolina, *creightoni* was extracted three times from leaf litter collected in mesic forest, and once from xeric forest (Carter 1962). In Florida, 14 collections of *creightoni* are from xeric forest, one from longleaf pine savanna with wiregrass, and one from a mesic pine and oak forest. Male specimens of *creightoni* are known (Brown 1964) but have not been described.

Name Derivation: Named by Marion Smith (1931a) for William S. Creighton, who collected the type series. Years later, Creighton's extraordinary manual of North American ants (1950) revolutionized the treatment and identification of the fauna, applying modern species concepts and sweeping away a chaotic mass of ill-defined forms. Creighton seized on the newly synthesized biological species concept with the enthusiasm of a stable hand offered a pitchfork in exchange for a garden trowel.

Strumigenys dietrichi M. R. Smith— Dietrich Mustache Ant (Plate 51)

Taxonomy and Similar Species: For approximately 75 years following its description, *dietrichi* was an absolutely unmistakable species, characterized by its elongate, pointed clypeus, from the tip of which protruded a cluster of long, slender hairs. In 2006, I intruded on this idyllic taxonomic situation with a new species, *boltoni*, which is almost identical to *dietrichi*. The two species differ in a set of small, but consistent features of the mandibles and clypeal hairs. The jaws of *dietrichi* protrude beyond the clypeus slightly more than those of *boltoni*, and there are six long hairs radiating from the tip of the clypeus of *dietrichi*, rather than the four found in *boltoni*. The two species have widely overlapping ranges, and occasionally occur in the same site, although never in the same colony.

Distribution: This species occurs from Maryland south through Florida, west into Illinois, and south into eastern Texas (Deyrup 2006). In Florida, *dietrichi* is the most widespread of the native *Strumigenys*, besides *louisianae*, occurring in the Florida Keys, with a few additional records in south Florida, and throughout the northern part of the state.

Natural History: There is no reason to believe that natural history information on *dietrichi* is confounded with that of *boltoni*, since published studies of *dietrichi* are from outside the known range of *boltoni*. Nests of *dietrichi* are generally in rotten wood, in contrast to those of the related species *ornata*, which are in leaf litter and soil (Brown

1953). Kennedy and Schramm (1933) report a colony in rotten wood with more than 80 workers. These authors also provide elegant drawings of the complex mouthparts of *dietrichi*. A captive colony observed by Wilson (1953) accepted various prey, including Entomobryiidae (*Entomobrya*, *Orchesella*, and *Pseudosinella*), Isotomidae (*Isotoma*), Sminthuridae (*Sminthurinus*), Japygidae (*Japyx*), and Scutigerellidae (*Scutigerella*). Foraging *dietrichi* were observed stalking Collembola, touching the potential prey lightly with the antennae and long clypeal hairs; there was no evidence that the prey were attracted to the hairs, but they did not take the evasive action that normally followed contact with another soil organism. In Florida, *dietrichi* is a relatively common species. It was collected 23 times in xeric forest, 15 times in mesic forest, and 11 times in swamp forest or low flatwoods. The male is undescribed, but males associated with workers are in the collection of the Archbold Biological Station (Zachary Prusak, collector).

Name Derivation: Named for Henry Dietrich, a coleopterist, who provided Marion Smith with the type series from a rotten log in Mississippi. Dietrich went on to become curator of the Entomological Museum at Cornell University, where he wrote a monograph on the click beetles of New York State. A clerid beetle, *Cymatodera dietrichi* Barr, is also named after Dietrich. His daughter Mary married, and collaborated with, the entomologist and natural history writer Howard Ensign Evans.

Strumigenys eggersi Emery— Eggers Pygmy Snapping Ant (Plate 54)

Taxonomy and Similar Species: This species is easily distinguished by its long, slender, forceps-like jaws, lined subapically with minute teeth. *Strumigenys gundlachi* is the only similar Florida species. Much rarer than *eggersi*, and confined to tropical areas, *gundlachi* differs in having a small but distinct spongiform appendage depending from the postpetiole, and a series of small teeth beginning about halfway up the jaws.

Distribution: This species appears to turn up almost everywhere it is sought in the neotropics, including the Caribbean. On a recent trip to Dominica, however, I found only a few specimens of *eggersi*, even in disturbed areas, where *Strumigenys subedentata* was the commonest species of *Strumigenys*. In North America, *eggersi* is reported only from

Florida, but the species is so well adapted to urban areas in warm climates that it is probably already established in cities like San Diego, Corpus Christi, Brownsville, and Tucson. In Florida, *eggersi* occurs from the Keys through the central Peninsula, with isolated records from north Florida. It is not clear whether *eggersi* is still expanding its range generally northward, or whether it is colonizing sheltered microclimates in such a way that it will eventually have a patchy distribution through the northern edge of its range. In either case, it will probably soon be found in coastal Georgia and along the Gulf of Mexico.

Natural History: Nests are small, usually with no more than 25 adults, including one or more dealate queens. The largest nest had 41 workers and one queen. The nests I have seen are in small hollow twigs or in nuts, such as acorns, buried in the leaf litter. One of these contained a set of dead Collembola: seven entomobryiids and one isotomid. This species also occurs where there is mulch but little leaf litter, such as plantings around shopping malls, and it may be that nests in such places are not in preformed cavities. The proportion of dealate queens appears unusually high in litter samples, and some apparently small nests have several alate queens. This suggests that *eggersi* may be adapted to rapidly colonize unstable disturbed sites. I have never seen a male in any nest, nor do Malaise traps in areas abounding in *eggersi* produce a large number of small male dacetines. Parthenogenesis has been reported in two unrelated dacetine species (Ito et al. 2010; Masuko 2013), but there is no direct evidence of parthenogenesis in *eggersi*.

Strumigenys eggersi qualifies as an invasive exotic species in Florida. It is probably the commonest litter-inhabiting ant through much of its Florida range. It is abundant in natural habitats, from swamp forest to xeric scrub, and also occurs dependably in urban, suburban, and other disturbed habitats. There is no conceivable way to slow or reverse the invasion of *eggersi*. It may well have altered the soil arthropod communities of Florida before they can even be studied, like the establishment of rats on some remote Pacific island a century before the arrival of the first biological explorer. Having said all this, there is no evidence that *eggersi* or other exotic dacetines have affected native species in any way. The first victims might be expected among the native dacetine ants, but the pre-*eggersi* records are not extensive. The first Florida record is in a paper by Brown (1960b), but the species was probably present well before that record. It seems to me that Walter Suter's litter samples collected from south and central Florida in the 1960s have more native *Strumigenys* and fewer *eggersi* than I find today, but Walter Suter used different sampling methods than the one I use. Moreover, many things besides the introduction of *eggersi* have changed in Florida since the 1960s. If *eggersi* continues to move beyond its present range, it may be practical to use the litter sample records from my survey to track gross changes in the ant fauna.

Interesting and unusual features of *eggersi* are the lack of spongiform appendages and the small number and minor development of the modified hairs on the head and body. If we knew more about the natural history of *eggersi*, and knew the significance of these features in other dacetines, it might be possible to present some clever evolutionary speculation about the morphology of *eggersi*.

Name Derivation: Carlo Emery named this species for the collector of the type series on St. Thomas: the Dane Henrik Franz Alexander Baron von Eggers (1844–1903). Baron von Eggers was a professional soldier posted to the Danish West Indies, where he pursued on the side a career as an explorer-naturalist, a fine tradition of the 19th century. His primary interest was in botany, and, on retiring from the military, he traveled around the Caribbean and South America collecting plants. Let me mention to any retired military personnel who may be reading this that entomology could still use an army of explorer-naturalists.

Strumigenys emmae (Emery)— Emma's Pygmy Snapping Ant (Plate 55)

Taxonomy and Similar Species: This species was in the genus *Epitritus* from 1890 to 1949, spent the next 50 years in *Quadristruma* (Brown 1949b), and ended up in *Strumigenys* in Bolton's 1999 revision of the dacetine genera. With its bowed jaws and its spoon-shaped hairs on the upper surface of its head, *S. emmae* is a highly distinctive species.

Distribution: The closest relatives of *emmae* are Australian, and its origin is probably Australia or nearby islands (Bolton 2000), even though the type specimens are from St. Thomas in the Virgin Islands. It is widespread in the Old World tropics and is established in the New World in Florida, the West Indies, Belize, and Suriname (Bolton 2000; Wetterer 2012b). Like some other "tropical tramp" species of ants, *emmae* is especially likely to be found on islands rather than mainland sites (Wetterer 2012b). This species probably occurs in irrigated urban areas in southern Arizona, southern Texas, and southern California. An extremely small (approximately 1.7 mm long), slow-moving ant that spends its entire life in hiding, this species is only likely to be detected by Berlese funnel extractions. In Florida, its range extends about two-thirds of the way up the Peninsula. The earliest collection record in Florida is 1945.

Natural History: *Strumigenys emmae* is a common species in a variety of mesic and xeric habitats, including old growth xeric habitats; this is based on 227 Berlese samples taken in various habitats (Deyrup and Deyrup 1999). A captive colony captured and killed entomobryiid Collembola (Deyrup and Deyrup 1999). This species is common in natural habitats as well as disturbed habitats, and could be competing with native dacetines, but there is no evidence that native species have been affected by this species. *Strumigenys emmae* often occurs with *S. eggersi*, frequently in the same small litter sample. Like *eggersi*, *emmae* is often in microhabitats that look perfect for some native *Strumigenys*, and it is difficult to avoid the suspicion that these exotics have excluded native species. Both *emmae* and *eggersi* also occur in extremely disturbed, often dry microhabitats, and it is easy to imagine that these two species will increasingly dominate Florida's dacetine fauna as natural habitats become increasingly embedded in a matrix of highly disturbed habitats. Unfortunately, the first extensive studies of Florida dacetines postdate the arrival of three invasive exotic dacetines. A presumed male is described by Deyrup and Deyrup (1999), but there appear to be no males associated with workers, and it is possible that this male is some other species. *Strumigenys emmae* might turn out to be a parthenogenetic species, but there is no good evidence of this other than an absence of the large number of males that one would expect to see in south Florida flight traps and the absence of males in numerous nests that I have examined.

Name Derivation: Carlo Emery and Auguste Forel were contemporaneous pioneers in ant taxonomy and occasionally named genera or species after each other. With *Strumigenys emmae*, Emery

(1890) carried this friendly reciprocity one step further: in the description of the species, he writes (here translated from Italian), "I dedicate this elegant species to Mrs. Emma Forel."

Strumigenys epinotalis Weber— Tree-Hole Mustache Ant (Plate 47)

Taxonomy and Similar Species: At first glance, this species looks rather similar to several other species with relatively sparse and evenly distributed spoon-shaped hairs on the head, such as *hyalina*, *creightoni*, *bunki*, and *margaritae*. First glances, however, are seldom diagnostic in *Strumigenys*. The most conspicuous diagnostic feature is the eye of the worker, which is relatively large, as wide as the antennal club; *margaritae* also has large eyes. In lateral view, the finely granulate, nonshining sides of the mesosoma separate *epinotalis* from *hyalina*, *creightoni*, and *bunki*, all of which have large smooth areas on the sides of the mesosoma. It shares granulate sides with *margaritae*, but has conspicuous spongiform bodies on the petiole and postpetiole, and the upper surface of the gaster is shining, whereas *margaritae* lacks spongiform bodies and the upper surface of its gaster is nonshining. In addition, the dorsal area of the pronotum has a distinctive curved row of short spoon-shaped hairs.

Distribution: Known from Mexico, Costa Rica, Brazil (Bolton 2000), Ecuador, and the southeastern United States (Louisiana and Florida) (Chen et al. 2012). Presumably, this species is widespread in Central and South America but has seldom been collected because of its arboreal habits. Florida records are from Highlands County, but this is not a likely point of entry into the state, and it is probably widespread in swamp forests.

Natural History: This species was long suspected of having arboreal habits, which was confirmed in 2009 by capturing specimens in floating pitfall traps in Louisiana cypress-tupelo swamps that lack terrestrial habitat (Chen et al. 2012). In 2013, two researchers working on laurel wilt, Paul Kendra and Wayne Montgomery, brought into the Archbold Invertebrate Laboratory the trunk of a dead red bay (*Persea borbonia*) that had a colony of *Strumigenys epinotalis* living in a fissure in the wood several meters above the ground. This colony had 121 individuals, including 13 dealate queens. The tree came from a seasonally flooded swamp forest beside a lake on the Archbold Biological Station. It is remarkable that this tropical arboreal dacetine has established itself in Louisiana and central Florida, areas subject to frost and even short-term freezes in the winter. The habitats where it has been found, however, are partially protected from cold temperatures by standing water and a tree canopy. *Strumigenys epinotalis* probably arrived at the Archbold Biological Station only recently. No specimens were captured in an extensive array of bowl traps in the swamp forest in 2001, while several specimens were collected in several bowl traps in a less extensive array in 2009.

While nobody would have predicted the establishment of *S. epinotalis* in southeastern North America, its transport to the region is unsurprising. Many years ago, a grower described to me collecting epiphytes in Costa Rica and Belize. "Back in those days," he said wistfully, "we would just fly down in our small plane and load it up. Not with rare stuff," he assured me virtuously, "but just the plants you might find loading down the branches of old citrus trees and such. We would give the farmer a hundred bucks, which was a lot." This was, apparently, standard practice for a number of purveyors of orchids and bromeliads, although the reminiscences might flow less freely today, following the accidental importation of the bromeliad weevil (*Metamasius callizona*), which has devastated Florida's larger native bromeliads. Such freewheeling importation no longer occurs, not so much because of increased vigilance by the U.S. Department of Agriculture as by that of the Drug Enforcement Administration, after whole fleets of small airplanes began bringing payloads with higher street value than bromeliads and even more destructive potential than weevils.

The large eyes of this species suggest that it is a diurnal and forages in the open, like *Strumigenys margaritae*, the other large-eyed Florida *Strumigenys*. Bolton (2000) places *epinotalis* and *margaritae* in the same species group.

The males of *S. epinotalis* are unknown. In Florida, alate queens have been captured in flight traps in June and August (two records).

Name Derivation: This species was described by Neal Weber (1934b) as a subspecies of another new species described at the same time, *Strumigenys studiosi*. One of the diagnostic features of this "subspecies" appears to be a higher lamina (narrow ridge) on each side of the propodeum (also called epinotum), hence the name *epinotalis*, directing

attention to this feature. William Brown later (1953) recognized *S. epinotalis* as a separate species.

Strumigenys gundlachi Roger— Gundlach Mustache Ant (Plate 54)

Taxonomy and Similar Species: There is a small group of neotropical species related to *gundlachi*, but the only other member of this group to reach Florida (so far) is *eggersi*. The presence of a small spongiform appendage on *gundlachi*, and its set of small, slender teeth that begin about halfway along the jaws separate this species from *eggersi*. Specimens in alcohol from litter extracts can often be distinguished from those of *eggersi* by their more parallel, less convergent, jaws.

Distribution: Widely distributed through the Neotropics, including the Caribbean (Bolton 2000). In Florida, *gundlachi* is apparently restricted to tropical areas of Miami-Dade and Monroe counties.

Natural History: Brown (1960b) found *gundlachi* in almost every leaf litter sample he took on Barro Colorado Island (Panama). Elsewhere, he found it in primary rain forest, second growth forest, and cacao plantations. In the Dominican Republic, I found *gundlachi* occasionally in dry scrub forest. In Florida, it has been collected about 20 times in tropical hardwood hammocks. A colony kept by Wilson captured and ate entomobryiid and sminthurid Collembola, usually stalking them and ambushing their prey, but sometimes approaching and attacking without waiting in ambush (Brown 1960b). The male of *gundlachi*, like that of *eggersi*, is apparently unknown.

As an exotic species in Florida, *gundlachi* has not prospered nearly as well as the closely related *eggersi*. It is possible that *gundlachi* is less cold-hardy and not as well adapted to highly disturbed situations. It is also possible that it has an unfavorable competitive relationship with *eggersi*. This species, like *eggersi*, was first reported from Florida in 1960 (Brown 1960b), but could have arrived much earlier.

Name Derivation: Julius Roger named this species for the collector of the type series, one of the great Caribbean naturalists, Johannes Christoph Gundlach. In 1839, when he was 28, Gundlach moved to Cuba, where he spent much of his long life (he died at the age of 85 in 1896) and eventually changed his name to Juan Cristobal Gundlach. He wrote an early treatise on the birds of Cuba and assembled a large collection of insects, birds, mammals, reptiles, and land shells, which he donated to the Museo Poey in Havana. About 60 species of animals were named after Gundlach, including *Mysoteles gundlachi*, a prehensile-tailed tree hutia from the Isla de Juventud.

Strumigenys hexamera (Brown)— Japanese Mustache Ant (Plate 54)

<u>Taxonomy and Similar Species:</u> With its large, disc-like hairs that even occur on the prominent bowed jaws, *hexamera* is an unmistakable species in the Florida fauna. This species, along with several somewhat similar species, were placed in the genus *Epitritus* until Bolton's 1999 generic revision of the dacetine ants, when it was removed to *Pyramica*, a genus later synonymized with *Strumigenys* (Baroni Urbani and De Andrade 2007).

<u>Distribution:</u> Korea, Japan, Taiwan, Florida, Louisiana (Bolton 2000), and Mississippi (MacGown and Wetterer 2012). It is undoubtedly an introduced species, as North American specimens are indistinguishable from Japanese specimens, and all closely related species are confined to the Old World. In Florida, it is known from a few specimens collected in Marion and Hernando counties, but there are many additional records from Mississippi and Louisiana, where *hexamera* appears to be more common (MacGown and Wetterer 2012).

<u>Natural History:</u> An excellent account of the natural history of *hexamera* is presented by Masuko (1985). This species is completely subterranean, with small monogynous colonies, usually with fewer than 50 individuals, normally found at a depth of approximately 10 cm below the surface. It seldom forages in leaf litter and probably confines most of its hunting to small cavities and tunnels in the soil. When a forager encounters a potential prey, the forager crouches, retracts its antennae, and becomes immobile. The head is lowered, and the jaws are not spread. If the prey steps on the head of the ant, the ant makes a sudden upward snap that impales the prey on the apical teeth. The ant then stings its prey, which may be a centipede or a dipluran considerably larger than the ant. This hunting technique is highly effective in narrow passages, much less so in an open arena. The strategy is well suited to capturing elongate prey such as Diplura, Symphyla, Geophilomorpha, and Lithobiomorpha, less suitable for Collembola. Mitsuko also examined nest chambers in the field, finding remains of five campodeids, one japygid, one cryptopid centipede, one isotomid, and one entomobryid. A curious

behavior that Masuko observed (1985) occasionally in *hexamera* and more frequently in several other dacetines was using the front legs to anoint or smear the head with a film of moisture or small soil particles scraped up from the floor of the tunnel. The significance of this is unclear, but Masuko suggests that it might help conceal the ants from prey that are using touch and scent to find their way through the dangerous interstices of the soil.

The subterranean habits of *hexamera* undoubtedly explain why it is so seldom encountered. Unlike many other introduced species, *hexamera* is not known from highly disturbed sites, but, for all we know, it could be living beneath the sidewalks of cities. Its point of introduction will probably remain a mystery. The introduction of a relatively rare, subterranean ant seems unlikely in the light of recent restrictions on nursery material, but once upon a time there was a massive, unsupervised trade from every exotic locale. Persimmons, for example, are fruit trees that do not breed true from seed and are therefore shipped in containers. Japanese persimmon trees were first brought to Washington, DC, by the Perry Expedition (1854), and in 1877, 5000 trees, representing 10 varieties, were imported from Japan (Hedrick 1919). These were probably only one of many opportunities for ants to move from the Far East to North America. There are several other ants that probably came to North America from Japan or Taiwan, including *Paratrechina flavipes*, *Vollenhovia emeryi*, and *Pachycondyla chinensis*. *Pyramica hexamera* is the only one, so far, that has a range extending as far south as Florida.

Recently, Masuko (2013) has made the remarkable discovery that at least one population of *hexamera* is able to reproduce parthenogenetically. Parthenogenesis appears to be rare in ants; among the dacetines, the only other known example is that of *membranifera* (Ito et al. 2010). As Masuko notes, if parthenogenetically reproducing populations are widespread in *hexamera*, this might help explain how this relatively uncommon species has managed to colonize new areas.

<u>Name Derivation:</u> From *hex* (Greek), meaning "six," and *meros* (Greek), meaning (in this case) "part," referring to the six-segmented antennae.

Strumigenys hyalina (Bolton)— Translucent Mustache Ant (Plate 47)

<u>Taxonomy and Similar Species:</u> This species is similar to *S. bunki*, *abdita*, and *creightoni*, but lacks

curved, spoon-shaped hairs on the dorsal area of the mesosoma; in *hyalina*, the dorsal hairs are fine and more or less erect. The clypeal hairs are much like those of *creightoni* or *bunki*, but are translucent, almost transparent under the right lighting conditions. Exactly why every species in this group seems compelled to make its own fashion statement with clypeal adornment is one of the great mysteries of myrmecology.

Distribution: Known from Ohio, Indiana, Mississippi (Bolton 2000), and Florida. The Florida specimens are from the western Panhandle; it is possible that *S. hyalina* is an example of a Mississippi Basin species that just makes its way into westernmost Florida. A few species of *Phyllophaga* (Scarabaeidae) seem to show this pattern (Woodruff and Beck 1989).

Natural History: The two known Florida specimens are from hardwood forest in a "steephead" or forested ravine on the Eglin Air Force Base. Males are unknown.

Name Derivation: Named for the hyaline (translucent or transparent) clypeal scales.

Strumigenys inopina (Deyrup and Cover)— Unexpected Mustache Ant (Plate 56)

Taxonomy and Similar Species: This species is noteworthy for a combination of features that I never would have expected to find in a dacetine ant. There are no strongly modified hairs of any sort, but an abundance of short, tapering suberect hairs of a sort not otherwise found on dacetines. The jaws are triangular, with regularly spaced, small teeth. Antennal scrobes are lacking, and the clypeus is relatively short and unmodified. There are no spongiform appendages. This species is like the most basic model of *Strumigenys* without any of the options and accessories. There is no reason, however, to believe that this species represents some primitive basal lineage of *Strumigenys*. In the heyday of erecting new genera for unusual dacetine ants, this species would probably have been placed in its own genus.

Distribution: Known from three queens, collected in Alachua, Putnam, and Marion counties, all sites in north-central Florida (Deyrup and Cover 1998).

Natural History: No workers are known, nor are there any local *Strumigenys* known only from workers that might be associated with these queens. All three specimens were extracted by Berlese

funnel from deep pine litter. Deyrup and Cover (1998) suggest that *inopina* might be parasitic in the nests of other dacetines. This is based on the following fragmentary and circumstantial evidence: (1) *Strumigenys inopina* appears to be extremely rare, like most species of parasitic ants. (2) It is known only from queens. (3) All three specimens were extracted from relatively small litter samples that also produced other native *Strumigenys*. (4) The lack of the usual kinds of dacetine specializations suggests that this species makes its living in some unusual way, and the pilosity of curved, tapering, suberect hairs resembles that of the parasitic genera *Stronglyognathus*, *Protomognathus*, and *Harpagoxenus*. All this, of course, is speculative, and it is possible that *inopina* is just a rare species that occurs in some unusual or subterranean microhabitat, has an atypical form of predation, and is represented in collections by stray queens that had not migrated to their normal microhabitat.

Name Derivation: From *inopina* (Latin), meaning "unexpected," referring to the novel set of features in this species.

Strumigenys laevinasis M. R. Smith— Smooth-Snouted Mustache Ant (Plate 56)

Taxonomy and Similar Species: This species is similar to *clypeata* and *pilinasis*, but, unlike those species, which have broadened clypeal hairs, the clypeal hairs of *laevinasis* are long, fine, and tapering, diverging from the midline of the clypeus to extend obliquely over the sides of the clypeus. Smith described *laevinasis* as a subspecies of *clypeata*, possibly because he had not seen enough specimens to be convinced that *laevinasis* was a distinct species, and not some aberrant form. Moreover, in 1931, when the species was described (Smith 1931a), both species concepts and geographic subspecies concepts were often different from those in general use today.

Distribution: Virginia south into Florida, west into Illinois, and south into eastern Texas (Bolton 2000). In Florida, *laevinasis* is known from one collection of two specimens in DeFuniak Springs, Walton County.

Natural History: This species seems to be generally uncommon, or not easily collected, and its biology has not been investigated. I have seen six collections, none with more than a few individuals, from leaf litter in hardwood forests. One dealate queen was found in a rotten log. The type series were from a cavity in a rotten log (Smith 1931a). Males are unknown.

Name Derivation: From *laevis* (Latin), meaning "smooth," and *nasus* (Latin), meaning "nose," referring to the smooth clypeus.

Strumigenys lanuginosa Wheeler—
Woolly Pygmy Snapping Ant (Plate 55)

Taxonomy and Similar Species: Easily distinguished from all other Florida species by the remarkable long hairs on the upper surface of the gaster.

Distribution: The Bahamas, Cuba, Mexico, Costa Rica, Panama, Colombia, and Venezuela (Bolton 2000). In Florida, where it was first found in 1987 (Deyrup et al. 2000), it is known from a few sites in Monroe, Miami-Dade, and Lee counties.

Natural History: This is an example of a rare exotic in Florida. In Florida, it is known from a few mesic sites, either tropical hardwood hammocks or old urban plantings. In the Bahamas, I found a large colony in a clump of sawgrass (*Cladium jamaicense*). It does not appear to be an "enclave exotic species" because the site records are reasonably dispersed. Nobody knows the function of the long hairs on the gaster. The male is apparently unknown.

Name Derivation: From *lanuginosa* (Latin), meaning "woolly."

Strumigenys louisianae Roger—
Louisiana Pygmy Snapping Ant (Plate 57)

Taxonomy and Similar Species: In the North American fauna, this species is easily recognized by its shining, spoon-shaped hairs on the dorsal surface of the head, and its long, somewhat swollen jaws. This widespread species varies from place to place in the Neotropics and might be a species complex (Bolton 2000).

Distribution: Southern United States, through Mexico and Central and South America, south into Paraguay and Argentina; also, the Greater Antilles and the Bahamas (Bolton 2000). In the United States, it occurs from southern South Carolina across the south into southern Texas. It appears to be commonest from South Georgia through the Gulf States to the Mississippi River. Specimens from Tucson, Arizona, may represent an introduction, as it has not been found in natural habitats in Arizona. This species is assumed to be native in North America. It occurs throughout Florida, but appears to be commoner in northern Florida.

Natural History: *Strumigenys louisianae* usually occurs in habitats that are reliably moist, but not subject to prolonged flooding. The predatory behavior of this species has been studied by Wilson (1950a, 1953). When a foraging worker senses a collembolan, usually at a distance of only a millimeter or so, the ant approaches slowly, opening its jaws and exposing the trigger hairs. When these hairs touch the intended prey, the jaws snap shut, impaling the prey, often with enough force to squeeze droplets of hemolymph from the punctures. The prey may then be raised into the air and the gaster bent forward between the ant's legs to deliver a sting. If the prey is quiescent after the bite, the stinging behavior may be omitted. Wilson (1953) also published a list of arthropods accepted and rejected by captive colonies of *louisianae*. Collembola of the families Entomobryidae, Isotomidae, and Sminthuridae were the preferred prey, but Symphyla of the family Scutigerellidae were also readily accepted. Collembola of the families Poduridae and Onychiuridae were rejected, as were a variety of other small arthropods. Some arthropods, such as termites and japygids, were attacked and killed, but not always consumed. Colonies are usually relatively small, but one colony had 181 workers, 119 larvae, and 1 queen (Wilson 1953). The male is known and described (Haug 1932), but without any indication of how to tell the difference between this and other species.

Name Derivation: The species epithet *louisianae* means "of Louisiana," referring to the source of the type specimens.

Strumigenys margaritae Forel—
Pearly Mustache Ant (Plate 48)

Taxonomy and Similar Species: This is the only Florida *Strumigenys* with the combination of short jaws, no spongiform appendages, and a nonshining gaster. Relatively large eyes and the absence of antennal scrobes are also distinctive features of *margaritae*, shared only with *epinotalis* and the rare *inopina*, of which there are no known workers.

Distribution: Northern South America (Colombia), Central America, Mexico, West Indies, Texas, Alabama, Georgia, and Florida (Bolton 2000). In Florida, there are a few records from Marion, Leon, and Bay counties.

Natural History: What little is known of *margaritae* suggests that it is an unusual dacetine. In

Trinidad in 1994, I found workers mixed in with a column of *Wasmannia auropunctata* walking along a small irrigation pipe in a garden. The two species are similar in size and color. Amazingly, two years later, Lloyd Davis found this mixed column of workers in exactly the same place. I have seen a few specimens taken by sweeping vegetation. The relatively large eyes and the lack of antennal scrobes also suggest that *margaritae* may forage primarily in the open. Colonies of *margaritae* may be relatively large for a dacetine: Forel (1893) reports that the colonies that supplied type material from St. Vincent included approximately 250 individuals, and two other colonies were each estimated to include approximately 200 individuals. Nests were under mats of vegetation on rocks. This information is from the accounts of a careful collector, H. H. Smith, who was sent on a faunal survey to the Antilles by the British Government and the Royal Society. Forel scrupulously included Smith's collecting notes in his review of the ants of St. Thomas.

In Florida, *margaritae* is a rare exotic. The first known Florida specimen is a queen collected in a Malaise trap in 1983, but the species was probably in Florida considerably earlier, as it had been collected in Alabama some time before 1953 (Brown 1953). It seems that there has been plenty of time for *margaritae* to become abundant and widespread in Florida, but this has not happened. It is unknown from subtropical and tropical Florida, where I might have expected it to join a series of well-established West Indian ants. Perhaps this and other mysteries surrounding *margaritae* will be explained when more is known about the biology of the species.

Name Derivation: August Forel does not include an explanation of the name in his description of the species (1893). The name might be derived from *margarita* (Latin), meaning "pearl," and referring to the numerous, shining, white scales on the head and body. It might, alternatively, be intended to honor a person named Margarita.

Strumigenys membranifera (Emery)— Bare Mustache Ant (Plate 48)

Taxonomy and Similar Species: This species is singularly hairless, with only two hairs on the dorsal surface of the head, a series on the antennal scapes, and a few around the tip of the gaster. The jaws are short, thick, and hooked at the tip, so that in lateral view, the head has a characteristic parrot-like look. For many years, *membranifera* was in the genus *Trichoscapa*, until that genus was synonymized with *Pyramica* in Bolton's revision of the dacetine genera (1999), later transferred to *Strumigenys* (Baroni Urbani and Andrade 2007). There are half a dozen old synonyms or subspecific names for *membranifera*, although variation in its morphology is small, and some of these forms seem identical (Brown 1948). Once Brown realized that *membranifera* is a widespread tramp species, its complex nomenclature collapsed. This is a minute ant and not easily found in many areas. This undoubtedly meant that only a few badly prepared specimens were available in most collections, and when the species was found in a new and isolated area, such as Hawaii, it was easy to imagine significant differences in the newly acquired specimens. I know from personal experience that when I am surveying the ants of some remote place, where I fantasize that no myrmecologist has ever rooted about, I am not happy to admit evidence that the area has been contaminated by exotic tramp species, probably for several hundred years.

Distribution: This appears to be an Old World tropical and subtropical species. At this point, it is probably impossible to determine whether *membranifera* originated in northern Africa or Asia, but the latter seems more likely, based on the large number of records. It has been found in the North Africa and the Mediterranean region, in South Africa, Nepal, Bhutan, Taiwan, various Pacific Islands, Indonesia, and New Guinea (Bolton 2000). Like some other "tropical tramp" ants, *membranifera* seems to thrive on tropical islands, but has seldom been recorded from continental South America or Africa (Wetterer 2011b). In the New World, it is known from several islands in the Caribbean, Venezuela, Costa Rica (Bolton 2000), and from the states of Virginia south through Florida and west in southern North America into California (Wetterer 2011b). Its distribution throughout its range is probably much more continuous than the actual collection records suggest. In Florida, *membranifera* has been collected throughout the state, but there are some large gaps in known distribution, especially in south Florida and in the Panhandle. It seems to be easiest to find in the central and north-central Peninsula.

Natural History: Smith (1931a) found *membranifera* nesting in small colonies (the largest with approximately 75–110 individuals) in soil beneath objects such as scraps of wood or pieces of concrete. Most nests were in somewhat dry soil. In western Japan, membranifera is usually associated with disturbed habitats, such as parks (Kitahiro et al. 2014). In Florida, membranifera is known from a wide variety of habitats. Habitat information is known for 82 collections: 19 from landscaped areas, roadsides, and highly disturbed sites; 5 from swamp forest; 4 from beach areas back of the dunes; 33 from mesic hardwood or hardwood and pine forests; 16 from xeric forests; and 5 from flatwoods. One specimen was found in the stomach of a toad, Bufo terrestris. Despite its apparent habitat flexibility, there are many places where membranifera is difficult to find. At the Archbold Biological Station, the most intensively sampled site in Florida, it has been collected only seven times. One mystery about this widespread and often common species has been the failure to discover males. Ito et al. (2010) provided a startling solution to this mystery: membranifera is parthenogenetic. Parthenogenesis is uncommon in ants, and this was the first example in a dacetine. More recently, Masuko (2013) has shown that hexamera is also parthenogenetic.

Wilson (1953) has studied the predatory behavior of membranifera in captive colonies. This species can move quickly for a defensive attack, seizing and stinging relatively large arthropods, such as termites and japygids. Offered the larvae of Monomorium and Solenopsis, foragers took both back to the nest, but the membranifera larvae accepted only the Monomorium larvae. It is unlikely that Strumigenys raid the nests of other ants under normal circumstances, but nobody really knows. Symphyla and campodeids were readily accepted as prey. When presented with Collembola, foragers switched to a specialized hunting mode. After a springtail is detected by an ant, which only happens when the prey is very close, the ant freezes and then very slowly orients toward the springtail. The ant then begins to creep toward the springtail, but so slowly that its motion is almost imperceptible. The trigger hairs must touch the springtail before the jaws snap shut, usually clamping onto an appendage. The springtail is then subdued with a sting. If the springtail wanders away from the ant, the ant does not follow, but resumes a slow foraging pattern. Masuko (1985) has observed foraging membranifera scraping moisture and organic matter from the substrate and applying it to the head; there is a photograph of this anointing behavior in the 1985 paper. The function of this behavior is unknown, but perhaps it makes the ant smell more like its surroundings.

I often lament that so little is known about the hunting behavior of dacetine ants, but I have not been moved to remedy this myself and can only admire the patient intensity of observers such as Wilson and Masuko. Aside from the initial problem of finding colonies and transferring them to observation nests, much of the activity of dacetines apparently occurs in slow motion and is frustrating to watch. These are small ants that generally attack relatively large and elusive prey. The only fast movement of a dacetine is its mandibular snap, and movements leading up to this snap need to be undetectable. As one watches dacetines, one can't help thinking that these ants invest relatively little energy to catch their large prey, so that if a forager is catching only one or two Collembola daily, it is being relatively productive. The idea that one might spend a whole day waiting for one or two events, each lasting a microsecond, is inevitably discouraging.

Name Derivation: From membranus (Latin), meaning "membrane," and fero (Latin), meaning to "carry" or "bear," referring to the large spongiform bodies on the postpetiole and at the base of the gaster.

Strumigenys metazytes (Bolton)— Middlegap Mustache Ant (Plate 53)

Taxonomy and Similar Species: This recently described species (Bolton 2000) is similar to pulchella and missouriensis, but metazytes lacks their conspicuous, backward-pointing enlarged hairs in the middle of the clypeal border. The gap where these hairs would occur has a series of very small, scale-like hairs. In other features, metazytes is almost identical to puchella, with slender jaws that show a prominent basal toothless gap when viewed from above, and a broad, scalloped clypeus with a series of forward-curving, enlarged hairs on each side.

Distribution: Kentucky, Tennessee (Bolton 2000), Alabama (MacGown and Forster 2005), Mississippi (MacGown and Brown 2006), and Florida. In Florida, metazytes is known from a single dealate queen from Walton County.

Natural History: This species seems to be known from a small number of stray specimens extracted from leaf litter. The specimens I have seen are from mixed pine and hardwood forest, including seven collections from north-facing slopes in the Ouachita Mountains of Arkansas and one from a Florida "steephead," a kind of seepage hollow with a relic population of more northern trees and shrubs. The male is unknown.

Name Derivation: From *metaxy* (Greek), meaning "between," and *ites* (Greek), meaning "characterized by," perhaps referring to the fact that *metazytes* is apparently intermediate between members of the *pulchella* group and those of the *rostrata* group.

Strumigenys missouriensis M. R. Smith— Missouri Mustache Ant (Plate 58)

Taxonomy and Similar Species: This species is distinguished from *pulchella* and *metazytes* by having jaws that are thickened apically (in lateral view), rather than uniformly slender. Both *Strumigenys missouriensis* and *reflexa* have apically thickened jaws, but all the lateral clypeal hairs of *reflexa* are directed backward, while those of *missouriensis* are directed forward. The median and submedian marginal clypeal hairs of *missouriensis* are variable, but are often directed backward.

Distribution: New York south into North Carolina, west into Iowa and Missouri, and south into Mississippi (Bolton 2000). I have collected specimens in Florida and South Carolina. In Florida, *missouriensis* is known from six collections, widely distributed around the state.

Natural History: The type series was from semi-abandoned galleries of *Aphaenogaster fulva* (Smith 1931a), but no subsequent collections have been found from nests of other ants. Specimens have also been found in rotten logs and stumps, and under objects on the ground (Smith 1931a). In North Carolina, Carter (1962) collected *missouriensis* from a wide variety of habitats, including mesic hardwood forest, mixed pine and hardwood forest, pine forest, open pine and shrub savannah, and xeric oak forest. Florida collections are from litter in swamp forest and mesic hardwood forest, and from a tree hollow in hardwood forest. A captive colony caught and consumed entomobryid Collembola and, on one occasion, a podurid Collembola; normally, dacetines, including *missouriensis*, reject podurids (Wilson 1953). The male of *missouriensis* is unknown.

Name Derivation: The name *missouriensis* (Latin) means "an inhabitant of Missouri."

Strumigenys nigrescens Wheeler—Neat Fringed Mustache Ant (Not Known from Florida)

Taxonomy and Similar Species: The clypeus of *nigrescens* is fringed with a uniform, evenly spaced,

row of long-ovate, scale-like hairs. In this feature, *nigrescens* resembles *carolinensis*, but the latter species has unusually large mandibular teeth and a small but distinct toothless gap that is visible when the closed jaws are viewed from above. There is no such gap visible in *nigrescens*, and its teeth are not unusually large. The clypeal margin of *nigrescens* is slightly turned up, although this is only apparent in clean specimens.

Distribution: This species is not known from Florida. It is included here because it is abundant and widely distributed in the Caribbean and Central America, and is suspiciously tramp-like (Brown 1953). Florida myrmecologists, who may be more numerous in the future, should keep an eye out for this species, not that we do not already have a surfeit of exotic ants.

Natural History: In Cuba, *nigrescens* occurs in both natural and cultivated habitats (Brown 1953). In the Bahamas, I found a nest in leaf litter of sea grape (*C. uvifera*) near the beach on Long Island.

Name Derivation: The name *nigrescens* (Latin) means "blackish," referring to the dark brown color of the type specimen (Wheeler 1911). Most specimens are paler and are often yellowish brown (Brown 1953).

Strumigenys ohioensis Kennedy and Schramm— Ohio Mustache Ant (Plate 56)

Taxonomy and Similar Species: The features that distinguish *ohioensis* are its wedge-shaped head with a rough or tuberculate clypeus fringed with short, fine, irregular, J-shaped hairs, and large-toothed jaws that show no basal toothless gap when viewed from above. It is somewhat similar to *laevinasis* and *pilinasis*, but both those species have a smooth clypeus with regularly oriented (combed-looking) hairs. In 1939, Wesson and Wesson described the species *manni*, which was eventually synonymized with *ohioensis*. According to Brown (1953), the Wessons did not have access to types or other identified specimens of *ohioensis*, and were misled by the original description and figure of *ohioensis* (Kennedy and Schramm 1933). It is easy to see how this could happen: the pen and ink portrait of *ohioensis* provided by Kennedy and Schramm is elegant and aesthetically attractive, but gives an inaccurate impression of the species. There are similar cases of mistaken identities in the entomological literature of the first half of the last century. The recent mass proliferation of high-quality automontage photographs of tiny insects on the Internet should prevent many such errors in the future. Myrmecologists have rushed to take advantage of this technology, producing hundreds of images of ants on a variety of sites.

Distribution: This is one of the more common and widespread species of Strumigenys. It occurs from New Jersey south into Florida, west into Ohio and Missouri, and south into Texas (Bolton 2000). In Florida, it is known from six northern counties, but has been found only eight times.

Natural History: This species is usually found in mesic forest or forest edges, including hardwood forest, pine forest, and mixed hardwood and pine (Brown 1953, 1964; Carter 1962). I have found ohioensis in an urban yard in Washington, DC, and in a highly disturbed forest buffer zone beside a busy highway in Georgia. In Florida, ohioensis is known from mesic hardwood or mixed hardwood and pine stands, often near water. The male was described by Brown (1953).

Name Derivation: The name ohioensis (Latin) means "an inhabitant of Ohio," the state where the type series was found.

Strumigenys ornata Mayr—
Ornate Mustache Ant (Plate 51)

Taxonomy and Similar Species: This species has an elongated, pointed, snout-like clypeus, attractively retroussé, ornamented at the tip with a set of eight long, radiating hairs, each expanded at the tip like stamens. This arrangement of clypeal hairs resembles that of dietrichi and boltoni, which are probably close relatives, but those species lack the knobs on the hairs found in ornata.

The type and arrangement of highly modified clypeal hairs on a species of Strumigenys are usually as consistent as they are mysterious. Occasionally, however, there are unusual aberrations, which cast a little doubt on our practice of blithely assigning a species name to every form of Strumigenys with distinctive hairs. In the case of ornata, there appears to be a population at Fort Clinch State Park in Nassau County on the northeast coast of Florida that is lacking the prominent pair of up-curving hairs that emerges from the anterior third of the clypeus. This pair of hairs also occurs in dietrichi and boltoni. I have not found any other differences in the Fort Clinch population. It is known from five collections, most of which were made by the Florida myrmecologist Clifford Johnson.

Distribution: This is a common and widely distributed species. It occurs from Virginia south into Florida, west into Missouri and Texas (Bolton 2000). In Florida, ornata is common in the northern half of the state, becoming much rarer in the south, as is usually the case with native Strumigenys species. There seem to be no records from the East Coast south of Volusia County, but it probably occurs at least as far south as southern Brevard County.

Natural History: In North Carolina, Carter (1962) found ornata to be the most common species of Strumigenys, occurring in all forest types, from xeric oak stands to bottomland forest. At higher elevations, ornata was usually in dryer and more open sites. In Florida, 112 collections include habitat information: 74 from mesic hardwood or hardwood and pine forest; 14 from swamp forest; 13 from xeric forest; 10 from pine forest, including pine flatwoods; and 1 from coastal marsh. Some of the mesic forest collections were from disturbed areas, such as urban and suburban lots.

Examining the clypeal hairs of ornata, it is easy to see why it has been suggested that the clypeal hairs of some species of Strumigenys might be tactile lures that attract Collembola, possibly resembling fruiting bodies of fungi or slime molds. I indulged in a brief fantasy that the hairs of ornata might be luminescent, which would solve some logistical problems of the lure theory. A short time in the darkroom with a colony of ornata convinced me that these ants have no glowing lures, or at least none with enough photons to register on my retinas.

Males of ornata have not been described, but there are associated males in the Archbold Biological Station collection

Name Derivation: From ornata (Latin), meaning "decorated," referring to the fancy clypeal hairs.

Strumigenys pilinasis Forel—
Hairy-Snouted Mustache Ant (Plate 49)

Taxonomy and Similar Species: The head of pilinasis in dorsal view is wedge shaped and lacks a basal gap in the mandibular teeth (diastema), much like clypeata and laevinasis. Unlike laevinasis, the clypeal hairs of pilinasis are broadened and shining. Unlike clypeata, the clypeal hairs of pilinasis stand up from the surface of the clypeus and are bent over at the ends, a difference that is not easily seen from above, but conspicuous in lateral view. There has been some taxonomic confusion surrounding this species. Two names have been synonymized with *pilinasis: medialis* (synonymy by Smith in the 1951 Catalog of Hymenoptera) and *brevisetosa* (synonymy by Brown 1964). There appear to be some puzzling variants of *pilinasis* that are somewhat intermediate with *laevinasis* (Brown 1953, 1964). Bolton (2000) did not see any intermediate variants, although he probably saw some of the same specimens examined by Brown. This is a relief, in light of Brown's dark hints (1964) about hybrids, regularly occurring mutants, and character displacement. Until Berlese funnels came into regular use in the 1960s, there were so few specimens of *pilinasis, laevinasis*, and many other dacetines that it was difficult to get a feel for variation within a species. Along with additional specimens came range extensions, showing that some pairs of apparently minor variants had broadly overlapping ranges, as is the case with *pilinasis* and *laevinasis*, and are not geographic variants or subspecies. Theoretically, a recurring, widely distributed form that is consistent within a nest series could be what Brown (1964) calls "a frequent mutant," but this situation is relatively rare in animals, and often seems to rely on the complexities of discontinuous natural selection. Occam's razor disallows this as a preliminary hypothesis to explain consistent, sympatric variation: it is simpler to postulate separate species. A final problem that bedevils the select group of dacetine enthusiasts is the variation caused by damage to specimens, or dirt and possibly secretions that dacetines apply to themselves (Masuko 1985). Under some conditions, *Strumigenys* specimens removed from fluid become coated with a fine deposit that can make hairs look thicker, or conceal their form.

Distribution: Pennsylvania south into Florida, west into Kansas and Louisiana (Bolton 2000). In Florida, *pilinasis* occurs from Highlands County through north Florida. It is not frequently collected.

Natural History: In North Carolina, *pilinasis* was collected one time in xeric oak forest, five times in mesic forest, and four times in dense pine forest (Carter 1962). At the Archbold Biological Station, there are 12 collections from north-facing hardwood sites in the Ouachita Mountains in Arkansas (Robison and Carlton, collectors). There are 15 Florida collections with habitat information: nine from mesic hardwood or mixed hardwood and conifer forest, five from swamp forest, and one from dense shrubs at the edge of a sawdust pile.

Nests have been found in soil (Brown 1964), in a fallen, rotten pine limb covered with leaf litter (Brown 1953), and in a rotten hickory log at the edge of a forest (Wesson and Wesson 1939). Wesson and Wesson (1939), observing a colony in an artificial nest, found that foraging workers remained inactive for long periods, with the head close to the substrate and the jaws closed. Dealate queens were seen foraging; this is a useful bit of information, as it helps explain why queens and workers show almost no differences in their jaws and modified hairs. In food preference tests, Wilson (1953) found that *pilinasis* ate entomobryid and sminthurid Collembola, and rejected podurid and onychiurid Collembola, as well as various small invertebrates. The male of *pilinasis* is unknown.

Name Derivation: From *pila* (Latin), meaning "hair," or "pile" (as in the standing nap of velvet), and *nasis* (Latin), meaning "nose," referring to the hairy clypeus.

Strumigenys pulchella Emery— Beautiful Little Mustache Ant (Plate 58)

Taxonomy and Similar Species: This is one of a series of species with a broad, somewhat scalloped clypeus, and a basal gap in the mandibular teeth (diastema) visible when the head is viewed from above. The jaws are slender apically in lateral view, unlike *missouriensis* and *reflexa*. There is a pair of prominent, backward-pointing, enlarged hairs on the anterior edge of the clypeus of *pulchella*; this pair of backward-pointing hairs is missing in *metazytes*. At one point, I was not convinced that *metazytes* was a distinct species, but rather an aberrant form of the much commoner *pulchella*; *metazytes* is now known from a good number of sites (although from relatively few specimens), all within the range of *pulchella*, and there do not seem to be intermediate forms. One can only wonder what, if anything, is so biologically important about those two backward-pointing hairs.

Distribution: New York and New Jersey, south into Florida, and west into Illinois, Kansas, and Louisiana (Bolton 2000). In Florida, *pulchella* is known from a relatively small number of sites, from Highlands County north.

Natural History: In North Carolina, Carter (1962) found *pulchella* one time in a pine forest, one time in a xeric oak and cedar forest, and several times in mesic hardwood forests. There are 12 Florida collections with habitat information: six in swamp forest, two in tree holes in hardwood forest, two in mesic hardwood forest, one in xeric forest, and one in a vacant lot in Tampa. Various myrmecologists have found colonies in moist, rotten logs or stumps, usually wood in a red-rot stage (Brown 1964; Kennedy and Schramm 1933; Smith 1931a; Wesson and Wesson 1939). Such logs with red rot are most likely to be pines. Smith (1931a) also found colonies in soil and under objects, not always in moist conditions. Colonies observed by Wesson and Wesson (1939) in artificial nests captured springtails by waiting for a prey to approach, turning slowly toward it, and waiting for the springtail to touch the mandibular-clypeal area, before striking, followed by stinging. This is one of the earliest accounts of predatory behavior in *Strumigenys*, preceded only by Wesson's earlier paper on *pergandei* (Wesson 1936).

The male was described by Brown (1953), who mentions that the sculpture and some other features are variable, even in nest series, while the genitalia resemble those of *talpa*, a species that does not appear to be closely related to *pulchella*. Male dacetines are something of a disappointment. For the most part, males that are not associated with workers cannot be identified to species, so males from flight traps are not, at this point, very useful as specimens.

Name Derivation: The name derives from the diminutive of *pulchra* (Latin), meaning "beautiful," or "beautiful little one."

Strumigenys reflexa Wesson and Wesson— Reflexed Mustache Ant (Plate 58)

Taxonomy and Similar Species: This is one of several species with a broad, scalloped clypeus and a prominent, basal gap in the mandibular teeth (diastema) in dorsal view. Unlike two somewhat similar species, *pulchella* and *metazytes*, *reflexa* has jaws that are broadened at the tip (in lateral view), rather than slender. This species differs from *missouriensis*, which also has thick jaws, in having all the hairs on the clypeal margin pointing backward; in *missouriensis*, only one or two pairs along the apical margin of the clypeus point backward. This feature is not as consistent as one might like: sometimes, in both species, one or two sets of hairs may appear to curve down, instead of backward or forward. All the clypeal fringe hairs of *reflexa* tend to be unusually large and spoon-shaped rather than spatulate.

Distribution: West Virginia south into Florida, west into Kansas and Texas (Bolton 2000). In Florida, *reflexa* is known from a few scattered sites, from Highlands County north.

Natural History: There are nine Florida records with habitat information: six from mesic hardwood or mixed hardwood and pine forest, two from swamp forest, and one from a pine buttress in an unknown habitat. I have collected specimens from a mesic pine and hardwood forest in Georgia and have seen three collections from north-facing slopes on the Ouachita Mountains of Arkansas. The type locality is a backyard in Jackson, Ohio, where Wesson and Wesson (1939) found a colony in a rotten, partially buried bit of planking. In an artificial nest, the foragers were "even more sluggish" than those of *pulchella*. This undoubtedly means that the foragers position themselves in a place where a collembolan might pass by and react slowly (except for the deadly mandibular strike) to the presence of prey. The male of *reflexa* was described by Brown (1953).

Name Derivation: From *reflexa* (Latin), meaning "bent back," referring to the position of the enlarged hairs of the clypeal fringe.

Strumigenys rogeri Emery— Roger's Pygmy Snapping Ant (Plate 57)

Taxonomy and Similar Species: The eyes of *rogeri* are on blunt tubercles on the sides of the head, distinguishing this species from other New World *Strumigenys*. The jaws are long and slender, there are no enlarged silvery hairs on the upper surface of the head (as in *emmae*, *louisianae*, and *silvestrii*), and the gaster lacks the dense array of elongate hairs found in *lanuginosa*.

Distribution: This species apparently originated in tropical Africa, where related species occur, but has been introduced into Malaysia, Indonesia, various Pacific islands (including Hawaii, Samoa, Tahiti, Fiji, and New Caledonia), the West Indies, the Bahamas, Mexico, Costa Rica, and Florida (Bolton 2000). The exodus from Africa must have begun early, as *rogeri* was described in 1890 from specimens collected on St. Thomas in the Virgin Islands (Bolton 2000). It was first reported from Florida in 1984 (Deyrup and Trager 1984), but earlier Florida specimens were found in litter samples collected and preserved by Walter Suter in 1965 (Deyrup et al. 2000). It is currently known from many sites in Florida, north into Orange, Seminole, and Osceola counties, well into areas that have winter frosts and short-term freezes. It is often extremely abundant, even in the northern part of its range, and it is quite possible that it is still expanding its range northward. It is probably transported in potted plants and should appear in irrigated areas in southern Arizona and southern California. It has been found in greenhouses in various northern sites where it would not be able to live outdoors (Wetterer 2012c).

Natural History: *Strumigenys rogeri* is found primarily in moist to wet wooded areas, such as well-irrigated landscaped areas, swamp forests, and bayheads. There is a good chance that *rogeri* is replacing some native dacetines in natural habitats, but there has not been any methodical before-and-after sampling to settle the matter. The problem of documenting ecological effects is compounded by the fact that native dacetines that favor wet areas are being affected by several environmental changes in south Florida, such as modification of local hydrology, introduction of wetland habitat-changing exotic plants (such as *Melaleuca*, *Casuarina*, and *Lygodium*), and additional ant species, such as *Pheidole moerens*. Nests of *rogeri* are usually in hollow twigs or sections of branches or nuts, often buried in the leaf litter. The prey of *rogeri* are probably Collembola and other small soil arthropods, but this has not been studied. Males are apparently undescribed. I have seen probable *rogeri* males from flight traps but have yet to make an association with workers.

Name Derivation: This species was named by Carlo Emery for the German ant taxonomist Julius Roger, who published a series of papers on ants between 1857 and 1863. James Trager informs me that the name "Roger" in *Strumigenys rogeri* should be given its Germanic pronunciation with a hard "g," as in (to take a random example) the name Trager. Roger that.

Strumigenys rostrata Emery—
Square-Snouted Mustache Ant (Plate 53)

Taxonomy and Similar Species: This species has a wide clypeus that is straight or very slightly concave across its apical margin. Like a number of other Florida species, *rostrata* has three or four forward-curving hairs along the sides of the clypeus, but the anterior edge of the clypeus has a set of three small hairs on each side that decrease in size toward the midline, so that the clypeal fringe is concave along the apical margin. Unlike several other non–wedge-headed Florida *Strumigenys*, *rostrata* has no basal gap in the mandibular teeth (diastema). This is a (relatively

speaking) large species, all of 2.5 mm in length, and usually blackish brown, so individuals can often be provisionally identified in the field.

The species most similar to *rostrata* is the aptly named *rostrataeformis*, but *rostrataeformis* should not concern North American myrmecologists, as it lives in Japan. This species is not usually available in North American collections, but can be seen as a scanning electron micrograph in Bolton's revision of the dacetines (2000), and there is a color photo on the ants of Japan website. Brown (1949c) supplies a series of small differences that separate the two species and suggests that this pair of species is an example of the biogeographic affinities between eastern North America and temperate eastern Asia. *Strumigenys rostrataeformis* is also an example of a species that could go undetected almost indefinitely if it became established in North America. With this in mind, and also thinking of the 3000 flowering cherry trees donated to Washington, DC, by Japan in 1912, I nervously inspected several series of *rostrata* that I had collected in urban and suburban sites around the city. I was relieved to find that these were normal-looking *rostrata*. I was further reassured to learn that the 1912 trees were actually the second such gift; the first, in 1910, was destroyed, amid general diplomatic consternation, because the trees were found on their arrival to be infested with various insect pests. Still, species of ants such as *Vollenhovia emeryi* and *Strumigenys hexamera* got here somehow.

Distribution: Pennsylvania south into Florida, west into Illinois, Missouri, and Texas (Bolton 2000). In Florida, *rostrata* is known only from the northern edge of the state.

Natural History: In North Carolina, *rostrata* is a common species that usually occurs in mesic hardwood or pine forests (Carter 1962). I have seen many collections from hardwood forest on north-facing slopes in the Ouachita Mountains of Arkansas (Robison and Carlton, collectors). From Florida, where *rostrata* is narrowly distributed in the extreme north, there are 20 habitat records: 14 in mesic hardwood forest, usually near a stream; three in mesic pine and hardwood forest; one in a pine plantation; one in pine flatwoods; and one in live oak litter in a dry depression in sandhill habitat. Nests are reported in rotten wood or in soil (Brown 1953). In the Great Smoky Mountains, I found many nests in hollow twigs and small rotten branches on the surface of the ground in moderately dense, moist forest. Colonies may be large, with more than 200 workers and three to five queens (Brown 1953).

Wesson and Wesson (1939) found that worker *rostrata* in artificial nests were, by *Strumigenys* standards, dynamic foragers. Instead of waiting in a likely spot, the foragers roam about, and, detecting a springtail 2–3 mm away, move carefully toward it, stopping just before the ant's mandibles make contact. If the springtail then blunders into the head of the ant, it is quickly seized and stung. Springtails that move away may be deliberately pursued. Wilson (1953) found that foragers in captive colonies would only take Collembola of the families Entomobryidae, Sminthuridae, and Isotomidae. Other Collembola and other small invertebrates were rejected.

The male of *rostrata* was described by Emery (1895) and redescribed by Brown (1953). Brown points out that the jaws of males have approximately eight conspicuous teeth, unlike other known males of North American *Strumigenys*. Their toothy jaws and large size allow one to provisionally identify those that appear in flight traps as *rostrata*, bearing in mind that some males in the *rostrata* group are still unknown, such as those of *bunki* and *carolinensis*.

Name Derivation: From *rostrata* (Latin), meaning "having a beak" or "having a snout."

Strumigenys silvestrii Emery— Silvestri's Pygmy Snapping Ant (Plate 57)

Taxonomy and Similar Species: This species might be confused with *louisianae*, but the jaws of *silvestrii*, viewed from above, are much more slender. The silvery hairs on the upper surface of the head of *silvestrii* are less expanded than those of *louisianae*.

Distribution: Argentina, Brazil, the West Indies, and the Bahamas; it is considered native to the Neotropics (Bolton 2000). *Strumigenys silvestrii* has also been reported from Madeira, Portugal, and Macau (MacGown et al. 2012). In the Southeast, it is known from Florida west through southern Alabama and Mississippi. In Florida, it is known from a small number of scattered sites. I consider this an introduced species in the Southeast, despite its wide distribution and apparent rarity. The isolation of the North American population and the occurrence of *silvestrii* in Nassau (Bahamas) suggest that this is a species that has been transported from place to place. It is possible, but less likely, that *silvestrii* reached Florida by island-hopping through the West Indies from northern South America.

Natural History: In my experience, this is a woodland species that is never abundant. I have extracted it from leaf litter and have not seen a nest. Its behavior, diet, and males are apparently unknown.

Name Derivation: Named in honor of the Italian Professor Filippo Silvestri, who is famous for his remarkable discovery that certain parasitic wasps can insert in a host larva a single egg that multiplies into numerous eggs (polyembriony). He wrote papers on inquilines in ant nests but did not describe any species of ants. Silvestri traveled in the early 1900s to China, Japan, Korea, Africa, and North and South America, collecting ant specimens that were described by other myrmecologists, such as Emery, Wheeler, Santschi, and Menozzi.

Strumigenys talpa Weber— Mole Mustache Ant (Plate 50)

Taxonomy and Similar Species: *Strumigenys talpa* has a clypeal fringe of long, slightly flattened, evenly spaced hairs that all curve forward. There are no clypeal hairs curving backward, as in *pulchella*. There is a conspicuous basal gap in the mandibular teeth (diastema). The enlarged hairs on the clypeus are elongate, but only slightly widened, not spoon-shaped as in some other some other species, such as *bunki* and *creightoni*. In northern and western populations, all the enlarged hairs on the antennal scapes are bent toward the apex of the scape. Unfortunately, it has become necessary to synonymize *Strumigenys deyrupi* (Bolton), which differs from typical *talpa* in having one or two enlarged hairs near the base of the antennal scape curved toward the base of the scape. I felt most honored to have this handsome ant bear my name, and it was with great chagrin that I began to notice evidence that *deyrupi* was a geographic form of *talpa*. Starting with individuals from the same collection as the type of *deyrupi*, I looked at several hundred specimens from peninsular Florida. I also examined specimens of *talpa* from the western Panhandle and from sites north and west of Florida. Beginning in the northwestern Peninsula, and extending into Leon and Liberty counties, there were sets of individuals with antennal scapes showing hairs in intermediate positions, some even having one antenna with the *talpa* configuration, the other with the *deyrupi* configuration. I was unsuccessful in a diligent search for any other character states separating the two species.

Distribution: Washington, DC, south into Florida (south into Highlands and Sarasota counties), and west into Illinois and Louisiana (Bolton 2000).

Natural History: In North Carolina, *talpa* was collected four times from pine forest, three times from hardwood forest, and three times from a grassy field with scattered pines and red cedars (Carter 1962). Nests may be in soil, grass tussocks, or leaf litter, and may be in open as well as forested sites (Wesson and Wesson 1939). Brown (1953) considers *talpa* one of the most common North American species of *Strumigenys*. This is one of the most abundant of the native *Strumigenys* in Florida and is represented in the Archbold Biological Station collection by more than 200 separate collections. Habitat information for 138 collections includes 64 collections from mesic forest (including oak and pine hammocks, hardwood forest, and urban forest), 57 from xeric forest (often long-unburned scrub and sandhill), 11 from low pine flatwoods, and 6 from swamp forests. These records are not based on methodical sampling in various habitats, but they show that *talpa* tolerates dry and mesic conditions and is relatively scarce in wet habitats. The survey of Florida ants includes large numbers of samples from wet habitats, especially various types of swamp forests. In feeding tests, Wilson (1953) found that *talpa* captured Collembola (Entomobryidae, Isotomidae, Sminthuridae), Japygidae, Campodeidae, and Scutigerellidae, but discarded the japygids and campodeids rather than feeding them to larvae. Foragers were "more active and aggressive" than those of related *Strumigenys* that Wilson studied. The male of *talpa* has been described by Brown (1953).

Name Derivation: From *talpa* (Latin), meaning "mole," as in the burrowing animal. The description of the species (Weber 1934a) does not mention what mole-like feature of morphology or behavior inspired the name. Perhaps *talpa* refers to subterranean habits; in Kentucky, I have found colonies by breaking open clods of clay.

Strumigenys wrayi (Brown)—
Wray Mustache Ant (Plate 50)

Taxonomy and Similar Species: In lateral view, the clypeus of *wrayi* is covered with short hairs that curve strongly backward; this distinguishes *wrayi* from all other Florida species. In dorsal view, *wrayi* has backward-pointing hairs along the sides of the clypeus, and jaws with a basal toothless gap

(diastema). These features are shared with *reflexa*, but *reflexa* has heavy jaws that are not tapered as in *wrayi*, and lacks the short, backward-pointing hairs on the dorsal surface of the clypeus. I have seen specimens from several sites that lack two characters supposedly characteristic of *wrayi*: one or more hairs on the antennal scape that are directed toward the base of the scape, and strongly enlarged tips on the hairs of the clypeal fringe (in these specimens, the hairs are very slightly enlarged at their tips). At first, I thought these might be an undescribed species, but eventually concluded that they were variants of *wrayi*. This is a very rare species in collections, and its variation is unknown. There are no extensive nest series, only a few stray individuals. The collection of the Archbold Biological Station has only 22 specimens, representing eight samples, but this little batch of *wrayi* may be among the largest available.

Distribution: New Jersey, North Carolina (Bolton 2000), Georgia, and Florida. The Florida records are from Leon and St. Johns counties.

Natural History: There are seven collections of *wrayi* with habitat information: one from xeric coastal hammock, one from pine forest, one from mesic pine and oak forest, one from hardwood forest at the edge of a swamp forest, one from a rotten log in mesic forest, one from hardwood litter in a ravine, and one from a yellow bowl trap in xeric scrub (it is unusual to find *Pyramica* workers in bowl traps, as few species forage in the open). It does not seem that the rarity of this species in collections is caused by its restriction to some rare habitat.

Name Derivation: Brown (1950) explains the name: "This species is named for the collector, Dr. David L. Wray, who has added very substantially to the knowledge of the Nearctic ant fauna through his zealous efforts to make known the microgenton of North Carolina." Wray wrote many papers on Collembola, not restricting himself exclusively to North Carolina.

Genus *Temnothorax*

Creeper Ants

Most of us myrmecologists over a certain age will go to our graves occasionally calling members of the genus *Temnothorax* (pronounced **Tem' nō thor" aks**) by their old genus name of *Leptothorax*. The genus name *Leptothorax* is still valid but applies to a relatively small Holarctic group of approximately

20 species (Bolton 2003), none of which occur anywhere near Florida. I am not at all opposed to this generic shift, proposed by Bolton in 2003, but it will take a while for myrmecology to catch up, for the extensive literature to be filed under Temnothorax rather than Leptothorax, and for museum specimens to get new labels on their trays and those trays shifted to the appropriate places. This might be a reasonable juncture for the comment, unnecessary, I am sure, that old determination labels on specimens should never be removed (another can be added), because they provide the only vouchered record of earlier ideas of the identity of a species. So, for example, if a specimen is labeled "Leptothorax wheeleri Smith, det. M. R. Smith 1932," that label is actually more valuable than a "correct" label, reading "Temnothorax smithi Baroni Urbani, det. M. Deyrup 2006."

Most species of Temnothorax move about close to the substrate, rather than standing high on their legs (like Pheidole, for example), and they tend to forage under leaf litter or in the concealment of bark fissures. These behaviors are responsible for the anglicized name applied here to the genus, the "Creeper Ants." While its representatives are small in size and retiring in habits, Temnothorax would nevertheless be a big winner in any biodiversity competition among ant genera. With approximately 315 described species (Bolton 2003), Temnothorax is only exceeded by the Big Three (Pheidole, Camponotus, and Crematogaster), Tetramorium (approximately 360 species), and Strumigenys (approximately 325 species). A final tally of Temnothorax awaits further work in the New World tropics and subtropics, and in China. This will take some time, because many species combine semicryptic habits with habitat specificity, so they are easily overlooked. The snag creeper ant, T. smithi, for example, is usually confined to large, standing dead trees in open areas, and may fail to appear at baits, even when a colony is known to occur in a particular tree.

In the Florida myrmicine fauna, Temnothorax is generally distinguished by its small size, short legs, propodeal spines, and heavy to moderate surface sculpture. Cardiocondyla species look like small Temnothorax, but Cardiocondyla species are almost completely without standing hairs (very unusual among ants). Small species of Tetramorium, the minors of small species of Pheidole, and Wasmannia auropunctata all look like little Temnothorax, but members of these genera all have antennal scrobes (the groove on the side of the head where the antennal scape can repose). Myrmecina has a short petiole, almost square in profile. The one Florida species of Myrmica is much larger than any local Temnothorax. Actually, there are hardly more species of Temnothorax in Florida than the number of character states needed to define the genus, so for the local naturalist, it may be easier to learn the species than to learn the generic diagnosis.

With only nine species, Florida is not particularly well endowed with Temnothorax. One or two species known from nearby states, especially tuscaloosae and subditivus, may yet turn up in Florida. There is the possibility of one or more undescribed species in north Florida, where some habitats have not been thoroughly explored for ants. Even with these possibilities, however, it is clear that Florida is not a center of diversity for Temnothorax. On a continental scale, there is a major Madrotertiary center of diversity in the Southwest, but this diversity quickly drops out to the east, with only texanus, palustris, and possibly pergandei as likely descendants of the western fauna. There is a remarkable radiation of species in the Greater Antilles (Wilson 1988), contributing to Florida the species allardycei and torrei. The remaining five Florida species may represent a small radiation of Appalachian species, but there is no center of diversity in the Appalachians, in contrast to the well-developed mid-elevation fauna of Eurasia. It may be that much of the Appalachians were too densely forested for Temnothorax, which, like many other ant groups, shows a generic preference for open and semiarid habitats.

In Florida, as elsewhere in the North America, identification of Temnothorax species is complicated by considerable intraspecific variability in color and sculpture. Temnothorax pergandei, for example, is golden orange in some upland areas of south Florida, and blackish, brown, or bicolored elsewhere. Temnothorax texanus is variable in both color and surface sculpture. All species are variable in size, and tend to show reduced sculpture in smaller individuals. (Incidentally, this reduction in surface sculpture in smaller specimens is such a consistent rule in the Hymenoptera that when one finds small individuals with accentuated sculpture, one should always look for evidence that the smaller individuals represent a distinct species.) Many species of Temnothorax have been described from small series, and the intraspecific variability has led to many synonyms. If nature had accommodated itself to the needs and limitations of taxonomists, there would be a rule that lineages with pronounced intraspecific variation, such as *Temnothorax*, would not include cryptic species as well. Unfortunately, there is no such rule; if anything, the reverse is true. One should be on the lookout for cryptic species of *Temnothorax*, although Florida is not the best place to look for such species.

Temnothorax, under the old name *Leptothorax*, was divided into subgenera, but these, according to Bolton (2003), are not sustainable when considering the worldwide fauna of *Temnothorax*. Some of these groups merge into each other, while there are other apparent unnamed groupings that are just as distinct as those that have been named. Eventually, it might be possible to recognize some subgenera or species groups for purposes of convenience. Mackay's revision of North American species (2000) sets up a series of "complexes," but these are rather arbitrary and inconvenient, and not easily applied to the Florida fauna. At the species level, William Mackay's revision is an excellent reference, with many valuable species accounts.

I have had some frustrating experiences baiting for *Temnothorax*, with the idea that foragers will lead me to the nest. On encountering a bait, such as a cookie crumb, a forager may spend a long time apparently licking the crumb to obtain sugar and oil. Foragers returning to the nest often have a trail that takes them under dead leaves, any one of which may conceal the nest. After waiting with increasing impatience, the observer may remove a leaf under which a forager has disappeared, only to find it "resting," with no nest entrance in sight. The nest entrance itself may be incredibly small. Although a successful forager may stimulate the appearance of additional foragers, these usually appear in small numbers and cannot be easily traced to the nest. Sugary liquid baits are often the most effective: the forager tanks up with remarkable speed and then drags its bloated gaster directly back to the nest.

Name Derivation: From *temno* (Greek), meaning "cut" or "divided," and *thorax* (Greek), meaning "chest," possibly referring to a notch in the mesosoma of the type species *Temnothorax recedens*.

Temnothorax allardycei (Mann)— Poisonwood Creeper Ant (Plate 59)

Taxonomy and Similar Species: Large propodeal spines, pale color, and reticulate sculpture

easily distinguish this species from all other ants in eastern North America. It is a member of a distinctive lineage that radiated extensively in the Greater Antilles. It was described under the name *Macromisha*, from specimens collected on the island of New Providence, Bahamas (Mann 1920). Wheeler (1931a) gave the name *Antillaemyrmex floridanus* to a Florida population.

Distribution: Known from tropical areas of south Florida. In the Bahamas, known from New Providence (type locality), North Andros (Deyrup et al. 1998), Gorda Cay (Zachary Prusak, collector), and Rum Cay (John Mangold, collector). It is not on lists of Cuban ants (Fontenla 1994, 1997), but it would be surprising if this species was absent from Cuba.

Natural History: In Florida, this species is found in dead, hollow vines, in dead culms of sawgrass (*C. jamaicense*), and in dead twigs on live trees, especially poisonwood (*Metopium toxiferum*). On North Andros, it is common in poisonwood twigs, and also in sawgrass culms where the sedge clumps grow up among shrubs. In my experience, it is not a true marsh species. Dead poisonwood twigs probably contain a smaller dose of skin irritants than leaves or live twigs, but myrmecologists sensitive to poison ivy or poison oak should avoid poisonwood. A species of *Orasema* (Eucharitidae) was reared from a colony on Key Largo.

Like many other ants that live in dead twigs and grass stems, *allardycei* is well suited to laboratory culture. It has been the subject of a series of studies by Blaine Cole. Colonies are generally small, approximately 60 workers and a queen (Cole et al. 1994). There is a dominance hierarchy among workers, and dominant workers are presented liquid food by subordinates (Cole 1981). Dominant workers often lay eggs that, being infertile, develop into males. These offspring of workers constitute a significant proportion of males produced by a colony (Cole 1981). Several additional papers by Cole (1986, 1992) and Cole et al. (1994) are intriguing examples of baseline colony dynamics, but their application to the field natural history of this species is unclear. Real-life colony dynamics are likely to be influenced by such factors as weather, the type and quantity of food available and its distance from the nest, and specific dangers affecting foragers. On the other hand, complex behavioral patterns, such as dominance hierarchies and reproduction by workers, are unlikely to be artifacts of artificial systems, and undoubtedly occur out among sawgrass stems and poisonwood twigs. The male of *allardycei* is known, but apparently not formally described.

As mentioned above, *allardycei* is one of the more widely distributed members of a large group of species of *Temnothorax* that evolved in the Greater Antilles, especially Cuba. This group shows more morphological divergence between species than in all the Nearctic *Temnothorax* put together, and it is probable that species also diverge in behavior and social ecology. We should not assume that social interactions of *allardycei* are typical of other Florida *Temnothorax*.

Name Derivation: The describer of this species, William Mann, was an adventurous ant collector who traveled widely and discovered many new species. His description of *allardycei* (1920) says that this species is "dedicated to Sir William Allardyce, Governor of the Bahamas, a naturalist himself, and a friend of naturalists." Despite this compliment, it appears that William Allardyce did not leave much of a direct legacy in natural history, perhaps because he was kept busy as a professional governor, apparently specializing in islands. He administered, in succession, Fiji, the Falklands, the Bahamas, Tasmania, and Newfoundland.

Temnothorax bradleyi (Wheeler)—
High Pine Creeper Ant (Plate 60)

Taxonomy and Similar Species: The most similar local species is *smithi*, until recently known as *wheeleri*, which has much longer propodeal spines and heavier sculpture. Another somewhat similar species is *schaumii*, which has even smaller propodeal spines than *bradleyi* and has delicate surface sculpture, usually absent entirely on parts of the dorsal surface of the head. It is tempting to suspect that the three species are a compact lineage of arboreal ants: *schaumii* adapted to life in hardwoods and pines in open forests; *bradleyi* specialized for life in fire-resistant pines in frequently burning high pine (also called sandhill) ecosystems; *smithi* also found in sandhills and prairie edges that burn frequently, but nesting in dead pines and large oaks that failed to survive a fire.

Distribution: Alabama, Georgia, and Florida (Wilson 1952). This species is difficult to find and probably occurs in some other states, such as North and South Carolina, that have remnants of a longleaf pine ecosystem that once covered much of the Coastal Plain of Southeastern North America. In Florida, *bradleyi* is known from scattered sites around the northern part of the state, with a few records from Highlands County, where it appears to be rare.

Natural History: From 1913, when the species was first described, through 1950, when Creighton released his ant manual, *bradleyi* was known only from a type specimen, collected in Georgia (Wheeler 1913a). It became possible to find specimens relatively easily when Wilson (1952) published his discovery of the nesting habitat of *bradleyi*: shallow cavities under thick bark of large standing pines in open habitats. Although it is difficult to obtain whole colonies of *bradleyi*, foragers can sometimes be lured to peanut butter or jelly baits. I have had difficulty attracting this species (and most other arboreal ants) during drought conditions; probably, these ants curtail foraging during droughts. The male of *bradleyi* was described by Wilson (1952).

Name Derivation: William Wheeler named this species for James Chester Bradley, who collected the type specimen in the Okefenokee Swamp. Bradley, a professor of entomology at Cornell from 1911 to 1952, was a beetle specialist with a broad general knowledge of entomology. He led several collecting expeditions, including the 1913 trip to South Georgia and the Okefenokee, which must have been a more arduous trip than it would be today. Many other species of insects were also named for Bradley.

Temnothorax curvispinosus (Mayr)—
Yellow Woodland Creeper Ant (Plate 61)

Taxonomy and Similar Species: Distinguished from other southeastern *Temnothorax* by its long, slender propodeal spines (almost as long as the petiole in lateral view) and pale color. The spines are longer than those of *allardycei* (another yellow species); *allardycei* also has coarse, reticulate sculpture on the head and mesosoma, and differs in many other ways from *curvispinosus*. The ranges of the two species do not overlap. In the northern part of its range, *curvispinosus* is sympatric with a similar species, *ambiguus*, but the range of *ambiguus* ends far north of Florida. In the field and in litter extractions, *curvispinosus* somewhat resembles *Cardiocondyla wroughtonii*, but the latter species has no standing hairs on its head or body, and has shorter propodeal spines.

Distribution: The distribution of this species is not as clear as one might expect. Some early records, especially northern records, may apply to *ambiguus* rather than to *curvispinosus*. For example, the distribution map in Mackay's revision has *curvispinosus* occurring in a broad band most of the way across North Dakota, but the Wheelers, who qualify as champion collectors, did not find this species in North Dakota (Wheeler and Wheeler 1963). Specimens I have seen provide a range from southern New England south into north Florida, west into Iowa and Oklahoma; it seems likely that it also occurs in eastern Texas (Mackay 2000). In Florida, *curvispinosus* is known from a few central Panhandle sites and from a site Nassau County. As far as I know, it does not occur in central peninsular Florida, as shown in the *Temnothorax* revision (Mackay 2000).

Natural History: The general habitat of *curvispinosus* is open woodlands and brushy areas. In Florida, it occurs in mature mesic forests that usually escape burning, such as ravines, steepheads, and mesic areas protected by adjacent wetlands. I have obtained specimens by sifting loose litter, but have not seen a nest. In the Northeastern and Midwestern states I have usually found nests on the ground in hollow twigs or pine cones, and in dead plant stalks and empty nuts. In the Appalachians I have found nests in dead twigs on living oaks, and dead shoots on fungus-infected dogwoods.

The behavioral ecology of *curvispinosus* is a pleasure to compile because it has been ably investigated by a number of myrmecologists (Fellers 1987, 1989; Headley 1943; Stuart 1987; Wesson 1939; Wilson 1974a). Mature colonies are large for the genus, with more than 200 workers. Colony fragments may be scattered about in several of the little hollow twigs and other nesting sites that this species prefers. There may be several queens, which display no animosity toward each other. There is an insidious rivalry, however, because each queen recognizes her own eggs, which she treats tenderly, but when she encounters eggs of another queen, she often handles them so roughly that she damages them. Workers are timid but persistent foragers, active during most of the summer. They are constantly moving about and are often among the first ants to find a bit of edible arthropod or plant material. When a larger resource is found, such as a sizeable dead insect, this species, like many other *Temnothorax* species, does not attempt to drag off the booty, or race about wildly in the manner of *Pheidole* or *Paratrechina*, but rather gnaws off a tiny portion. It does so with so little movement of legs and antennae that it is easily overlooked by an observing myrmecologist, distracted by more active species at the resource. If approached by another species of ant, *curvispinosus* is likely to withdraw, moving slowly back when the coast is clear. Foragers returning home from a large resource do not lay down a chemical trail that guides a foraging column but may return to the resource accompanied by a single individual following close behind, a behavior called "tandem running." All this adds up to a way of life that minimizes aggressive or defensive encounters with ants of other species, while guaranteeing a good share of small food morsels and an unobtrusive place at a larger banquet. This is one of the species of *Temnothorax* that is parasitized by other ants (Wesson 1939), but it is unlikely that any of these parasites occur in Florida, where *curvispinosus* is probably too rare to support populations of parasitic ants. The male of *curvispinosus* somehow escaped description for more than 100 years; it was described by Mackay in 2000.

Name Derivation: From *curvus* (Latin), meaning "bent" or "bowed," and *spinosus* (Latin), meaning "spined," referring to the unusually long, curved propodeal spines.

Temnothorax palustris (Deyrup and Cover)— Flatwoods Creeper Ant (Plate 59)

Taxonomy and Similar Species: Similar in size and structure to *texanus*, but easily distinguished by color (*palustris* is yellowish, *texanus* is blackish) and by the shape of the postpetiole in dorsal view (postpetiole almost as long as wide in *palustris*, conspicuously wider than long in *texanus*). In the field, it might be possible to mistake this species for *Tetramorium caldarium* or *T. simillimum*, but it is unlikely that these species would be found in natural wetlands of northern Florida where *palustris* occurs. In the field, the contrast between the darker head and more yellowish body of *palustris* would also be diagnostic, as no other Florida myrmicines share this color combination.

Distribution: Known only from northern Florida: Liberty, Leon, and Columbia counties (Deyrup and Cover 2004a). There is a good chance that *palustris* also occurs in flatwoods habitat in the western Panhandle of Florida and in southern Georgia and Alabama. This is obviously a species that is easily overlooked, as it escaped detection until recently.

Natural History: The colonies that are the basis of the type series (Deyrup and Cover 2004a) were found on small ridges of sandy soil in a marsh in the Apalachicola National Forest. Foragers were easily baited with pecan sandy cookie crumbs, which they carried to minute open nest holes that had no surrounding excavated dirt. Entrances were so small that it was not unusual for a crumb-carrying forager to run right over the entrance, turn around, run over it again, and finally set down the crumb in order to locate the hole. Nests, at least at this site, are shallow, only a few centimeters deep. Stray specimens were also captured in pitfall traps in low flatwoods at several sites, and this seems a good way to sample for *palustris*. This species might be a wet-site species of the same lineage as the dry-site species *texanus*, which it strongly resembles in size, pilosity, and general morphology (Deyrup and Cover 2004a). The male of *palustris* was described by Deyrup and Cover (2004a).

Name Derivation: From *palustris* (Latin), meaning "marshy," referring to the habitat where this species lives.

Temnothorax pergandei (Emery)— Pergande's Creeper Ant (Plate 61)

Taxonomy and Similar Species: The smooth, rounded propodeum with its little, pointed spines is diagnostic for this species, distinguishing it from other Florida ants. The conspicuous groove between the mesonotum and the propodeum (in lateral view) is unique among North American *Temnothorax*. This groove was the main reason why *pergandei* has generally been put in its own subgenus, *Dichothorax* (meaning, approximately, "with the thorax divided into two sections"), occasionally used as a full genus. Males of *pergandei* are also aberrant relative to other Nearctic *Temnothorax*, with a strongly swollen mesonotum and scutellum, and an elongate, flattened head with huge eyes and no malar space. Compared, however, with some West Indian *Temnothorax*, *pergandei* workers look relatively "normal." Clearly, *pergandei* is in its own species group of *Temnothorax*, but the extent of its phyletic divergence from other species groups is currently unknown. In the field this species can be confused with minors of *Pheidole dentata*, and, in areas where the dark form of *pergandei* occurs, with *Solenopsis invicta*.

Color forms of *pergandei* include blackish brown, brown, golden yellow, and bicolored. These have been the bases of species and subspecies names, most of which were rejected by Creighton (1950). A remaining subspecies, *pergandei floridanus*, was synonymized by Mackay (1993a). Nobody has investigated what the color variation might be telling us about the ecology and biogeography of *pergandei*. In Florida, a golden yellow variant is the only color form on the Lake Wales Ridge in the center of the state, while other upland areas that are just as dry and have similar habitat types are usually inhabited by a brown form. When one is in the scrub forest of the southern Lake Wales Ridge, it is noticeable that *Pheidole dentata*, *Dorymyrmex bossuta*, and *T. pergandei* all have the same coloration and are easily mistaken for each other in the field. All three of these species are darker elsewhere in Florida. Arnold Van Pelt, who observed the dark form in north Florida, states (1958), "Its mannerisms and appearance in the field are much like those of *P. dentata*, and it is sometimes necessary to examine closely a wandering individual before a determination can be made." The significance of these convergent colorations is unknown.

Distribution: New Jersey south into Florida, west to Nebraska and Texas (Smith 1979), with apparently isolated populations in the desert of Arizona and New Mexico (Heinze et al. 1995; Mackay 1995). This extensive distribution makes more sense in the light of the remarkable habitat range of the species, discussed below. In Florida, *pergandei* occurs throughout the state, but there are some large gaps in the known distribution, especially in the southern part of the state.

Natural History: This species shows great flexibility in its habitat, which ranges from moist bottomland forest to desert. It occurs in the foothills of the Appalachians up to 2800 feet (Carter 1962). In Florida, it is most abundant, or at least most easily found, in sandy uplands, such as Florida scrub, sandhills (high pine), and former sandhills that have become xeric or mesic forest. It also occurs in pine flatwoods and, surprisingly, in coastal salt marshes. I have not found *pergandei* in strongly modified habitats such as gardens, lawns, improved pastures, and grassy roadsides. Nests in Florida are usually in hollow twigs, nuts or pine cones, partially or completely buried in leaf litter, but sometimes in cavities in leaf litter, in grass tussocks, rotten wood, or sand. Florida populations are likely to be strongly affected by fires that consume leaf litter. I have not found nests in bare soil in the open, as they occur in Arizona (Heinze et al. 1995). At the Archbold Biological Station, this species is monogynous, based on a sample of 35 colonies. Colonies usually included approximately 20–100 workers. In Arizona, *pergandei* produces swarms of males during the day, flying approximately 1.5 meters above the ground (Heinze et al. 1995). Queens fly into the swarm, connect with a male, and mate in the air; each female mates only once (Heinze et al. 1995). The male of *pergandei* was described by Wheeler (1903).

Name Derivation: Named by Carlo Emery in honor of Theodore Pergande (1840–1916), discussed under *Proceratium pergandei*. Thirteen species of North American Hymenoptera, seven of them ants, were named for Pergande, as well as some flies and a couple of Heteroptera.

Temnothorax schaumii (Roger)— Schaum's Arboreal Creeper Ant (Plate 60)

Taxonomy and Similar Species: Delicate sculpture of the head and mesosoma distinguish this species from the arboreal species *bradleyi* and *smithi*, both of which have coarse ridges on the head and mesosoma. In the field, the head sometimes appears smooth and shining. Like *pergandei*, this species occurs in different color forms, ranging from golden yellow, through yellowish brown, to blackish brown. The yellow and blackish forms of *schaumii* may occur in the same area, and have never been considered geographic variants (as in the case of yellow and blackish *pergandei*), but have sometimes been mistaken for separate species. As nest series accumulated in collections, however, mixed colonies of pale and dark workers appeared with disturbing regularity. In 1940, Laurence and Robert Wesson published an account of eight mixed nest series, including one in which both color forms were produced in an artificial nest over the term of a year. This resulted in the synonymy of the former species *fortinodis* (established for the dark form).

Distribution: Maine, south into Georgia, west into Iowa and Texas (Smith 1979). In Florida, there are scattered records in the northern third of the state and records from Highlands County in the south.

Natural History: In Florida, this species occurs in open forests, less commonly in closed canopy forests, and may occur in shade trees of urban and suburban areas. Nests, which occur in live trees,

are in dead branches, hollow twigs, in patches of dead bark, and occasionally under mats of moss. In Ohio, *schaumii* occurs dependably in large oaks (Wesson and Wesson 1940). Nests may have more than 100 workers (Wesson and Wesson 1940). Yellow queens may produce both black and yellow workers, suggesting to Wesson and Wesson that the black coloration was dominant, and the yellow workers must be produced parthenogenetically, but this seems improbable, given the reproductive behavior of other ants.

Temnothorax schaumii is probably a predator on small insects and probably feeds on honeydew and sweet secretions produced by some cynipid galls. At baits on tree trunks, this species is furtive and difficult to see and catch, often hiding in crevices or under bark flakes adjacent to the bait in order to feed without being exposed. The male of *schaumii* was described by Mayr (1886).

Name Derivation: Julius Roger named this species for a Professor Schaum, who collected the type specimen in Pennsylvania. There are several other insects, primarily beetles, which were named schaumii or schaumi around the same time, so Schaum may have been an important collector back in the 1860s through the 1880s.

Temnothorax smithi (Baroni Urbani)— Snag-Creeper Ant (Plate 60)

Taxonomy and Similar Species: This is a reddish brown arboreal species that could easily be mistaken for *bradleyi* in the field, but under the microscope, it is distinguished by its long, sharp propodeal spines, and much stronger sculpture of ridges along the sides of the mesosoma. Unlike *bradleyi*, *smithi* usually occurs in dead trees.

Distribution: North Carolina south into Florida, west into Ohio and Mississippi. Locality records appear to be highly disjunct, but this is probably because it is difficult to find specimens without baiting a large number of dead trees. There are only a few Florida records of *smithi*, but it probably occurs through most of the central and northern parts of the state.

Natural History: Nests are usually in large, weathered dead trees. In the formerly extensive longleaf pine ecosystem of the Southeast, *smithi* was probably a common species living in long-standing dead pines (locally known as standing snags) that had been struck by lightning or killed by fire. In the western part of its range, it may have been a species of corridor forests and transition zones of prairie ecosystems where fires occasionally killed large trees. Although there are many species of arboreal ants in eastern North America, almost all of them are confined to dead portions of live trees, including rotten heartwood, dead branches and twigs, the dead outer layers of bark, and hollow nuts, cones, and galls. Few ant species inhabit dead trees, probably because the resources available in such trees are strictly limited. The diet of smithi is probably wood-eating insect larvae, such as cerambycids, buprestids, anobiids, and their predators, such as wasps and predatory beetles. Many dead trees are inhabited by solitary bees and wasps, whose larvae or prey might be attacked by ants, but this might require resistance to whatever factors (presumably chemical) solitary bees and wasps use to protect their larvae and prey in other situations, such as burrows and sandy soil. Wesson and Wesson (1940) found this species common in hardened, weathered, exposed logs of old cabins exposed to the sun, a habitat that might be equivalent to a pile of fallen snags. Two colonies were found in dead parts of large oaks, and on one occasion workers were seen "feeding" on the cast larval skins of wood-boring beetles. The male of smithi is undescribed, apparently unknown.

Name Derivation: This species was originally described as Leptothorax wheeleri by Marion Smith (1929a). When the genus Macromisha was synonymized with Leptothorax by Baroni Urbani (1978), an earlier-described species, Macromisha wheeleri (Mann 1920) became a senior synonym of Smith's L. wheeleri. Baroni Urbani was therefore forced to rename the North American species and chose to honor its original describer, Marion Smith. Both species were later transferred to the genus Temnothorax. Marion Smith was an important myrmecologist who published a large number of papers between 1923 and 1967. Many of these papers are still very useful, and his key to the genera of North American male ants (1943a) is still the only comprehensive reference on the subject. He also gave myrmecology a major boost by strongly encouraging E. O. Wilson to pursue an interest in ants.

Temnothorax texanus (Wheeler)— Sand-Loving Creeper Ant (Plate 59)

Taxonomy and Similar Species: This species is most similar to palustris but is distinguished by its broader postpetiole (that of palustris is almost as long as broad), darker color (texanus is black to dark reddish brown, while palustris is yellowish), and habitat, which is sandy uplands (palustris lives in low sandy flatwoods and sandy hummocks in marshes). An eastern population of texanus was originally described as a separate subspecies, texanus davisi, raised to species rank by Mackay in 2000. Although the types of davisi from New Jersey look different from the types of texanus from Texas, specimens from the intervening area show every combination of the character states that were supposed to be useful for distinguishing the species, sometimes with striking variation in a single colony (Deyrup and Cover 2004a). As is often the case in the taxonomy of variable ants, the problem in understanding the morphological diversity in texanus seems to have arisen from the limited number of specimens available to myrmecologists attempting to define the species. Temnothorax texanus does have a remarkably extensive range, and there could be two or more undetected species occupying this range, but there is no evidence that this is the case (Deyrup and Cover 2004a). In the field, this small, dark, shining species could be confused with Monomorium viridum or one of the three species of small, sand-loving Pheidole that have black minors.

Distribution: Massachusetts south into Florida, west into Minnesota, Utah, and Arizona (Mackay 2000). In Florida, texanus is known from sandy uplands in Collier County northwards. The lack of records from the Atlantic Coastal Ridge south of Volusia County is probably attributed to insufficient collecting. This species is often difficult to find because it forages in small numbers, often moving about under a thin covering of dry oak leaves or pine needles.

Natural History: Temnothorax texanus is associated in Florida with open sandy areas with scattered trees. It is also known from sandy sites in Ohio (Wesson and Wesson 1940) and Michigan (Wheeler et al. 1994); in North Carolina, most collections by Carter (1962) were from dry, open sites, but two were from mesic forest. In Florida, these ants respond readily to crumbs of pecan sandy cookies, often returning directly to the nest with a tiny crumb, rather than licking it for long periods. Nests are often near the surface in Florida scrub habitat. In interior peninsular Florida, the males are black, and the queens may be either black or brick red. I have never seen more than one queen in a nest. The male of texanus was described by Wheeler (1903).

Name Derivation: The type series was collected in Texas, in the town of Milano.

Temnothorax torrei (Aguayo)—
Minute Tropical Creeper Ant (Plate 61)

Taxonomy and Similar Species: Small size, the lack of any raised surface sculpture, nonshining head and mesosoma, yellow color, and extremely broad postpetiole (about three-fourths the width of the gaster) distinguish *torrei* from all other eastern *Temnothorax*. *Cardiocondyla wroughtonii* is also small and yellow, with an expanded postpetiole, but, like other *Cardiocondyla*, lacks standing hairs on the head and body, represented in *torrei* by small but distinct spatulate hairs on the head and mesosoma. This species was originally described in the genus *Macromischa*, then transferred to *Leptothorax* (Baroni Urbani 1978), and then to *Temnothorax* (Bolton 2003).

Distribution: Cuba, the Bahamas (North Andros, Gorda Cay), the Florida Keys, and tropical Atlantic Coast. It is not known from the Everglades or tropical coastal areas of the Gulf Coast, but it could occur there. This is a tiny ant that probably never leaves the leaf litter and seems to occur in low densities, so it would be easy to miss. There is no good reason to suspect that *torrei* is not native to Florida, and it occurs on nearby islands of the Great Bahama Bank and the Little Bahama Bank.

Natural History: Workers and queens have been extracted from leaf litter in tropical hammocks. A nest with males was found at the base of a large *Pinus elliottii* on Big Pine Key. Naturalists still know almost nothing about the natural history of the many species of *Temnothorax* that are members of the remarkable evolutionary radiation of species in the Greater Antilles, especially Cuba. It seems probable that many of these species have specialized ecological roles, or roles that are different from those of most mainland *Temnothorax*. Since *torrei* is a species that belongs to this Cuban group, its behavior might be especially interesting. Males of *torrei* are known but have not been described.

Name Derivation: This species was named by C. G. Aguayo, a Cuban naturalist specializing in the study of mollusks, in honor of Carlos de la Torre y la Huerta, an eminent Cuban naturalist of the previous generation, also with a special expertise in the land mollusks of Cuba. The island has produced a diversity of endemic land snails that is at least as noteworthy as its diversity of *Temnothorax*. Aguayo (1931) writes, "Described from one specimen found alive among land shells given to me by Dr. Carlos de la Torre, and it is a great honor to me to associate his name with this interesting species."

Temnothorax tuscaloosae (Wilson)—
Alabama Creeper Ant (Not Known from Florida)

Taxonomy and Similar Species: The dark brown color of the head and body, contrasting with the pale appendages, separates this species from other eastern *Temnothorax*. The smooth and shining head and mesosoma are also unique among the eastern species, although there are western species with these features. The shininess, small size, and general shape are similar to those of various species of *Monomorium*, but the large propodeal spines of *tuscaloosae* make it impossible to mistake for *Monomorium*.

Distribution: Known from the type series collected in Tuscaloosa and Elrod, Alabama (Wilson 1950), from several sites in North Carolina (Carter 1962), and from Mississippi (MacGown and Brown 2006). It is not known from Florida, but probably occurs in hardwood forests of the Panhandle.

Natural History: Alabama specimens (Wilson 1950b) were collected from ground nests in forested areas; one was at the edge of a swamp. Workers were also found during the day on low bushes near the nests. North Carolina specimens (Carter 1962) are from hardwood forests, primarily mixed stands of oak and beech; some specimens were extracted from litter and humus taken at the bases of beech trees. A Mississippi specimen (MacGown and Brown 2006) was collected at a peanut butter bait on the ground in a hardwood forest near a creek. The male of *tuscaloosae* is unknown.

Name Derivation: Named for Tuscaloosa County, Alabama, where the type series was found.

Genus *Tetramorium*

Pennant Ants

Species of the genus *Tetramorium* (pronounced **Tet' ra mō" ri um**) have a high petiolar node that is usually squarish or blocky in profile, rising rather abruptly from the narrow basal part of the petiole. This is a large genus of almost 360 species (Bolton 1995), almost all of which are native to the Old World. There is a distinctive group of four species native to southwestern North America, but the four species found in Florida are all introduced from the Old World tropics. Another introduced species, *Tetramorium caespitum*, which has the common name of the "pavement ant," is a European species common in urban areas, occurring as far south as North Carolina. It is not likely to move much further south, considering that it has probably had a couple of centuries to do so if that were an option (Wheeler 1927). The pavement ant introduced me to myrmecology, as I grew up in New York City, which is mostly paved. As a very young child in my small cemented yard, I would play The God of Crumbs for my tiny friends. I discovered that when I put out a little crumb, it would eventually be discovered by an aimlessly wandering ant, which would then head straight back to its nest. I concluded that every ant had memorized as landmarks every pebble protruding from the surface of the rough concrete. It was not until I was in college that I read how bees and ants navigate in a vast and cluttered world, a process that is both simpler and far more sophisticated than I could have imagined. The pavement ant is also known to gather around open burrows of solitary bees, dropping bits of dirt down until the bee comes up to investigate, only to be seized and killed by the ants (Schultz 1982). Fortunately, as a small child, I never observed anything of this sort. If I had, my suspicion that pavement ants were smarter than I was would have turned into an indelible certainty.

Although this genus includes a few dietary specialists, such as *Tetramorium sericeiventris*, which preys on other ants, the great majority of species appear to be generalists, feeding on honeydew, live or dead insects, and sometimes seeds (Bolton 1976). Most species of *Tetramorium* nest in the ground or in rotten wood and hollow twigs on the ground (Bolton 1976). Although *Tetramorium* workers are able to sting and may be abundant, they do not, as far as I know, rush out to defend their nests against humans.

The end of the stinger has a peculiar flat triangular or spatulate dorsal flange, often resembling a microscopic pennant on the end of the stinger. This feature induced Andersen (2002) to invent the English name "Pennant Ants." Although this structure looks as if it would interfere with operation of the stinger, it "nevertheless retains piercing ability" (Kugler 1978), as I, along with many other myrmecologists, can attest. The function of this conspicuous appendage is apparently unknown, although from its position and structure, it looks like it might be used in some way

to deliver or disperse chemicals from the stinger. The stinger of *Crematogaster* species has evolved into a spatulate structure that holds and dispenses a drop of venom. Perhaps the *Tetramorium* lineage has somehow managed to combine venom-holding and stinging functions. As Kugler (1978) puts it, the stinger has "a unique morphology, for which there may be unique behavioral correlates."

Name Derivation: From *tetra* (Greek), meaning "four," and *morion* (Greek), meaning "section," referring to the four-segmented maxillary palps found in almost all species.

Tetramorium bicarinatum (Nylander)— Larger Imported Pennant Ant (Plate 62)

Taxonomy and Similar Species: *Tetramorium bicarinatum* is easily recognized by its square petiolar node (in profile) combined with conspicuous parallel ridges on the front of the head. In the field, this species could be confused with the fire ant *Solenopsis invicta*, which it resembles in size, color, and gait, but even at low magnification, *T. bicarinatum* can be distinguished by its conspicuous sculpture and large propodeal spines. Another tramp species, *pacificum*, is somewhat similar to *bicarinatum*, but is uniformly dark brown, not reddish with a much darker gaster like *bicarinatum* (Bolton 1979). *Tetramorium pacificum* is not known from Florida, but is possibly established in California (Bolton 1979).

Distribution: *Tetramorium bicarinatum*, like some other tramp species of ants, is so widely distributed by human commerce that it is difficult to pinpoint its place of origin. Long thought to be a native of Africa, it now appears to have originated somewhere in Southeast Asia, where it has some close relatives (Bolton 1977). It now occurs in coastal areas and on islands throughout most of the tropics and subtropics, with the exception of parts of coastal Africa, where it is apparently excluded by other species of *Tetramorium* (Bolton 1977; Wetterer 2009a). *Tetramorium bicarinatum* also readily colonizes greenhouses in cold climates, providing many records of the species from well outside the area where it could live outdoors (Wetterer 2009a).

Natural History: In Florida, I have found nests of *bicarinatum* in soil, in rotten wood, and occasionally in dead stalks of large weeds or in rotten branches on trees. The natural diet is living or dead insects and honeydew from sap-sucking insects; occasionally, *bicarinatum* invades houses, where it has been found to be "almost omnivorous" (Smith

1965). When present in large numbers, *bicarinatum* can be an effective predator of insect pests of some tropical crops, but their role in fostering sap-sucking insects can be injurious to crops and to native vegetation (Wetterer 2009a).

When a worker of *bicarinatum* finds a large attractive resource, it returns to the nest and recruits a group of workers (De Biseau et al. 1994). The scout that has found the resource appears to lay down a chemical trail back to the nest and then enters the nest and excites a series of workers by rushing at them and making short runs in the nest. A group of workers then leaves, guided by the scout; the path taken does not match that made by the scout on the way back to the nest but seems to depend on visual cues. It is possible for some followers to use the original trail to find the resource, but they greatly prefer to follow the scout. If the scout is removed on the way to the resource, the follower ants are confused and are unlikely to find the resource. De Biseau et al. (1994) suggest that it may be more effective for recruited workers to follow a scout rather than a trail because it is possible for workers in the nest to respond appropriately to the level of excitement displayed by the scout, while a scent trail provides less information.

One reason for the success of *bicarinatum* as a tramp and colonizer species is a lack of aggression between separate colonies, even if the colonies have been brought together from different continents (Astruc et al. 2001). This means that colonies, which have multiple queens, can fuse or form new colonies by budding, making large populations that easily respond to the level of available resources. Astruc et al. (2001) propose that a species-wide chemical profile is responsible for this lack of intraspecific aggression, and even suggest a chemical mechanism in the form of relatively simple straight-chain alkanes and alkenes. As in the case of the polygynous phase of *Solenopsis invicta*, the ability to create a single population rapidly expanding outward may come at the price of reduced long-range dispersal by single founding queens and, in the long run, reduced genetic diversity and evolutionary plasticity.

Even tramp ants may have something to offer humanity. Rifflet et al. (2012) have discovered a powerful new antimicrobial peptide in the venom of *bicarinatum*, one that combats some worrisome pathogens, including strains of *Staphylococcus aureus*.

Name Derivation: From *bi-* (Latin), meaning "two," and *carinatum* (Latin), meaning "ridged,"

probably referring to the large ridges bordering the upper borders of the antennal grooves.

Tetramorium caldarium (Roger)— Confusing Yellow Pennant Ant (Plate 62)

Taxonomy and Similar Species: *Tetramorium caldarium* strongly resembles *T. simillimum*. Roger, who described *caldarium* in 1857, synonymized it with *simillimum* five years later; it was not recognized again as a distinct species for more than a century (Bolton 1979). Both species are approximately 2 mm long, yellowish or reddish yellow, with heavy sculpturing on the head and body, except for the gaster. The petiolar node is squarish above in profile. The frontal carinae of *caldarium* fade out toward the back of the head, but remain strong in *simillimum*. The ground sculpture between the fine ridges on the front of the head is faint and unclear in the case of *caldarium*, distinct in *simillimum*. Both species are common in southern Florida, so the local naturalist can readily obtain specimens of the two species for comparison. *Tetramorium caldarium* and *simillimum* are structurally similar to the even smaller *Wasmannia auropunctata*, but that species has long, sharp propodeal spines, in contrast to the short, triangular spines of *caldarium* and *simillimum*.

Distribution: *Tetramorium caldarium* is native to Africa, but has been dispersed by commerce to tropical and subtropical areas in the New World and Africa, occurring more rarely in Southeast Asia and the Indo-Australian region (Bolton 1979). There are records from greenhouses in Europe (Bolton 1979), and it has probably been introduced into greenhouses and indoor atria in various places in North America. Published records of *caldarium* before Bolton's 1979 revival of the status of *caldarium* are under the name *simillimum*. In Florida, *caldarium* occurs in disturbed areas at least as far north as Alachua County, but the northern records are mostly from urban areas that are warmer than surrounding countryside.

Natural History: I have found *caldarium* most frequently in paved areas or at the edge of paved areas, but it also occurs in open disturbed areas away from buildings. Nests are in soil, or under debris on the ground; one colony was found in a hollow twig on the ground. Foragers are attracted to dead insects and cookie crumbs. This species appears to be polygynous. Alates (six specimens) were found in June, July, October, and December.

Name Derivation: From *caldarium* (Latin), meaning "hot room," undoubtedly referring to the type collection site, a hothouse for growing pineapples in Germany (Roger 1857).

Tetramorium lanuginosum Mayr— Downy Pennant Ant (Plate 63)

Taxonomy and Similar Species: *Tetramorium lanuginosum* is an unmistakable species, its head and body densely and evenly covered with long, soft, erect down. In lateral view, this ant appears to be enveloped in a kind of pale aura. Closer inspection of this unique fur coat reveals that most hairs divide near the base into two hairs, sometimes three. This is most easily seen on the gaster, where hairs are less dense and can be seen against the dark background of the gaster. For just over a century, species with these weird hairs were grouped into a separate genus, *Triglyphothrix*, but in 1985, Bolton added these species to *Temnothorax*, partly because there are species of *Tetramorium* that have simple hairs while other features suggest their closest relatives are species placed in *Triglyphothrix*. Moreover, in reviewing the diversity within *Tetramorium*, Bolton (1985) found various sorts of strange types of hairs, so segregating a group of species on the basis of unusual pubescence makes little sense in this particular genus.

Distribution: This species is related to species in tropical India and Burma, and probably originated in that area (Bolton 1976), possibly spreading naturally throughout tropical Asia, Malaysia, New Guinea, and northern Australia (Wetterer 2010). It is also possible that it was distributed by commerce through much of that range, as it is a tramp species, recorded from many Pacific islands, from Madagascar, the western Caribbean, and southeastern North America (Wetterer 2010). Outside its presumed native range, *lanuginosum* appears to be especially well adapted to islands, and less well adapted to mainland tropical areas of Africa and Central and South America (Wetterer 2010). It appears to be rare but widely distributed in southeastern North America. In 30 years of collecting, I have never found *lanuginosum*, but there is a recent collection from the Florida Keys.

Natural History: Members of the group of *Tetramorium* species formerly placed in *Triglyphothrix* are general predators and scavengers, foraging slowly, and not known to recruit to baits (Bolton 1976). A nest was found under a rock in the Florida Keys; otherwise, nests are probably in the soil or leaf litter. In the West Indies, I have found this species in disturbed areas, in both open and well-shaded sites. It is nowhere a pest species (Wetterer 2010). The function of the long, dense, erect down is unknown. One might expect this layer of fur to collect fine debris, but all specimens I have seen are completely clean.

Tetramorium lanuginosum is an inconspicuous, small, slow-moving species and one would not expect it to be easily found by casual collecting unless it was reasonably abundant. It is known from an extensive area of southeastern North America, from South Carolina into Florida, west to Louisiana (Smith 1979). The first North American specimens were discovered in 1913, from Louisiana (Wheeler 1916). It appears that *lanuginosum* is now rare or difficult to find; this may be an example of a species that became relatively common and widespread following its introduction and then became much scarcer.

Name Derivation: From *lanuginosum* (Latin), meaning "covered with down." In an earlier publication (Deyrup et al. 2000), I called this species "wooly," but that was a mistake. The Latin word for wooly is *lanatum*, not *lanuginosum*.

Tetramorium simillimum (F. Smith)— Similar Yellow Pennant Ant (Plate 62)

Taxonomy and Similar Species: *Tetramorium simillimum* is easily confused with *caldarium*, as discussed above under that species. The two species occur together in many places, so it is necessary to check the identity of every sample.

Distribution: *Tetramorium simillimum* originated in Africa and now has a pantropical distribution (Bolton 1977). Before Bolton's 1979 paper reviving *caldarium*, records of *caldarium* and *simillimum* were intermingled in the published literature. In Florida, *simillimum* is known from the Keys into Marion County; it probably occurs farther north in urban areas.

Natural History: *Tetramorium simillimum*, like *caldarium*, lives in open disturbed areas, but I have an impression that *caldarium* is more likely to be around buildings and pavement, while *simillimum* is usually in grassy areas or near the beach, often nesting in grass tussocks, as well as under debris and in rotten wood. Working out the ecological differences (if any) between these two similar species might be a relatively easy project that could be done in an urban/suburban area. This species comes to baits, such as cookie crumbs, but seems to recruit slowly. Perhaps recruitment is primarily by following a leader, as in *bicarinatum*. This species appears to be polygynous. I have seen many alate queens of *simillimum*, some taken in flight, many collected from nests. I have seen only one male, which, on closer inspection, proved to be a gynandromorph, with only its right side male. It seems that there is an unusual aspect to reproduction in Florida *simillimum*: either males leave the nest immediately, or few males are produced, or the local population is parthenogenetic. Queens fly at dusk, May–July (six specimens).

Name Derivation: From *similis* (Latin), meaning "similar," because the describer, Fredrick Smith, thought this species was similar to another species now in a different genus, *Temnothorax tuberum*. Fredrick Smith (1805–1879) appears to have been rather like a taxonomically myopic Mr. Magoo of myrmecology, placidly describing species after species without reference to the work of other myrmecologists, his descriptions generally lacking any diagnostic value. Creighton (1950) records some irritated comments by other early specialists forced to deal with Smith's species. In a friendly obituary (Dunning 1879) of Fredrick Smith's "peaceful and uneventful life," one nonetheless gets a hint of the obliviousness that infuriated his fellow myrmecologists: "Regular and methodical in his habits, patient and persevering, laborious and industrious,—like his favorite ants and bees,—he plodded on, piling fact upon fact, and adding to his ever-increasing store of knowledge."

Genus *Trachymyrmex*

Tuberculate Fungus Ants

The genus *Trachymyrmex* (usually pronounced **Trak' ē mur" meks**) is easily identified by conspicuous tubercles of various sizes on the head and body. If there were a term *compound tuberculate*, it would apply to *Trachymyrmex*, as the larger tubercles are themselves covered with tubercles. The only obvious function of these tubercles is to break up both the visual outline and surface of the ant, so that when an alarmed *Trachymyrmex* freezes, it vanishes among soil particles and debris. The tubercles, combined with a hard exoskeleton, might also deter some small predators. I have found,

for example, that antlions usually reject worker *Trachymyrmex septentrionalis*, although the immobility of the ants thrown into the antlion pits may also have a role in rejection. Foraging *Trachymyrmex* workers gather bits of vegetation and caterpillar feces that have fallen from above onto the surface of the ground, so they never need to crawl through leaf litter where large tubercles would impede passage.

All species of *Trachymyrmex* are fungus gardeners, using a compost of plant material to grow filamentous fungi in underground chambers (Brandão and Mayhé-Nunes 2007). *Trachymyrmex* fungal gardens require constant care, including collecting and preparing the compost, setting up new garden chambers, harvesting edible hyphal nodules of fungus, and weeding out competing fungi. Given the complexity of tasks challenging attines, it is not surprising that fungus gardening seems to have evolved in ants a single time (Mueller et al. 2001). The fungal gardens of *Trachymyrmex*, like those of *Cyphomyrmex*, are usually susceptible to destruction by a group of weed fungi that can be controlled by bacteria growing on the head and body of the ants (Brandão and Mayhé-Nunes 2007). This bacterial culture is often seen as a grayish film on Florida species of *Trachymyrmex*.

The three-way mutualism between ant, fungus, and bacteria has excited the interest of many evolutionary ecologists, with the predictable discovery of increasingly complex permutations of a basic story. *Trachymyrmex* species control weed fungi in several ways. They pull out weed fungi, they place pieces of their own fungal cultivar on top of patches of weed fungi, they kill weed fungi with secretions of their metapleural gland, and they have large numbers of secretory pockets or pits on the body that nourish bacterial colonies that specifically attack weed fungi (Fernández-Marín et al. 2013). A study of weed control by four *Trachymyrmex* species indicates that there may be trade-offs in types of weed control (Fernández-Marín et al. 2013). This study found less reliance on physical weeding in species that make extensive use of chemical and bacterial control, and less abundant bacterial cultures in species that spend more time applying chemicals from the metapleural gland (Fernández-Marín et al. 2013). It is suggested (Fernández-Marín et al. 2013) that adaptive management through physical weeding and implanting the edible fungal cultivar may be more compatible with adoption of new fungal strains than chemical control. There is good evidence that some attines adopt new fungal cultivars relatively frequently (Mueller et al. 2005). The human species may not be alone in benefiting from genetic diversity in essential crops.

Trachymyrmex is a genus exclusive to the New World, like other fungus-growing ants, all of which are attines (tribe Attini). Most species of *Trachymyrmex* are tropical, but several have adapted to temperate or south-temperate zones in North America. There are approximately 45 described species, with others waiting to be described in an ongoing series of revisions by Brandão and Mayhé-Nunes. While every worker *Trachymyrmex* looks as if it were covered with taxonomically useful morphological features, the conspicuous tubercles and ridges can vary within species, and it has been difficult in some cases to pick out reliable diagnostic features. North American species have been reviewed by Rabeling et al. (2007). There are only two Florida species of *Trachymyrmex*.

Trachymyrmex jamaicensis (André)— Caribbean Tuberculate Fungus Ant (Plate 64)

Taxonomy and Similar Species: *Trachymyrmex jamaicensis* is similar in general morphology to *septentrionalis*, but conspicuously larger (approximately 4.5 mm rather than 3 mm long) with relatively longer legs. Under the microscope, *jamaicensis* has four narrow tuberculate ridges on the gaster; these ridges are absent in *septentrionalis*. *Trachymyrmex jamaicensis* in Florida is restricted to tropical hardwood hammocks and tropical coastal hammocks, habitats where *septentrionalis* is absent.

It is fortunate that *jamaicensis* from Florida were not examined too carefully at the time when species and subspecies names were being awarded to local variants of *jamaicensis*, or the Florida population might have been given its own name. Florida specimens are distinguished by stouter mesosomal projections (Mayhé-Nunes and Brandão 2007) and especially by expanded, thickened frontal lobes, with a strong bend in the antennal scapes where they fit into these lobes, and unusually wide antennal scapes distal their basal constriction. If these features are restricted to the Florida population, they would suggest that *jamaicensis* has been in Florida long enough to evolve noticeable morphological differences.

Distribution: This species is widely distributed in the Caribbean, including Jamaica, the Greater Antilles, the Bahamas, a few of the islands of the Lesser Antilles, and tropical areas in the Florida Keys and the Atlantic Coast of Florida (Rabeling et al. 2007). Smith (1954), in his study of the ants of Bimini (Bahamas), suggested that the Florida population might have been introduced, but since *jamaicensis* has apparently moved throughout the Bahamas, there is no reason to believe it would not be able to reach Florida independently. Bimini itself is closer to Florida than it is to any Bahamian island. Queen *jamaicensis* have a relatively thick integument and a nonswollen gaster; they should be able to disperse widely, as they have through the Caribbean.

Natural History: *Trachymyrmex jamaicensis* nests on beaches and in dry Caribbean scrub forest (Torres et al. 1999; Wheeler 1907). In the Florida Keys, I have usually found this species in dry tropical hammocks, the nests in pockets of soil amid chunks of limestone. The nests usually have a distinctive turret of plant material, a feature not reported in other populations. Wheeler (1907) suggests that nest turrets in the western species *turrifex* prevent rain from washing into the nest, and it is possible that turrets of Florida *jamaicensis* keep rain temporarily pooling in limestone pockets from pouring down the nest entrance. Nests I have seen in the Florida Keys were in cracks or cavities in limestone. Although there seems to be plenty of habitat in the Upper and Middle Keys, it could be difficult for a nest-founding queen to find an accessible cavity of suitable size that does not flood in wet weather. Nests in the Bahamas excavated by Wheeler (1907) and Weber (1967) had fungus gardens suspended from small roots growing into the chamber. Weber (1967) noticed that foragers were usually active at night and early morning, probably to avoid excessive daytime temperatures. Avoidance of diurnal predators may also be important: when Weber was excavating a nest, a small lizard (probably an anole) came over and ate some of the ants. The foragers I have seen returning to the nest were carrying small bits of vegetation, such as dead flowers, but fragments of woody material and fragments of green leaf are also collected (Weber 1967). On Andros Island (Bahamas), Wheeler (1907) observed foragers cutting bits out of corn leaves in the manner of leafcutter ants of the genus *Atta*. In Puerto Rico, Torres et al. (1999) found the pulp of fleshy fruits to be the usual substrate for fungal gardens. At the Puerto Rico study site, *jamaicensis* preferred to nest in deeper soil of ravines, avoiding shallow soils. Torres et al. (1999) initially hypothesized that *jamaicensis* might affect species composition of dry forests by discarding seeds in nutrient-rich refuse piles, but discovered instead that while seeds of some species germinated better in refuse piles, the porous debris of these piles retained water more poorly than normal soil, so seedling survival was lower.

Trachymyrmex jamaicensis in Florida may be a species of conservation concern. It belongs to a group of West Indian species of plants and animals whose North American populations are confined to the Florida Keys, or the Keys and small nearby areas of the Mainland (Moreau et al. 2014). It appears to be uncommon or absent on most Keys, with the exception of Elliott Key. The Florida population of *jamaicensis* is likely to be affected by continuing development of the Keys, rising sea level, which is already changing some plant communities (Ross et al. 2009), and possibly by pesticide residue from spraying to control mosquitoes.

Name Derivation: From Jamaica, where the type specimens were collected, and the suffix *-ensis* (Latin), meaning "a resident of."

Trachymyrmex septentrionalis (McCook)— Northern Tuberculate Fungus Ant (Plate 64)

Taxonomy and Similar Species: *Trachymyrmex septentrionalis* is easily distinguished from *jamaicensis* by smaller size (approximately 3 rather than 4.5 mm long) and the lack of narrow four tuberculate ridges on the gaster. *Trachymyrmex septentrionalis* lives in sandy areas throughout Florida (except for the Keys), while *jamaicensis* is found in tropical hammocks in southernmost Florida, primarily in the Keys. West of the Mississippi, *septentrionalis* occurs together with the similar species *turrifex* in some areas; Rabeling et al. (2007) present convenient ways to distinguish between the two species. Like other species of *Trachymyrmex*, *septentrionalis* is somewhat variable, leading to subspecific and varietal names that have been abandoned (Rabeling et al. 2007).

Distribution: *Trachymyrmex septentrionalis* occurs from southern New York south through Florida, west into Illinois and southwest through most of eastern Texas (Rabeling et al. 2007). Inland from the sandy Coastal Plain, *septentrionalis* is generally confined to sandy areas along rivers or seacoast, so its occurrence is sporadic, especially in the northern part of its range. In Florida, *septentrionalis* is found wherever there are natural habitats with well-drained sandy soil, even on small ridges surrounded by lowlands. It is absent from most of the interior Peninsula south of Lake Okeechobee, an area of marshes and cultivated muck soils.

Natural History: *Trachymyrmex septentrionalis* is the best-known species of *Trachymyrmex*, primarily because it is common within the home range of the ever-curious Walter Tschinkel and his students and colleagues in the Tallahassee area. Foraging workers collect plant debris, such as oak catkins and bits of leaves, as well as caterpillar droppings (Wheeler 1907). One might expect dense populations of *septentrionalis* in forested sites, such as a multilayered closed-canopy woodland in Florida scrub habitat, with plenty of fallout of compostable materials. This is not the case; colonies usually occur in open sites (Wheeler 1907), with the highest numbers, sometimes more than 1000 colonies per hectare, recorded from frequently burned sandhill areas with well-separated longleaf pines (Seal and Tschinkel 2006a). The warmer temperature in open areas appears to increase worker activity and fungal growth, although the most barren areas appear to be too hot for nest-founding queens (Seal and Tschinkel 2006a). In prime sandhill pinelands, *septentrionalis* may excavate more than a metric ton of soil per hectare each year (Seal and Tschinkel 2006a).

Trachymyrmex septentrionalis in Florida appears to be well adapted to upland habitats that burn frequently or occasionally. It is a member of a suite of species that thrive in longleaf pine stands with summer fires every two to five years (Seal and Tschinkel 2006a), a fire regime ideal for management of sandhill pinelands (Myers 1990). The eastern range of *septentrionalis*, up through the coastal sandhill areas of the Southeast, north into the pine barrens of New Jersey and New York, corresponds with habitats maintained by fire. Even in Ohio, where *septentrionalis* is confined to a small area in the southern part of the state, it is found associated with pitch pine, a fire-adapted species (Coovert 2005). While fires, especially summer fires, open up habitats and promote growth of fire-adapted plants (Myers 1990), they also cause a temporary scarcity of compostable materials falling from above. At the Archbold Biological Station, and probably elsewhere, *septentrionalis* deals with a dearth of plant material and caterpillar droppings by becoming a leaf-cutter, emerging in groups from the nest, trailing to plants that resprout days after fire, climbing the stems, and cutting notches from leaves in the manner of obligate tropical leaf-cutting ants. Facultative leaf-cutting has been observed in other species of *Trachymyrmex* (Brandão and Mayhé-Nunes 2007; Mehdiabadi and Schultz 2009), although not in the context of fire adaptation. Facultative leaf-cutting in response to periodic disturbances such as fire might have evolved in a common ancestor of *Trachymyrmex* and the related genus *Acromyrmex*, becoming obligate in *Acromyrmex*.

Population surveys of *septentrionalis*, such as the studies by Seal and Tschinkel (2006a, 2008), are relatively easy because of the unique and conspicuous nest mounds, which form a high crescent on one side of the nest. There is only one such mound per colony. In fact, *septentrionalis* is one of the few Florida ants whose colony density can be estimated quickly and directly. Numerous ecology students responsible for projects that can be finished in a week have independently chosen to study orientation of *septentrionalis* mounds, intrigued by the fact that there seems to be a consistent orientation in some spots, more random in others. This was studied methodically by Tschinkel and Bhatkar (1974), who found that compass orientations were random, but where there was a slope, the mound was normally downslope; altering the slope caused a gradual shift toward the new downslope position. This does not answer the researchers' other question: how does a worker emerging with a pellet of sand know where to leave it? Some kind of visual cue seems to be involved, but its nature is unclear. If the mound is removed, as it might be by a heavy rain, it is replaced in the same position, as if individual workers had learned the appropriate dumping site, which might well be the case. The shape of the nest mound crescent is usually symmetrical, and it is easy to observe that this is not caused by pellets of sand tumbling down from the top, but rather by directional deposition that resembles a bell-shaped curve, or as if excavating workers fanned out from the nest entrance. With current technology, one could easily set up cameras that would allow one to speed up the construction of the nest mound to provide a clearer picture of behavior of workers. The adaptive significance, if any, of a one-sided mound is also unknown. At a time when so many ecological studies, no matter how small, seem to tout themselves as revealing the linchpin of the Universe,

one can only applaud the intentional humor of the final sentence of Tschinkel and Bhatkar's 1974 publication: "We are continuing to work on this fascinating and unimportant problem."

For years, I blithely impressed visiting naturalists with the co-evolution story of *Trachymyrmex*, its mutualistic garden fungus, and its mutualistic fungicidal bacteria nurtured in glandular pits on the bodies of the ants. With a good hand lens, it is even possible to demonstrate the bacterial film on many workers. Unfortunately, I was casually extrapolating from work on other *Trachymyrmex* species. When microbiologists turned their attention to *septentrionalis*, it turned out that the weed fungus specifically targeted by the fungicidal bacteria does not occur in gardens of *septentrionalis*, at least in Texas, where multiple nests were sampled throughout the year (Ishak et al. 2011). There were plenty of types and strains of bacteria, but their function is unknown. Almost inevitably, the story of *Trachymyrmex* microbiology seems to be reaching a stage of terminal complexity, as far as research is concerned. Both the cuticular bacterial films and bacterial residents of fungus gardens appear to constitute separate "microbiomes," elaborate and dynamic bacterial communities (Ishak et al. 2011). It is difficult to read about this without a guilty nostalgia, so familiar to modern biologists, for clearer scenarios presented by less generous information. Apparently, the multiple roles of numerous microbes might be context dependent, shifting in ecological importance, including temporary associations, with the most abundant species not necessarily key players (Ishak et al. 2011). The edifying little tale of three-way co-evolution may have to be replaced by a massive opus that could have been written by a microbial Dostoyevsky. The somewhat discouraging conclusion of Ishak et al. (2011): "A pluralistic conceptual approach acknowledging diverse functions of microbiomes is therefore most likely to advance understanding of the attine-associated microbes." As a simple scientist, however, I am drawn to simpler hypotheses. Perhaps the major weed fungi are poorly adapted to the ecology of *septentrionalis*, either its seasonal cycle, or its habitat of sandy well-aerated soil with rapid and drastic changes in moisture. There is no reason to assume speedy evolution to dispense with the normal *Trachymyrmex* bacterial symbionts, even if the relationship is no longer vital.

Turning with relief to a grosser consideration of *septentrionalis* fungus gardens, the dense distribution of *septentrionalis* colonies in certain habitats caused Seal and Tschinkel (2008) to wonder whether compostable materials might sometimes be a limiting resource. Accordingly, they supplemented colonies with frass donated by native caterpillars feeding on native plants. Supplementation increased growth and biomass of symbiotic fungus, resulting in larger numbers of ants. This study shows that *septentrionalis* populations may be limited by available substrate for fungal gardens, meaning that colonies may be competing for a vital resource. Despite this, there is no evidence of territoriality in *septentrionalis*. One reason for this may be that the kind of extreme crowding that would favor strong territoriality does not occur in *septentrionalis* because founding queens, which must forage extensively for their beginning fungus garden, are unable to establish in areas with numerous foraging workers (Seal and Tschinkel 2008). Since a founding *septentrionalis* queen depends on fungus to feed her young, she does not need the internal resources, primarily fat, needed by a queen of a species that raises her first set of workers on secretions from her body. While this lack of internal resources may be a problem for founding queens where foraging is difficult, it must be an advantage in dispersal to carry 50%–75% less fat than queens of species that do not forage to feed their first set of workers (Seal and Tschinkel 2006b). Perhaps more importantly, queens are cheaper to produce, so the real test of colony productivity, the number of reproductives, is theoretically more efficient in fungus-gardening ants.

Trachymyrmex septentrionalis, like other fungus-growing ants, probably has its share of interesting natural enemies and insect associates, but there are only hints of this in the literature, and this would be a good subject of natural history research. The only proven predator is the eastern narrow-mouthed toad, *Gastrophryne carolinensis*, a specialized predator of nocturnal ants (Deyrup et al. 2013). A study of stomach contents of 146 toads revealed 71 *septentrionalis*, distributed among 27 toads (Deyrup et al. 2013). A tantalizing paper on a rare firefly, *Pleotomodes needhami*, reports 23 larval and 2 pupal fireflies in the fungus gardens of *septentrionalis*, but in the laboratory, these larvae ignored adult and larval ants, and the diet of *P. needhami* is still unknown (Sivinski et al. 1998). Adult fireflies were found on the surface by the entrances of *septentrionalis* and also *Odontomachus relictus* (Sivinski et al. 1998). Some diapriid wasps (Diapriinae) are associated with attines (Masner and García 2002), and it would not be surprising to find a species in nests of *septentrionalis*. Unidentified milichiid flies also occur in nests of *septentrionalis*; their diet is unknown, but they are attacked and killed by the ants (Sivinski et al. 1998). In southwestern North America, the army ant *Neivamyrmex rugulosus* attacks *Trachymyrmex arizonensis* (LaPolla et al. 2002), and it would seem that *septentrionalis*, which nests in sand easily tunneled by army ants, might also be vulnerable to local species of *Neivamyrmex*.

Alate *septentrionalis* in Florida are produced March–May, departing in locally synchronized flights following the first summer rains, usually in June (Seal and Tschinkel 2006b). Flight trap records indicate that there may be additional minor flights in October and November. Male size varies conspicuously; it is possible that small males are less handicapped than small queens, and workers apportion limited resources accordingly. Although there is no evidence of sex-based resource provision in *Trachymyrmex*, it occurs in parasitoid ichneumonids and braconids (Deyrup 1975).

Name Derivation: From *septentrionalis* (Latin), meaning "northern," referring to the New Jersey locality where this species was first collected, far north of the range of any other fungus-gardening ant. The *septen* part of the name is derived from the Latin for "seven," which makes little sense except to a scholar of mythology, like at least one of the authors of the review of North American *Trachymyrmex* (Rabeling et al. 2007). "The species name *septentrionalis* refers to the seven plowing oxen, the brightest stars of the Great Bear constellation, which dominate the skies of the Northern Hemisphere."

Genus *Wasmannia*

Little Fire Ants

The genus *Wasmannia* (pronounced **Was man" i uh**) is a neotropical group of 10 described species, 9 of which are relatively rare, while the 10th is all too common in many disturbed sites through the Neotropics (Longino and Fernández 2007). This same species, *auropunctata*, has also been introduced in parts of the Old World tropics, sometimes with devastating results (Wetterer 2013a). This is the only species of *Wasmannia* occurring in Florida. Similar Florida ants are covered below under

W. auropunctata. A few references, such as Smith (1979), place this species in *Ochetomyrmex*, but it has been returned to *Wasmannia* (Bolton 1995). The little fire ants are not closely related to fire ants in the genus *Solenopsis*, but both have a fiery venom.

Name Derivation: August Forel named this genus in honor of the Austrian myrmecologist Erich Wasmann (1859–1931). Wasmann wrote many important papers on ants and termites, but he is best known for his fascination with arthropods that live with ants. He wrote more than 150 papers on this subject, publishing a list of 1246 species in 1894 (Wheeler 1910a). Wasmann described many of the weird morphological traits found in myrmecophiles and discussed the adaptive value of these traits. The little-used term *Wasmannian mimicry* refers to species of insects and spiders that resemble the ants or termites with which they live. Insects and spiders that live with ants and also resemble ants may be protected from general predators by whatever defenses the ants may have, to which they may add their own defenses, such as jumping, that protect them from predators of ants. Even entomologists who are unfamiliar with Wasmann's research often repeat his name, thanks to the notoriety of *Wasmannia auropunctata*, the little fire ant. Erich Wasmann has a status resembling that of researchers whose names are associated with some dreaded medical condition.

Wasmannia auropunctata (Roger)— Little Fire Ant (Plate 63)

Taxonomy and Similar Species: *Wasmannia auropunctata* is similar to other species of *Wasmannia*, but none of these occur in Florida, as far as anybody knows. In the Florida, ant fauna *auropunctata* can easily be mistaken for minor workers of the small, reddish yellow species of *Pheidole*, such as *moerens*, *dentigula*, *floridana*, and *flavens*. In the field, colonies of *auropunctata* appear as a dense cluster of slow-moving, yellowish, tiny ants, with no major workers but frequently with one or more large queens. A tiny, reddish yellow ant that delivers a painful sting is definitely *auropunctata*. Under the microscope, *auropunctata* shows a distinctive hatchet-shaped petiole (in lateral view) and longer, sharper propodeal teeth than those of local species of *Pheidole*. Minor *Pheidole* may have obvious frontal carinae curving back from the clypeus, but these are much more strongly developed in *auropunctata* and can be used to identify specimens in mixed samples of small

ants. *Tetramorium simillimum* and *caldarium* are structurally similar to *auropunctata*, but are larger, with short, triangular propodeal spines.

Distribution: *Wasmannia auropunctata* is native to the tropical and subtropical mainland of the New World, introduced into Florida and the Galapagos (Wetterer 2013a). It might be either native or introduced in the West Indies (Wetterer 2013a). In the Old World, there are introduced populations in Africa, Israel, Papua New Guinea, New Caledonia, Australia, and a series of Pacific Islands (Wetterer 2013a). It is currently absent from parts of Africa and all of India and Southeast Asia, but this could change, as colonization of some Old World sites is recent and progressing (Wetterer 2013a). Expansion in outdoor habitats in North America is evidently limited by cold weather. Northern Florida is apparently unsuitable for *auropunctata*, but populations may well occur in protected sites around buildings in cities as far north as Jacksonville.

Natural History: In Florida, nests may be in leaf litter, under objects on the ground, or arboreal in dead wood or under bark. It is unusual to find subterranean nests in dry, sandy habitats, but I have occasionally found such nests associated with sap-sucking Homoptera on roots of grass clumps. Workers are easily observed visiting extrafloral nectaries, for example, those on *Passiflora incarnata*, and also visit honeydew-producing insects. Honeydew is sometimes reported as the primary diet of *auropunctata* (Spencer 1941). Foragers are easily attracted to dead insects offered as bait.

Understanding of the insidious nature of *auropunctata* has grown slowly. On the one hand, this species has long been known as an invasive species able to displace other ants (Smith 1937). On the other hand, everybody who has observed the behavior of *auropunctata* has been struck by the apparent passivity of its workers. Individual workers move slowly, are timid at baits, freeze or retreat when threatened, and are inefficient at retrieving food (Spencer 1941; Vonshak et al. 2012). In Florida, even in the field, foragers can be identified as *auropunctata* rather than similar-sized *Pheidole* on the basis of their leisurely gait. Nicolas Kusnezov (1951) remarks that the workers are not combative, which is certainly true compared with the larger species of *Solenopsis* fire ants infesting the region of Argentina that Kusnezov called home. Even more striking, when a nest is uncovered, for example, a nest under bark or under a rock, the workers do not rush with their brood to shelter, nor do they mobilize to attack the intruder. They remain in a bright yellow patch of densely crowded workers, which, as Kusnezov states (1951), stands out brightly against any dark background. Reasoning that the imperturbable attitude of *auropunctata* must be justified by strong defenses, Howard et al. (1982) investigated both the sting and mandibular gland secretions of *auropunctata*. The sting was shown to kill *Monomorium minimum* workers within seconds. Alkylpyrazines produced by mandibular glands repelled other ants, including fire ants (*S. invicta*), which staggered off rubbing their appendages. This same chemical attracted and alerted *auropunctata* workers. Workers are also defended by long spines that protect the vulnerable petiole area and deep antennal scrobes that allow the antennae to be stowed flat against the head. The aggressive mode of *auropunctata* is more difficult to document with casual observation. To date, it is best shown by the work of Vonshak et al. (2012), who set up arenas for interactions between *auropunctata* and two other species of ants, *Monomorium subopacum* and *Pheidole teneriffana*. As I interpret the experiments, small numbers of *auropunctata* workers slowly infiltrate both the foraging areas and nests of other ants, going into a defensive posture when challenged. Over time, however, numbers of *auropunctata* workers can build up to a level that allows them to band together and use their sting as well as their repellent and irritating chemical defense to drive other ants from the foraging arena and kill the inhabitants of the infiltrated nests. The larvae of the conquered colony are gathered together and probably fed to the *auropunctata* larvae.

Wasmannia auropunctata is able displace a great variety of other ants, according to field studies that seem to document disappearance of native ant fauna following invasion by *auropunctata* (Clark et al. 1982; Le Breton et al. 2003; Lubin 1984). It is not known whether *auropunctata* displaces all species in the same way as seen in the laboratory experiments of Vonshak et al. (2012). In the Galapagos, *auropunctata* can eliminate a much larger species, *Camponotus planus* (Clark et al. 1982), a species that should be able to fight back effectively. The shift to attack mode remains unclear and would make an interesting study. It might be that achieving a certain number of workers in or around the nest of another ant triggers a behavioral shift. Alternatively, it might be that *auropunctata* makes its move when members of the target colony become habituated to the presence of *auropunctata* workers. Whatever the method by which *auropunctata* overcomes neighboring colonies of ants, an initial rapid population growth of invading *auropunctata* might be fueled in part by killing and consuming the colonies of other ants.

The sting of *auropunctata* is painful to humans, out of all proportion to the size of the ant, as Creighton puts it (1950). Kusnezov (1951) disputes this, but may not have put his theory to the test; it is not easy to get these ants to sting, unless they have been trapped between the skin and a sleeve or a collar. In my experience, they often sting if they are caught in a film of sweat on the body. The pain of the sting is persistent, lasting, in my experience, for approximately 15 minutes, but other victims have told me it lasts longer. Attacks on humans almost invariably occur when there are great numbers of ants visiting sap-sucking insects in overhanging vegetation. A small proportion of the ants tumble out of the trees or shrubs, some landing on any humans below. These ants are so tiny that their landing is not detected, but those that get caught in sweat or trapped under clothing defend themselves by stinging. Shaking an infested tree increases the rain of ants; fruit pickers yanking citrus from infested trees are especially susceptible, and in extreme cases may be unable to harvest (Spencer 1941). *Wasmannia auropunctata* is small enough to pass through screening, sometimes entering houses to forage. I encountered a family who complained they were frequently stung when sitting on the sofa watching TV. The ants were probably gathering small crumbs from snacks. When there is a major infestation near a swimming pool, swimmers emerge with a batch of floating ants adhering to their skin and are stung when lounging back against the cushions of poolside recliners. Ants can also be brought indoors on the fur of pets that have been moving through infested vegetation.

Other vertebrates are also affected by the stings of *auropunctata*. There is a strong correlation between occurrences of gradual, cataract-like blindness and outbreaks of *auropunctata* (Wetterer and Porter 2003). This has been observed in dogs and cats, as well as some livestock. A direct link between blindness and *auropunctata* has not been established experimentally; such experiments would obviously be unethical. There is a correlation between infestation by *auropunctata* and reduction or disappearance of populations of endemic lizards in New Caledonia, although the reason for this correlation is unknown (Jourdan et al. 2001). There will probably be similar reports as *auropunctata* continues its expansion through the Old World tropics. There is no reason to believe that the powerful sting of *auropunctata* evolved to deter vertebrates. Unlike the fire ant *Solenopsis invicta*, the little fire ant does not swarm out to protect its nest. Perhaps, *auropunctata* can repel birds and lizards that come upon colonies under bark or in leaf litter, but I have not seen reports of this.

Wasmannia auropunctata, like some other invasive species of ants, can produce populations that are "supercolonies" or "unicolonial," with no aggression between what look like separate colonies. In this case, it appears that single queens may found invasive populations, so even individuals taken from distant parts of the introduced range are able to accept each other (Mikheyev and Mueller 2007; Mikheyev et al. 2009). In parts of its native range in Brazil, individuals from separate colonies maintain their aggressiveness (Errard et al. 2005). The absence of aggression in unicolonial populations is apparently the result of close genetic relationship throughout the population, as queens in such populations reproduce parthenogenetically (Fournier et al. 2005). Unlike most other parthenogenetically reproducing ants, *auropunctata* also produces males, which mate with queens to produce workers that are more genetically diverse, as they share genes with both parents (Fournier et al. 2005). There is however, a strange genetic anomaly in this species. Males, which are haploid as in all male ants, are produced by the queens, but when a male mates with a queen, his genome somehow prevents the female from producing haploid eggs other than those with his own genome (Fournier et al. 2005). Therefore, males, although they require a female to produce more males, are also clones of their male parent. There is an enjoyable analysis of the evolutionary factors that allow this situation to appear and persist (Queller 2005), but unicolonial populations of *auropunctata* may have no more long-term stability than other parthenogenetic populations. While workers acquire genetic diversity through recombination, there is no efficient way to get favorable recombinants back into the reproductive lineages. If, for example, a disease evolves to afflict a typically densely packed population of *auropunctata*, a mutation that conferred resistance would need to appear simultaneously in both male and female lineages, since both are required to produce workers. While a genetic system in which males can reproduce themselves clonally may tickle the fancy of some male biologists, in evolutionary terms, the cloned males are likely to find themselves on the receiving end of Princess Leia's classic comment in the movie Star Wars: "Into the garbage chute, Flyboy!"

In undisturbed habitats in its native range, *auropunctata* is a minor member of a diverse ant fauna (Longino and Fernández 2007; Wetterer and Porter 2003). This suggests that natural enemies or other factors can control populations of *auropunctata*; Wetterer and Porter (2003) suggest a search for useful biological controls. In Florida, I have seen several impressive but highly restricted outbreaks, usually associated with plants with abundant extrafloral nectaries, such as candle bush (*Senna alata*). *Wasmannia auropunctata* has been in southern Florida since 1924 or earlier (Wheeler 1929) and is always found with numerous other species of ants, some of them also introduced. It is possible that *auropunctata* went through an explosive phase in Florida before encountering some kind of natural enemy, or it might have never been a dominant species.

Name Derivation: Apparently from *aurum* (Latin), meaning "gold," and *punctata* (Latin), meaning "spotted" or "covered with small pits." The antennal scrobes and much of the mesosoma are covered with tiny, regular reticulations, which might be interpreted as punctures or tiny pits, but they are the same golden color as the rest of the body; they are not golden pits. The name might mean, "Golden, and covered with small pits."

Genus *Xenomyrmex*

Minute Flattened Tree Ants

The genus *Xenomyrmex* (pronounced **Zee' no mur" meks**) includes only three species and is restricted to Central America, Mexico, the West Indies, and southern Florida. These tiny ants are easily mistaken in the field for small species of *Solenopsis* such as *zeteki*, *corticalis*, and *picta*, which may inhabit the same tree. They also resemble the introduced *Monomorium floricola*, although to those whose myopia or hand lens allows close observation, *floricola* can be seen to be bicolored orange and blackish. Under the microscope, species of *Xenomyrmex* are easily distinguished from similar ants in other genera by the cylindrical petiole with a greatly reduced dorsal node. The large petiolar node found in most ants of the subfamily Myrmicinae is probably associated with muscles that raise and lower the gaster, an activity that may be less practical for ants that live in shallow galleries in dead wood and hug the surface of the bark or wood when out foraging.

Name Derivation: From *xeno* (Greek), meaning in this case "guest," and *myrmex* (Greek), meaning "ant." The first specimens seen by August Forel appeared to be associated with a species of *Camponotus* in an enormous oak gall in Guatemala (Forel 1885). This led Forel to believe that these ants were parasitic, like the European genus *Formicoxenus*. Wheeler (1931b) pointed out that "this opinion has received no support from subsequent observations. The generic name is therefore a misnomer." It is possible, however, that *Xenomyrmex* species are unwelcome guests in the form of facultative predators of ants or termites living in the same piece of dead wood, acting somewhat like thief ants such as *Solenopsis molesta* that sneak into the galleries of larger ants.

Xenomyrmex floridanus Emery— Florida Minute Flattened Tree Ant (Plate 42)

Taxonomy and Similar Species: *Xenomyrmex floridanus* is the only member of its genus in Florida. Its color varies from dark brown to yellow, including many colonies whose members are yellowish with a darker gaster. Color varies from place to place with no obvious geographic pattern, and members of individual colonies have the same coloration. This variation inspired a number of subspecific names through the range of the species in the West Indies and Florida, but most of these were synonymized by Creighton (1957). In the field, X. *floridanus* can be mistaken for some other small arboreal Florida ants: the light form of X. *floridanus* resembles *Solenopsis zeteki*, and the dark form resembles *Solenopsis picta* and *Monomorium floricola*. Experience permits field identification of *floridana* by its unusually flattened appearance. Under the microscope, the absence of a conspicuous petiolar node separates *floridanus* from similar species.

Distribution: Southern Florida, the West Indies, Mexico, at least from the states of Tamaulipas south into Veracruz (Creighton 1957). Creighton (1957) recognized the Mexican population as a separate subspecies (*skwarrae*) on the grounds that its color is consistently yellow rather than variable. This is taxonomically permissible, but one could use the same logic to name subspecies on various Caribbean islands that seem to have consistently colored populations of *floridanus*; such names would only revive the former taxonomic confusion surrounding this species.

In Florida, *floridanus* occurs as far north as Alachua County, but seems commoner farther south. It is not known from the Panhandle.

Natural History: *Xenomyrmex floridanus* is an arboreal species, most reliably found in dead twigs of red mangrove. It probably arrived in Florida from the West Indies by island-hopping from one coastal site to another. In southern Florida, it has moved inland, but is still primarily associated with trees and shrubs in wet areas. Like almost all arboreal ants, this species lives in dead twigs and branches on living trees. Although workers can move rapidly, they can also flatten themselves against the bark when alarmed, the tibiae fitting into unusual grooves in the surfaces of the swollen femora. Like other arboreal ants, *floridanus* could easily be transported from place to place, for example, in dead hollow flowering stems of orchids, but there is no reason to believe that dispersal has been helped by humans. It is preadapted for movement around the Caribbean, as it often colonizes red mangrove islets in shallow water. *Xenomyrmex floridanus* returned within two years to mangrove islands from which all ants had been extirpated (Simberloff and Wilson 1970). A captive colony fed on sweet liquids and termites, and it is possible that *floridanus* attacks the arboreal termites found in dead branches of mangroves and other tropical trees (Creighton 1957). Like many ants living in dead branches, *floridanus* appears to be drought resistant (Creighton 1957), and it should be relatively easy to keep colonies in the laboratory for further study. The poison gland of alate female *floridanus* produces a pheromone that is highly attractive to males during mating flights (Hölldobler 1971a); efficient aggregation for mating should be especially important for a species that often disperses over water and whose gnat-like males could be swept away by a sea breeze. At the Archbold Biological Station, there were large flights in late May and early June, with a few specimens found in flight traps July–December. The Dufour's gland of workers produces a trail-marking substance, and worker mandibular glands produce a chemical that appears to be a weak alarm pheromone (Duffield et al. 1980; Hölldobler 1971a). There is minor size variation in workers, with occasional workers much larger than normal.

Name Derivation: Described as a subspecies, *Xenomyrmex stolli floridanus*, the specimens sent from Florida.

SUBFAMILY DOLICHODERINAE

Genus *Dolichoderus*

Tongue-and-Groove Ants

North American members of the genus *Dolichoderus* (usually pronounced **Dōl' I kō" der us**, sometimes **Dōl' I kō der" us**) are small, sleek, shiny ants that are usually uncommon. These species have a peculiarly shaped propodeum that is strongly overhanging and concave posteriorly, forming a space into which the petiole fits in a kind of tongue-and-groove arrangement. This might allow these ants to clamp the petiole against the propodeum, thus protecting the vulnerable anterior joint of the petiole. The petiolar spines found on many myrmicine ants have a similar function. The major of *Camponotus riehlii* has a somewhat similar petiolar structure, but the cork-shaped head of this species immediately distinguishes it from any species of *Dolichoderus*. Nearctic species of *Dolichoderus* were reviewed by Clifford Johnson (1989a) and the complete New World fauna by William Mackay (1993b).

Dolichoderus is primarily tropical, but there are a few species, apparently closely related, in north temperate areas, one in Eurasia and four in eastern North America. The genus as a whole includes more than 100 species and gives the impression of a diverse and evolutionarily dynamic lineage, judging by present species. It is probable, however, that *Dolichoderus* was once a more dominant genus, as there are approximately 30 described fossil species (Bolton 1995), an unusually large proportion of extinct species. Several extinct species preserved in Oligocene and early Miocene amber represent part of the "remarkable retreat of the Dolichoderinae from the West Indies since Dominican amber times…" (Wilson 1988).

Name Derivation: From *dolichos* (Greek), meaning "long," and *dere* (Greek), meaning "neck," referring to the type species D. *attelaboides*, a long-necked tropical species that does not superficially resemble our species.

Dolichoderus mariae Forel—
Mary's Tongue-and-Groove Ant (Plate 65)

Taxonomy and Similar Species: *Dolichoderus mariae* resembles *pustulatus*, but lacks standing dorsal hairs on the mesosoma (there are several in *pustulatus*) and has only a few standing hairs on the head (many in *pustulatus*). In the field, *mariae*, unlike *pustulatus*, lives in large terrestrial nests and often forages in large columns. At first glance in the field, *mariae* could be mistaken for one of the terrestrial species of *Crematogaster*.

Distribution: Massachusetts south into Florida, west into Minnesota and Oklahoma (Johnson 1989a; Mackay 1993b). Within this large area, populations often seem to be scattered and localized, and some early records are apparently based on misidentified *pustulatus* (Johnson 1989a). Some blame goes to Wheeler, who insisted on synonymizing various species to fit his view of the evolution of the genus; this was grumpily corrected by Creighton (1950). The controversy is of interest today as a very early example of how a dogged pursuit of phylogeny can sometimes obscure reality. In Florida, the only reliable records of *mariae* are from Leon and Taylor counties. This species has huge colonies with large numbers of workers moving about in the open, so it is not a common species that has usually been overlooked.

Natural History: This species lives in big colonies in open areas, and the nests are usually beside or under grass tussocks (Wheeler 1905b). Workers form extensive foraging columns and collect both honeydew and arthropods (Wheeler 1905b). A mound of plant debris may be built above the nest (Talbot 1956). Flights of queens and males occur in August and are dependent on temperature in Michigan (Talbot 1956), but it is unlikely that cool temperatures are a factor in Florida. Workers facilitate both initiating and ending flights (Talbot 1956).

Despite the relative rarity of *mariae* in Florida and elsewhere, its natural history is the best understood of any North American *Dolichoderus*, thanks to an amazingly detailed study by Laskis and Tschinkel (2008) of a series of colonies in the Apalachicola National Forest. The Florida habitat is open pine flatwoods with numerous large grass tussocks, primarily wiregrass (*A. stricta*). This habitat is maintained by relatively frequent fires that suppress development of a dense shrub canopy. Nests are large irregular cavities under grass clumps, sheltered by the roots of the grass and by debris piled over the nest. Colonies have multiple nests that grow during the summer to contain large numbers, often hundreds, of fertile, egg-laying queens, with streams of workers moving freely from nest to nest. Such a large number of queens, combined with dispersal of nests over the resource area, allow explosive growth of the colony, with worker numbers rising to as much as one or two million by fall. Colonies contract during the winter, with an associated die-off of most workers. This type of colony structure and seasonal cycle is similar to that found in the dolichoderines *Linepithema humile* and *Tapinoma sessile*, and in the formicine *Lasius neoniger*. Laskis and Tschinkel (2008) suggest that the remarkable rate of seasonal colony growth might allow strategic exploitation of rapidly changing populations of various homopterans that provide honeydew essential for *mariae*. Laskis and Tschinkel also suggest that flatwoods fires, while benefiting *mariae* in the long term, might be devastating in the short term, another reason why rapid colony growth might be adaptive.

Name Derivation: August Forel (1885) named this species in honor of Mary Treat, who sent him the type series from her home in Vineland, New Jersey. A thumbnail biography of Mary Treat appears under *Aphaenogaster mariae*.

Dolichoderus pustulatus Mayr— Variable Tongue-and-Groove Ant (Plate 65)

Taxonomy and Similar Species: Similar to *mariae* in size and color, but *pustulatus* has several erect dorsal hairs on the mesosoma, more than a few erect hairs on the head, and the head is covered with large, shallow, oval impressions. Florida specimens of *pustulatus* are reddish with a mostly dark gaster, but northern specimens are highly variable in color (Johnson 1989a). In Florida, *pustulatus* lives in dead trees and branches, unlike the terrestrial *mariae*. In the field, *pustulatus* is easily confused with minors of *Camponotus impressus*. Johnson (1989a) suggested that the two species are convergently mimetic and that they might even have a tendency to cohabit the same twigs and branches. In the field, *D. pustulatus* colonies can be distinguished by the almost uniform size of the workers, while workers of *C. impressus* are highly variable in size.

Distribution: Nova Scotia south through Florida, west into Illinois, Oklahoma, Mississippi, and southern Texas (Johnson 1989a). In Florida, *pustulatus* is known from scattered sites in the Peninsula south into the Keys. I have not seen specimens from the western Panhandle, but *pustulatus* probably occurs there.

Natural History: As indicated by Johnson (1989a), the most strikingly variable characteristic of *pustulatus* is its choice of nesting sites. In Florida, *pustulatus* lives in dead twigs and branches on live trees and shrubs, usually at the edges of swamp forests or in marsh-dwelling shrubs such as buttonbush (*Cephalanthus occidentalis*). In northern North America, this same species is apparently terrestrial, although the nests may be in debris or tussocks on the ground rather than subterranean (Johnson 1989a). There should somewhere be a transition zone between these two nesting behaviors, but such a zone is unknown. Most habitats of the southeastern Coastal Plain are strongly influenced by relatively frequent fires, with many species of plants and animals resistant to or dependent on frequent fires. Species that are vulnerable to fire may have Coastal Plain populations confined to habitats that are less fire prone. *Aphaenogaster lamellidens*, whose Florida populations are usually restricted to swamp forests, might be an example of an ant that avoids upland areas in Florida, but is less discriminating in the North. Florida populations of *pustulatus* might have evolved from a population whose superficial nests at the bases of trees and shrubs extended up into dead stems. It is possible that a divergence in habits and habitat occurred so long ago that the southeastern Coastal Plain form has diverged genetically to become a distinct species. This would be an interesting little puzzle for the right naturalist living in the right location. To the north, *pustulatus* and its relatives apparently subsist on honeydew and arthropods, such as caterpillars (Wheeler 1905b), and Florida populations presumably do the same.

Name Derivation: From *pustulata* (Latin), meaning "full of blisters or pimples." Mayr's description (1886) mentions no pimples or blisters, but takes note of the large, shallow punctures on the head and mesosoma. If one thinks of these depressions as acne scars, the name seems appropriate.

Genus *Dorymyrmex*

Cone Ants

North American species of the genus *Dorymyrmex* (pronounced **Dōr' ee mur" meks**) were placed in the genus *Conomyrma* until Shattuck's revision of the Dolichoderinae (1992) synonymized *Conomyrma* with the South American genus *Dorymyrmex*. The cone-shaped projection on the propodeum (lateral view) provides an easy way to recognize the genus and is the derivation of the old name *Conomyrma*. The function of this unusual projection is, of course, unknown. The name "cone ants" is also applicable to the nest mound, which is usually a beautifully symmetrical little volcano of sand. The traffic pattern of *Dorymyrmex* bringing up sand appears to be almost deliberately randomized. These ants can race over sand so fleetly that they have no problem going in and out of the nest, and the steepness of the cone may keep other insects from blundering in, or perhaps alert the ants to other insects floundering around the nest entrance. The sand-running ability of *Dorymyrmex* workers allows them to sprint out of the pits of antlion larvae, but one of the non–pit-making species, *Brachynemurus nebulosus*, sometimes takes up a troll-like residence in the nest mound itself, darting out to seize and drag back the occasional worker (Bahls and Deyrup 1988).

Dorymyrmex includes approximately 55 described species, most of which (42) are from Central and South America, especially Argentina. In the United States, there are six described southwestern species and seven described southeastern species. All these numbers are preliminary, as *Dorymyrmex* includes difficult species complexes that have not been thoroughly studied. Species of *Dorymyrmex* have little integumental sculpture and most species have the same number of large hairs identically positioned. Florida species are identified by differences in color, head shape, length of appendages, and mesosomal profile. Even in Florida, where there is a good revision of species (Trager 1988), there are a few residual problems. Remarkably, considering that only five or so people are actually interested in the topic, there is controversy about some species definitions in Florida (Johnson 1989b; Snelling 1995). I have chosen to adhere to the treatment of Trager (1988) because this is the only careful revision available, and because it agrees with subsequent field work in the Southeast. There is no clear evidence showing that any southeastern species are the same as southwestern species. It is clear, even from my own haphazard western collections, that there are more than six southwestern species, and the group needs revision on a continental basis, preferably with supportive chemical and genetic evidence. *Dorymyrmex* is not an appropriate genus for summary synonymy of strongly disjunct populations on the basis of series that more or less resemble each other.

While the taxonomy of *Dorymyrmex* is messy, the ants themselves are neat and elegant, built for fleetness and agility. Secure in their ability to retreat rapidly, and well protected by copious chemical defenses and sharp jaws, these are among the most venturesome of diurnal ants. They are often the first to arrive at baits and recruit rapidly. A student observing ants at the Archbold Biological Station had one of his toes nibbled by a wandering *Dorymyrmex bureni* that was exploring the student's sandal. This ant appeared to withdraw, but soon after, a foraging column emerged purposefully from the nest with the evident intention of dismembering this enormous resource for transport to the waiting larvae.

Some *Dorymyrmex* species, especially those in a widespread complex of wide-headed species, are unusually creative in their exploitation of other ants. At least one species, *reginicula*, is a monodomous temporary nest parasite of other *Dorymyrmex* (Trager 1988). Another species, *medeis*, is initially a temporary nest parasite and is later in its colony cycle a usurper or predator of host nests (Buren et al. 1975). The southwestern *bicolor* has been observed taking over the nests of *insana* (Martinez 1995) and, more weirdly, using workers of *Myrmecocystus kennedeyi* as slaves (Bernstein 1978). A species identified as *insana*, although it could have been some other species such as *smithi*, was seen using *Crematogaster emeryana* as slaves (Bernstein 1978).

Dorymyrmex is a genus of well-drained open areas. Some species tolerate a light shade, as one might find along sand roads that penetrate forested areas. Several Florida species are completely restricted to Florida scrub or sandhill habitats with deep, well-drained sand and patches that are bare of vegetation. The genus as a whole is partially defined by a sparse "psammophore" of curving hairs on the ventral side of the head, supposedly useful in transporting dry sand and soil. Not all species of *Dorymyrmex* live in sandy areas. *Dorymyrmex antillana*, for example, often occurs in rocky sites with little sand, and the same is true of some southwestern species. Nevertheless, North American *Dorymyrmex* seem most common and diverse in open sandy habitats, and there is evidence from Florida that species can be so dependent on excessively drained and sparsely vegetated habitats that they may be geographically restricted because of this requirement. *Dorymyrmex elegans*, *flavopectus*, and, to a lesser extent, *bossutus* are examples of such species.

This suggests that there might also be geographically isolated species in the Southwest.

Name Derivation: Gustav Mayr (1866a) probably derived the name from *doré* (French), meaning "golden," and *myrmex* (Greek), meaning "ant." The type species, *D. flavescens*, is yellow. It is also possible to derive *dory-* from *doratos* (Greek), meaning "spear," but this does not seem to have any application to either the ants or, to the extent I can translate it from Latin and German, Mayr's description of the genus.

Dorymyrmex bossutus (Trager)— Hump-Backed Cone Ant (Plate 66)

Taxonomy and Similar Species: This is a small species, about the size of a minor of *Pheidole dentata* or *morrisi*, with a shining dark gaster that is almost devoid of microsculpture and with short, sparse, appressed hairs. *Dorymyrmex grandula* is similar in size but is not bicolored and has a nonshiny gaster. Through most of Florida, *bossutus* is orange with a black gaster; the head may be dark or orange like the mesosoma. At the south end of the Lake Wales Ridge, *bossutus* is unusually pale and the gaster is grayish rather than black. There are various additional organisms that show evidence of isolation and divergence on the Lake Wales Ridge. The posterior area of the mesonotum of *bossutus* drops off strongly, with a slightly concave declivity, giving this species a hump-backed profile.

Distribution: Sandy uplands; Collier and Brevard counties north into southern Georgia. Walton County is the westernmost record, but this species probably occurs across the Panhandle and possibly into coastal Georgia.

Natural History: *Dorymyrmex bossutus* lives in well-drained, open sandy uplands, primarily Florida scrub and sandhill habitats. Nests are in patches of open sand. To the naked eye, *bossutus* foragers resemble those of *Pheidole dentata*, often found in the same site, but the volcano-shaped nest mounds betray the presence of *Dorymyrmex*. In my experience, this species does not occur in urban or heavily populated suburban areas.

Name Derivation: "The name *bossuta* is a Latinized (and presumably vulgar Latin form) of the French adjective *bossu*, meaning humpbacked" (Trager 1988). James Trager is a versatile myrmecologist and naturalist who has done revisions of several important groups of ants, including the *saevissima* group of fire ants. These revisions are notable for their careful taxonomic procedure and for their attention to ecological information and field observation.

Dorymyrmex bureni (Trager)— Buren's Cone Ant (Plate 67)

Taxonomy and Similar Species: The pronotum and mesonotum of *bureni* are smoothly convex in profile, with no blunt angle or change in outline on the posterior part of the mesonotum. This separates *bureni* from the eastern species *bossutus*, *grandula*, *medeis*, and *reginicula*, and from the western species *flavus*, which has been confused with *bureni*. There is a strong obtuse notch between the mesonotum and the propodeum, as in most Florida *Dorymyrmex*, but lacking in *elegans*, which otherwise shares the even pronotal–mesonotal profile of *bureni*. The mesosomal profiles of *bureni* and *flavopectus* are similar, but the latter species is distinctively bicolored: orange mesosoma with blackish head, legs, and gaster. There is a dark form of *bureni* along parts of Florida's Atlantic Coast, but this form is never bicolored like *flavopectus*. There is a dark coastal Caribbean species, *antillana*, but this species is much more densely punctate and pubescent than *bureni*. *Dorymyrmex antillana* has not (yet) appeared in Florida.

Soon after its description (Trager 1988), *bureni* was synonymized with *flavus* (Johnson 1989b), but was reinstated by Snelling in 1995. It now appears that *bureni* and *flavus* have overlapping ranges, with *bureni* possibly more strongly restricted to sandy sites.

Distribution: Coastal Plain of Maryland and Virginia south through Florida, west into Louisiana and Texas, with an isolated, possibly introduced population in California (Snelling 1995).

Natural History: The following paragraph is summarized from Trager (1988). This species is widely distributed in open, sandy areas, including both natural habitats, such as sandhill and Florida scrub, and disturbed habitats, such as roadsides, lawns, and cultivated fields. A versatile predator, *bureni* may be a significant biological control agent in some field crops, orchards, and gardens. Foraging occurs during much of the day during summer, except in periods of extreme heat, and may also occur at night, especially in the case of young colonies. This species is generally monodomous, but colonies may temporarily occupy more than one nest, particularly in spring and fall, when nest emigrations are more frequent. Mating flights are crepuscular or nocturnal and take place from spring through fall, usually correlated with rain. Males have large ocelli typical of night-flying ants, in contrast to the small ocelli seen in a small series of specimens of *flavus*.

Unlike most native Florida ants, *bureni* is a frequent beneficiary of destruction and modification of natural habitats by humans. Clearing of forests and draining of lowlands are both helpful for *bureni*, and this species is often one of the few abundant native ants in habitats otherwise dominated by exotics such *Solenopsis invicta* and *Pheidole moerens*. In primeval Florida, *bureni* was probably restricted to beaches and frequently burned habitats. Its success in lawns and landscaped areas earn it pest status in the perverted values of insect control enthusiasts, who condemn it for constructing "unsightly" mounds and running about on sidewalks.

Name Derivation: "*C. bureni* is named after the late William F. Buren, who first recognized it as a species distinct from *C. flavopectus* and *C. flava*" (Trager 1988). William Buren was an important Midwestern myrmecologist who published on ants (especially *Formica* species) in Iowa and did pioneering preliminary studies of North American *Crematogaster*. Later, he moved to Florida, where he began projects on *Dorymyrmex*, *Crematogaster*, *Pheidole*, and *Solenopsis*, as well as a general survey of Florida ants. His early death from cancer ended his career at the height of its productivity, but he inspired several of his graduate students to finish myrmecological projects after his death.

Dorymyrmex elegans (Trager)— Elegant Cone Ant (Plate 67)

Taxonomy and Similar Species: This racehorse among the cone ants is unusually slender, with a flattened mesosomal profile, a greatly reduced propodeal cone or tubercle, an elongate head, and relatively long appendages. These character states are less extreme in specimens from the southern Brooksville Ridge and could be mistaken for unusually slender *bureni* but comparison of specimens (the two species occur sympatrically) reveals the elongate head and antennal scapes of *elegans*.

Distribution: Confined to Florida, where it is restricted to upland habitats on the Lake Wales and Southern Brooksville Ridges. This is one of the most narrowly distributed of Florida's endemic ants. There are other endemic species that are known from only a few specimens (*Solenopsis phoretica*

is known from one queen), but there is no reason to assume that such species only occur in a small area. As mentioned above, specimens from the Brooksville Ridge can be distinguished from those from the Lake Wales Ridge and could be recognized as a separate species or subspecies, if one wished to do so, on the basis of its relatively shorter legs and antennal scapes. To date, genetic evidence separating these two forms is ambiguous. An unrelated ant, *Odontomachus relictus*, is also confined to these two ridges. Aside from these two ants, there is no evidence of a historical connection between the ridges.

Natural History: This species is confined to sandhill and Florida scrub habitats. Nests are in open patches or along sand roads, usually but not always in yellow sand rather than white sand. In these habitats, *elegans* and *bureni* may occur together, but *bureni* moves readily into disturbed areas that appear to be unsuitable for *elegans*. In the field, *elegans* often appears slightly paler than *bureni* and, after a bit of practice, the more slender shape and longer appendages can be recognized without magnification. Behaviorally, *elegans* often moves with an unusual hesitant gait, but foraging workers can run rapidly.

Name Derivation: According to Trager (1988), "*C. elegans* (Latin for exquisite or graceful) was the name originally selected for this species by Buren (personal communication), referring to the elegant appearance of this gracile, yellow ant, especially when alive."

Dorymyrmex flavopectus M. R. Smith— Yellow-Chested Cone Ant (Plate 67)

Taxonomy and Similar Species: This species is easily identified, even in the field and without magnification, by its color pattern: pale orange in the middle, black at both ends. It is larger than *bossutus*, our other orange and black species, and has a nonshining gaster, unlike the conspicuously shining gaster of *bossutus*. This species, like *medeis*, is polydomus, often another useful field character.

This species was collected by Schneirla (Archbold Biological Station is the type locality) and described by M. R. Smith (1944) as a subspecies of "*Dorymyrmex pyramicus*." Creighton (1950), apparently distressed by what he termed "the maze of color variants" in *Dorymyrmex*, ignored the distinctive color emphasized in Smith's description, using instead structural character states that were certain to cause confusion between *flavopectus* and the species we now call *bureni*. There are plenty of instances in which this kind of decision by Creighton worked well, but in this case, several studies of "*flavopectus*" were published (Buren et al. 1975; McGurk et al. 1968; Nickerson et al. 1975a) that refer to some other species, probably *bureni*.

Distribution: Restricted to peninsular Florida, where it occurs on the Lake Wales Ridge and in a few known sites in the Ocala National Forest in both Lake and Marion counties.

Natural History: This species is found in well-drained, open sandy areas, especially "rosemary balds," a type of Florida scrub characterized by bushes of scrub rosemary (*Ceratiola ericoides*) interspersed with patches of bare sand. Even within its narrow geographic range, *flavopectus* only occurs in a small proportion of rosemary balds. When one encounters this species, there are often hundreds of workers racing about between a series of nest entrances, giving the impression of a robust and abundant species. On a larger scale, however, this is one of Florida's rarer ants. Moreover, the northern populations in the Ocala National Forest depend on managers resisting the urge to transform rosemary scrub into sand pine plantations.

Name Derivation: From *flavus* (Latin), meaning "yellow," and *pectus* (Latin), meaning "chest" or "breast," referring to the contrasting yellow mesosoma.

Dorymyrmex grandula (Forel)— Mousy Cone Ant (Plate 66)

Taxonomy and Similar Species: This small species is somewhat variable in color, but is usually light brown and covered with a grayish pubescence that gives it a "mousy" look (Trager 1988). It is smaller than the other two brownish Florida species, *reginicula* and *medeis*. The heads of the latter two species in frontal view are broad with a concave occipital border, while the head of *grandula* has a convex occipital border. The mesonotum usually has a rounded angle posteriorly and is not smoothly convex as in the dark coastal forms of *bureni*. In the field, this is a remarkably nondescript ant, and in the field, I have a tendency to pass over workers until alerted by the cone-shaped nest mound. In this inattention, I am in good company: Forel, surprisingly, described *grandula* as a subspecies of *Paratrechina parvula*, and Creighton consigned it, probably sight unseen, to synonymy with *P. parvula*. When Trager was revising *Paratrechina* (now *Nylanderia*) of the United States

(1984), he borrowed the types of *grandula* and recognized them as a widespread species of *Dorymyrmex* that had previously appeared to be an undescribed species (Trager 1988).

Distribution: New Jersey south into Florida (Trager 1988), west into Michigan (Wheeler et al. 1994), Ohio (Coovert 2005), and Missouri (Trager 1997). The distribution of *grandula* in the northern part of its range appears to be correlated with areas of sandy soil. This may be a Coastal Plain species that followed glacial and riverine deposits into the interior. In Florida, *grandula* is known from Citrus County northward, west through the Panhandle.

Natural History: This species is usually found in forest clearings and edges, sometimes in more open sandhill habitat. In Florida, it is restricted to sandy uplands. Many open habitats in Florida, especially sandhill and Florida scrub, have become increasingly overgrown by trees and shrubs because of a reduction in fire frequency. One might expect this trend to strongly favor *grandula*, but this species remains somewhat rare and sporadically distributed in Florida. It is also odd that *grandula* does not occur in uplands of central and southern Florida.

Name Derivation: *Grandula* (Latin) is the diminutive of *grandis*, meaning "large," and can be roughly translated as "a little larger," referring to its slightly larger size relative to *Nylanderia parvula*, of which this species was supposedly a subspecies. It is hard to see how this taxonomic confusion arose, as *D. grandula* is completely lacking the conspicuous enlarged hairs characteristic of *P. parvula*.

Dorymyrmex medeis (Trager)— Usurper Cone Ant (Plate 66)

Taxonomy and Similar Species: There are two Florida *Dorymyrmex* (*medeis* and *reginicula*) that are blackish with broad heads and a concave occipital margin in frontal view. The two species differ in queen size (*reginicula* has tiny queens) and colony structure (*medeis* forms large extended colonies). When deprived of these indicators, I am not confident of my ability to distinguish between workers of the two species, but *reginicula* tends to be redder than *medeis*, especially the head and mesosoma.

Distribution: North Carolina south into Florida (Trager 1988). The distribution of this species remains unclear. Southeastern records of *smithi* and some records of *insanus* probably refer to this species or to *reginicula*. If one considers *medeis* a junior synonym of *smithi*, as proposed by Snelling (1995), its range would be much larger, extending west into Colorado, New Mexico, and Texas. The types of *smithi* are from western Nebraska (Cole 1938). I am, however, rejecting Snelling's synonymy of *medeis* under *smithi* as premature for obvious reasons. We already know that even in the Southeast, workers within this species complex can be almost indistinguishable, thanks to Trager's study of *medeis* and *reginicula*. Snelling, incidentally, did not attempt to include workers of *reginicula* in his 1995 key to Nearctic *Dorymyrmex*. The taxonomy of western species is in a chaotic state and includes more than one blackish species. We also know that there are separate centers of *Dorymyrmex* species diversity in the Southeast and Southwest, and *smithi* as defined by Snelling would be the only transcontinental species. We have nothing to lose by considering *medeis* a separate species until somebody is willing to study the southwestern species as carefully as Trager studied the southeastern species, perhaps supplementing with some biochemical work. I am, in general, a proponent of old-fashioned taxonomy, that is, looking at large series of specimens until my eyes glaze over, but I admit that there are limitations to this approach.

Natural History: Colony founding and organization of this remarkable species have been the subjects of several studies. *Dorymyrmex medeis* depends on the usurpation of *bureni* colonies and is restricted to areas where *bureni* occurs (Trager 1988). New colonies are founded by solitary queens that take over a *bureni* nest, killing and replacing the resident queen (Buren et al. 1975). This results in a mixed colony that eventually becomes a pure colony of *medeis* (Buren et al. 1975). Colony founding by solitary usurping queens may be risky: Trager (1988) reports that several small mixed colonies that he had been tracking eventually reverted to pure *bureni*. Once the colony is established, workers begin to invade and take over nearby colonies of *bureni*, forming an expanding group of satellite colonies with multiple queens (Buren et al. 1975; Nickerson et al. 1975a). It is not known whether queens move out with a battalion of workers, or whether the nest is taken over and one or more queens move in. The *bureni* workers tend to disappear within a few weeks (Buren et al. 1975). This suggests that few, if any, *bureni* larvae survive to adulthood under the new regime, and adults are probably also dying off unusually quickly. Encounters between workers of *medeis* and *bureni* result in a "panic retreat" by the latter species (Buren et al. 1975). While this reaction may be useful in nest usurpation, readily panicked slaves may not be highly productive or well adjusted. It appears that satellite colonies can form without the benefit of a host colony (Buren et al. 1975), constituting a third method of colony formation.

Except on a very local scale, however, *medeis* remains much less common than *bureni*, so there must be some unknown limit to colony expansion. This limit might be the availability of nearby colonies of *bureni*. *Dorymyrmex medeis* is an unusually aggressive predator (Nickerson et al. 1975a; Whitcomb et al. 1972a), but it is possible that foraging is not sufficient by itself to maintain the large teams of hyperactive workers characteristic of *medeis*. With its rapid and sequential usurpation of neighboring *bureni* colonies, *medeis* seems to be edging toward an early stage of the army ant lifestyle.

In a study of ants in north Florida soybean fields, Whitcomb et al. (1972) found that *medeis* was one of the most important predaceous ants. While *bureni* tends to be a scavenger, *medeis* captures live caterpillars, leafhoppers, small beetles, and other insects. It also readily overwhelms and kills queen *Solenopsis invicta* that are seeking nesting sites following the nuptial flight (Nickerson et al. 1975b). The effectiveness of *medeis* as a predator is limited, however, by the patchy distribution of its colonies (Whitcomb et al. 1972a). This ant also collects honeydew from a wide variety of sap-sucking insects both aboveground and underground, as well as nectar from flowers and from extrafloral nectaries (Nickerson and Whitcomb 1988). I have seen several massive colonies in Gainesville, including one that has persisted for eight years. Such colonies are remarkable for their streams of workers racing along the curb and their teams of avid explorers moving into vehicles and over one's shoes. I have often been nibbled by these enterprising ants, but they do not seem to have a coordinated defense of their nests against intruding humans. There seems to be an effort to hype *medeis* into an important pest (Vazquez et al. 2008). I would hesitate to build a career on the pest potential of this ant, especially since it may deter *Solenopsis invicta*. Replacing *invicta* with a nonstinging, native species that is a dedicated predator of herbivorous insects should occasion more celebration than panic. In any case, colonies of *medeis*, as far as is known, are self-limiting, unlike those of polygyne fire ants.

Name Derivation: Trager considerately explains the source of this name: "the adjective *medeis* derives from a sorceress in Greek mythology, Medea, who successively dominated several households, sometimes killing their head persons, to serve her own selfish ends. The figurative analogy of Medea's behavior to the social parasitic behavior of this ant lead to the choice of this name."

Dorymyrmex reginicula (Trager)— Dwarf Queen Cone Ant (Plate 66)

Taxonomy and Similar Species: Its miniature queens and monodomous colony structure separate this species from *medeis*. The workers of the two species are similar in structure, but the head and mesosoma of *reginicula* are usually reddish, whereas those of *medeis* are usually blackish. The pubescence of *medeis* is described as denser than that of *reginicula* (Trager 1988), but it may be that the pubescence is more visible on a darker background. I have seen queens and workers of *reginicula* or a very similar species collected by Stefan Cover in Colorado, but I am not sure what to call these at this point. *Dorymyrmex* provides enough taxonomic and biogeographical puzzles to keep myrmecologists happily engaged for decades.

Distribution: Scattered sites from Hendry County north into Alachua County. If the Colorado specimens really are *reginicula*, they would appear to represent a remarkably disjunct population, but it is always possible that there are intervening populations that have been identified as *medeis* or *smithi* or something else.

Natural History: Like many parasitic species, *reginicula* is relatively rare, but it is easy for the ambling myrmecologist to pick out the mixed *bureni/reginicula* nests among hundreds of yellow or orange *bureni* nests. I have only seen *reginicula* colonies parasitizing *bureni*, but this species also parasitizes *bossutus* (Trager 1988). Alates are produced in mixed nests, and I have seldom seen a pure colony of *reginicula*, so it is possible that *reginicula* is unable to live independently for any length of time. The exact nature of the parasitism is not known. Temporary nest parasites with miniature queens usually kill the host queen (at least in the genus *Formica*), but *reginicula* may follow its own rules.

Name Derivation: From *reginicula* (Latin), meaning "little queen," referring to the small size of the queen.

Genus *Forelius*

Asbestos Ants

Members of the genus *Forelius* (pronounced **Fō rel" ee us**) occupy open, usually semiarid habitats, where they show amazing resistance to daytime heat. The genus includes approximately 18 species (Cuezzo 2000), divided into two apparently disjunct lineages, one in North America and the Caribbean, the other in dry areas of southern South America (Shattuck 1992). These are small, drab ants, generally lacking distinctive features other than some rather unexciting hairs, and most active during the heat of the day when sensible myrmecologists have retired for a glass of iced tea. Perhaps in consequence of these factors, North American *Forelius* have operated below the taxonomic radar, despite their ecological importance in many habitats. Only three Nearctic species are currently recognized (Cuezzo 2000), but there is at least one undescribed species in Florida (Fisher and Cover 2007). I hesitate to venture into comparative biogeography of different genera, but, in general, ground-foraging ants found in open habitats are represented by separate species in southeastern and southwestern North America. This makes me suspect there might be at least five or six North American species.

Florida species of *Forelius* are superficially similar to *Tapinoma sessile* and *Linepithema humile*. One way to distinguish *Forelius* is by the presence of at least two conspicuous standing hairs on the pronotum. One might imagine that this is yet another obscure and possibly ambiguous character beloved of ant taxonomists, but these hairs are actually rather easy to spot (when the pronotum is observed in lateral view under the microscope), and they are not easily dislodged or matted down on the pronotum. Our *Forelius* species are also smaller than their two look-alikes, and differ in mandibular structure, especially in the size and placement of teeth.

Name Derivation: Carlo Emery (1888) named this genus for the great Swiss myrmecologist and psychologist, August Forel (1848–1931). As a pioneering clinical psychologist, Forel championed more humane treatment of the insane, and strongly advocated replacing the horrendous "madhouses" of the early 19th century with clinics where patients received curative treatment. Forel's taxonomic work includes descriptions of approximately 3000 species of ants and a large number of genera. He must have been gifted with a remarkable memory, as his descriptions seldom include keys or other diagnostic guides that would allow a normal entomologist to identify a series of species belonging to the same genus. As a behavioral myrmecologist, Forel helped overturn the ancient belief that ant colonies were guided by intelligence and logic.

Forelius pruinosus (Roger)— Frosty Asbestos Ant (Plates 68 and 69)

Taxonomy and Similar Species: This little, gray, fast-moving species is usually easy to identify, but in north Florida, it may occur together with the undescribed "*Forelius* species A." The latter species is generously endowed with bristling hairs on the head and body. A pair of strong hairs on the pronotum distinguish F. *pruinosus* from the similar *Tapinoma sessile* and *Linepithema humile*. It is possible that *pruinosus* will prove to be a complex of two or more cryptic species.

Distribution: Cuba and the Bahamas, Mexico, Guatemala (Cuezzo 2000), New York west into Wisconsin, south through Florida, and west into New Mexico (Smith 1979). In Florida, this species occurs throughout the state.

Natural History: In Florida, *pruinosus* is most abundant and conspicuous in open natural habitats, especially sandhill and Florida scrub. It does not seem well adapted to man-modified habitats, although it frequently occupies the edges of sand roads traversing natural habitats. In the Florida Keys, *pruinosus* occurs in open rocky areas as well as in sandy sites. I have collected a series from seasonally flooded prairie in Kissimmee Prairie State Park (Okeechobee County) and seen specimens from seasonally flooded areas in Collier County. Queens from the Kissimmee Prairie site are blackish, unlike the bicolored reddish and gray queens found in upland habitats throughout Florida. It is possible that there is a cryptic species associated with seasonally flooded sandy areas in Florida, or perhaps queen color varies by habitat. Nests in Florida uplands are in bare sand, sometimes at the base of a small plant. Nest excavation is an organized operation, apparently involving many workers, each of whom rushes out of the nest entrance with a few grains of sand, drops the load a few centimeters from the entrance, and dashes back into the nest. There may be more than one nest entrance, as this species is often polydomous and usually polygynous. During mating flights, queens appear to mate with males from the same colony and then return to the nest (Hölldobler 1982).

In Florida and elsewhere, *pruinosus* is notable for its ability to forage under hot conditions when other ants have retreated underground. Foraging also occurs during cooler parts of the day, but appears to cease at night. In Florida, *pruinosus* rapidly recruits to baits and can exclude other ants by forming a circle around the bait, their heads inward and feeding, their gasters laden with chemical repellent directed outward. Hölldobler (1982) provides details of foraging behavior at a site in New Mexico. Foragers recruit to dead insects in large numbers in response to chemicals produced by the sternal gland. They do not use group transport, but cut insect prey into tiny pieces that are carried by individual workers. A concentration of repellent-bearing ants on and roaming around the prey keeps at bay other ants during the laborious and time-consuming process of prey dissection. Swarms of *pruinosus* are even able to displace much larger *Myrmecocystus* from baits of dead insects. Several other species of small ants in the genera *Forelius*, *Solenopsis*, and *Monomorium* also protect resources by chemical means, but the desert *Forelius* have an additional trick that is more unusual. When a dead insect or nectar source is near a nest entrance of *Myrmecocystus*, the *pruinosus* foragers mount a preemptive strategy, gathering around the nest entrance and using their chemicals to prevent the *Myrmecocystus* from emerging to forage. A chemical involved in these pheromonal encounters is 2-heptanone, which serves both as an alarm pheromone to attract and excite large numbers of workers and as a repellent to other species (Amoore et al. 1969; Blum et al. 1966).

Name Derivation: From *pruinosus* (Latin), meaning "covered with frost," referring to the silvery appearance derived from the covering of fine, pale, appressed pubescence on the dark background of the head and body.

Forelius Species A, Undescribed Species (Plate 68)

Taxonomy and Similar Species: Strangely enough, this undescribed species is the least taxonomically ambiguous of Nearctic *Forelius*. It is distinguished by its short, fine, bristling, erect hairs on the head, body, and appendages. *Forelius mccooki*, which occurs in the West and east into Mississippi, also has erect pubescence, but it is not as abundant as that of species A; *mccooki* also has a wider head, with a more concave occipital margin, and is yellowish or yellowish brown in color.

Distribution: Known from a site in Georgia (Emanuel County) and several sites in north Florida: Citrus, Alachua, Putnam, and Walton counties.

Natural History: This species inhabits open sandhill habitat. It usually, perhaps always, occurs together with *pruinosus*, but is much rarer. How much rarer is open to question, as it is difficult to identify species A in the field. I think that in the field, species A appears very slightly larger and darker than *pruinosus*, but it is easy to convince oneself that every colony has slightly larger, darker inhabitants, or that none of them do. The distinctive erect hairs of species A, although short, are silvery and conspicuous, and can actually be seen with a good hand lens in the field if the ant is held against a dark background. Unfortunately, this species is too rare, and *pruinosus* is too common, to make this method of survey very practical. Considering its status as a rare associate of *pruinosus*, it is reasonable to wonder whether species A might be a temporary nest parasite of *pruinosus*. This species was discovered by Lloyd Davis.

Forelius Species B, Taxonomic Status Unclear (Plate 68)

Taxonomy and Similar Species: There is a common yellow species of *Forelius* in the Southeast that barely enters Florida in the western Panhandle—there is a single record from north of Pensacola. This is a species that I would have preferred to ignore. It has been sometimes treated as a subspecies: *Forelius pruinosus analis*. It is definitely not a subspecies of *pruinosus*. Cuezzo (2000) considers *analis* a western species, mentioning no eastern populations. If the *analis*-like populations are all one species, this species would seem to have an unusual distribution: an enormous arc from western North Dakota and southern Idaho, sweeping south into Mexico, then up around the Gulf of Mexico (but usually somewhat inland), and up the eastern Coastal Plain and Piedmont into Virginia. I could call the southeastern form *analis* by default, but this would only be projecting an illusion of taxonomic certainty over my doubts and suspicions. Features distinguishing Species B from other Florida species are as follows: yellowish or yellowish brown color; head wide, with slightly concave occipital border (in frontal view);

no erect hairs on the legs or antennal scapes; and only a few erect hairs on the pronotum.

Distribution: I have seen specimens from Virginia south into northwestern Florida, west into Mississippi. *Forelius analis*–like ants are known from sites through the southeastern Coastal Plain, and from western North Dakota (Wheeler and Wheeler 1963) west into southern Idaho (Creighton 1950) and south into Texas and Mexico (Cuezzo 2000). In Florida, this species is known from a single series collected by Lloyd Davis north of Pensacola.

Natural History: This species, in my experience in the Southeast, is found in open areas, including somewhat disturbed open sites. Nests are usually in soil that has some clay content, but occasionally in pure sand. Although this seems to be a common southeastern species, natural history accounts are sparse, and some may refer to *pruinosus* or some other species.

Genus *Linepithema*

Argentine Ant and Relatives

I have heard various pronunciations of the genus name *Linepithema*, including **Lin' eh pith" eh muh**, **Lin' ē pith ē" muh**, **Līn' epi thē" muh**, and **Lin' eh pith" ē muh**. Although there are supposed rules for pronouncing scientific names, common usage always wins over rules. Common usage is often dependent on the rules of pronunciation of the speaker's native language; the four variants listed above are English pronunciations, while Spanish speakers have their own pronunciation. The name *Linepithema* gets plenty of use, as the species *L. humile* is a major pest that has invaded many regions, but I have not observed that differences in pronunciation of its name lead to problems in communication when myrmecologists get together. Pronunciation was simpler when *L. humile* (the Argentine ant) was in the genus *Iridomyrmex*, but that changed in 1992 when Steven Shattuck demonstrated that *Iridomyrmex* had become a dumping ground for a big batch of unrelated dolichoderine ants.

Linepithema is a genus of 19 species native to tropical and subtropical regions of South and Central America and the Caribbean, with one species, *humile*, which has been carried to warm-temperate areas around the world (Wild 2007). *Linepithema humile* is established in Florida, where it might be confused with species of *Forelius* and *Tapinoma*, as discussed below. Like most dolichoderine ants, species of *Linepithema* are general scavengers and predators, but also avidly consume honeydew from sap-sucking insects and nectar, both from flowers and from extrafloral nectaries (Wild 2007). Species that have been studied are active day and night, quickly recruiting to new resources using chemical trails that become the basis of busy foraging freeways.

Name Derivation: The derivation of *Linepithema* is unclear. The genus was described from the male of L. *fuscum* only (Mayr 1866a), and I am unable to find anything in the description that would explain the name, which can be broken down into various word roots, none of which seem to be relevant. The name might be a Latin translation of some German term relating to insect genitalia, the most distinctive and carefully described features in Mayr's description of the type specimen.

Linepithema humile (Mayr)— Argentine Ant (Plate 69)

Taxonomy and Similar Species: *Linepithema humile* is a grayish black, rather nondescript species with no conspicuous hairs or sculpturing. It is similar to *Tapinoma sessile*, and the two species might be mistaken for each other, either in the field or in the laboratory. *Linepthema humile* has a conspicuous petiolar scale, lacking in T. *sessile*, and the mandibular teeth of L. *humile* are large and uneven, whereas those of T. *sessile* are relatively small and even, except for those near the mandibular apex. *Forelius pruinosus*, another nondescript grayish black Florida species, has two to four pairs of conspicuous hairs on the mesosoma. *Technomyrmex difficilis* might also be mistaken for L. *humile* in the field, although it is darker without the grayish look of L. *humile*. Under the microscope, T. *difficilis* shows some distinctive features, including conspicuous sculpturing on the sides of the mesosoma and large, erect hairs on the mesosoma and gaster. In its native range, which may be confined to the Paraná River drainage in Paraguay, Uruguay, Argentina, and Brazil, *humile* shows some confusing morphological variation (Wild 2004), but variation does not seem to occur in southeastern North America, probably because the species was imported from a small geographic area (Tsutsui and Case 2001). A somewhat similar species, *iniquum*, which lives in hollow stems and branches, widely distributed in Central and South America and parts of the Caribbean, has been reported as an exotic in greenhouses (Smith 1929b; Wild 2007). It is surprising that *iniquum* has not become established outdoors in the Miami area, although not every possible exotic ant resides in Miami; it just seems that way.

Distribution: The distribution of *humile* has been thoroughly researched by Wetterer et al. (2009). *Linepithema humile* is native to the Paraná River basin of subtropical Brazil, Paraguay, Uruguay, and Argentina, the same region that gave us *Solenopsis invicta*. It was reported from Louisiana in 1891 and from Florida in 1914. It occurs outdoors across southern North America, and in subtropical areas around the world. It appears to be especially well adapted to regions with a Mediterranean climate: cool wet winters, warm dry summers. It has become a major pest in California, South Africa, southern Australia, and parts of the western Mediterranean region. *Linepithema humile* is a household pest in some areas, with many northern records referring to indoor infestations.

Natural History: In its native range, *humile* is associated with large rivers in the Paraná River system and with low areas within this drainage (Wild 2004). Like *Solenopsis invicta* and *saevissima*, the Argentine ant is apparently adapted to habitats disturbed by flooding, even displaying a S. *invicta*-like ability to form balls of ants that float on flood water, with larvae and pupae in the center of the ball of ants (Barber 1916). Like the imported fire ant, the Argentine ant seems to have parlayed its winning strategies in naturally disturbed areas into an ability to occupy areas disturbed by humans. These strategies include complete acceptance of individuals from nearby colonies, multiple queens, and what appears to be dispersed central place foraging, with large groups of workers and queens moving quickly to position themselves near concentrations of resources (Barber 1916; Newell 1909). Early studies of *humile* in North America state that scattered colonies unite in the fall in a dry, protected location (Barber 1916; Newell 1909). This may involve movement of a local population from one habitat to another. In southern California, Markin (1970) found that large populations found in summer in irrigated citrus groves emigrated in fall to aggregation sites on nearby dry hillsides. Newell (1909) accidently attracted such an aggregation to a sheltered box of garden compost; he estimated that this aggregation had more than 1000 fertilized queens. In spring, the ants moved out to form small colonies, which quickly grew to overrun the premises and become "intolerable nuisances." Newell performed, at least on a small scale, the experiment of setting out overwintering containers of compost and then killing the aggregation before spring; he claimed this worked well until colonies invaded from neighboring properties. Newell also took advantage of dispersed central point foraging, which he described as "colonies moving into close proximity to any constant source of food." A suitable nest site, such a piece of decaying log, was set up in a shaded location with a nearby jar of honey or sugar covered with coarse mesh. The trap nests were dropped into boiling water when occupied. These early attempts at control of humile revealed a pattern of seasonal colony structure similar to that of Tapinoma sessile, reported almost a century later (Buczkowski and Bennett 2006). Both these species aggregate in dry winter headquarters in wet areas and then spread out in spring to take advantage of areas that might have been affected by winter flooding.

Outdoor populations of humile are most persistent in a climate similar to that of its native range, with cool, moist winters and warm, dry summers (Wetterer et al. 2009). Where there is little rain in the summer, as in the Sacramento Valley of California, humile requires a permanent source of water and is confined to sites near streams or rivers or artificially irrigated sites (Ward 1987). Since invasive populations of humile spread by budding off groups of workers accompanied by flightless queens, humile depends largely on humans for long-range dispersal (Suarez et al. 2001). This means that humile continues to expand its range as it is carried to new areas that have a suitable climate and to isolated stream systems or irrigated habitats in arid areas (Wetterer et al. 2009). Humid tropical areas and warm-temperate areas with wet summers are apparently unfavorable to humile, although the reason for this is unknown. Shortly after its introduction into southeastern North America, humile spread like a plague through the Southeast, as described by Barber in 1920: "Ants are common and annoyance from them is by no means unknown, yet the Argentine ant has distinguished itself by greatly exceeding other species in its injury. Other ants may make themselves troublesome, but the Argentine ant goes so far as to cause homes to be vacated in an infested neighborhood." Abandoning real estate

to Argentine ants may seem an extreme reaction, but before air conditioning was available, southern homes were necessarily much more accessible to ants than today. Inside the home, as a last resort, legs of kitchen tables and other furniture were painted with a band of bichloride of mercury, or placed in saucers of moth balls (Barber 1920), measures completely unacceptable these days. Today, in the Southeast, the Argentine ant is primarily an urban problem (Wetterer et al. 2009). On a smaller scale, in the archipelago of Madeira, humile has retreated from areas of humid climate where it was formerly abundant but remains common in drier areas (Wetterer et al. 2009). The most likely explanation for these declines is that certain pathogens gradually build up in warm, humid conditions.

A curious feature of humile natural history is the springtime slaughter of the great majority of queens by groups of workers. This phenomenon, first described in a California population (Markin 1970), has been examined in more detail by researchers in France (Keller et al. 1989; Reuter et al. 2001). It does not appear that workers are targeting weak, infertile, or senescent queens, which might be less productive as the colony begins to bud off groups of workers and queens in the summer (Keller et al. 1989). Workers are not, on average, less closely related to the queens they assassinate than the queens they spare, so it does not seem to be a case of workers favoring their own genotype by killing less-related queens (Reuter et al. 2001). It is still possible that groups of workers more related to each other gather to kill less-related queens, although demonstrating this would demand the near-impossible task of catching in the act individual queen-killers for genetic testing (Reuter et al. 2001). Markin's own explanation (1970) is that workers are able to regulate queen numbers, with the implication that there is some adaptive ratio of queens to workers when the colony fragments in spring. It might be advantageous to produce a surplus of queens in the fall if there is often disproportionate winter queen mortality, for example, if South American birds or lizards attacking winter aggregations tend to target the large queens. There is, however, no evidence of this, and it is not even known whether queen executions occur in colonies within the native range of humile.

Under suitable conditions, invasive populations of humile become large enough to reduce or even extirpate native species of ants. Erickson (1971) observed an invasion of *humile* as it progressively displaced several native seed-eating species of California ants. *Linepithema humile* is an efficient forager, but in this case, the eradication of native species seemed to be less by competition than by direct aggression in the foraging area. In contrast, in a study of interactions between *humile* and other ants at baits, Human and Gordon (1999) found that when an ant of one species approaches an ant of another species, the ant that is challenged tends to retreat. This was true of both native ants and *humile*, but since *humile* greatly outnumbered all other ants and recruited workers to the baits quickly and in large numbers, a much greater proportion of native ants retreated from the baits. In this competitive situation, *humile* was winning by force of numbers alone. The unusual way that a group of *humile* expands its foraging area helps ensure that it will immediately have enough workers to competitively dominate new territory. Exploration of new foraging areas somewhat resembles a small-scale advance by army ants, with workers moving rapidly back and forth from the nest to the head of the exploratory front, laying down chemical trails that recruit increasingly large numbers of workers (Aron et al. 1990). In California, humans facilitate the process of domination by *humile*. This species thrives in disturbed habitats if there is a water supply; residential areas and croplands allow this species to invade surrounding areas of more natural habitat (Ward 1987). As natural areas become more fragmented, *humile* displaces more species of native ants, occupying areas up to 200 meters from the edges of disturbed areas (Suarez et al. 1998). California is gradually being divided up by supercolonies of the Argentine ant. There are five such colonies, one of which is much bigger than the others (Thomas et al. 2007). These supercolonies are aggressive toward each other (Thomas et al. 2007), but this probably provides no more relief to native species than the territorial rivalry of criminal gangs benefits the residents of a city.

Linepithema humile in Florida has not received much scientific attention. On a regional scale, it does not appear to be an important species. Where it occurs, in the town of Niceville for example, it is abundant, but there are other ant species conspicuously present. I know of two colonies in south-central Florida that have apparently disappeared. Judging from the scarcity of Florida specimens in collections, *humile* may have never

gone through an outbreak phase in Florida. Some of the records on Florida maps of distribution of *humile* represent very old records.

Name Derivation: From *humilis* (Latin), meaning "low" or "flattened," possibly referring to the head, which is flattened relative to that of somewhat similar European ants, especially the common species now in the genus *Lasius*. When applied to character, *humilis* means "humble," but this would only apply ironically to the Argentine ant.

Genus *Ochetellus*

Tongue-Node Ants

The genus *Ochetellus* (pronounced **Ō' kē tel" us**) is an Australasian group containing seven species (Shattuck 1992). These were originally included in the genus *Iridomyrmex*, but were moved to a new genus on the basis of several features, including concave clypeal border, concave or flat propodeal declivity, and the unusual form of the petiole, which is large, flattened from front to back, and slightly expanded from side to side in its apical third, giving a tongue-shaped appearance (Shattuck 1992). The propodeum is set off from the rest of the mesosoma by a prominent groove. These ants forage, often in columns, on trees or on the ground, sometimes venturing into houses for sweets (Shattuck 1992).

Name Derivation: From *ochetos* (Greek), meaning "channel," and the diminutive *-ellus*, hence, "little one with a groove." This refers to the conspicuous groove separating the propodeum from the remainder of the mesosoma, and the small size of the species relative to most other members of the Dolichoderinae.

Ochetellus glaber (Mayr)— Bicolored Tongue-Node Ant (Plate 65)

Taxonomy and Similar Species: This is a highly distinctive small, shiny, almost hairless ant that is usually black with an orange or dark red mesosoma. The petiole is high and somewhat flattened; from an anterior view, it is somewhat broadened in its apical third, giving a tongue-like appearance. Members of some Old World populations are completely black.

Distribution: Native to Australia and New Guinea, introduced into Hawaii, New Zealand, Japan, Taiwan, the Philippines, and Florida (McGlynn 1999). In Florida, this species was first reported in 1979 from Winter Park in Orange

County (Smith 1979) and is now known from a cluster of nearby localities in Orange and Lake counties. It is unlikely that we will ever be sure of how this ant got to North America, but a possible inadvertent importer is Henry L. Mead (1852–1931), an avid horticulturalist who traveled to Australia and many other regions to collect butterflies and nursery stock. His garden of exotic plants is still preserved in Winter Park.

Natural History: I have found this species in dead, dry branches and also in large tussocks of *Andropogon* grass. Most nests were in moist, semidisturbed habitats. There is no obvious reason why this species should not have attained a much more extensive range in Florida. As humans continue to move potted plants around the state and nation, *O. glaber* will probably gradually expand its range. In Hawaii, where it was first reported in 1977, *O. glaber* lives in both dry and mesic lowland areas (Reimer 1994), and is considered an urban pest (Cornelius and Grace 1997). Colonies are polygyne and can reproduce by budding (Cornelius and Grace 1997). Like other tropical and subtropical ants that nest in dead wood, *O. glaber* probably competes with termites for nesting sites, and may take over galleries abandoned by termites, or possibly evict termites. Cornelius and Grace (1996) discovered that secretions of *O. glaber* can repel, or even kill, Formosan subterranean termites.

Name Derivation: From *glaber* (Latin), meaning "hairless," referring to the general scarcity of standing hairs or pubescence on this species.

Genus *Tapinoma*

Odorous Ants

The genus *Tapinoma* (pronounced **Tap' i nō" muh**) is distinguished by the absence of a scale or knob on the petiole (the petiole is a simple flattened cylinder), combined with the absence of erect hairs on the head and body. Dried specimens often shrivel in a way that hides the petiole under the overhanging gaster, but they can still be separated from similar species of *Forelius* by a lack of erect hairs. *Technomyrmex difficilis*, introduced in Florida, also lacks a scale on the petiole, but has erect hairs on the mesosoma and gaster, fine granulate sculpture on the sides of the mesosoma, and the propodeum in profile is angulate above, unlike the rounded profile of *Tapinoma*. Like other ants of the subfamily Dolichoderinae, species of *Tapinoma* are

primarily defended by speedy retreat and strong repellent chemicals. *Tapinoma* species are especially well endowed with the latter defense, which has been described as "a characteristic disagreeable, rotten-coconut like odor" (Smith 1965). Few of us have much exposure to rotten coconuts; to me, the *Tapinoma* odor is strong but does not resemble any reference odor. As far as I am concerned, *Tapinoma* has invented its own smell. This penetrating smell is a combination of several cyclopentanoid monoterpenes (Blum 1981), earning the name of the Odorous House Ant for the widespread North American species *sessile* (Smith 1965).

Tapinoma presently includes approximately 60 species, the majority of which live in the Old World tropics, with approximately 12 found in Old World temperate areas, 10 in the New World tropics, and one (*sessile*) with a remarkably wide range through North America. There are three species in Florida, one introduced from tropical Asia, a native Caribbean species, and the native *sessile*. *Tapinoma* species collect honeydew from Homoptera, often tending scales or mealybugs; they are also general scavengers (Shattuck 1992).

Name Derivation: From *tapeinos* (Greek), meaning "low," and *-oma* (Greek), meaning "tumor," presumably referring to the undeveloped node of the petiole.

Tapinoma litorale Wheeler— Odorous Tropical Twig Ant (Plate 70)

Taxonomy and Similar Species: *Tapinoma litorale*, despite its small size (approximately 1.5 mm), is easily identified in the field by myrmecologists seeking ants in dead wood. It is the only pale yellow species that comes dashing out in surprising numbers from a slender broken twig and darts about so rapidly that it is almost impossible to capture individual specimens. It is easy, however, to get plenty of specimens by blowing on a section of hollow twig held over a vial; this method also provides an additional identifying feature, the distinctive flavor of *litorale*. Preserved specimens, lacking these convenient field marks, are distinguished from yellow species of *Brachymyrmex* by the absence of a scale on the petiole and absence of long hairs on the gaster. *Brachymyrmex* species have only 9 antennal segments, whereas *Tapinoma* species have 12. *Plagiolepis alluaudi*, a widespread Old World species that might someday appear in Florida, is another tiny yellow species but has a petiolar scale and conspicuous hairs on the head and gaster.

Distribution: The Greater Antilles, the Bahamas, and southern Florida. In Florida, records of *litorale* are from as far north as Pinellas, Highlands, and Brevard counties, but it might occur farther north at the edges of coastal hammocks. An outlier record from southern Mexico (Del Toro et al. 2009) might be some other species of *Tapinoma*, but it is also possible that *litorale* has been moved about by commerce, for example, in the dead flowering stems of orchids, a habitat where I have found *litorale* in the Bahamas.

Natural History: Nests occur in relatively open areas, often at the edge of a hammock or brushy area. If I were on a mission to find colonies, I would look first in small dead twigs on living red mangroves, but there are other less predictable sites, such as the culms of sawgrass, small dead vines on living trees, small dead twigs on trees and shrubs, such as Brazilian pepper, and dead hollow flowering stems of orchids. Several dealate queens often occur in a single nest.

Name Derivation: From *litoralis* (Latin), meaning "of the shore," referring to the coastal site in the Bahamas where this species was first found by William Wheeler (1905a).

Tapinoma melanocephalum (Fabricius)— Ghost Ant (Plate 70)

Taxonomy and Similar Species: Despite its minute size (approximately 1.5 mm), there is no ant easier to identify in the field than *Tapinoma melanocephalum*. The head and mesosoma are black, the gaster whitish, and on a pale background such as a kitchen counter, a group of these ants appear as black dots running erratically over the surface on the way to harvest a few grains of spilled sugar. There are no similar species in Florida. I see nothing in the appearance or behavior of these ants to justify the commonly used English name of "ghost ant," but I do not claim familiarity with the field marks of ghosts.

Distribution: *Tapinoma melanocephalum* occurs through most tropical areas and often becomes established in greenhouses or zoos in temperate areas (Wetterer 2009b). Similar species live on Pacific islands and on the Indian Subcontinent, and it is highly probable that *melanocephalum* is native to the Indo-Pacific Region (Wetterer 2009b). In Florida, I have seen outdoor colonies as far north as Pasco County.

Natural History: In Florida, *melanocephalum* usually lives in dry microhabitats near buildings. I have not seen colonies that appeared to be established indoors, probably because most Florida homes and businesses are kept cool and dry most of the year. This species thrives in outdoor dining areas and can also enter buildings through tiny cracks or through coarse window screens. In relatively natural areas, I have seen large outdoor colonies under bark, in dry sticks, in dead mangrove branches, under debris or rocks, and in bases of dead palm fronds. In Florida, it is not so common in any of these kinds of nesting sites that I would be able to depend on finding a colony if I wished to do so. Colonies can be found in cardboard boxes left outdoors, and James Wetterer (2009b) mentions a colony that moved overnight into his luggage in Martinique. It is easy to see how *melanocephalum* could have achieved its wide distribution, as it has been found on ships and frequently intercepted at ports (Wetterer 2009b). Colonies have multiple queens, which would facilitate dispersal by groups of workers accompanied by one or more queens (Smith 1965). Workers from different colonies are not antagonistic (Smith 1965); absence of aggression between colonies of the same species is a regular characteristic of "tramp" ants (Passera 1994).

Tapinoma melanocephalum tends sap-sucking insects and readily invades houses where any kind of sweets are available (Smith 1965), and I have seen *melanocephalum* visiting extrafloral nectaries. Many of the species of small ants that can be seen running about in the open are probably micro-predators, but this has seldom been documented. In the case of *melanocephalum*, Osborne et al. (1995) managed to observe workers carrying off mites, adult whiteflies, thrips, and immature aphids. In further greenhouse studies, Osborne et al. (1995) showed that *melanocephalum*, under some circumstances, could be a significant predator of two-spotted spider mites on bean plants. This may be the first known example of an ant that eats spider mites. Larger prey have also been observed, including small larvae of moths and beetles, hemipteran eggs (Osborne et al. 1995), and housefly eggs (Smith 1965). *Tapinoma melanocephalum* is probably defended by repellent chemicals, which are strong enough that the human nose can easily detect the odor of two or three squashed ants, despite their tiny size.

Tapinoma melanocephalum is so well adapted to man-made habitats that it often becomes a pest, despite its tiny size and its inability to bite or sting humans. In Florida, it is a common problem in outdoor dining areas, but it is a more severe problem in the tropics where buildings usually depend largely on open ventilation rather than on air conditioners operating in a sealed environment. In Brazil, *melanocephalum* is one of several ants often found in hospitals, potentially carrying pathogenic bacteria while traipsing through designated sterile areas (Bueno and Fowler 1994).

Name Derivation: From *melano* (Greek), meaning "black," and *cephalo* (Greek), meaning "head," referring to the black head of this species.

Tapinoma sessile (Say)—
Odorous House Ant (Plate 69)

Taxonomy and Similar Species: *Tapinoma sessile* might be described as nondescript. It is a dull blackish brown species with no sculpturing of the head or body, no standing hairs on the mesosoma, no development of the petiolar node, and the mandibular teeth small and almost even in size except for the apical and subapical pair. In the field, it is almost indistinguishable from *Linepithema humile* (Argentine ant), but is slightly larger and darker, and more likely to be found in wet areas, at least in Florida. Under the microscope, *L. humile* can be seen to have a conspicuous node on the petiole, and the mandibular teeth are large and uneven in size. *Tapinoma sessile* also resembles *Technomyrmex difficilis* in the field, but that species has a petiolar node, two pairs of upright hairs on the mesosoma, and fine granular sculpture on the sides of the mesosoma. *Forelius pruinosus* is smaller, with two to four upright hairs on the mesosoma, a node present on the petiole, and large jagged mandibular teeth.

Distribution: *Tapinoma sessile* occurs from Nova Scotia south through Florida, west to Washington, south into California and Mexico (Smith 1979). It has the widest distribution of any North American ant, with the possible exception of *Brachymyrmex depilis*. This enormous range is partially explained by the great variety of habitats accepted by *sessile*, ranging from tropical to north temperate, and from dry, rocky sites to the edges of marshes. In desert areas unsuitable for *sessile*, it occurs along the edges of streams and rivers, which may provide corridors between larger mesic areas. Shifting climatic zones of the late Pleistocene may have allowed *sessile* to reach mesic habitats that are isolated today. This does not preclude the possibility that there are cryptic species lumped in *sessile*. A bicolored form in California has recently been recognized as a separate species (Hamm 2010); as the author of this species notes, it was only the relatively small but consistent difference in coloration that stimulated a careful examination of this population.

Tapinoma sessile is like *Brachymyrmex depilis* in that it seems odd that such an adaptable, common, and widespread North American species has not been accidentally spread far outside its native range. There is a recent report of a localized population in Hawaii, but this seems to be the first such record (Buczkowski and Krushlycky 2012).

Natural History: On a continental scale, *sessile* lives in many different habitats, although in my experience in arid habitats, it only occurs near watercourses or in irrigated areas. In Florida, I have found *sessile* in marshes (including salt marshes) and open edges of swamps, never in scrub or sandhill areas. It is not a common Florida species, and I have not been sent specimens that were infesting buildings. North of Florida, *sessile* can be found in upland as well as wet areas, and is considered a major household pest (Hedges 1998; Smith 1965). Colonies have multiple queens and establish satellite colonies near prime resources, such as groups of honeydew-producing Homoptera, usually retracting into a centralized colony as resources diminish in the fall (Buczkowski and Bennett 2006, 2008). This system of resource exploitation, called dispersed central place foraging, allows much more efficient use of scattered concentrations of resources (Buczkowski and Bennett 2006); *Lasius neoniger* has a similar system of satellite nests that expand in summer and contract in fall, but without extra queens in the satellite nests. Buczkowski and Bennett (2008) state that in natural habitats, *sessile* forms relatively small colonies with single queens, but it is not clear that these are not single-queen satellite nests of a larger colony. Queens and males may mate in the nest, but there are also flights of alates that presumably mate elsewhere (Smith 1965). The ability to mate in the nest facilitates the formation of huge "supercolonies" in prime locations, but to date, there is no indication that these colonies have given up the option of nest founding by dispersing alate queens.

Tapinoma sessile was predestined to become a household pest because any indoor resource, such as a sugar bowl left on the kitchen counter, can immediately stimulate the establishment of a satellite colony in proximity. Such colonies may be outside, or they may be indoors in damp areas such as the area around water pipes in wall voids (Hedges 1998). Dispersed central area foraging implies that workers from a colony explore a relatively large area, increasing the probability that a worker will find its way into a kitchen. In some residential areas, *sessile* can form enormous "supercolonies" (Buczkowski and Bennett 2008). These can occur where humans have created a kind of *sessile* paradise consisting of nesting sites (heavy mulch, piles of debris, damp enclosed spaces around buildings), outdoor feeding stations for workers (infestations of Homoptera, often exotic, on plants that are also often exotic), plenty of live insect prey in a disturbed and ecologically unbalanced system for the larvae, dead insects around outdoor light fixtures, additional protein scavenged from garbage cans, and frequent access to the careless opulence of the kitchen.

Tapinoma sessile is the host of two rare parasitic species of *Tapinoma* in northern and western North America (Fisher and Cover 2007), but it is unlikely that these species occur in Florida, which is a long way from the known incidence of these parasitic species.

The pungent odor of cyclopentanoid monoterpenes produced by *sessile* (Blum 1981) explains the name of "odorous house ant" used by Smith (1965) and other entomologists.

Name Derivation: From *sessilis* (Latin), meaning "sitting," referring to the way the overhanging gaster appears to be sitting on the mesosoma, concealing the simple petiole to which the gaster is attached. Thomas Say's description of this species (1836) is brief and the type specimen is missing, so Christopher Hamm (2010) provided a new description and designated a new type specimen (neotype) collected from the same site as Say's specimens. There was never any doubt about the identity of Say's species, as he clearly describes the petiole, but the new description will be useful if additional similar species are found in North America.

Genus *Technomyrmex*

Techno Ants

The genus *Technomyrmex* (pronounced **Tek' nō mur" meks**) includes approximately 90 species, almost all of which are native to the Old World tropics, although several species have been introduced elsewhere (Bolton 2007). Amazingly, there are two native species in Central America and South America (Fernández and Guerrero 2008); these might be relict representatives from before the "retreat" of some dolichoderine lineages in the New World (Brown 1973; Wilson 1988). There are four "tramp" species that have been spread around the world, but only *difficilis* is established in Florida (Bolton 2007). Most species of *Technomyrmex* are arboreal and strongly dependent on the honeydew of sap-sucking insects.

Many *Technomyrmex* species are similar to species of *Tapinoma*, and some *Technomyrmex* species were originally placed in *Tapinoma*. In both genera, the petiole is a simple flattened tube and the gaster overhangs the propodeum. *Technomyrmex* has five visible tergites rather than the four visible in lateral view in *Tapinoma*, but in some specimens, the gaster has telescoped or collapsed, making it difficult to see the structure of the gaster. In the Florida ant fauna, the one species of *Technomyrmex* can be distinguished from local *Tapinoma* by the sculptured sides and angled crest of the propodeum and by the presence of several erect hairs on the mesosoma.

Name Derivation: Possibly from *techno* (Greek), meaning "art or skill," and *myrmex* (Greek) meaning "ant." There is no derivation provided in the description of the genus or the type species (Mayr 1872), or any indication why this ant might be considered skillful. Another possible derivation might be from the Latin word *techna*, meaning "a trick," perhaps referring to the difficulty of distinguishing between *Technomyrmex* and *Tapinoma*. In the absence of any convincing derivation of the name *Technomyrmex*, it is reasonable to accept James Wetterer's catchy name, "techno ants" (Wetterer 2008b).

Technomyrmex difficilis Forel—
Difficult Techno Ant (Plate 70)

Taxonomy and Similar Species: *Technomyrmex difficilis* is a small grayish black ant that could be confused in the field with *Nylanderia bourbonica* or possibly *Tapinoma sessile* or *Linepithema humilis*. In the field, *Nylanderia bourbonica*, which often occurs in the same disturbed sites as *T. difficilis*, is noticeably shiny. Under the microscope, *difficilis* is easily distinguished by its angulate and granulate propodeum (in lateral view), yellowish white tarsi, and peculiar erect hairs on the gaster. When *difficilis* was first found to be established in Florida (Deyrup 1991b), it was identified as *albipes*, a similar species that has also been transported around the Old World tropics and subtropics (Bolton 2007). There is a good chance that *albipes* or another tramp species of *Technomyrmex* may eventually appear in Florida. Bolton (2007) provides a key to the four tramp species known to occur in the New World or likely to be introduced. *Technomyrmex difficilis* is distinguished by the combination of one or two pairs of setae on the posterior part of the head, and yellowish white tarsi that are much paler than the tibiae (Bolton 2007).

Distribution: *Technomyrmex difficilis* is probably native to Madagascar, but is now widely distributed in the Melanesian Region (Malaysia and the Philippines south through Indonesia and New Guinea), Australia, various Pacific Islands, parts of the West Indies, and Peninsular Florida (Wetterer 2013b). In Florida, it occurs outdoors from Key West north into Brevard and Hillsborough counties, and it may be expected around the outside of buildings in urban and coastal areas through the Peninsula. It has been found in greenhouses and zoos far north of its outdoor zone (Wetterer 2013b).

Natural History: *Technomyrmex difficilis* is commonest in disturbed habitats, including urban habitats and some East Coast tropical hammocks that are naturally disturbed by storms. In my experience, it is rare in natural habitats inland and is not common in disturbed habitats in south-central Florida. In this region, I have not seen outbreaks such as those that occur in coastal southeast Florida. Honeydew produced by sap-sucking Homoptera is apparently an important part of the diet of *difficilis*; like many other ants, *difficilis* may increase this food supply by tending and protecting various kinds of sap-sucking Homoptera (Warner et al. 2013). This positive feedback loop may explain outbreaks of *difficilis*, outbreaks that might be most likely in a disturbed system with a variety of introduced plants and introduced sap-sucking insects. Nests are concealed in mulch or debris, or above ground in trees and buildings. These ants are difficult to control because colonies are dispersed and not confined to the ground where insecticides are usually applied. Control is by persistent use of slow-acting baits containing boric acid (Warner et al. 2013). When residents complain of ants invading from outside, exterminators often concentrate pesticides on the ground and around the base of the house. This is likely to reduce populations of ground-nesting ants in the genera *Pheidole*, *Nylanderia*, *Solenopsis*, and *Dorymyrmex*

that compete with *difficilis* for honeydew. In the West Indies, *difficilis* may occur in forested areas (Wetterer 2008b).

Populations of *difficilis* might increase in two ways. Winged male and queen *difficilis* can be captured in flight traps, suggesting independent nest founding. When one looks at the kind massive population typical of a *difficilis*, it seems likely that large, multiqueen nests of *difficilis* are budding off new colonies, but I have not seen studies demonstrating that this is happening. Both forms of colony reproduction occur in the related species *brunneus*, which produces numerous fertile wingless females that mate with wingless males in the same colony (Yamauchi et al. 1991). The study of *brunneus* was published using the name *albipes*, and it appears that the observations of Yamauchi et al. may have been applied to *difficilis*, which was also believed to be *albipes*. The work on *brunneus* (Yamauchi et al. 1991) revealed other remarkable features: relatively huge numbers (sometimes over half the colony) of fertile, wingless "intercaste" females; the normal worker-to-worker sharing of food such as honeydew replaced by consumption of specialized eggs produced by workers; and colony dependence on a worker force comprising less than half the colony. Perhaps these weirdities are widespread in the *albipes* group of *Technomyrmex*, or they might be confined to *brunneus*.

Name Derivation: The name *difficilis* (Latin) might best be translated as "problematic." Forel described *difficilis* as a subspecies of *mayri*, but was unsure of its status, and even thought that it might be a hybrid between *mayri* and *albipes* (Wetterer 2008b). Both *mayri* and *albipes* are now considered distinct species (Bolton 2007). As James Wetterer remarks (2008b), *difficilis* remains difficult in the sense of a pest that is difficult to control.

SUBFAMILY FORMICINAE

Genus *Brachymyrmex*

Rover Ants

The genus *Brachymyrmex* (pronounced **Brak' ē myr" meks**) is native to the New World and most common in the tropics and subtropics, with the remarkable exception of the hardy species *depilis*. These common little ants are easily recognized by their nine-segmented antennae and their plump physique. Almost 40 species of *Brachymyrmex* have been described (Bolton 1995), and there is plenty of evidence that there are many undescribed species. The review of this genus by Santschi (1923) is not useful in Florida and is probably not very useful anywhere.

There appear to be at least eight species of *Brachymyrmex* in Florida, making the state the North American hot spot for what Creighton (1950) called this "miserable little genus." Creighton's frustration is understandable. Small and soft-bodied, *Brachymyrmex* tend to make poor specimens, especially if stored in alcohol for some time before mounting. In extreme cases, they collapse and crumple into tiny nubbins at the end of their paper points, often with a few hairs and deformed appendages sticking out at unnatural angles. In life, however, *Brachymyrmex* species are small but cute. In his description of *heeri*, Forel (1874) calls it, "Cette charmante petite espèce…." (This charming little species….). *Brachymyrmex* specimens preserved in alcohol are far easier to study than dry, pinned specimens.

Viewed from another perspective, this genus of small, chubby ants, weak-jawed and stingless, presents an interesting challenge to ecologists. How do these little ants, common throughout Florida, survive in a world filled with fiercely stinging *Solenopsis* and *Wasmannia*, the serried ranks of *Pheidole* with their scissoring jaws, burly *Camponotus* that can preempt honeydew sources and, with a single spray, drown a *Brachymyrmex* in formic acid? The probable generic answer is that *Brachymyrmex* species have potent defensive chemicals. Herpetologists have become interested in this topic because some *Brachymyrmex* are sources of pumiliotoxin alkaloids found in the skins of frogs and toads (Saporito et al. 2004). These compounds, however, appear to be produced inconsistently, suggesting that the ants derive them from yet another source, or that pumiliotoxin production is environmentally induced (Smith and Jones 2004). Oribatid mites seem to be a prime source for poison frog toxins (Saporito et al. 2007), but nobody has investigated whether species of *Brachymyrmex* eat oribatids. While the frog research shows that *Brachymyrmex* species may be defended by significant amounts of toxic alkaloids, defenses against aggressive ants must involve compounds that are constantly ready to be deployed. A particularly useful class of compounds for small ants might be chemicals that disrupt or corrupt communication, as such compounds should be effective in tiny quantities, and might be manufactured along pathways already used for communicatory chemicals.

Name Derivation: From *Brachys* (Greek), meaning "short," and *myrmex* (Greek), meaning "ant." These ants are relatively compact, and, in addition, often contract distressingly as dried specimens.

Brachymyrmex depilis Emery— Hairless Rover Ant (Plate 71)

Taxonomy and Similar Species: Distinguished from most other yellowish Florida species by the absence of any large standing hairs on the mesosoma (check in lateral view with a dark background). An undescribed yellow Florida species has much smaller eyes with only about a dozen facets.

This widespread species might be a species complex. It is highly unusual for a North American ant species to have the huge geographic range of *depilis*. One can, however, imagine a scenario in which a subset of the *depilis* lineage makes an evolutionary breakthrough in cold hardiness and then spreads rapidly through the newly suitable territory. One would expect eventual genetic divergence through founder effects, genetic drift, and natural selection. As in other widely distributed species, genetic differences in widely separated populations do not necessarily mean that these populations have become separate species, as there might be gradual (clinal) change from one area to another. Specimens I have seen from northern New England are noticeably larger and somewhat browner than specimens from southeastern North America, but I see no structural differences. Even in Florida, some specimens are a clear yellow, whereas others have a slightly brown tint.

Distribution: Records of *depilis* appear to be scattered across most of North America from southern Canada into northern Mexico. It occurs throughout Florida.

Natural History: *Brachymyrmex depilis* has an enormous habitat range. I have found colonies in swampy areas, usually in rotten wood, and in the driest Florida scrub, often among roots of grass tussocks. Heavily wooded and open sites are both acceptable. I have extracted specimens from leaf litter in the tropical Florida Keys, from random scoops of sand in excessively drained, open sandhill, from soil of a deciduous forest in central Maine, and have encountered a colony under a rock in southern Arizona at 1700 feet. This species and *Tapinoma sessile* may be the most widespread species of North American ants; both must owe their extraordinary ranges to their ability to make a living in almost any habitat. It is remarkable that *depilis* has never become established as an exotic in temperate or subtropical regions of the Old World.

Brachymyrmex depilis appears to be primarily subterranean, but I regularly catch workers in bowl traps set flush with soil surface, so workers probably forage in the open at night. Colonies in grass tussocks are often associated with sap-sucking insects; ecologically, *depilis* might resemble a miniature *Lasius*. Male *depilis* are minuscule compared with females of their species, so small that two or three are able to mate with a female simultaneously (Page 1982). Alates are sometimes attracted to lighted windows in Florida, where this multiple mating can occasionally be observed.

Name Derivation: From *depilis* (Latin), meaning "without hair," referring to the absence of enlarged standing hairs on the head and body.

Brachymyrmex minutus Forel— Hairy Yellow Rover Ant (Plate 71)

Taxonomy and Similar Species: In Florida, *minutus* is the only known *Brachymyrmex* species that is yellow with large standing hairs on the mesosoma. *Brachymyrmex minutus* is similar to a number of yellowish Neotropical species. One of these might be *heeri*, a species that must get moved around, as the type series is from a botanical garden in Zurich (Forel 1874). For some years, I thought that the yellow Florida species might be *heeri*, but Forel's comprehensive description of *heeri* states that *heeri* has no ocelli and the first abdominal tergite is strongly pubescent. This description would not fit the Florida species, which has ocelli and a shining first abdominal tergite with very sparse pubescence, both features included in the description of *minutus* (Forel 1892). The types of *minutus* are from St. Vincent and St. Thomas. While it seems likely that *minutus* is the species found in Florida, most identifications of Neotropical *Brachymyrmex* will be somewhat provisional until species of *Brachymyrmex* have been investigated more carefully. The differences between Forel's species *minutus* and *heeri* are emphasized here because it is reasonable to think that *heeri* might also have been brought to Florida, although the Florida species we call *obscurior* might really be *heeri*, as discussed below. Forel (1893) provides additional differences between *minutus*

and *heeri*, but these are comparative features that are best seen, as Forel remarks, if specimens of the two species are examined side by side.

Distribution: The types of *minutus* are from two widely separated sites, the islands of St. Vincent and St. Thomas (Forel 1893). I have seen specimens that I believe to be *minutus* from Florida, the Bahamas, and Bermuda, comparing them with specimens that I believe to be *heeri* from Puerto Rico and Hispaniola. Wheeler (1905a) found both *minutus* and *heeri* in the Bahamas.

Natural History: *Brachymyrmex minutus* appears to be a cold-sensitive species, in Florida confined to the southern tip of the Peninsula, extending north along the East Coast into Broward County. Within this area, it is much less common than *depilis*; there are fewer than 50 Florida specimens in the Archbold Biological Station collection, and it is probable that there are no Florida specimens in any other collection. Specimens were extracted from leaf litter, mostly in disturbed areas.

Name Derivation: From *minutus* (Latin), meaning "small," because this species is even smaller than *heeri*.

Brachymyrmex obscurior Forel— Seaside Rover Ant (Plate 72)

Taxonomy and Similar Species: *Brachymyrmex obscurior* is easily differentiated from other Florida *Brachymyrmex* by the combination of pale to dark brown color and dense, appressed pubescence, especially on the gaster. In the field, it can be confused with *patagonicus*, another common brown species. Under the microscope, however, the eye of *obscurior* (in lateral view) is shorter than the distance between the eye and the base of the mandible, while the eye of *patagonicus* is at least as long as the distance to the base of the mandible. The gaster of *obscurior* is only feebly shiny, because of the dense pubescence, while that of *patagonicus* has sparse, appressed pubescence on a strongly shining background.

The taxonomy of *obscurior* is unsettled. Forel (1893) described *obscurior* as a dark-colored variant of *heeri*, and there is no reason to question his judgment, especially as the species currently going under the name *obscurior* varies from very pale brown to dark brown. Wilson and Taylor (1967) raised *obscurior* to species level as a "provisional measure, contingent upon a fuller revision of the large and difficult genus to which it belongs." It is possible that the long-awaited "fuller revision" will synonymize *obscurior* with *heeri*. It is somewhat worrisome that workers of both *obscurior* and *heeri* are described as lacking ocelli (Forel 1874, 1893), whereas the species that numerous myrmecologists (including myself) have identified as *obscurior* typically has ocelli, easily seen in larger specimens. In unusually small specimens of *obscurior*, it is difficult to decide with dry specimens whether there are functional ocelli, or just tiny indentations in the head among the hairs and punctures. Ocelli occur in some species of *Brachymyrmex*, not in others. In *minutus*, they are "parfois visibles" (Forel 1893). Perhaps there is variation in the occurrence or visibility of ocelli between populations of *obscurior*.

Distribution: *Brachymyrmex obscurior* occurs in Florida, southern Texas, Central America, and the West Indies (Smith 1979). It has been imported into Samoa and Hawaii (Wilson and Taylor 1967) and could appear almost anywhere. I have often found colonies of *obscurior* in potted plants. I saw a batch of queens fly into the open door of a small plane in Fort Lauderdale, crawl about the carpet for an hour or so, and then head for the door when the plane landed on San Salvador in the eastern Bahamas. Had I not collected the specimens, they would have joined a resident population already on the island. In the British Virgin Islands, *obscurior* were found as stowaways on a barge filled with nursery plants that originated in southern Florida (Miller 1994). Even though *obscurior* may have been imported into Florida many times, it is probably native here. This species seems especially at home in sandy areas near the seacoast, so island-hopping would be a good mode of dispersal. They are apparently expert at short-distance island-hopping, as they are the most abundant ants on small islets (under 1000 m²) in the Bahamas (Morrison 2002, 2006).

Natural History: *Brachymyrmex obscurior* in Florida is usually found in open sandy areas with sparse vegetation. It is one of the commonest ants in disturbed areas, including scruffy lawns, edges of mall parking lots, scuffed vegetation by the curbs of sidewalks, drought-stricken highway medians, and among the litter of gum wrappers and cigarette butts along the fences of playgrounds. It is seldom found in natural upland habitats such as sandhill and Florida scrub. All this fits the profile of an introduced species with a masochistic preference for the worst kind of disturbed areas, but *obscurior* is also abundant in beach and other coastal habitats. These natural habitats are as extreme and highly disturbed as the man-made habitats where

obscurior occurs, and it is reasonable to assume that *obscurior* evolved as a coastal species.

The extreme hardiness and resilience of *obscurior* was shown by Lloyd Morrison (2002, 2006), who demonstrated his own endurance by conducting a 14-year survey of the ants on more than 260 tiny cays in the Bahamas. Morrison found ants on almost all cays that had vegetation, even if the islets were easily washed over by storms. *Brachymyrmex obscurior* was the most prevalent of these ants. On islets that were so inhospitable that they had only one ant species, that ant was *obscurior* in 75 out of 81 instances. These cays are composed of porous limestone that is filled with cracks and holes and chambers in the rock, within which *obscurior* can survive at high tide or when the cay is submerged during storms. In the course of the 14-year study, occupancy by *obscurior* was stable, with rates of extinction or immigration under 5%. The other common terrestrial ant on these keys was a species of *Dorymyrmex*, a species much larger and faster than *obscurior*, arriving earlier at baits but eventually displaced by large numbers of *obscurior*.

The diet of *obscurior* appears to be primarily honeydew and dead arthropods. Workers readily visit extrafloral nectaries. Despite its abundance, *obscurior* seldom seems to be a significant pest. Large numbers of queens may be attracted to lights; I have been brought batches of these queens by alarmed homeowners who thought their houses were being invaded by ants (sometimes misidentified as termites) that would take up residence indoors. *Brachymyrmex obscurior* is one of the ants that tends the Asian citrus psyllid (*Diaphorina citri*), apparently reducing rates of parasitization by the eulophid wasp *Tamarixia radiata* (Navarrete et al. 2013).

Morrison (2006) reported slight variations in color and eye size, and suggested the possibility of a species complex on the Bahamian cays. At some sites in Florida, *obscurior* are slightly larger or smaller than normal. I have assumed that this reflects prevailing environmental conditions, preferring this explanation to the prospect of a swarm of ineluctably cryptic species.

Name Derivation: From *obscurior* (Latin), meaning "darker," referring to the coloration, a darker brown than the similar *heeri*.

Brachymyrmex patagonicus Mayr— Patagonian Rover Ant (Plate 72)

Taxonomy and Similar Species: *Brachymyrmex patagonicus* is a common brown species with large erect hairs on the mesosoma, much as in *obscurior*. In the field, the two species may be difficult to distinguish, although, in Florida, *obscurior* is often a paler brown than *patagonicus*. Under the microscope, *patagonicus* is distinguished by its larger eyes, which are longer than the space between the eye and the mandible in lateral view, and by the sparsely pubescent, strongly shining first gastral tergite, which in *obscurior* is densely covered with appressed hairs and feebly shining. In southeastern North America, *patagonicus* was at first erroneously identified as the Central American species B. *musculus* (Deyrup et al. 2000; Wheeler and Wheeler 1978). *Brachymyrmex patagonicus* was redescribed by Quirán et al. (2004), who examined Mayr's type material. Quirán later identified as *patagonicus* specimens sent from southeastern North America (MacGown et al. 2007b).

Distribution: The South American distribution is Argentina, Paraguay, Bolivia, Brazil, Venezuela, and the Guianas (Quirán et al. 2004). In North America, *patagonicus* occurs from South Carolina through Florida, in a more or less continuous band through Louisiana, with sporadic occurrences in Texas, Arkansas, Arizona (MacGown et al. 2010), and California (Martinez et al. 2011). It was established, at least temporarily, in a zoo in The Netherlands (Boer and Vierbergen 2008). As long as *patagonicus* remains extremely abundant in southeastern North America, it will probably continue to spread through commerce and could appear almost anywhere in warm-temperate climatic zones.

Natural History: *Brachymyrmex patagonicus* occurs in a variety of natural and disturbed habitats, usually in sites that are relatively open, such as pine flatwoods, grasslands, fields, and coastal sites, as well as urban and suburban areas. In my experience, it is usually found in less arid sites than those favored by *obscurior*. For example, in the Florida Keys, where *patagonicus* is a recent arrival (Moreau et al. 2014), it can be found in irrigated landscaped areas, but has not replaced *obscurior* in grass tussocks along the shore. As *patagonicus* becomes more common in south Florida where *obscurior* is a dominant ant in open disturbed sites, it might displace *obscurior*, but interactions between these two species are unknown. Both species seem to subsist primarily on honeydew, extrafloral nectar, and dead insects. It seems to me that *patagonicus* prefers superficial nesting sites, such as landscape mulch, leaf litter, thatch found in grass tussocks, and loose bark at the base of trees. In contrast, *obscurior* is more subterranean, with extensive galleries at least a centimeter below the surface. Whenever a newly introduced species of ant begins to spread rapidly through Florida, there is another opportunity for research on the effects of an exotic species on the preexisting fauna, with special reference to closely related species.

At some sites, there have been outbreaks of *patagonicus*, with large numbers coming indoors to forage (MacGown et al. 2007b). Since nests are outside, it would probably be easy in most cases to exclude these ants, but many homeowners seem to expect and demand that their homes should be surrounded by a kind of large, ant-free bubble, requiring perpetual vigilance and drastic control measures. Although I am a coauthor of a paper labeling *patagonicus* an "emerging pest species" (MacGown et al. 2007b), nobody knows whether *patagonicus* will become a long-term pest (in some situations) or whether its populations will subside. This could happen if some organism or group of organisms begin to exploit the species as an abundant resource.

Name Derivation: Named for the Patagonia region of Argentina, where the type specimens were collected.

Brachymyrmex Species A— Small-Eyed Rover Ant (Plate 71)

Taxonomy and Similar Species: *Brachymyrmex* species A is distinguished from *depilis* by its small eyes, which have only approximately 12 facets, sometimes fewer. In the survey of Florida ants, Species A has been collected 25 times, as opposed to hundreds of collections of *depilis*. I have seen no specimens that show intermediate eye size between Species A and *depilis*. Litter samples with Species A usually had only a few specimens. When I first collected this species, I assumed it was an aberrant form of *depilis*, but with 25 collections from 10 Florida counties, this seems unlikely. Eye size is the only way I know to distinguish Species A from *depilis*, although it seems to me that the short pubescence on Species A is a little more erect and abundant than that of *depilis*.

Distribution: Known only from Peninsular Florida, from Levy County south into Okeechobee County.

Natural History: Almost all specimens are from scrub or sandhill areas, sometimes heavily overgrown. A series of specimens from the

Archbold Biological Station were collected from heavily overgrown scrub after persistent rains had brought the water table near the surface. It is possible that this is a subterranean species occasionally forced up near the surface, or found as small numbers of strays in leaf litter.

Brachymyrmex Species B— Banded Florida Rover Ant (Plate 72)

Taxonomy and Similar Species: *Brachymyrmex* Species B lacks standing hairs on the mesosoma, like *depilis* and Species A, but is dark brown like *patagonicus* or *obscurior*. The eye length is slightly shorter than the distance between the eye and the base of the mandible in lateral view. The femora are broadly banded with brown, and the bases and apices are pale, unlike other Florida species of *Brachymyrmex*. The first gastral tergite is densely covered with long appressed hairs, with enough space between to see the shining underlying surface. Tergites 2–3 are more sparsely haired, with spaces between hairs a little less than half the length of the hairs. Tergites 1–4 have a sparse subapical fringe of long pale hairs. The single known queen is conspicuously smaller than queens of *patagonicus* or *obscurior* and lacks long hairs on the head.

Distribution: This apparently rare species has been collected only three times, in Columbia, Clay, and Alachua counties, all in northern Peninsular Florida.

Natural History: Only 17 specimens are known. One series of 12 specimens was collected by Lloyd Davis, who found a column of these ants running along the ground, including a male and a queen. One collection was made in mid-November, and the other two were made in January. The Alachua County series (four specimens) was collected in a shaded slope with disturbed vegetation on the edge of a catchment basin with clay soil. It is possible that this species is more active in winter.

Brachymyrmex Species C— Fuzzy Queen Rover Ant (Plate 73)

Taxonomy and Similar Species: *Brachymyrmex* Species C, known from three queens, is another apparently undescribed species. The queen of Species C is dark brown with pale brown appendages, approximately 3 mm long, with no long, standing hairs on the head or body. The head, mesosoma, gaster, and legs are remarkable for their dense covering of short, even pubescence. The pubescence on the propodeum and petiole is erect, whereas that of the rest of the body is reclinate or proclinate, not appressed. These queens, taken in flight traps, have the moderately swollen gaster typical of founding queens; there is no reason to believe that this species is parasitic. Workers are unknown, but are probably dark colored like the queens and might have unusual pubescence.

Distribution: Known from Polk County, one from the Nature Conservancy Tiger Creek Preserve, and two from the Florida Wildlife Commission Sunray property.

Natural History: The queens were collected in flight traps in Florida scrub habitat in May and July. The sites where the queens were found got some intensive sampling for ants and other ground-inhabiting insects, but this yielded no dark *Brachymyrmex* workers. They are out there somewhere, but possibly in some unusual microhabitat.

Brachymyrmex Species D— Tiny Queen Rover Ant (Plate 73)

Taxonomy and Similar Species: *Brachymyrmex* Species D is known from 31 queens in the Archbold Biological Station collection. All were captured in flight traps; males and workers (if the latter exist) are unknown. This yellow species is easily distinguished from other queen *Brachymyrmex* by its small size (under 2 mm long), nonswollen gaster, and somewhat reduced forewing venation: the radial sector and medial veins are not joined beyond the first segment of the radial sector (these veins form a long V rather than a Y with a short stem). There is a pair of elongate curved hairs on each side of the scutellum, usually two to four erect mesonotal hairs (apparently easily dislodged), and the gastral tergites are covered with long, appressed hairs.

Distribution: *Brachymyrmex* Species D is known from flight traps in Highlands and Polk counties. This distribution may primarily reflect the sites where I had flight traps rather than the true distribution of this species. I did not find any specimens among a large collection of flight trap ants from Alachua and Leon counties, but have no way to compare sampling efforts.

Natural History: Every myrmecologist who looks at Species D is convinced that this species must be parasitic on some other species of ant, probably a species of *Brachymyrmex*. This is partly because some parasitic ants have queens that are small relative to other species in their genus, as in Species D. The convincing feature in Species D, however, is not size but the relatively small gaster, in marked contrast to that of queens of other species of *Brachymyrmex*. In alate queens of the subfamilies Formicinae (like *Brachymyrmex*), a small gaster, proportionally similar to that of a worker, is strongly associated with parasitism. Currently, however, no species of *Brachymyrmex* has been shown to have a parasitic lifestyle. A consistent feature of this species is its reduced wing venation relative to that of other queen *Brachymyrmex*.

Although there is a good series of queen Species D, they were accumulated from hundreds of flight trap samples, so this species is probably relatively rare. One might need to look in hundreds of *Brachymyrmex* nests to find Species D, if it is actually a parasitic species. The best chance to find this species is when alate queens are present, probably in July. Males might be flightless and extremely small. Revealing the natural history of the various rare species of *Brachymyrmex* may require a large number of dedicated myrmecologists, or a smaller number of very lucky ones.

The queen Species D in the Archbold Biological Station collection were collected in scrub habitat, but the traps were intended to sample the insects of Florida scrub, and this species could be in other habitats, especially if it is associated with the omnipresent *B. depilis*. Queens fly primarily in July and August (28 out of 31 specimens).

Genus *Camponotus*

Carpenter Ants

The genus *Camponotus* (pronounced **Cam' pō nō" tus**) is the only genus of Florida Formicinae that has major and minor workers, an excellent field character that also works well with museum series. Most species have a distinctive smooth, arching mesosomal profile. The exceptions are *sexguttatus*, which has a notched mesosoma, and three small Florida species whose majors have enlarged, cork-like heads that are used to block nest entrances. These last species have small minors that can easily be mistaken for *Dolichoderus* in the field.

Camponotus is a major worldwide genus found in almost every terrestrial habitat, from boreal forest to desert to the wet tropics. This is a huge genus: Bolton (1995) lists 932 species and 566 subspecies.

Camponotus has grown so unwieldy that its revision would be a myrmecological Labor of Hercules. A good number of supposed species, and a larger proportion of subspecies, are probably synonyms. Some subspecies are probably distinct species. There is every reason to believe that there are many undescribed species. These are large, often conspicuous ants, and many species were described long before the advent of the biological species concept or any clear understanding of the biogeography of ants. Many species are variable in color, and most species have conspicuously different major and minor forms, usually with a continuum of intermediate forms. Add to this the problem that many species were described from a few specimens taken from isolated sites, and the wealth of opportunities for taxonomic confusion is all too obvious. The North American fauna, which one might expect to be well understood, has not been immune from the confusions of *Camponotus*, eliciting frequent exasperated comments from Creighton (1950), such as, "It is much to be regretted that Wheeler elected to describe the variety that he called *osceola*."

Putting aside the moot question of how William Wheeler (1910b) would have fared had he our conceptual tools, or how we would have fared with his, *Camponotus* remains a taxonomically challenging genus. There are several difficult species complexes in North America, and I am not sure how to deal with some southeastern species until there have been a few studies that incorporate ecology, genetics, and collecting transects across geographic ranges. I doubt that we can achieve real understanding of these species by rooting about in old literature and museum collections. A useful example of such a problem is provided by *nearcticus* and *decipiens*, which cannot be separated by any known structural features, but differ in color and ecology, and occur together at a number of sites in Florida. With such cases in mind, I have no interest in synonymizing Florida species for the convenience of museum taxonomists, and the treatment below is generally conservative.

As their English name implies, *Camponotus* species are usually associated with wood, both stumps and dead wood lying on the ground, or dead branches and twigs on live trees. In pre-European times, Florida's upland habitats were structured by frequent fires, and Florida populations of *Camponotus* species reflect this history. Florida *nearcticus*, for example, usually nest in branches in crowns of fire-resistant slash and longleaf pines. To the north, this species often occurs in hardwoods, and even in vines. *Camponotus castaneus* is usually a ground-nesting species in Florida, but in the Mid-Atlantic States, it usually lives in rotten wood. Several species of carpenter ants regularly appear in buildings, but they seem overrated as structural pests, except possibly *pennsylvanicus*. More damage may be done by colonies moving into outdoor pump housings and fuse boxes.

Some species of *Camponotus* are likely to be transported from place to place. These species live in small branches and twigs that can be moved with nursery stock, or they may be in timbers or containers that are shipped from other regions. I have even seen specimens of the boreal *novoboricensis* that were spending the winter in a travel home that had been brought from New England to Lee County. *Camponotus sexguttatus* is the most obvious example of an exotic established in Florida, but *planatus* is probably also exotic, and there is some doubt about the native status of *inaequalis* and *floridanus*. On the other hand, *conspicuus* and *ramulorum*, which are apparently common in Cuba and the Bahamas, have not been found in Florida. These and other exotic *Camponotus* could have small populations established in Florida for some time before attracting attention, because *Camponotus* species tend to be nocturnal, especially if they come from areas where ant-eating lizards patrol tree trunks during the day.

Most species of *Camponotus* that live in Florida also live in Mississippi, and there is an excellent guide to these species (MacGown et al. 2007a).

Name Derivation: From *kampe* (Greek), meaning "a bend," and *notos*, meaning "back," referring to the strongly arched mesosomal profile seen (in lateral view) in most species of the genus.

Camponotus americanus Mayr—
Black-Headed Carpenter Ant (Plate 74)

Taxonomy and Similar Species: *Camponotus americanus* resembles *castaneus* but has a black or brownish black head and large, elongate punctures in the area between the eye and mandible in lateral view. Wheeler (1910b) considered *americanus* and *castaneus* subspecies with broadly overlapping ranges. He was writing before there were clear concepts of species and geographic subspecies, so this probably did not seem strange to him. In a kind of over-reaction, Creighton claimed that the two species were only distantly related, but Coovert (2005)

considered them distinct but closely related species. In Florida, *castaneus* sometimes has a dark brown or even blackish head, so the most reliable difference between the two species is the collection of elongate punctures between the eye and mandible. There are no other Florida *Camponotus* that would be easy to confuse with *americanus* and *castaneus*. The four other large Florida *Camponotus* have obviously different coloration: *pennsylvanicus* is all black, *socius* is red with a yellowish black-banded gaster, *floridanus* and *inaequalis* are reddish with a black gaster, sometimes yellow-banded in the case of *inaequalis*. In Florida, *americanus* is also a rare species, known only from a few collections in the Panhandle.

Distribution: Ontario south just into Florida, west into Michigan, Oklahoma, and Texas (Coovert 2005). In Florida, good series of *americanus*, along with some *castaneus*, were collected by Peter Kovarik and Paul Skelley in the ravines of the Apalachicola River in Liberty County. This area is well known for its large number of isolated populations of northern plants and animals, many of which are normally associated with the southern Appalachians (Whitney et al. 2004). Other northern ants with apparently isolated populations in Apalachicola River ravines are *Formica subsericea* (now possibly extinct in Florida) and *Lasius interjectus*. An old record of *americanus* from Quincy (Wheeler 1910b) might refer to specimens taken in ravines near the Apalachicola River.

Natural History: *Camponotus americanus* is a woodland species, usually found in relatively open woodlands (Carter 1962). Nests are subterranean, usually under rocks or dead wood (Smith 1979). Like *castaneus*, *americanus* is nocturnal (Ellison et al. 2012).

Name Derivation: Gustav Mayr named this species *americanus* because of its origin (New Orleans); it was not the first *Camponotus* to be named from North America, but one of the first to be named after the genus itself was recognized as distinct from *Formica*. This species was named *americanus* in 1862, at a time when the country the world knew as America appeared as if it might have a temporary existence, but there is no reason to believe that Mayr was thinking of the national rather than geographic homeland of this ant.

Camponotus caryae (Fitch)—
Hickory Carpenter Ant (Plate 75)

Taxonomy and Similar Species: This is a black species closely resembling *nearcticus*, but *caryae* has large, conspicuous punctures in the malar area (side of the head between the eye and jaw insertion). *Camponotus pennsylvanicus* is another black species, but is larger and much hairier, especially the gaster, which is covered with long, pale, almost contiguous hairs lying flat on the surface. In structural morphology, *caryae* seems almost indistinguishable from *discolor*, but *discolor* has a red head, mesosoma, and appendages. It is possible that *caryae* and *discolor* are the same species, but the two seem to overlap in distribution without intergrading (Snelling 1988).

Distribution: New York south into Florida, west into Ohio and Iowa (Snelling 1988). A purely speculative hypothesis explaining the distribution of *caryae* and *discolor* is offered in the discussion of *discolor*. In Florida, *caryae* is known only from Liberty (Snelling 1988) and Columbia counties.

Natural History: This species appears to be rare throughout its range (Creighton 1950), which is certainly the case in Florida. I have collected *caryae* only once, a single worker taken on the trunk of *Quercus laevis*. It seems likely that nests are in dead branches of living hardwoods. Foraging probably occurs in the canopy, with only stray workers chancing into the foraging zone of myrmecologists, who are themselves relatively rare.

Name Derivation: From *karya* (Greek), meaning "walnut," but now used as the scientific genus name for hickory. Hickories have hard, resilient wood, but produce twigs and small branches with a soft pith that can be hollowed out relatively easily by a variety of insects. Hickories produce powerful repellent chemicals, which are easily detected from the odor of a crushed fresh leaf. The leaves, bark, and wood of hickories tend to be exploited by *Carya* specialists, but there is no evidence that *caryae* is one of these specialists.

Camponotus castaneus (Latreille)— Chestnut-Colored Carpenter Ant (Plate 74)

Taxonomy and Similar Species: This is the only shiny, chestnut-colored carpenter ant in south and central Florida. The queens and males share this coloration, and are easily identified when they appear in large numbers at lights in winter and spring. Sometimes, it seems that there are never any simple taxonomic situations in the genus *Camponotus*; *castaneus* provides another example. In the North, *castaneus* and *americanus* are easily distinguished species. *Camponotus castaneus* has a brown head, its malar area (side of head between eye and

mandible) lacks large, elongated punctures, and it usually nests in rotten logs or stumps. *Camponotus americanus* has a black head, and its malar area has large, elongate punctures. In the Florida Panhandle, there is a brown-headed form of *castaneus* with a few to substantial numbers of elongate punctures. Even in south Florida, queens may or may not have elongate malar punctures. There are specimens that are "classic" *americanus* from the Apalachicola River ravines (see *americanus*, above), but it also seems possible that there might have been some interbreeding between *castaneus* and *americanus* at some point in the past. Perhaps, however, elongate malar punctures are spontaneously appearing in Panhandle specimens. These elongated punctures, each with a bristling hair, seem to be important features in distinguishing between species of *Camponotus*, for example, between *discolor* and *decipiens*, and between *nearcticus* and *caryae*. I doubt that anybody knows the biological significance of these punctures, much less why they should start to appear in Florida *castaneus*. It is inspiring to think how many interesting taxonomic/ecological questions remain in the genus *Camponotus*.

Distribution: New York south through Florida, west into Iowa, Oklahoma and Texas (Smith 1979). In Florida, *castaneus* has been collected from Collier County northward.

Natural History: In North Carolina, *castaneus* lives in both xeric and mesic forests, nesting in stumps and rotten logs, the colonies sometimes extending into the soil (Carter 1962). In Welaka, Putnam County, Florida, *castaneus* lives in xeric, mesic and wet forests, the nests in well-decayed hardwood logs or stumps (Van Pelt 1958). In upland sites in Florida, I have usually found *castaneus* nests associated with dead wood or saw palmetto trunks, but partly or completely subterranean. I have often found incipient colonies, a queen with a few workers, in rotten wood. My guess is that colonies are founded in rotten wood and then spread facultatively into the soil, depending on size and moisture content of the wood. In upland Florida habitats, periodic fires may also completely burn up dry rotten logs, along with any insects living therein. Mating flights and presumably colony founding are during the dry season in Florida (December through April) when choice of suitable moist nesting sites is limited, but any site that is moist at that time of year probably remains damp throughout the year. There seem to be large, synchronized emergences of alates, which are strongly attracted to lights. Workers of most species of Florida *Camponotus* are primarily nocturnal, but *castaneus* seems extreme in this respect, and it is rare to see workers during the day even where honeydew-producing Homoptera or extrafloral nectaries have lured out species such as *floridanus* and *inaequalis*.

Name Derivation: From *castaneus* (Latin), meaning "chestnut colored."

Camponotus decipiens Emery— Deceptive Arboreal Carpenter Ant (Plate 75)

Taxonomy and Similar Species: As far as I can tell, this species is structurally identical to *nearcticus*, another arboreal species, but is red with a black or partially black gaster, while *nearcticus* is completely black, at least in Florida. In Florida, *nearcticus* is usually found in large, open-grown pine trees. The red-and-black *discolor* differs from the similar *decipiens* by its large, elongate punctures in the malar area (side of head between eye and mandible). This leaves the very similar *snellingi*. I have sometimes despaired of definitively distinguishing between *decipiens* and *snellingi*, and have been tempted to call them the same species. In much of its range, *decipiens* is dark red with a black gaster, while *snellingi* is a paler or more yellowish red, with red on the basal tergites of the gaster. In Florida, there is a dark red species with a black gaster in the southern Peninsula, and no species with red on the gaster. In north Florida, all specimens I have seen have at least some red on the gaster. In Highlands and Polk counties, in the center of the Peninsula, both forms occur. It is unlikely that *snellingi* is replacing *decipiens* in north Florida, as *decipiens* is known from west and north of Florida. It seems more likely that the two species converge in coloration. My guess is that the yellowish red individuals are *snellingi* and the dark red individuals are *decipiens*. This problem obviously requires some quality time in the field and laboratory, not a guy scratching his head over a few trays of specimens.

Distribution: Georgia into Florida, west into North Dakota and mountains of western Texas (Snelling 1988). If one were to map this distribution, its peculiarity would be obvious. Why should a species that is at home in both Florida and North Dakota be absent from the East north of Georgia? A plausible answer is that *decipiens* is the species that used to be known as *rasilis* Wheeler or *sayi* Emery. *Camponotus snellingi* might also occur north of Georgia.

In Florida, *decipiens* probably occurs throughout the state, probably occurring together with *snellingi* from the central Peninsula northward.

Natural History: Like most other arboreal ants, *decipiens* is absent from habitats that burn frequently, such as Florida scrub and sandhills. It occurs in oak–hickory hammocks and in swamp forests. Replacement of natural habitats by suburbs with large trees has probably benefited *decipiens*. Nests are usually in dead branches and twigs in the canopy, and individual workers are most likely to be found on trunks of large trees or running along wooden railings of boardwalks. It is probable that dense populations of exotic *Pseudomyrmex gracilis* have usurped most potential nest sites of *decipiens*, and it is not certain that the two species can coexist indefinitely.

Name Derivation: From *decipiens* (Latin), meaning "deceiving." Carlo Emery, in his description of this species, mentioned that it is deceptively similar at first glance to *discolor*. Fortunately for Emery's peace of mind, he did not know that identification of *decipiens* is even trickier with respect to another species.

Camponotus discolor (Buckley)— Bicolored Arboreal Carpenter Ant (Plate 75)

Taxonomy and Similar Species: This species resembles *decipiens* in color, but has conspicuous, elongated, hair-bearing punctures in the malar area (side of head between eye and mandible). *Camponotus discolor* appears to be structurally identical to the all-black *caryae*, and there is some doubt that the two are separate species (Snelling 1988). The two forms, however, seem to have overlapping ranges in the Southeast, so it is reasonable to consider them distinct species until the situation has been studied.

Distribution: South Carolina into Florida, west into North Dakota and Texas (Snelling 1988). As in the case of *decipiens*, something seems to be strange or missing in this distribution. One possibility is that *caryae* is a species that spread north and south from a Pleistocene refuge in the Appalachians, while *discolor* spread north, east, and west from a Pleistocene refuge in the Mississippi Basin. In Florida, *discolor* appears to be a rare species, known from only a few collections in the central and northern Peninsula.

Natural History: This species is seldom collected in Florida, and I have only seen one nest, in a large, dead live oak branch that must have fallen out of the tree shortly before I chanced by. Individuals are occasionally seen on trunks of large oaks, and nests must be in dead branches or twigs in the canopy of such trees. This species is apparently common in Texas, where it is usually found on oaks, but also occurs on hickory, willow, and cottonwood (Snelling 1988). Wheeler (1902, 1910b) reports small colonies in oak galls. In his description of the species, Buckley (1866) states that these ants "dwell beneath stones and logs, having cells a few inches beneath the surface of the ground." This observation must refer to some other ant than the one we call *discolor*.

Name Derivation: From *discolor* (Latin), meaning "of different colors," presumably referring to the red and black coloration of the species. The describer of this ant, S. B. Buckley (1809–1884), was the State Geologist for Texas, with an avocational interest in plants and ants. He described 67 forms of ants, but the types of most have been lost, and his descriptions are curiously useless. William Wheeler remarked (1902), "With a persistency, which at times seems almost intentional, the author selects for description the worthless, insignificant features of the ant's body, and passes without a word over the important, distinctive characters. His conception of generic characters is even more nebulous than his appreciation of specific differences. Sometimes he mistakes the sex of the form he is describing, and at other times confounds several very distinct forms in a single description." Nevertheless, Wheeler (1902) set himself the task of figuring out the identities of Buckley's species, even going to Buckley's preferred collecting areas in Texas. Despite these efforts, it is not clear that the species we call "discolor" is the same species that Buckley described as "*Formica discolor*," but the name has achieved some validation through usage.

Camponotus floridanus (Buckley)—
Florida Carpenter Ant (Plate 76)

Taxonomy and Similar Species: This large, red and black species can be easily distinguished by its long bristling hairs, especially on the legs. This hairiness can often be seen even without a microscope, especially if the viewer is a bit myopic.

The application of names to this common and ecologically important species is complicated and poorly resolved, but it may well eventually have some name other than *floridanus*. I apply the name

floridanus to the Florida population of the widespread and variable Neotropical *C. atriceps* complex, until quite recently known as the *abdominalis* complex. In 1973, Hashmi revised the *abdominalis* complex, clearing up several annoying problems, such as species names resulting from separate descriptions of minors and majors of the same species. Hashmi then reviewed the subspecies of *abdominalis*, using the techniques of phenetics. These techniques may be useful, but not as useful as they seemed when they were first invented. They are not easily applied to some groups of *Camponotus* that show great intraspecific variation and often negligible morphological differences between species. This is true of many other groups of ants; one wonders, for example, how the *Formica fusca* group would have fared if Francoeur, who also published his revision in 1973, had used pure phenetics, rather than an intimate knowledge of not only morphology, but also field ecology and biogeography. Hashmi (1973) refused to use subspecific names and seemed to have a strong interest in minimizing geographic variation. In the long run, *atriceps* (formerly *abdominalis*) is likely to be split into two or more species or subspecies. The fact that *floridanus* and *atriceps* (the latter described from Brazil) have radically different trail pheromones (Haak et al. 1996) suggests that these forms are separated by biology as well as coloration. The situation is complicated by such forms as the southwestern *transvectus*, which strongly resembles *floridanus* in morphology and coloration. It seems most conservative to continue using the name *floridanus* for the southeastern population of the *atriceps* complex, pending a thoughtful review of this complex. In this way, even if the name *floridanus* is eventually replaced, all the information that has accumulated about it can be conveniently attributed to an actual population, rather than disentangled from a mass of information filed under the name *atriceps*. It is better to impress posterity with our consideration than our contemporaries with our certitude.

Distribution: North Carolina south through Florida, west into southern Mississippi (Smith 1979). It seems less common in the western part of its range. It occurs throughout the state in Florida, but there are relatively few records from the western Panhandle. The tropical and subtropical distribution of the *atriceps* complex made Creighton question whether *floridanus* was native to the southeast, and he noted that the species is absent from the Greater Antilles, the source of many tropical species that have probably made their way to Florida unassisted. These ants move their nests readily and can, for example, take up residence in a stack of flower pots in my yard overnight, so it is possible that *floridanus* was been brought by commerce from distant ports. Nests of species of the *atriceps* complex have been found in containers of orchids and bunches of bananas in England (Donisthorpe 1915) and bananas in New York City (Bequaert 1923). It is also plausible that *floridanus* and the similar southwestern form *transvectus* once occurred as a single population around the Gulf of Mexico, later becoming separated into eastern and western populations. Representatives of various southwestern lineages, such as *Pogonomyrmex badius*, have isolated populations east of the Mississippi. They are probably part of the eastward faunal movement during dry climatic periods in the late Pliocene or early Pleistocene (Webb 1990). It might be possible to settle the matter by looking at genetic relatedness of various populations of the *atriceps* complex, concentrating on red and black populations. Something of this sort has been done with *floridanus* on a very small scale in southern Florida (Gadau et al. 1996).

Natural History: This species occurs in a wide variety of natural habitats, from seasonally flooded forests to Florida scrub, as well as in disturbed habitats, including urban vacant lots and landscaped areas, the baking gravel of exposed railroad embankments, and piles of trash dumped along country roads. Nests are usually under objects on the ground, in rotten logs, or in manmade containers and structures. Subterranean colonies are not uncommon, usually at the base of a large tree trunk. In south Florida, *floridanus* seems less common in attics and wall voids than *inaequalis*, but I am brought occasional specimens that probably wandered from these hideouts into a kitchen or bathroom. Like *inaequalis*, *floridanus* is common around the outside of houses, moving into pump housings, piles of garden refuse, large potted plants, piles of flower pots, lumber, shingles, and junk of every description. When a cold snap hits Florida, causing gardeners to bring in their tender plants, whole nests may be brought indoors, where workers immediately begin exploratory foraging. Like *inaequalis*, *floridanus* seldom excavates solid, hard wood in the manner of *pennsylvanicus*. While this species gets considerable attention because of its large size and excitable

workers, it is not an especially threatening pest. It is relatively easy to control because the nests are compact, with a single queen.

The defensive behavior of *floridanus* is often called "aggressive" by those entomologists who lack any appreciation of an ant's point of view (Nickerson and Harris 1985). These ants, when their homes are under attack, rush out in an impressive and often effective display of force that deserves applause from the residents of Florida, a state that zealously guards the legal right of homeowners to shoot presumptive intruders dead (meter-readers deserve hazard pay). Natural selection may have promoted rapid defensive mobilization and unusually large and powerful majors in this species because nests are often in piles of leaf litter or friable rotten wood. It is, of course, possible that evolution worked in the opposite direction: highly developed defenses were a preadaptation for a behavioral shift toward occupying readily available but vulnerable nesting sites. Large majors are sometimes able to pierce human skin, making a tiny cut that is easily ignored until formic acid is sprayed into it. There is some evidence that biting and spraying occur in an obligatory sequence. When an antlion larva seizes a worker and drags it down into the sand, the struggling ant is unable to bite, and also fails to spray, as shown by the replete acid sacs of ants that have been sucked dry and tossed up out of the pit (Eisner et al. 1993). In laboratory experiments, ants that are grasped with forceps refrain from spraying until given the opportunity to grasp an object with the jaws, whereupon the gaster is bent forward beneath the ant to deliver a massive spray of formic acid (Eisner et al. 1993). The myrmecological ethologist T. C. Schneirla (1944) examined large numbers of nests occurring under heavy paper fertilizer sacks discarded in orange groves. Schneirla, who must have been examining disturbed nests at close range (without ever calling these ants "aggressive"), noted that alarmed workers move about in short, jerky bursts of activity, accompanied by a trembling of the body. This could be a form of alarm drumming, common in ants that nest in wood (Hölldobler and Wilson 1990).

Camponotus floridanus, like most other members of its genus, is a generalist predator of arthropods and also feeds extensively on honeydew and nectar. Small rodents caught in live traps for scientific surveys are sometimes killed by this species; at some sites, researchers are forced to surround each trap with a ring of insecticidal dust. It is possible that the ants first recruit strongly to the bait and then attack the mammal when it is captured. I have seen a few instances of nestling birds attacked by *floridanus*, but the condition of the nestlings before ant attack is not known. This ant is so abundant and ubiquitous in peninsular Florida that predation on nestlings would be widely reported if it were a normal activity. I have not seen this species at road kills, which are strongly attractive to *Solenopsis invicta*.

Workers feed on nectar and honeydew both day and night, jealously guarding their resources. The use of extrafloral nectaries on ceasarweed (*Urena lobata*) has been studied by Dreisig (2000). Two or three ants on a plant visit the nectaries systematically, keeping the supply low and increasing visitation rate in response to increased nectar flow. They chase off other ants, such as *Pseudomyrmex gracilis*, when these intruders approach a nectary. *Camponotus floridanus* is the primary ant species guarding the endangered Miami blue butterfly, *Cyclargus thomasi bethunebakeri* (Saarinen and Daniels 2006). The Miami blue caterpillar has a dorsal nectary gland on the seventh abdominal segment, and a pair of eversible, tentacle-like glands on the sides of the eighth abdominal segment; secretions of the latter gland apparently make attending ants more excited or agitated (Saarinen and Daniels 2006). Aggregations of certain aphids and scale insects that produce honeydew are also guarded, although I sometimes see larvae of syrphids and coccinellids in apparently ant-protected aggregations of honeydew producers, as well as the mummies of aphidiine braconids. The anti-ant defenses of these predators might be relatively easy to study using *C. floridanus*, as the workers are relatively large and active.

Hardy and adaptable, *floridanus* is well suited to laboratory culture, and has been the object of several experimental studies (Carlin and Hölldobler 1986, 1987, 1988; Heinze et al. 1994; Sauer et al. 2002). Studies by Norman Carlin and Bert Hölldobler tease apart the genetic and environmental factors that permit kin recognition in ants. One important discovery is that fully functional, prolific queens appear to produce a unique chemical label that is transferred to workers, allowing them to distinguish between nestmate and nonnestmate workers of their own species. This is useful for abundant species with wide-ranging foragers that frequently meet up with possible competitors or aggressors of their own species.

Since this chemical trademark has both genetic and environmental components, and is being constantly produced by the queen, it is not susceptible to easy forgery or replacement. If this system is widespread in the genus, it may help explain why social parasitism by congeners or other ants is so rare in the enormous genus *Camponotus*. This system, however, might have a limiting effect on reproductive options, as it should inhibit the development of polygyny, which appears to be rare in *Camponotus*. Polygyny could be a benefit to arboreal species that nest in dead twigs and branches, because loss of the piece of dead wood that holds the queen could doom the colony. To pile on yet another speculation, it would be adaptive for monogyne lineages that live in hazardous situations to emphasize production of alates.

Camponotus floridanus is an abundant species that frequently develops colonies of thousands of individuals (Schneirla 1944), providing resources for various inquilines. A eupelmid wasp, *Alachua floridensis*, is a gregarious parasitoid (several emerge from each host) that attacks the pupa of *floridensis* (Schauff and Bouček 1987). *Obeza floridana* (Eucharitidae) is also a pupal parasitoid (Davis and Jouvenaz 1990). Other inquilines include *Microdon fulgens* (Syrphidae) (Davis and Jouvenaz 1990), *Myrmecophila* sp. (Gryllidae), and an unidentified thysanuran, probably *Battigrassiella* sp. (Nicoletiidae). An unusual myrmecophilous cockroach, *Myrmecoblatta wheeleri* (Polyphagidae), also occurs occasionally in large colonies of *floridanus* (Deyrup and Fisk 1984).

Name Derivation: The name *floridanus* (Latin) indicates the state where the type specimens were collected. This is another species named by S. B. Buckley (1866), who is discussed above under *C. discolor*. Carlo Emery saw a type specimen and recognized it as a subspecies of *abdominalis*, so there can be no doubt about the identity of the species Buckley described (Emery 1893). This is fortunate, because Buckley's description gives no character states, or combination of them, that are specific to this species, but provides many unhelpful features, such as "...eyes of medium size, circular..." and "...mandibles reddish brown, curved inwards and toothed...."

Camponotus impressus (Roger)— Common Stopper Ant (Plate 77)

Taxonomy and Similar Species: This is one of several Florida species whose queens and majors have a remarkable stopper-like or cork-like head, used to block entrances to the nest or prevent passage of intruders through narrow tunnels. These ants are often placed in the subgenus *Colobopsis*, a small and distinctive group of species living in southern North America, Central America, the West Indies (the Bahamas and Cuba), and Europe. Most species are identified by features of the head of the major. Three of the four southeastern species tend to live up in the canopy of trees and are generally difficult to obtain. North American *Colobopsis* were reviewed by Wheeler in 1904. *Camponotus impressus* is the commonest and best known species. I have looked at types of Wheeler's *pylartes*, and believe they are the same as *impressus*; *fraxinicola* is probably also *impressus* (Wilson 1974b).

Camponotus impressus is somewhat variable in coloration: the faces of majors seem paler in north Florida than in south Florida, and the amount or pale or red marking on the first gastral tergite is also variable. *Camponotus obliquus* has sharply raised reticulations on the face; the sides of the face of *obliquus* have sharp rims, like the rim on the upper side of the face (rim usually but not always rounded on the sides in *impressus*); the face of *obliquus* is not conspicuously paler than the rest of the head; majors and queens of *obliquus* are slightly smaller than those of *impressus*, so in an unsorted box of southeastern *Colobopsis*, one can often pick out majors of *obliquus* initially by their smaller size and dark heads. Majors of *mississippiensis* that I have seen have a sharp rim on the sides of the face, the face in profile is scoop-shaped rather than more or less vertical, the clypeus scarcely overlaps onto the upper surface of the head, and there are only faint traces of shallow punctures on the upper surface of the head between the frontal carinae. The mangrove-inhabiting species *riehlii* has a deep notch in the mesosomal profile of the major, a feature found only in minors of other species.

Distribution: Assuming that *pylartes* is a synonym of *impressus*, the distribution extends from Maryland south through Florida and into the Bahamas, west into southern Missouri, Arkansas, and Texas. In Florida, *impressus* occurs throughout the state.

Natural History: This species occurs in a wide range of habitats, from shrub-filled marshes and riverine forests to sand pine scrub. Nests are in hollow twigs and vines, only occasionally in weed stems. Originally, *impressus* in Florida might have been largely restricted to wet areas, including the edges of swamp and riverine forests that escaped the frequent fires of the southeastern Coastal Plain. More opportunistic, perhaps, than our other *Colobopsis*, it appears to spread readily into habitats that have not burned in recent decades, such as the patches of scrub hickories where it was studied by Walker (see below).

Wilson (1974b) used laboratory colonies in glass tubes to study the function of majors of *impressus*. While majors respond to threats or alarm odors by moving to block entrances, they are no more effective than minors when there is mandible-to-mandible fighting in the nest. Majors also serve to store food, which they distribute to minors in times of scarcity. Majors do not take part in foraging or nest construction.

The natural history of *impressus* was comprehensively studied by Janet Walker, who worked on colonies at the Archbold Biological Station. I have pleasant memories of Janet, sometimes with her infant in a backpack and often assisted by her mother, recording the activities of *impressus* in the canopies of short scrub hickories. The following information is from her PhD thesis (Walker 1984) and resulting paper (Walker and Stamps 1986).

Colonies may only persist for a few years, judging by the size of field colonies and the longevity of laboratory colonies. Suitable nesting sites may be limited, and dead twigs are subject to breakage, circumstances that may give an evolutionary advantage to lineages that invest heavily in reproduction and less in long-term maintenance. The number of reproductives is correlated in this study with the number of majors, possibly because the food-storing majors maintain food supplies during the time reproductives are larvae. All this may mean that colonies end up somewhat understaffed with foragers.

Single, newly mated queens establish colonies in spring, guarding the nest entrance themselves until a major appears, usually when there are 8–12 minors. Larvae destined to become reproductives appear the following spring. These grow to maturity in April and May, generally flying the night after the first heavy afternoon rain of the south Florida rainy season. Leaves and flowers of trees also appear well before the start of the rainy season, and the entire ecosystem may be under drought stress when reproductives are being reared. During one especially dry year, even foraging by minors (at least during the day) was suppressed. The seasonality of colonies might be somewhat different in other parts of the range of *impressus*.

Colonies of *impressus* are relatively small, averaging 146 ± 76 minors and 34 ± 15 majors. The optimal number of majors is difficult to guess because the importance of their food-storage function depends on the late spring food abundance, and the importance of their entrance-guarding function depends on the number of separate twigs that house the colony, which varies from 1 up to 10. Walker showed the importance of the latter role by removing the majors or breaking the nest twig so there was no defensible entrance. In these experiments, *impressus* was sometimes attacked or supplanted by *Crematogaster ashmeadi* or *Pseudomyrmex gracilis*. These species might be expected to compete with *impressus* for food or nest twigs, but there was no correlation between the size and success of *impressus* colonies and presence or absence of the other two species.

Founding queens often move into small twigs or galls that must quickly become too small for the colony, but it is probably adaptive for the queen to find shelter as rapidly as possible. Sometimes, however, the situation is reversed: a spacious gallery has been left by a leaf-cutter bee or longhorn beetle larva, but it has a large entrance that would be difficult to defend. In such a case, the queen may block the entrance with wood fibers glued with saliva and then cut a new entrance through the plug (Tynes 1964).

In north Florida, there appears to be a remarkable facultative association between *impressus* and the similar-appearing *Dolichoderus pustulatus* (Johnson 1989a). The two species apparently share nest cavities in tree limbs, such as those of sweetgum (*Liquidambar styraciflua*), and it seems probable that workers of *D. pustulatus* are able to induce doorkeeper majors of *impressus* to allow them to enter the combined nest, although this has not been observed (Johnson 1989a). *Dolichoderus pustulatus* also occurs in its own nests without *Camponotus*. A somewhat parallel situation between the European species *Camponotus* (*Colobopsis*) *truncata* and *Dolichoderus quadripunctatus* was reported by Forel and discussed by Wheeler (1904). These two species are often associated in the same tree, and may have a mimetic association, but are not known to share nests.

Name Derivation: From *impressus* (Latin), meaning "engraved" or "imprinted," probably referring to the conspicuous punctures on the face of the major.

Camponotus inaequalis Roger—
Bare-Cheeked Caribbean Carpenter Ant
(Plate 74) (Probably = *tortuganus* Emery)

Taxonomy and Similar Species: This species is distinguished from the superficially similar *floridanus* by its lack of long, standing hairs on the legs, antennal scapes and malar area (cheeks). Both *floridanus* and *inaequalis* frequently appear in houses, especially when alates are being produced, so the general entomologist would do well to learn to recognize these species for the occasion when a neighbor stops by with one of these ants in a pill bottle. The most important distinction, of course, is that between ants and termites. A third large *Camponotus* whose range overlaps with *inaequalis* in Florida is *castaneus*, but this species has a shining head and body, both of which are chestnut colored, at least in south Florida.

Camponotus inaequalis in Florida has usually gone under the name of *tortuganus*. Its taxonomy is both tedious and unresolved, and I present it in the briefest possible form. 1863: Roger describes *inaequalis* from Cuba. 1895: Emery describes *maculatus*, subspecies *tortuganus* from a specimen collected in the Dry Tortugas by Pergande. 1923: Wheeler raises *tortuganus* to species. 1923 onward: the name *tortuganus* was used by various authors for Florida specimens. 1965 onward: Florida specimens accumulate with coloration like that of *inaequalis* or intermediate between *tortuganus* and *inaequalis* coloration. 1988: Deyrup et al. report specimens from the Florida Keys resembling museum specimens of *inaequalis* in coloration; typical *tortuganus* also in the Keys. 2002: Wetter and O'Hara claim, contrary to Emery's description, that the "holotype" of *tortuganus* is Bahamian, representing a Bahamian species not found in Florida. Names, if any, previously applied to the Bahamian species were not provided. The name *zonatus* was applied to the Florida species, without explanation. *Camponotus zonatus* (Emery 1894), described from Costa Rica, may be a synonym of *inaequalis* (Wheeler 1913c). Current: in desperation, I am reverting to the oldest plausible name: *inaequalis* Roger (1863). Perhaps this designation will encourage future taxonomic work to start at the beginning of the story.

Distribution: Unclear, because of taxonomic uncertainty. Described from Cuba, and there is an *inaequalis*-like species in the Bahamas, as well as Florida. This species is probably a good stowaway and is likely to occur in coastal Texas in the Brownsville area. In Florida, *inaequalis* occurs from the Dry Tortugas north into Brevard, Polk, and Hernando counties.

Natural History: This species occurs along beaches, in open Florida scrub, and in disturbed areas. Nests are usually under objects on the ground, making *inaequalis* a frequent beneficiary of illegal dumping. *Camponotus inaequalis* readily moves into enclosed spaces, including stacks of flower pots, pump housings, fuse boxes, and miscellaneous machinery left outdoors. It also takes advantage of wall voids and attics, but does not seem to cause structural damage by excavating chambers in sound wood. Foraging, which is primarily nocturnal, often involves collecting honeydew from sap-sucking insects, and it is rare to find extensive foraging within buildings. When alates are produced by a building-infesting colony, workers cut new exit holes, and these often open into the interior of the building, alarming the human occupants. Foragers may also start to patrol the kitchen or bathroom, usually at night and in small numbers. Quite often, I am brought workers in medicine vials, with the comment, "There are always one or two of these in the bathroom every day or so." I have never known what this means, except that it seems that the ants have access to the interior, but usually prefer to go outside, and may be attracted to the bathroom by its higher humidity. When large numbers of ants are emerging into the house, it is sometimes possible to find their emergence hole, especially at night when the ants are active. The hole or crack can then be blocked with plastic wood. This, of course, means that there is still a colony somewhere in the walls, but it is not necessarily doing any damage.

The tendency of *inaequalis* to move into containers means that this species could have been transported anywhere, especially during the unregulated days of commerce during several centuries following the European invasion of Florida. On the other hand, *inaequalis* might have come to Florida from Cuba, possibly via the Bahamas during the Pleistocene or earlier. Well adapted to coastal habitats, *inaequalis* might be able to island-hop relatively easily. If the red-and-black form occurs only in Florida, this would suggest that the species has been here long enough to converge on the general red-and-black coloration of some other Florida ants and should be regarded as a native species. The yellow-banded or spotted color form has also been in Florida for many years (Smith 1965), possibly representing a separate introduction. It may never be possible to determine whether Florida *inaequalis* is native, or exotic, or both.

Name Derivation: From *inaequalis* (Latin), meaning "unequal," or "different," perhaps referring to the variation in the extent and pattern of yellow markings on the gaster, even within colonies.

Camponotus mississippiensis M. R. Smith— Mississippi Stopper Ant (Plate 78)

Taxonomy and Similar Species: Generally similar to the other Florida species of stopper ants (*Camponotus*, subgenus *Colobopsis*), but the modified anterior part of the head of the major extends farther back, so that the head in lateral view is scoop-shaped. The area between the frontal carinae has only a few punctures, which are faint. There are only a few hairs on the top of the head (in lateral view). The rim around the face of the "stopper" is sharp all the way around.

Distribution: Maryland south into Florida, west into Illinois, Louisiana, and Oklahoma (Smith 1979). I have not seen any Florida specimens, but the Florida record is probably from specimens identified by M. R. Smith himself. This species probably occurs in north Florida in *Fraxinus americana* and similar species.

Natural History: Details of the natural history of this species first appeared in an unpublished PhD thesis (Tynes 1964); some of this information was summarized in a guide to Mississippi *Camponotus* (MacGown et al. 2007a). This species makes galleries in living twigs and small limbs of white ash, *Fraxinus americana*, a tree that has relatively soft live wood and thick, corky pith. Galleries may be several centimeters to almost 2 meters long. Queens disperse in summer, moving into young ash growth from that same year. The entrance hole is on the lower side of horizontal branches or on the sunny side of vertical branches. Foundress queens are common in new nests in July and August. Queens do not appear to forage but raise a set of small workers claustrally. The tissue of the twig tends to grow over the entrance and must be cut back. The diet appears to be primarily honeydew, and Tynes seems to have maintained colonies in the laboratory on honey alone. Majors accepted and stored excess honey in their crops, dispensing it to other workers on demand.

Other southeastern *Colobopsis* do not, as far as known, nest in live twigs, although nests are almost always in living trees and shrubs. Tynes (1964) went so far as to attempt to measure the effect of an infestation of *mississippiensis* on the growth of its host trees; he found no effect. On the one hand, it seems that, among southeastern twig-and-branch-nesting ants, *mississippiensis* has achieved something of an ecological breakthrough. It is able to choose among abundant nesting sites appropriate to its own needs and without facing competition from other species. On the other hand, it appears to be restricted to a single type of tree, possibly because it must deal with the defenses and phenology of a living host. This is just another example of the ecological rule that the specialist uses resources more efficiently, while the generalist has a larger resource base. This species probably avoids the frequent fires of the southeastern Coastal Plain by having a habitat restricted to mesic forests. If *mississippiensis* in Florida is dependent on white ash (*Fraxinus americana*) or the similar green ash (*F. pennsylvanica*), its distribution is probably restricted to a few areas in northern Florida. If, however, it is able to live in pop ash (*F. caroliniana*), it might be able to live in swamp forests through much of Florida.

Name Derivation: The name *mississippiensis* means "inhabiting Mississippi," "…so named because it is the most common species of the genus in Mississippi" (Smith 1923).

Camponotus nearcticus Emery—
Pine Branch Carpenter Ant (Plate 75)

Taxonomy and Similar Species: This is a small, black, shining species, similar to *caryae*, but lacking the large pits in the malar area (cheeks). It is smaller than *pennsylvanicus*, which is also distinguished by numerous coarse hairs on the head and body and a nonshining appearance, especially the gaster. *Camponotus nearcticus* is very similar (identical, as far as I can tell) in structure to *decipiens*, but the latter species is red and black, while Florida populations of *nearcticus* are completely black (there are reddish markings on some populations). I have no concern that *nearcticus* and *decipiens* might be color forms of the same species because the two occur together at many sites in Florida without intergrading and differ in nesting preferences. The structural similarity between these two species is a reminder that we take on faith the identifications of some widespread and behaviorally diverse ant species. If *decipiens* were black like *nearcticus*, we would have no idea that we were looking at two species.

Distribution: This may be the most widespread North American *Camponotus*. Its northern range is transcontinental, from Ontario into British Columbia and California; its range extends south from Ontario into Florida and west into North Dakota south into Texas (Smith 1979). It might be simpler to say that *nearcticus* occurs, at least sporadically, throughout the United States and southern Canada, except for desert habitats of the Southwest. In Florida, *nearcticus* is recorded from scattered sites throughout the state. It probably occurs in almost every open flatwoods and sandhill site, but is often difficult to find because workers tend to be up in the pine canopy. Stray workers on the trunks of trees are tricky to catch, as they run speedily and dart into bark fissures, and, as a last resort, fling themselves from the tree trunk to the ground.

Natural History: In Florida, *nearcticus* almost always lives in dead branches, twigs, and hollow cones in the tops of open-grown pine trees. These sites are relatively safe from the frequent fires that once dictated the structure of upland habitats in Florida and much of the southeastern Coastal Plain. *Crematogaster pinicola* and *Temnothorax bradleyi* are other members of this little group of ants that were able to flourish over vast stretches of primeval native pineland, thanks to their selection of fire-resistant pines for their homes. The red-cockaded woodpecker and the brown-headed nuthatch are two birds that share the southeastern open pineland habitat. These birds have become rare because they require much more extensive pine stands than required by ants, and because they are not adapted to other habitats off the Coastal Plain. In areas to the north and west, away from the sandy, open pine forests, *nearcticus* occupies a greater variety of nest sites. In Maine, for example, I find this species in cedar posts, dead trees, dead grape vines and rustic structures that are close to sources of honeydew and small insects. Nests may also be in dead branches of hardwood trees such as oak and hickory, as well as in conifers (Smith 1965). In North Carolina, as in New England, nests in trees are in open woodland or forest edges (Carter 1962). Nests occasionally occur in the timbers of roofs or porches, and workers may wander into buildings, but this species is not generally a persistent pest and does not cause significant structural damage.

Name Derivation: Described by Carlo Emery as the Nearctic form of the European species *Camponotus marginatus* (now *aethiops*).

Camponotus novogranadensis Mayr—
Black Compact Carpenter Ant (Plate 79)

Taxonomy and Similar Species: This species is similar to *planatus*, but can be distinguished by its black coloration; *planatus* is usually, not always, dark red with a black gaster. Under the microscope, *novogranadensis* is seen to have sparse, long, curved hairs, rather than abundant, short, straight hairs like *planatus*. The sides of the face above the mandibles are brownish.

Distribution: Mexico through Central America, with scattered records in Ecuador, Peru, and Brazil (Deyrup and Belmont 2013). It is possible that *novogranadensis* has been introduced into part of its Neotropical range, and it is also possible that there is a complex of similar species in the Neotropics. In North America, *novogranadensis* is known only from Lee County in the vicinity of Koreshan State Park (Deyrup and Belmont 2013).

Natural History: Nests of *novogranadensis* are in dead wood or hollow stems, usually in disturbed areas (Longino 2002; Ribas et al. 2012; Vasconcelos 1999), but also in the canopy of primary forest (Longino 2002). Workers forage for honeydew produced by Membracidae and Aetalionidae (Letourneau and Choe 1987) and collect honeydew from extrafloral nectaries (Damon and Perez-Soriano 2005). There is no way to know how long this species has been established in Florida. Koreshan State Park was the site of a tropical plant nursery more than a century ago, when there was little regulation of plant imports. On the other hand, this species might have been recently imported, for example, in a shipment of orchids, the mode of transportation for *novogranadensis* to Kew Gardens in England long ago (Donisthorpe 1915). If this species begins to spread rapidly in Florida, that would strengthen the hypothesis of a recent introduction. *Camponotus novogranadensis* does not seem to be regarded as a significant pest in its native range.

Name Derivation: The name *novogranadensis* is a Latinization of Nueva Granada, the Spanish Empire name for northern South America.

Camponotus obliquus M. R. Smith— Oblique Stopper Ant (Plate 77)

Taxonomy and Similar Species: The punctures of *impressus* are shallow and vague, often not in contact, while those of *obliquus* are deep, with well-raised rims that are usually in contact and sharing borders. In lateral view, *obliquus* has many more hairs with enlarged tips on the top of the head (the function of these hairs is unknown). *Camponotus mississippiensis* has a scoop-shaped head (lateral view), and the mangrove stopper ant has a conspicuous metanotal groove.

Distribution: Alabama, Mississippi (Smith 1979), Georgia, and Florida. This species is easy to overlook and may be more widespread, not to mention the fact that it may have been misidentified, as in my 2003 list of Florida ants (Deyrup 2003). Queens are readily attracted to lights, so examination of light trap catches should provide more records. In Florida, *obliquus* is known from Highlands, Clay, Orange, Baker, Lake, Volusia, and Alachua counties.

Natural History: *Camponotus obliquus*, like most arboreal *Camponotus* in Florida, has probably been forced to adapt its lifestyle to the frequent fires that used to typify forests of the southeastern Coastal Plain. Most Florida colonies that I have seen are from large twigs or small branches in large, fire-resistant trees in low flatwoods or at the edges of swamp forests. Zachary Prusak has found colonies high up in bald cypress trees. One collection from a *Quercus laevis* in a long-unburned sandhill was near a riverine mesic forest, and it is possible that *obliquus*, like *impressus*, may move out from its usual lowland habitat into uplands where fire has been suppressed. This is the least-studied southeastern *Colobopsis*, in part because of the inaccessibility of nests, but it seems to be genuinely rare in some areas, such as Mississippi, where there have been two surveys of *Colobopsis* (MacGown et al. 2007a; Tynes 1964), as well as Marion Smith's pioneering work (1923, 1930b).

Name Derivation: The name *obliquus* (Latin) means "slanting" or "oblique," referring to the obliquely truncate head (Smith 1930b).

Camponotus pennsylvanicus (DeGeer)— Pennsylvania Black Carpenter Ant (Plate 76)

Taxonomy and Similar Species: This is a burly, hirsute species that is easily distinguished from other Florida *Camponotus* by the combination of black coloration, large size, general hairiness, and nonshining body surface, especially the gaster.

Distribution: New Brunswick and Quebec, south into Florida, west into North Dakota and Texas (Smith 1979). In Florida, *pennsylvanicus* occurs from Alachua County north, and west through the Panhandle. The southern limits of its range might be associated with warmer temperatures or with the southern limits of remnants of Appalachian hardwood forests.

Natural History: This species is usually associated with forests that have large trees and stumps for nesting. In Florida, it occurs in mature hardwood forests and in towns and cities that have large trees. I have never found it in southern pines, although it often excavates nests in the heartwood of pines and other conifers in northern North America. Unlike other large Florida carpenter ants, such as floridanus and inaequalis, pennsylvanicus regularly excavates extensive galleries in sound wood, both heartwood of live trees and wood of dead trees, stumps, and structural timbers. Galleries run with the grain in both vertical and horizontal tree trunks and timbers, giving the nest a layered appearance, resembling nests of many termites. This probably reflects a number of factors: excavation is easier with the grain, structural integrity of the wood is less affected, and there is much more vertical space than horizontal in a tree trunk.

Much of what is known about the biology of this species is owed to John Pricer (1908), whose monograph is currently available online at the website for the Biological Bulletin of Woods Hole Marine Laboratory. A colony is founded by a single queen, who walls herself into an old beetle gallery or other cavity under bark or in dead wood. There, she raises a small number of miniature workers, using resources stored in her body. The colony grows slowly at first and does not become large enough to produce reproductives until there are approximately 2000 workers, a process that requires at least three years. Colonies may exceed 3000 individuals, including a queen, workers, reproductive, and brood. In the laboratory, queens of several species of Camponotus can live more than five years (Hölldobler and Wilson 1990), and the known demographics of pennsylvanicus suggest that there would be strong selection for queen longevity in this species.

Pricer (1908) found that the diet of pennsylvanicus is primarily liquids, especially honeydew, but also fluids of dead insects, and occasionally sap of plants or juices of fruits. Despite its dependence on sap-sucking insects, pennsylvanicus does not carry them into its nest or move them from place to place on plants. Head capsules of dead insects may be removed and taken to the nest, where they are apparently emptied out by larvae. The nutritional ecology of pennsylvanicus has been studied in amazing depth by Colleen Cannon (1998) in a dissertation currently available online. Cannon analyzed crop contents of foragers, finding primarily honeydew, much smaller levels of proteinaceous material, and almost negligible amounts of lipids.

She discovered that, given a choice of baits, pennsylvanicus greatly prefers protein-rich baits to carbohydrates. This suggests that protein is limited in the natural diet. It is probable that other arboreal Florida Camponotus have feeding habits and nutrition somewhat similar to those of pennsylvanicus.

Pricer (1908) was perhaps the first myrmecologist to realize that some Camponotus species store substantial food supplies within their bodies. In preliminary experiments, he showed that colonies deprived of all food in spring are able to rear large numbers of small overwintering larvae to adulthood. Pricer suggests that this is an adaptation to reduce the effects of fluctuations in quantity or quality of food available in spring. Similar food storage is mentioned above in the account of C. impressus, a species that appears to be only distantly related to pennsylvanicus. A capacity for food storage may be one of several keys to the success of the genus Camponotus, along with an advanced ability to excavate nests in wood and pith of living and dead plants, swift locomotion, heavy investment in chemical defense, and adaptations for nocturnal foraging when many predators are less active.

Foraging and spatial orientation of pennsylvanicus have been studied by several researchers. Hartwick et al. (1977) showed that a worker returning successfully from a bait left an odor trail that was followed by other foragers emerging from the nest. The forager returning from a bait often left an indirect trail, and subsequent foragers eventually made a shorter and more direct trail, using visual cues. Scout workers returning from a food source leave trails composed of formic acid, which stimulates rapid recruitment, and chemicals from the hindgut, which make more residual trails (Traniello 1977). Scouts returning from a rich food source also have a recruitment display, whose strength and effectiveness increases if the colony has been deprived of food (Traniello 1977). Like many other species of Camponotus, pennsylvanicus forages primarily at night, but also forages by day (Fowler and Roberts 1980; Klotz and Reid 1993; Pricer 1908). Nocturnal orientation usually depends on light sources that provide directional cues; originally, this would be the moon, but artificial lights such as street lights are readily accepted modern conveniences (Klotz and Reid 1993). In the absence of light beacons, ants use visual landmarks, including patterns made by the canopy of tree branches against the sky (Klotz and Reid 1993). An interesting feature of foraging noted briefly by Pricer (1908) and Fowler and Roberts (1980) is the positioning of some large workers in a kind of sheltered bivouac near a rich source of honeydew. Smaller workers collecting honeydew bring the proceeds to deposit with the waiting large workers. Cannon (1998) calls these large workers "tankers." This system should reduce the number of trips to the nest per unit of liquid, thereby decreasing exposure to predators and saving time and effort.

Among the species of *Camponotus*, *pennsylvanicus* is the principal pest species in eastern North America (Akre and Hansen 1990; Fowler 1990). It not only damages structural timbers, but also landscape, forest, and fruit trees (Fowler 1990). Nests in buildings usually begin in damaged or partially decayed wood, and expand into sound wood. Damaged wood usually has a higher moisture content than normally found in waterproof, heated structures (Smith 1965). Consequently, this ant is most likely to be a problem in summer cottages, porches, barns, and buildings with leaking roofs or siding. Collapse of portions of trees overhanging buildings may be facilitated by this species (Fowler and Roberts 1982). Preventive control in structures includes injecting insecticide dust (boric acid or various other chemicals) into wall voids, preventing persistent moisture in boards or timber, basing support beams on concrete rather than soil, and avoiding flat roof construction in areas with high risk of infestation by *C. pennsylvanicus* (Akre and Hansen 1990). Control of infestations should first focus on precisely locating the colony, which is sometimes based in a woodpile, stump, or tree outside the structure, where the colony can be treated with insecticide relatively easily (Hedges 1998). If the main body of the colony is within a building, it is usually treated with insecticides injected into walls after the colony has been accurately located (Hedges 1998). Although baits are often effective in controlling ants living in buildings, they often fail with carpenter ants, affecting only a few workers and competing poorly with natural foods (Akre and Hansen 1990). For details on detection and control of carpenter ants in structures, see Hedges (1998).

Name Derivation: Named for the Colony of Pennsylvania in 1773 by Carl DeGeer. This is the first endemic ant to be described from North America (Smith 1979). Carl (or Charles) DeGeer (1720–1778) was a Swedish industrialist and naturalist who published a seven-volume treatise called *Memoirs Pour Servir a l'Histoire des Insectes*, presumably in homage to the work by the same name published earlier by Reaumur. It seems to me that DeGeer's description of *pennsylvanicus* is insufficiently diagnostic to separate it from one or two other species, but, fortunately, his collection was preserved, so there is little doubt about most of his names.

Camponotus planatus Roger—
Compact Bicolored Carpenter Ant (Plate 79)

Taxonomy and Similar Species: *Camponotus planatus* can be distinguished from most other Florida *Camponotus* by the combination of small size, compact body plan, abundant silvery hairs, and the lack of any shining areas on the head or body. In the field, it could be confused with *novogranadensis* (currently known from one site in Florida), although *novogranadensis* is completely black, while the head and mesosoma of *planatus* is usually (not always) dark red. Under the microscope, the head and mesosoma of *planatus* is seen to have abundant, short, straight, erect, silvery hairs, while *novogranadensis* has long, curving, dark hairs.

Distribution: Southern Florida and southern coastal Texas, south through Mexico and Central America into Colombia; in the West Indies on Cuba and some smaller islands, including Gorda Key (Bahamas) and Dominica. In Florida, this species has long been known from Monroe and Miami-Dade counties, and seems to have spread up the coast recently into Brevard County to the east and Pinellas and Hillsborough counties to the west. Although *planatus* was described from Cuban specimens in 1863 (Bolton 1995) and has been in Florida since before 1910 (Wheeler 1910b), it is probably not native to either place, but imported from Mexico or Central America. It is highly unlikely that *planatus* would be native to Cuba and Florida, but absent from Puerto Rico, Hispaniola, and all but the most northern Bahamian Islands. Its recent spread in Florida (Deyrup 1991a) also suggests an exotic species that is being moved around by humans.

Natural History: In Costa Rica, *planatus* is often abundant in open, cut-over forest, around homes, and in other relatively open disturbed areas (Longino 2004). In the Florida Keys, this is a common ant in tropical forest and disturbed areas, a diurnal forager often seen tending honeydew-producing insects. Nests are in hollow twigs, dead branches, sawgrass culms, standing dead wood, and, occasionally, in dead wood on the ground. Unlike other Florida

Camponotus, this species has multiple queens (Carlin et al. 1993). This probably promotes dispersal by humans of viable colony fragments. Like many polygynous ants, *planatus* sometimes occurs in large extended colonies that dominate a patch of habitat.

The abundance of *planatus* and its predilection for sweets may make it a minor pest around the patio in some areas (Hansen and Klotz 2005) and also may lead to interesting mutualistic relationships with other organisms. This is one of the ants associated with the extremely endangered Miami blue butterfly, although *planatus* is less assiduous than *C. floridanus* (Saarinen and Daniels 2006). It visits the extrafloral nectaries on the floral bracts of the Panamanian wild ginger *Costus woodsonii*, protecting the seed arils from damage by larva of a fly, *Euxesta* sp. (Schemske 1980). It visits extrafloral nectaries and may protect flowers and fruits of the orchid *Schomburgkia tibicinis* (Rico-Gray and Sternberg 1991; Rico-Gray and Thien 1989). This same orchid has hollow pseudobulbs that *planatus* occupies, partially filling them with a midden that provides nutrients absorbed by the orchid (Rico-Gray et al. 1989). Finally, *planatus* appears to be the model for a small mimetic complex at a site in Honduras (Jackson and Drummond 1974).

Despite its sometimes benign interactions with other species, *planatus* seems to have the potential to invade semidisturbed tropical habitats. It already occurs on Hawaii and in the Galapagos (McGlynn 1999). While probably most dangerous on oceanic islands such as Samoa, it would be an unwelcome addition to the fauna anywhere in the Old World tropics.

<u>Name Derivation:</u> From *planus* (Latin), meaning "flat," referring to the broad, flattened pronotum (in dorsal view), whose dorsal edges are not curved downward gradually like those of most *Camponotus*, but flattened, then abruptly turned down, producing an angle along the dorsal sides of the pronotum. The whole word *planatus* means "flattened," as if the dorsum of the pronotum had been slightly squashed.

Camponotus riehlii Roger— Mangrove Stopper Ant (Plate 77)

<u>Taxonomy and Similar Species:</u> *Camponotus riehlii* was described in 1863 from a single queen collected in Cuba. This specimen apparently no longer exists, prompting the romantic question, "Who is the heir to the lost queen of Cuba?" One likely candidate is *baronii* Alayo and Zayas, which is probably synonymous with *riehlii*, unless Cuba has more than one species of *Camponotus* in the subgenus *Colobopsis*. Judging from the illustration in Alayo and Zayas (1977) and Zayas (1981), the species described as *baronii* is probably our mangrove-inhabiting species.

The mangrove stopper ant is easily distinguished from other Florida *Colobopsis* by its strong metanotal groove in the major, and in the minor by its peculiar propodeum, which extends backward toward the gaster in a blunt triangle.

<u>Distribution:</u> Florida, the Bahamas (North Andros), and Cuba. All Florida specimens are from Monroe County, but it might occur in mangroves in Miami-Dade and Collier counties as well.

<u>Natural History:</u> In Florida, *riehlii* appears to be almost completely restricted to dead twigs on living red mangroves (*Rhizophora mangle*). This species is among the ants studied by Daniel Simberloff and Edward Wilson in their famous project on recolonization of mangrove islets from which all arthropods had been removed (1969). *Camponotus riehlii* was the ant slowest to return. Following this demonstration of the convenience of mangrove islands for ecological studies, Blaine Cole (1980, 1983a, 1983b) studied *riehlii* along with the other common mangrove ants, *Crematogaster ashmeadi*, *Pseudomyrmex elongatus*, *Xenomyrmex floridanus*, and *Cephalotes varians*. *Camponotus riehlii* requires larger islands than the other species and is usually absent from islands less than $25.4 \, m^3$. Nests are in dead branches in, or just below, the leaf canopy, not in a zone of dead branches near the trunk. Peripheral location of nests may reduce interactions between this "secondary" species and more aggressive, more rapidly colonizing species *P. elongatus*, *C. ashmeadi*, and *X. floridanus*. This species was, by far, the least abundant of the five principal species of mangrove ants. In a study of the time spent by workers in various activities (ethogram), Cole (1980) discovered that major workers have fewer types of activities (15) than minors (23), although they do assist somewhat with brood care. Majors seem to function as food repositories, as in *impressus*.

<u>Name Derivation:</u> Julius Roger (1863) probably named this species for a colleague or friend with the last name of Riehl.

Camponotus sexguttatus Emery— Six-Spotted Carpenter Ant (Plate 78)

<u>Taxonomy and Similar Species:</u> In a lateral view, *sexguttatus* has a strongly notched mesosomal profile,

completely unlike the strongly arched profile of most *Camponotus*. Members of the subgenus *Colobopsis* ("stopper ants") also lack a strongly arched mesosoma, but the small size of these species, to say nothing of the cork-like head of majors, easily distinguish them from *sexguttatus*. The variable pale spots on the gaster are similar to those of some populations of *C. impressus* and *inaequalis*. At first glance, both in the field and in a collection, *sexguttatus* might be mistaken for *Formica archboldi*.

Distribution: Known from Paraguay and Brazil north through Central America; also on many islands of the Caribbean (Kempf 1972). A population has also appeared in Hawaii (McGlynn 1999). It was introduced into Florida, where it was first collected in 1993 (Deyrup et al. 2000). It is well established in Miami-Dade and Broward counties, and probably also occurs in eastern Monroe County. It may well spread throughout tropical Florida. There is no reason to believe that *sexguttatus* has been imported elsewhere in the Neotropics. In the Caribbean and some mainland sites, there are some distinctive color forms (named subspecies) that suggest some degree of isolation rather than recent importation as stowaways.

Natural History: I have usually found this species in semiopen habitats. In Florida, it occurs in disturbed sites, but also on "tree islands" in the eastern Everglades. Nests are in hollow twigs and branches, including material on the ground. On the island of Dominica, small dead bamboo stems are a favored habitat. In Florida, I have found colonies in hollow sawgrass culms (*Cladium jamaicense*). Foragers visit sap-sucking insects. As *sexguttatus* becomes more abundant in Florida, there should be opportunities to study its colony structure and other aspects of its natural history.

Name Derivation: From *sex* (Latin), meaning "six," and *guttatus* (Latin), meaning "spotted," referring to the six pale spots on the gaster of specimens from the type locality.

Camponotus snellingi Bolton—
Snelling's Confusing Carpenter Ant (Plate 75)

Taxonomy and Similar Species: The name *snellingi* is Bolton's replacement name for *C. pavidus* Wheeler, a name that had been applied earlier to another species. As discussed under *decipiens*, the species *nearcticus*, *decipiens*, and *snellingi* seem to lack consistent morphological differences. In the case of *decipiens* and *snellingi*, even differences in color and in the density of

decumbent hairs on the gaster are ambiguous, apparently varying from place to place in Florida. The differences in color seem to be clearer in Mississippi, where there is a form with a completely black gaster (*decipiens*) and a sympatric form with red on the gaster (*snellingi*) (MacGown et al. 2007a). This same distinction seems to occur in the central area of the Florida Peninsula, although *snellingi* is known from only a few specimens in this region. One could, obviously, simplify the taxonomic situation by combining *snellingi* with *decipiens*, but concealing a problem does not make it go away, and generally delays a constructive solution. I here present *snellingi* and *decipiens* as a challenge to current and future myrmecologists who are more focused and skilled than I.

Distribution: Central Texas east through Georgia and Florida (Snelling 1988). In Florida, *snellingi* probably occurs from Highlands County north, although everything becomes unclear in north Florida, where there may well be color convergence between *snellingi* and *decipiens*. The range of one or both species apparently extends into North Carolina under the former name *rasilis*.

Natural History: This species occurs in mesic habitats with tall shrubs or trees. It is probably unable to persist in sites that are frequently burned. I have also wondered whether it has been excluded by *Pseudomyrmex gracilis* in central Florida, as I have not seen recent specimens from the Archbold Biological Station or nearby areas. In Mississippi, colonies identified as *snellingi* occur in dead twigs and branches, in logs and stumps, and in standing dead trees (MacGown et al. 2007a). In Florida, it is not common to find *Camponotus* of any species in standing dead trees, while stumps and rotten logs are occupied by *C. floridanus*, or occasionally by *castaneus* or *pennsylvanicus*.

Name Derivation: Named for Roy Snelling, an important contemporary hymenopterist specializing in ants and bees. An ardent collector, under his care the ant collection of the Los Angeles County Museum has become one of the most important ant collections in the world. Snelling has a particular interest in the genus *Camponotus*, and the name *snellingi* specifically draws attention to his work on the subgenus *Myrmentoma* in North America.

Camponotus socius Roger—
Sandhill Carpenter Ant (Plate 76)

Taxonomy and Similar Species: This species is reddish brown; the gaster is black with yellowish or pale brownish markings that are sometimes restricted to the first two gastral segments. The gaster is completely matte (nonshining), a characteristic that does not sound especially distinctive but which immediately distinguishes *socius* in the field from the other large, reddish and black Florida species *inaequalis* and *floridanus*. The legs and antennal scapes of *socius* lack the conspicuous long hairs found on *floridanus*. The clypeus of *socius* lacks the sharply raised median ridge of *inaequalis*.

Distribution: North Carolina south through Florida, west into Louisiana (Smith 1979). In Florida, *socius* occurs as far south as Glades County inland, and south into Collier and Broward counties in coastal scrub habitat.

The type specimens of *socius* were supposedly collected in Brazil, and the species has been listed as introduced (Smith 1979). Creighton, however, remarks (1950) that *socius* seems so at home in southeastern North America that he would be as willing to believe the species had been introduced into Brazil. My feeling is that the species is unlikely to have been introduced anywhere. Nests are large, subterranean, in deep sand, and restricted to natural habitats, usually sandhill or scrub. More than 50 species of ants are known to have been introduced into the Southeast (Deyrup et al. 2000), all of which have nests that are either superficial, or in dead wood, hollow twigs, or opportunistically in relatively small man-made containers. Almost all of these species are commonest in disturbed habitats, although many also occur in natural habitats. I would not, of course, want to attempt to prove that a particular ant does not occur anywhere in Brazil. Still, I would go so far as to say that if the types really were collected in Brazil, and not the objects of some labeling glitch, it is probable that our species is not actually *socius*, but a similar-appearing undescribed species.

Natural History: *Camponotus socius* is most closely associated with open sandhill (high pine) habitats with scattered oaks and pines (Carter 1962; Van Pelt 1958; Van Pelt and Gentry 1985). It also lives in open Florida scrub habitats, but I have not found it in scrub where there is a closed canopy of oaks and pines. Nests extend down as much as 60 cm, with a series of more or less horizontal chambers coming off the main shaft at various levels. This nest architecture is extensively described and illustrated by Walter Tschinkel (2005), who even provides stereo images for three dimensional viewing. It appears that *socius* tends to be weakly polydomous, with one to three nearby nests (Tschinkel 2005). Both the habitat preferences and nest structure of *socius* suggest that it is adapted for open landscapes with frequent fires. It most clearly fire-adapted terrestrial *Camponotus* in the Southeast. In pre-Columbian times, *socius* and *nearcticus* were probably the only common *Camponotus* species over a huge swath of uplands in the southeastern Coastal Plain.

The recruitment behavior of *socius* was studied by Hölldobler (1971b). Recruitment is a three-stage process: a scout that has found a bait leaves a chemical trail back to the nest, performs a "waggle" display that includes vibrations of the head and mesosoma, and then leads the excited recruits back to the food source. Similar recruitment behavior guides nestmates to a new nest site. The scent trail is composed of chemicals from the hindgut, while formic acid released in the nest stimulates recruitment (Kohl et al. 2001).

Name Derivation: From *socius* (Latin), meaning "companion," or as an adjective, "allied." Roger did not explain the significance of this name.

Genus *Formica*

Classic Northern Ants

Members of the genus *Formica* (pronounced **For mī" kuh** or **For" mi kuh**) are, according to Donisthorpe (1915), "robust and intelligent ants which live an open-air life, hunting insects and other prey, and attending plant-lice on trees and shrubs." Aside from this supposedly admirable lifestyle, the genus is characterized by moderately large size (length, approximately 6 mm), conspicuous ocelli, and a clear notch in the mesosomal profile. In the Florida fauna, one might confuse species of *Formica* with its associated slave-raiders in the genus *Polyergus*, but members of the latter genus have long, sickle-shaped jaws; *Polyergus* species are rare in Florida. *Camponotus sexguttatus* is somewhat similar to *F. archboldi* but lacks the prominent trio of ocelli characteristic of *Formica* species.

This is a Northern Hemisphere genus that is most dominant in cool climates with strong topography. It is one of the most speciose genera of North American ants, the other being the more southern *Pheidole*. There is a bewildering array of species in the western mountains of North America, an impressive showing in the Great Plains states, and a strong representation in the Northeast. Like some northern European contemporaries of Donisthorpe, whose vigor and intelligence might flag in tropical colonies, Formica fades out quickly in the Southeast, where there are relatively few species. The ecological reasons for this are unclear, but might include the following: (1) requirement for a cold season to induce dormancy or structure the phenology of the species; (2) a scarcity of reliable, large-scale outbreaks of honeydew-producing insects; (3) an inability to change from diurnal to nocturnal foraging; and (4) inability to adapt to periodic severe droughts.

Northern latitudes have until recently produced a disproportionate number of myrmecologists, resulting in a huge literature about the behavior and ecology of various species of Formica, ants that can be found roaming the backyards of most northern naturalists. There is a particularly rich literature on the ecological effects of the mound-building species that prey on forest insects, and on the behavior of the slave-raiding species. So numerous are the species, however, that there are still species that have not been extensively studied, including many rare or localized species. As one might expect for a genus of conspicuous diurnal ants, the taxonomy of the group in North America is fairly well known, in the sense that there are not large numbers of species waiting to be discovered, although some new species continue to turn up. There is, however, no modern guide to the genus as a whole, and several species groups have not been revised in a long time. Within species groups, distinguishing features are often subtle, but are usually accompanied by ecological differences, so that field observations provide valuable clues for the taxonomist. New England myrmecologists are fortunate to have the ability to identify Formica species by using A Field Guide to the Ants of New England (Ellison et al. 2012).

Successful Florida species all belong to the small pallidefulva group of slender species with an evenly rounded propodeum, long legs, and antennae. The antennal scape is clearly longer than the length of the head measured from the anterior edge of the clypeus to the posterior border of the head in frontal view. The long legs promote extreme speed and agility; these ants never walk when they can run. Capturing individual foragers often involves grabbing the ant along with the leaf litter it is traversing and then tossing the whole handful into a tray or insect net. The pallidefulva group has recently been revised in Trager et al. (2007), including extensive natural history accounts.

In addition to species in the pallidefulva group, there is a widespread species of the fusca group that has been collected a single time in Florida.

Name Derivation: From Formica (Latin), meaning "ant."

Formica archboldi M. R. Smith— Archbold's Fleet Formica (Plate 80)

Taxonomy and Similar Species: This dark brown to blackish species is similar in structure to the yellowish Florida species of the pallidefulva complex, but is distinctively darker in color. Formica subsericea, with one known Florida population, is black with a conspicuous covering of flattened silvery hairs. Camponotus sexguttatus lacks the conspicuous ocelli found in Formica species.

Distribution: Florida, Georgia, and Alabama; a record from Virginia is a mistake (Trager et al. 2007). The Georgia record is from Tanner County in northern Georgia, not from near the Florida border (Ipser et al. 2004). It is probable that the range of archboldi contracted after the conversion of sandhill habitat in Georgia and Alabama. In Florida, archboldi occurs through the northern and central Peninsula, and into the Panhandle to just west of the Apalachicola River, but there seem to be no records from the western Panhandle. It occurs south into Highlands and Okeechobee counties, with disjunct records in Collier County, but there are no East Coast records from Broward County southward. Even in the northern Peninsula, there are no records from many counties. This species is fast and wary, often living in dense stands of grass, so it is easily overlooked.

Natural History: Formica archboldi often lives in grass tussocks around seasonal ponds; in north Florida, it also occurs in wiregrass clumps in sandhill habitat. The following information is summarized from Trager et al. (2007).

Although archboldi requires disturbance, in the form of fairly frequent fires, to maintain its open, grassy habitat, it does not persist in areas disturbed by humans. In this respect, it differs from the related species pallidefulva and (farther north) incerta, which are common in fields and can even nest in lawns. The nest entrance of archboldi is usually at the base of a grass tussock and may be surrounded by plant fragments brought in by foragers and carefully placed around the entrance. Foraging for food is concentrated between 8 a.m. and noon, and between 4 p.m. and dusk. Colonies of honeydew-producing

Homoptera may be guarded constantly. Seasonal activity is from March through October, longer in southern Florida. The diet is primarily honeydew, scavenged dead insects, and insects that are killed by the ants themselves. *Cinara* aphids and *Toumeyella* scales on young pines are common honeydew hosts. Remains of worker *Odontomachus* are often abundant around nest entrances, and there is some evidence that *archboldi* is a specialized predator that uses a chemical spray to subdue *Odontomachus* (King and Trager 2007), although there are no published reports describing this unique behavior in any detail. Alates are in nests from late April through June. They fly around dawn and do not appear to be attracted to light. Some nests are monogynous, whereas others are polygynous. The tiny cricket *Myrmecophila pergandei* may occur in nests of *archboldi*, as well as in the nests of many other ant species.

In the northern part of its range, *archboldi* is the host of a host-specific species of slave-maker ant, *Polyergus oligergus* (King and Trager 2007; Trager 2013; Trager and Johnson 1985). This slave-maker appears to be much rarer than its host and is probably dependent on denser populations of *archboldi* than are found in many areas. Worker *archboldi* immediately flee when attacked by *P. oligergus*; there is a report (Trager 2013) of a successful raid by four worker *P. oligergus* on a colony of more than 100 *archboldi*.

The type series of *archboldi* consists of specimens sent from the Archbold Biological Station by T. C. Schneirla. Schneirla is best known for his discoveries about army ants, but he was interested in social organization in all ants. He carefully described a nest of *archboldi* from a seasonal pond at the biological station. The entire nest was in and around a grass tussock, and restricted to an area approximately 12 inches in diameter and 7 inches deep (the original excavation presumed a much larger nest). The water table was only an inch below the lower galleries. Schneirla collected the entire nest, consisting of 1 queen, 1210 workers, and 414 large larvae and pupae. Small larvae and eggs were not counted. Only a small proportion of the pupae were in cocoons. The prevalence of cocoons might be variable in this species, as naked pupae have been considered more characteristic of *pallidefulva* than *archboldi* (Trager et al. 2007). Surrounding Schneirla's nest were five colonies of *Solenopsis pergandei*, each with numerous queens and with galleries connecting with those of *F. archboldi*.

Name Derivation: "This species is named in honor of Richard Archbold, the owner of the Archbold Biological Station, who not only encouraged Dr. Schneirla in a study of the ants of the station but who showed a special interest in the habits of this particular ant" (Smith 1944). Richard Archbold had his workers dig a big trench around the nest that Schneirla wanted to study and then remove a block of soil with the entire nest. For years afterward, we could see the hole left by this excavation at the Archbold Biological Station. The nest was set up in one of the laboratories and persisted for some time. For some reason, Schneirla removed the queen for a month, and she was killed by the workers after reintroduction.

It may seem strange that Marion Smith named *archboldi* as a subspecies, since I know that Schneirla also sent Smith specimens of typical *pallidefulva* from the same site. In 1944, however, the biological species concept and the geographic subspecies concept were not consistently applied to ants. Ernst Mayr's major work on the biological species concept appeared in 1942 (Mayr 1942); it was quickly embraced by William Creighton for his book *Ants of North America*, but did not percolate instantly through the taxonomic community.

Formica biophilica Trager— Wilson's Fleet Formica (Plate 81)

Taxonomy and Similar Species: This species was described recently (Trager et al. 2007) as part of the untangling of the *pallidefulva* group, whose taxonomy had become almost impossibly compromised, considering the small number of species involved. As usual in this genus, habitat differences provided some valuable cues. This species is one of three yellow Florida *Formica*: *biophilica*, *dolosa*, and *pallidefulva*. It has sparse but conspicuous and tapering standing hairs on the mesosoma, unlike *pallidefulva*, which has only a few, blunt hairs on the mesosoma, usually concentrated on the mesonotum. The appressed hairs on the gaster are somewhat longer and more abundant than those of *pallidefulva*, but not nearly as dense as those of *dolosa*. The gastral hairs of *dolosa* are consistently longer than the distance between the hairs. Unusually small and less hairy specimens of *biophilica* can be confused with *pallidefulva* (Trager et al. 2007), another good reason to collect series of specimens whenever possible. Florida males, in contrast to workers, are brown.

Distribution: North Carolina west into Missouri, south into north Florida, and central Texas (Trager et al. 2007). In Florida, *biophilica* is known from scattered sites from Polk County north and through the Panhandle (Trager et al. 2007).

Natural History: Unlike *pallidefulva*, *biophilica* often occurs in wet areas such as low flatwoods, seasonally wet prairies, marshes, and fields (Trager et al. 2007). Nests are usually at the base of a clump of grass or sedge, and may be built up into a mound of soil and plant fragments during cold or wet weather (Trager et al. 2007). In its range north of Florida, *biophilica* is the host of the slave-maker ant *Polyergus ruber* (Trager 2013). This association has not been seen in Florida, where populations of *biophilica* might not be dense enough to support *P. ruber*, but it is still possible that *P. ruber* occurs somewhere in northern Florida.

Name Derivation: The name *biophilica* is an allusion to Edward Wilson's inspirational neologism "biophilia," meaning a love of other species inherent in human nature. "Specimens from Alabama, Dr. Wilson's home state, were chosen as the type series to further honor his contributions to myrmecology, conservation and behavioral biology" (Trager et al. 2007).

Formica dolosa Buren—
Wily Fleet Formica (Plate 81)

Taxonomy and Similar Species: Until recently, the name *schaufussi* was generally used for the species now known as *dolosa* (Trager et al. 2007). This is a distinctive member of the Florida trio of yellow *Formica*: *biophilica*, *dolosa*, and *pallidefulva*. This species is conspicuously hairy: the gaster is covered with long hairs that are consistently longer than the distance between them. It is still useful to have the other species on hand for comparison. In some specimens of *dolosa*, the gaster is a little darker than the head and mesosoma. Florida males, in contrast to workers, are brown.

Distribution: New England west into Wisconsin, south into north Florida and Texas (Trager et al. 2007). In Florida, *dolosa* is known from scattered sites from Lake County north and through the Panhandle.

Natural History: The following information is summarized from Trager et al. (2007). This species lives in dry, open habitats. In Florida, it may occur together with *pallidefulva* and *archboldi*, but the latter species are also found in mesic areas, or, in the case of *archboldi*, sites with a high water table. In Florida, *dolosa* is probably primarily a sandhill species that requires frequent fires to maintain its open habitat.

Farther north, *dolosa* occurs in barrens, dry prairies, and dry, open woodlands. It does not nest in mesic woodlands or in moist soil. Nests are usually at the base of a grass clump or other plant. When the nest entrance is in bare soil, it is usually surrounded by bits of plant material or charcoal. Like other species of *Formica*, *dolosa* collects insects and honeydew. Trager observed *dolosa* displacing *archboldi* in colonies of *Toumeyella* scales on young pines. This species is a host of the slave-maker *Polyergus longicornis* (Trager 2013).

Name Derivation: From *dolosa* (Latin), meaning "wily" or "cunning." The author of *dolosa* is Buren (1944) because he was the first (in 1944) to publish the name as a "trinomial": *Formica schaufussi dolosa*. The "*dolosa*" part is taken as a subspecies, a valid taxonomic use. Originally, however, the name was coined by Wheeler (1912) as a "quadrinomial": *Formica pallidefulva schaufussi dolosa*. The "*dolosa*" part is a "variety," which has no taxonomic standing in zoology, so Wheeler's name was not a valid name until used as a trinomial by Buren. Wheeler (1913b) does not explain the meaning of "*dolosa*," but he states the habits of this form are similar to those of *pallidefulva schaufussi*, which he claims is "an extremely timid ant, usually fleeing with great precipitation when its nest is disturbed, never stopping to defend itself and returning to secure its brood in a furtive and hesitating manner." I suppose this behavior might be considered cunning and wily.

Formica pallidefulva Latreille—
Variable Fleet Formica (Plate 81)

Taxonomy and Similar Species: This common and widespread species is yellow in Florida, but darker, sometimes dark brown in some northern populations (Trager et al. 2007). There is a wide transition zone, roughly following the Mason–Dixon Line, in which coloration is variable (Trager et al. 2007). In Florida and elsewhere, *pallidefulva* is distinguished by its extremely short and appressed pubescence on the gaster (the hairs are shorter than the distance between them) and the sparse standing hairs on the mesosoma; these hairs are short and blunt rather than tapering. Florida males are yellow.

Distribution: Southeastern Canada into Florida, west across the Great Plains and in the lower elevations of the Rocky Mountains from Wyoming into New Mexico (Trager et al. 2007). In Florida, *pallidefulva* occurs through the northern and central parts of the state, extending south in remnant scrub and sandhill uplands in Highlands, Martin, and Collier counties. Much of south Florida is too low and wet for *pallidefulva*.

Natural History: In Florida, *pallidefulva* occurs in a variety of upland habitats, including scrub, sandhill, fields, open forests, and lawns. I have not seen *pallidefulva* in mesic forests, probably because such forests in Florida are usually associated with a high water table in the summer. Farther north, *pallidefulva* is the only member of its group that often occurs in mesic forests, and may be restricted to riparian forests in dry prairie regions (Trager et al. 2007). Nest entrances in Florida are often concealed and protected by a clump of grass or by the low grass of a mown field or lawn. Like other Florida *Formica*, *pallidefulva* has major nest chambers close to the surface where they would be easily accessible to digging animals (such as skunks) if they were in open sand. In addition to foraging for live and dead arthropods, *pallidefulva* gathers honeydew, such as that produced by the membracid *Idioderma virescens* on palmetto flowering stalks, and also collects nectar. According to Trager et al. (2007), *pallidefulva* does not tend or defend sap-sucking insects in the manner of *archboldi* and *biophilica*. In northern Florida (Leon County), *pallidefulva* is host to the slave-maker ant *Polyergus montivagus* (King and Trager 2007; Trager 2013).

Name Derivation: From *pallida* (Latin), meaning "pale," and *fulva* (Latin), meaning "reddish yellow," referring to the color of the type series, presumably from one of the yellow southern populations. This ant was described by Latreille in 1802 (Latreille 1802), the locality "Les États Unis." Pierre-André Latreille (1762–1832) was one of the most important early entomologists. He described great numbers of insect species, but his most significant work may have been the organization of species into genera and other groups. Westwood (1838) calls Latreille "One of the most distinguished of modern entomologists, whose writings for nearly half a century have tended in the highest degree to improve the science he so ardently loved." From his earliest publications "… he ceased not to labour toward the accomplishment of a natural classification of insects, and to a perfect investigation of their general structure."

Formica subsericea Say—
Silky Wood Ant (Plate 80)

Taxonomy and Similar Species: In Florida and adjacent states, *subsericea* is the only dark *Formica*

with appressed silvery pubescence. This species and its close relatives to the north are robust with short appendages relative to other southern *Formica* such as *archboldi* and *pallidefulva*.

Distribution: New Brunswick west into Manitoba, south into South Carolina, with scattered records in northern Mississippi and northern Arkansas (Francoeur 1973). An isolated population was discovered in Florida 10.4 km north of Bristol in Liberty County, approximately 300 km south of the next nearest population in Georgia (Wilson and Francoeur 1974). I have not been able to find this population, and it is possible that it has been eradicated by habitat modification. Another northern ant that has been collected only once in Florida is *Lasius interjectus*, from Torreya State Park, near the collection site of *F. subsericea*.

Natural History: This species is usually found in open or semiopen forest, either broadleaf or conifer, and also occurs in landscaped areas around buildings (Francoeur 1973). Nests are usually in well-drained soil (Francoeur 1973) and often under rocks, which serve to warm the adults and brood on sunny days in the north. I have often observed this species in northern New England, where, like most species of *Formica*, *subsericea* avidly visits honeydew-producing Homoptera, and also collects insect prey, especially caterpillars.

Name Derivation: From *sub* (Latin), in this case meaning "somewhat," and *sericeus* (Latin), meaning "silky." This refers to the shining, silvery, appressed pubescence covering most of the head and body.

Genus *Lasius*

Fuzzy Ants and Yellow Underground Ants

The genus *Lasius* (pronounced **Lā" si us**) is a large Holarctic genus of approximately 150 species worldwide (Ellison et al. 2012) and approximately 56 North American species (Fisher and Cover 2007), with additional species awaiting description. A dominant group in the northern United States, *Lasius* has fewer species and is less abundant in the Southeast and is a minor genus in Florida, with only five species, three of which are rare. The most common and widespread Florida species, *Lasius alienus*, barely gets as far south as Orange County. The genus *Lasius* is a good reminder that while ants in general flourish in warm humid climates, there are some big exceptions to this rule, namely, the important genera *Lasius*, *Formica*, and *Myrmica*. There are no known southeastern endemic species in these large genera, with the single exception of *F. archboldi*. In the States bordering the Gulf of Mexico, species of *Nylanderia* might be ecological replacements of *Lasius* species.

Now included in *Lasius* is a set of North American species until recently placed in a separate genus, *Acanthomyops*. These species are distinguished by their much reduced maxillary palpi, and strong citronella odor. They are probably all temporary nest parasites of other *Lasius* species, although this has not been demonstrated for all species. This change in nomenclature reflects important new phylogenetic analyses (Janda et al. 2004; Maruyama et al. 2008); these studies confirm the conclusion of Wilson (1955) that *Acanthomyops* is a recent offshoot of *Lasius*, phylogenetically embedded within *Lasius*. The name *Acanthomyops* remains as a subgenus of *Lasius* and is useful for discussing this well-differentiated lineage.

Among Florida ants, *Lasius* is a distinctive genus. Some species might be mistaken for species of *Nylanderia*, but they lack large hairs on the head and body typical of *Nylanderia*. In the field, *L. alienus* and *neoniger* could be mistaken for *Tapinoma sessile* or *Technomyrmex difficilis*, but closer examination of those species will show that they lack an elevated petiolar scale found in *Lasius* species. *Lasius alienus* and *neoniger* somewhat resemble *Linepithema humile*, but they are much chunkier, while *L. humile* is a slender racehorse ant, built for speed. These *Lasius* species also have an angulate propodeum (rounded in *Linepithema*) and the distance between eye and jaw in lateral view is obviously longer than the length of the eye (shorter in *Linepithema*). Florida species formerly in the genus *Acanthomyops* are yellow and shiny, with much reduced eyes and a strong citronella odor; they resemble no other Florida species.

A remarkable biological feature of *Lasius* is the high percentage of temporary nest parasites, species whose queen usurps the nest of another species of *Lasius*. Eleven of 17 *Lasius* species living in northeastern North America are known or suspected to be temporary nest parasites (Ellison et al. 2012). *Lasius* species in the *niger* group, especially the superabundant *alienus* and *neoniger*, are the most common victims. A number of factors have probably favored this proliferation of temporary nest parasites: the abundance of their hosts; the strong seasonality of northern climates, allowing synchronization of life histories; the monogynous nature of the hosts, which apparently produce colonies only by solitary founding queens; and a possible lack of strong competition between the primarily subterranean parasitic species and their more open-foraging hosts, so that hosts and parasites can coexist in large numbers on the same site. In addition, colonies of *neoniger* and perhaps *alienus* often divide up into smaller groups tending colonies of root aphids, so an invading parasitic queen may seldom need to face an entire colony, enabling her to insert herself into a relatively small group of workers and absorb their nest odor. In Florida, there are two host species and three temporary nest parasites, but the hosts, *alienus* and *neoniger*, are seldom abundant, and two of the nest parasites have only been collected once in Florida.

Species of *Lasius* are normally associated with root-feeding aphids or mealybugs, which are treated as livestock. Some other Florida ants, such as some *Brachymyrmex* and *Dorymyrmex*, are often associated with root-feeding Homoptera, but seem less dependent on this resource than species of *Lasius*.

Species of *Lasius* in the *alienus/neoniger* group were revised by Edward Wilson (1955), and the species formerly placed in the genus *Acanthomyops* were revised by Merle Wing (1968). The challenge of matching the scope of these revisions has perhaps discouraged myrmecologists from attempting a new revision of North American *Lasius* species. Such a revision is overdue as there are known undescribed species, and some species, especially *neoniger*, could be species complexes.

Name Derivation: From *lasios* (Greek), meaning "hairy" or "wooly," referring to the hairiness of the type species, *niger*, which is covered with dense appressed pubescence and also has a generous supply of erect hairs. The English name "Fuzzy Ants" was invented by Ellison et al. (2012) and fits some species. Most of the more subterranean species are shiny yellow with relatively sparse hairs.

Lasius alienus (Foerster)— Woodland Fuzzy Ant (Plate 82)

Taxonomy and Similar Species: *Lasius alienus* is similar to *neoniger* but lacks the fine erect hairs on the antennal scapes found in *neoniger*. To the north, there are additional species in the *niger* group similar to *alienus*.

Distribution: While many species of ants have been spread around the globe by human commerce, *Lasius alienus* seems to have occupied its enormous range through its own powers of dispersal.

In Eurasia, it occurs from Scotland across into Japan, south into the Middle East, while in North America, it occurs from British Columbia to Nova Scotia, south along the western mountain ranges into Mexico, and in eastern North America as far south as northern Florida (Wilson 1955). The lack of geographic differentiation in *alienus* suggests that this species had a more or less contiguous Holarctic range rather recently; the Late Pleistocene must have been a good time for *alienus*. Although *alienus* is native in North America, it is likely that it was also imported from Europe in Colonial times or after, and it is possible that some populations in northeastern North America have a mixed genetic heritage.

Natural History: In Florida, *alienus* is usually found in rotten wood in swamp forests and is not an abundant woodland species as it is farther north. Wilson (1955) observed that in Europe, *alienus* lives in open areas, woodlands being occupied by *niger*, whereas in North America, *alienus* lives in woodlands while open areas are occupied by *neoniger*. Wilson suggests that competition between local *Lasius* species has resulted in the contrasting habitat preferences in *alienus*. This means that European studies of *alienus* ecology may not be applicable to North American populations. In North America, nest sites are sometimes in soil, but typically in stumps and in rotten wood. In New England, I have often found large colonies of *alienus* in rotten wood without any visible root aphids or other subterranean Homoptera, while aboveground foragers are common around sap-sucking insects and also appear at baits and collect dead insects. In contrast, a population of *alienus* in a Danish heath was found to be primarily supported by secretions of Pseudococcidae on roots of heath plants, although insect prey was also collected (Nielsen and Jensen 1975). Nielsen and Jensen (1975) have also illustrated a spectacular cast of a nest system of *alienus* in soil, but this probably has little relevance to the North American population of *alienus*, except that both populations have dispersed rather than compact nests. In a study of the alarm and defense system of *alienus*, Regnier and Wilson (1969) discovered that alarm substances produced by *alienus* make workers run erratically about and evacuate a section of the nest rather than congregate around the source of disturbance.

In cool climates, *alienus* adults emerge in large, well-synchronized swarms, in which individual females may be seen mating multiple times (Bartels 1985). One swarm was large enough that its fallout was large enough to change ammonium concentrations of a subalpine lake (Carlton and Goldman 1984). In Florida, where *alienus* is not very common, no large swarms have been reported.

The abundance of *alienus* may help account for its large number of parasites and inquilines, at least in Europe, where the species has been studied in detail (Hölldobler and Wilson 1990). In North America, *alienus* is host for the temporary nest parasites *Lasius claviger*, *latipes*, *minutus*, and *umbratus*; of these species, *claviger* and *umbratus* occur in Florida.

Name Derivation: From *alienus* (Latin), meaning "strange." This species was first described in 1850 as a strange species of *Formica* at a time when there were only a few recognized genera of ants into which early myrmecologists attempted to fit every new species they found. *Formica* was established as a genus in 1758; *Lasius* was not established as a separate genus until 1804 (Bolton 1995).

Lasius claviger (Roger)— Hairy Yellow Underground Ant (Plate 83)

Taxonomy and Similar Species: *Lasius claviger* is a shiny yellow ant with tiny eyes that are about the width of the antennal scape. It is shinier and larger than the pale species of *Nylanderia*, *arenivaga*, and *phantasma*, and lacks the outsized erect hairs of *Nylanderia* species. It is very similar to another *Lasius* in the subgenus *Acanthomyops* that is also rare in Florida, *L. interjectus*, but segments 2–3 of the gaster of *interjectus* lack hairs except for those on the posterior margins, while segments 2–3 of *claviger* have hairs distributed over their dorsal surfaces. *Lasius umbratus*, also a rare species in Florida, is less shiny, brownish yellow in color, and has short, bristling, erect hairs on the gaster.

Distribution: New England west to Minnesota, south into Mississippi and Florida (Wing 1968). Wing (1968) examined one series from Florida, without locality information; I have seen a single Florida specimen, from Walton County. Wing (1968) saw specimens from Escambia County, Alabama, which is adjacent to Escambia County, Florida.

Natural History: *Lasius claviger* occurs in a wide variety of habitats, from forest to open fields (Wing 1968). It is difficult to get information on the temporary hosts of *claviger* because the action is all underground and host workers may not persist for long in the colony. There is convincing evidence that *alienus* is a temporary host of *claviger* (Raczkowski and Luque 2011). I have not seen published evidence of *neoniger* as a host (as listed by Ellison et al. 2012), although *neoniger* seems a likely host, especially since *claviger* can occur in habitats not favored by *alienus*. In Florida, *alienus* is more common than *neoniger* and seems the most likely host.

Lasius claviger is a subterranean species living (as far as is known) on the exudations of Homoptera that suck sap from the roots of plants (Wing 1968). Workers appear briefly on the surface to open exits for alates but are otherwise found under rocks (not available in Florida) or in logs or stumps (Wing 1968). Wing (1968) lured specimens to the surface by setting out bones from which *claviger* workers collected fat. This suggests that *claviger* and perhaps other subterranean *Lasius* species are less pastoral than they seem, and might supplement their diet by preying on subterranean insects. It would be easy to test this in a captive colony, although it might be difficult to set up such a colony.

In the northern part of the range of *claviger*, flights of queens and males generally occur in autumn, with dealate queens apparently spending the winter in hibernation outside any nest (Wing 1968). This engendered the hypothesis that queens invade host nests in early spring when hosts are still groggy with cold, but work by Raczkowski and Luque (2011) does not support this idea. If *claviger* queens usurp nests by moving into outposts of their host species (there is no evidence of this), it would be advantageous to do so when outposts are expanding in the spring rather than retracting in the fall. There are major gaps in our understanding of *claviger*, but Florida is not the place to study this species, since it has only been found twice.

Name Derivation: From *clava* (Latin), meaning "club," and *-ger* (Latin), meaning "carrying," referring to the club-shaped antennal scape of the type specimen, a queen.

Lasius interjectus Mayr—
Less Hairy Yellow Underground Ant (Plate 83)

Taxonomy and Similar Species: *Lasius interjectus* is similar to *claviger*, also a member of the subgenus *Acanthomyops*, but the gaster of *interjectus* has long hairs on tergites 2–4 confined to posterior margins, not scattered about on the tergites as in *claviger*.

Distribution: New England west through North Dakota, scattered locations in the west into Idaho, Utah, and New Mexico, in eastern North America south into Georgia and Mississippi (Wing 1968). In Florida, *interjectus* is known from a single large colony found in Torreya State Park (Liberty County) by Lloyd Davis as part of surveys for the Ants of Florida project. This Florida site is far south of other known localities, a distributional anomaly similar to that of the widespread species *Formica subsericea*, known in Florida from a single collection just south of Torreya State Park.

Natural History: *Lasius interjectus* is usually found in forested areas or at forest edges, but most North Dakota records are from grasslands (Wing 1968). In acceptance tests with potential host *Lasius*, Raczkowski and Luque (2011) found that *interjectus* queens were readily accepted by workers of *claviger*, sometimes accepted by *alienus*, *latipes*, *murphyi*, *umbratus*, and *minutus*, but not accepted by *neoniger*. In the field, dealate queens were seen entering a mature nest of *latipes*. If the laboratory tests reflect the situation in the field, *interjectus* may have a wide host range and can act as a hyperparasite, taking over the nests of five other species of temporary nest parasites. Of the hosts tested by Raczkowski and Luque (2011), *alienus*, *umbratus*, and *claviger* are the only Florida species compatible with *interjectus*, but *claviger* is so rare that it is unlikely to be a host, and *alienus* is the only common potential host.

In the northern part of the range of *interjectus*, alates usually fly from June into August (Wing 1968). The single nest seen at Torreya State Park was active in February, producing a large mound. The single specimen of *clavipes* was collected in December. These two observations do not mean that *Lasius* (*Acanthomyops*) species are only active in winter in Florida, but any targeted search for these species might as well be undertaken in winter. There should, at least, be fewer ticks and mosquitos about. Members of the subgenus *Acanthomyops* are easily recognized, so one would not need to be an ant expert to add substantially to knowledge of these ants in Florida.

Name Derivation: From *interjectus* (Latin), meaning "lying between," because Mayr considered this species appeared to be somewhat transitional between the subgenus *Acanthomyops* and the subgenus *Lasius*. Mayr (1866b) had only a queen to work with; if he had described the species from workers, he would probably have considered *interjectus* an obvious representative of *Acanthomyops*.

Lasius neoniger Emery—
Field Fuzzy Ant (Plate 82)

Taxonomy and Similar Species: *Lasius neoniger* is similar to *alienus* but has a few to many fine erect hairs on the tibiae and antennal scapes. North of Florida, there are other similar species of *Lasius*.

Distribution: *Lasius neoniger* is a common, sometimes dominant species in open areas in northern and midwestern North America, decreasing in abundance through the lower Midwest and the Southeast, local or rare in the southeast around the Gulf of Mexico (Wilson 1955). Wheeler (1905c) considered it the most abundant of New Jersey ants, "hence of all our insects," a typically myrmecocentric view. It has been found in a few sites in northern Florida.

Natural History: In the Northeast, I have often found *neoniger* in disturbed open areas, including fields, lawns, and roadway medians, as well as in naturally open rocky barrens and sandy areas. In Florida, open areas as well as mesic woodlands are usually occupied by one or more species of *Nylanderia*, so it is tempting to suggest that this genus of similar-sized ants has largely replaced *Lasius* in Florida. The importation of *Nylanderia vividula* and *Solenopsis invicta* might have also contributed to the rarity of *neoniger* in Florida. Farther north, *alienus* has apparently made an ecological claim to woodlands, and *neoniger* is the open-site species, where foragers may be seen collecting dead insects, honeydew, and nectar (Wilson 1955).

An amazingly comprehensive study of the life history of *neoniger* in Illinois was published by Stephen Forbes in 1908 (Forbes 1908). Forbes called this ant *Lasius niger americanus*, but it is clear from his account and from his illustrations that *neoniger* was the species studied. In Illinois, *neoniger* is, as one might expect, strongly seasonal, with flights of alates in June to September, producing foundresses that begin to lay a few eggs in summer, or wait until spring. In warm weather, it takes about two months to produce the first adult workers, and colonies appear to grow relatively slowly, probably requiring more than two years to produce a colony large enough to invest in alates. Older nests are large, with more than 1000 individuals, including larvae. Flourishing colonies of *neoniger* seem to get much of their sustenance from the secretions of root aphids, which the ants move about from plant to plant. In winter, aphid eggs are stored and protected, to be placed on roots in spring. *Lasius neoniger* are able to take advantage of annual crops, such as corn, moving aphids onto roots in numbers large enough to reduce crop yield. In early spring, the aphids are brought to roots near the surface where warmth makes these livestock grow quickly, while the ant larvae are kept in lower, cooler chambers until a good supply of honeydew is available. When a young worker is ready to emerge, older workers bite off the tip of the cocoon and help the callow worker emerge.

Around their nests, workers recognize and attack ants from other colonies; in 1908, this might have been considered simple territoriality based on intercolony competition. Forbes, however, sought a deeper explanation. He noted that foragers wander extensively, and if they were free to work with members of another colony, their own colony would suffer. Forbes anticipated later sociobiological theory in stating that protecting and providing for "the family" is the single goal of every worker, and hostility toward members of other colonies derives from this. "Clannishness in ants has the same justification that it has among savage men. It is a means of maintaining the necessary concentration of the group for the care of the young, and hence for the preservation of the race." More recently, Traniello (1980) showed that there is colony specificity in the trail pheromone of *neoniger*. Traniello suggested that this specificity not only leads workers to and from their own colony but also acts as a "spacing mechanism," allowing colonies to avoid conflict by laying claim to a specific resource or arena of resources.

The complexities of territoriality in *neoniger* have been further studied by Traniello (1980, 1983) and Traniello and Levings (1986). Their work might serve as a model for studies of polydomus ants that are more common in Florida, especially species of *Nylanderia*. Colonies of neoniger not only have multiple entrances, but workers have a tendency to use a particular entrance, implying some autonomy of the underground chambers associated with each entrance. Approximately 30–60 workers show some fidelity to each entrance. This provides a relatively large team that can be rapidly recruited to claim and retrieve insect prey, the recruitment involving a colony-specific trail pheromone. When more prey is available above ground, a colony expands the area claimed by its colony by opening new entrances. Seasonal expansions in colony territory probably also reflect increasing numbers and dispersal of root aphids tended by each colony. As prey become scarcer in fall, colonies tend to retract their colony fragments; this might also be correlated with root aphids producing overwintering eggs. The work by Traniello (1980, 1983) and by Traniello and Levings (1986) tells a remarkable story of adaptive territoriality. *Lasius neoniger* is able to live in densely aggregated groups of colonies that can take advantage of patches of plants that support root aphids. Conflict is largely avoided by restricting hostilities to areas immediately adjacent to nest areas. Disputes over prey are largely avoided by quick teamwork to carry off prey by the nearest colony outpost. Foragers moving from the defended area of the colony are guided back to their colony and away from other colonies by colony-specific pheromones. Colonies can strategically expand or retract territory by positioning colony outposts with separate nest entrances. In terms of territoriality, *neoniger* shows dispersed central place foraging, with temporary satellite groups of workers established in response to distribution of resources.

Like *alienus*, *neoniger* is host to other *Lasius* that are temporary nest parasites: *claviger*, *latipes*, *murpheyi*, and *umbratus* (Ellison et al. 2012). Of these, only *claviger* and *umbratus* occur in Florida, but *neoniger* might not be common enough in Florida to support these parasites. *Lasius neoniger* may also be too rare in Florida to support populations of the wasp *Pseudometagea schwarzii* (Eucharitidae), a larval parasite that is not known from southeastern North America (Ayre 1962).

Name Derivation: From *neos* (Greek), meaning "new," and *niger* (Latin), meaning "black," referring to the similarity between this New World species and the blackish Old World species, *L. niger*.

Lasius umbratus (Nylander)— Dusky Fuzzy Ant (Plate 82)

Taxonomy and Similar Species: *Lasius umbratus* somewhat resembles species in the subgenus *Acanthomyops* such as the Florida species *claviger* and *interjectus*, but has larger eyes (they are still relatively small), is much less shiny, and is brownish yellow rather than clear yellow. It has groups of conspicuous curved erect hairs on the back of the head and on the pronotum and mesonotum. The gaster is evenly covered with short, semierect,

bristling hairs, a noticeable and distinctive trait. *Lasius umbratus*, which belongs to the unpronounceable subgenus *Chthonolasius*, could not easily be mistaken for any other Florida species.

Distribution: Like *Lasius alienus*, *umbratus* is a Holarctic species, commonest in the northern part of its range, but with populations in southern temperate regions as well. In North America, *umbratus* occurs from Nova Scotia into Florida, west into Idaho, with scattered populations in southwestern mountains, including Arizona (Smith 1979). *Lasius umbratus* has been collected a few times in the western Panhandle of Florida.

Natural History: *Lasius umbratus* in North America is a temporary nest parasite of *alienus*, as shown by mixed colonies found in the Northeast (S. Cover 2015, personal communication). It is found in both open and shaded areas (Ellison et al. 2012), so there is a good chance that it is also a parasite of *neoniger*, which is more likely to occur in open sites than *alienus*. In Europe, *umbratus* is also a temporary nest parasite of *alienus* (Sciaky and Rigato 1987). In a possible wolf-in-sheepskin scenario, queens carrying a dead worker of *alienus* have been seen approaching the nest entrances of their host (Sciaky and Rigato 1987). Nest usurpation in North America has not been studied. *Lasius umbratus*, like other members of its genus, seems to be dependent on the secretions of sap-sucking Homoptera on roots, but workers also forage for insects above ground (Talbot 1965; Wilson 1955). Nests are underground or in rotten logs and stumps (Talbot 1965; Wilson 1955). The few Florida colonies I have seen were in rotten wood.

Name Derivation: From *umbratus* (Latin), meaning "shady" or "dusky," probably referring to the subdued yellowish color of this species relative to yellow species such as the European *flavus*.

Genus *Myrmelachista*

Short-Antenna Stem Ants

The genus *Myrmelachista* (pronounced **Mur' a kis" tuh**) is a Neotropical genus of 63 described species, with more probably awaiting discovery (Longino 2006). The antennae are relatively short, often with only nine segments; the scape may be almost as long as the remaining segments combined.

Some species of *Myrmelachista* live in dead twigs and stems, but many species live in hollow living stems, including species that have specialized relationships with certain plant genera or species (Longino 2006). This contrasts with the twig- and stem-inhabiting ants of Florida, which live in dead material, with the single known exception of *Camponotus mississippiensis* (Tynes 1964). At least one species of *Myrmelachista* has a remarkable specialized mutualistic relationship with their host plants, getting food and lodging from the plants while killing surrounding vegetation by spraying toxins into wounds made in buds and leaf veins (Morawetz et al. 1992). The single Florida species of *Myrmelachista* does nothing so amazing but lives in dead twigs on living trees and shrubs. Although sound dead twigs in the productive upper canopy of trees and shrubs must often be scarce, there is no reason to suspect that any species of *Myrmelachista* use toxins to augment the supply. It is not even clear that species living in dead twigs and stems can hollow out their own nest cavities rather than occupying abandoned galleries of twig-boring insects (Longino 2006).

Name Derivation: From *myrmex* (Greek), meaning "ant," and *elachistos* (Greek), meaning "smallest." Species of *Myrmelachista* are small, but not the smallest ants. Roger (1863) knew this when he erected the genus for a species that is 2 mm long, while describing in the same paper species of *Plagiolepis* that are 1.5 mm long. He might have used *elachistos* in an emphatic rather than a comparative sense: "very small ant." For the *Myrmelachista* myrmecologist, who is dealing with difficult complexes of similar species (Longino 2006), many of which are probably hiding out undescribed in jungle treetops, these ants are plenty small.

Myrmelachista ramulorum Wheeler— Common Short-Antenna Stem Ant (Plate 83)

Taxonomy and Similar Species: *Myrmelachista ramulorum* is a shining bicolored species, with a black head and gaster, and orange mesosoma. In Florida, it might be confused with *Ochetellus glaber*, but *Myrmelachista* species have a prominent triangular petiole, while that of *Ochetellus* species is more or less cylindrical, with no dorsal node. In the field, this species might possibly be confused with three other small, black or bicolored, shining arboreal ants: *Monomorium floricola*, *Solenopsis picta*, and dark forms of *Xenomyrmex floridana*. All of these have the myrmicine two-segmented petiole. Behavior might be useful for field identification of *ramulorum*: "In life the head and gaster appear blue black, and the ants will be at once recognized by their slow and deliberate but purposeful movements, as their files move slowly up and down the trunks of coffee and coffee shade trees in fine weather" (Wolcott 1948). Questions of its identity in Florida may never arise, as it has not been seen since a single collection in 1967.

Distribution: *Myrmelachista ramulorum* is widely distributed in the West Indies (Smith 1979). In Florida, it is known from a 1967 collection on an orange tree in Highland City, Polk County (Smith 1979). On a trip to Highland City, I was unable to find this species, but that does not mean that it has disappeared. There have been, however, some cold winters in Polk County since 1968, and in recent years, more powerful insecticides have been used more frequently on commercial citrus, so the Florida population might well be extinct. The best place to look for *ramulorum* in Florida might be in sea grape twigs (*Coccoloba uvifera*) along the coast of the southern Peninsula.

Natural History: *Myrmelachista ramulorum* occurs in various habitats, including grassland, agricultural land, and forest (Torres 1984a), provided there are suitable host trees. I found this species once in the Bahamas, in dead twigs in the canopy of a sea grape growing in the open, but in Puerto Rico, it is also reported from densely forested areas (Lavigne 1977), and at one time, it was the most serious pest in Puerto Rican coffee plantations (Smith 1937; Wolcott 1948). The reports of Smith (1937) and Wolcott (1948) are especially interesting for several reasons. Wolcott considered this the most destructive native ant in Puerto Rico, and his observations, as well as the earlier report by Smith, suggest that there might have been an outbreak of *ramulorum* while they were doing their surveys of the ants of Puerto Rico. This species tends and shelters mealy bugs and scale insects, and it is possible that introduction of new homopteran hosts, combined with ideal conditions in the coffee plantations, led to this possible outbreak. While the ants fostered homopterans on coffee and other trees, two species of the genus *Inga* that were grown as shade trees seemed to be especially favored, with tens of thousands of ants per tree (Smith 1937). The ants do much more than carry sap-suckers from place to place; they dig channels into the pith of coffee shoots, introducing mealy bugs and scale insects into these grooves (Hooker 1913), or bore directly into twigs and small branches, bringing homopterans into the interior. Working at the junction of branch and tree trunk, the ants also induce swollen galls in which they live with their livestock (Smith 1937; Wolcott 1948). These galls become zones of weakness, causing extensive branch drop during hurricanes, during coffee harvest, or even when there is little wind (Smith 1937; Wolcott 1948). The induction of galls suggests a nonmutualistic form of plant manipulation through chemistry, although the phenomenon has apparently never been studied, and it is possible that physical damage or the actions of the homopterans caused the growths. The ants repeatedly caused the death of all the new shoots of a large Ficus (Wolcott 1948), an anecdote reminiscent of the methodical tree-killing by the Amazonian species studied by Morawetz et al. (1992). From an evolutionary standpoint, ramulorum might be a rewarding species to study, as it appears to be an adaptable generalist species with behaviors that indicate origins of more specialized relationships of some other species. As might be expected from its habits, ramulorum is polygynous and polydomus (Smith 1937). Blum and Wilson (1964) did a preliminary study of the trail pheromone, suggesting that the pheromone is not produced by the poison gland but by a source emptying into the hind gut.

In some contrast to the early reports of conspicuous trails of ramulorum on tree trunks (Smith 1937; Wolcott 1948), Lavigne (1977) found ramulorum "secretive" in the Luquillo Forest of Puerto Rico, and only deduced its abundance from the large numbers in stomachs of three species of lizards and six species of frogs.

Name Derivation: From ramulorum (Latin), meaning "of branches," referring to the arboreal habits of the species.

Genus Nylanderia

Crazy Ants

The recent adoption of the genus name Nylanderia is another change in the genera of Florida ants that tests the mental flexibility of aging myrmecologists such as myself, even though the change seems well justified and overdue. Members of this large genus of abundant ants were long included in the genus Paratrechina, until careful work by LaPolla et al. (2010) showed that the type species of Paratrechina, Paratrechina longicornis, was not even closely related to the other supposed species of Paratrechina. The name Nylanderia, which was invented to designate a subgenus of Prenolepis and later of Paratrechina when that genus was separated

from Prenolepis, became the default name for the genus. The only good reason to remember all this is that most literature on Nylanderia species was published using the name Paratrechina. Nylanderia includes approximately 130 described species and subspecies (LaPolla et al. 2011), with many species waiting to be described.

Nylanderia is a difficult genus, with many species that closely resemble each other, especially when one is working with dead specimens, which are often slightly deformed because of drying, with some fading of color, which is often not distinctive to begin with. It is, however, worth getting to know these ants, as they are among the most abundant and conspicuous of Florida ants. In Florida, there are 13 species of Nylanderia, including one Palaeotropical exotic and three Neotropical exotics. This wins Florida the prize for Nylanderia diversity among the United States. Nearctic species of Nylanderia were rescued from a dismal state of taxonomic confusion by James Trager in 1984. This involved redescribing the known species, describing five new species, and, equally important, establishing the identity of N. faisonensis, the most abundant native species in much of eastern North America. The next major taxonomic advance was the revision by Robert Kallal and John LaPolla (2012), which cleared up some outstanding problems and provided descriptions of three new species. At least three rare parasitic species remain to be described.

In Florida, these ants are easily identified, at least to genus, by their small size (2–3 mm), combined with long legs and, especially, by thick, enlarged hairs ("macrochaetae") that stick up from the head and body. From their position and socketed attachment, it is reasonable to suppose that they are sensory hairs. These hairs are minutely barbed (the barbs invisible under a normal dissection scope) and might provide a ratchet-like stimulus when pulled across an obstacle, functioning rather like the microscopic ridges on human fingertips. These hairs also occur in P. longicornis. Nylanderia species have a characteristic fast, zigzag way of running that allows the experienced naturalist to recognize them in the field without resorting to a microscope. These ants appear to follow diffuse or vaporizing odor trails rather than tracing a chemical trail along a substrate. Individual ants follow slightly different paths, angling from one side to the other of the chemical roadway. A diagram of such a path for a species of Lasius appears in Hölldobler and Wilson (1990). Although this style of trail-following increases the distance traversed, it incorporates an evasive function lacking in more precise trail-followers in such genera as *Solenopsis* and *Monomorium*. When alarmed, these ants race about in a "crazy" way that is actually an effective dodging technique; they are "crazy like a fox," as the saying goes. This frenzied alarm behavior is even more conspicuous in the species remaining in *Paratrechina, P. longicornis,* and can also be seen in some species of *Forelius*.

Nylanderia species are predators and scavengers, and also dependent on sugary fluids, especially honeydew. Species of *Nylanderia* tend to be strongly opportunistic, with superficial or diffuse nests that can be easily moved to take advantage of a new habitat or resources. While these relocation events might make the colony vulnerable to nest predators, they should make individual foragers less vulnerable and more efficient because they have a shorter trip to and from resources. This opportunistic lifestyle preadapted several species to invade and exploit the chronically disturbed habitats created by humans. Four such species have invaded Florida and are either pests or possible ecological threats.

There are three known parasitic species of *Nylanderia* in eastern North America, and it is likely that there are additional parasitic species elsewhere. The three eastern species are associated with hosts that are more or less polydomus, with small aggregations of brood scattered through pockets in the leaf litter. This might facilitate evolution of a form of parasitism in which dispersing queens take over isolated groups of workers and brood, eventually resulting in a specialized parasitic species.

Name Derivation: Carlo Emery established *Nylanderia* as a subgenus of *Prenolepis* in 1906 (Emery 1906). This subgenus was intended to accommodate the ant then known as *Prenolepis vividula*, which was described by the Finnish botanist and entomologist William Nylander (1822–1899). The restriction of the name *Paratrechina* to *P. longicornis* and the elevation of *Nylanderia* from subgenus to genus bring to prominence the name of a long-ago naturalist little known on this side of the Atlantic.

Nylanderia arenivaga (Wheeler)—
Sand-Loving Crazy Ant (Plate 84)

Taxonomy and Similar Species: Most species of Florida *Nylanderia* are brown or blackish brown, but *arenivaga* is yellow, often with a slightly darker gaster. The other pale Florida species is *phantasma*, which is yellowish in north Florida, almost white on the Lake Wales Ridge. The enlarged hairs on *arenivaga* are dark, whereas those of *phantasma* are almost as pale as the body; there are usually 5–17 enlarged hairs on the antennal scapes of *arenivaga*, 1–3 in the case of *phantasma* (Trager 1984).

Distribution: Sporadically distributed from New Jersey south through Florida, west into Nebraska, and south into east Texas (Trager 1984). It appears to have colonized widely separated sandy areas, possibly by following riverine deposits inland from sandy coastal plain areas. In Florida, *arenivaga* occurs in sandy uplands through most of the state. Records are scarce for the southern third of Florida, where there are few sandy uplands and even fewer with remnants of natural habitat.

Natural History: *Nylanderia arenivaga* is a species of open sandy areas that lack an over story of trees or shrubs, or a dense covering of herbaceous vegetation. In primeval Florida, it must have been confined to gaps, including small gaps, left after fires in sandhill and Florida scrub habitats that once covered much of the state. Today, this species also occupies edges of fire lanes and sand roads traversing natural habitats, but, unlike *Dorymyrmex bureni*, does not usually occur in more drastically altered habitats such as borders of cultivated fields, shoulders of highways, or urban and suburban vacant lots.

This species is polydomus, in the sense of having a somewhat dispersed colony. As many as 20 entrances may occur, separated by distances ranging from a few centimeters to approximately 1 meter (Trager 1984). Some of these appear to provide convenient access to an underground tunnel, and one often fails to find ants by digging up a "nest entrance." Foraging takes place both day and night (Van Pelt 1958), with reduced or interrupted foraging during the middle of the day. Queens and males appear in nests in late summer or in fall, but do not fly until early spring (Trager 1984). The diet is dead insects and probably honeydew; *arenivaga* is often associated with Homoptera feeding on roots (Trager 1984). A species of phorid fly, *Pseudacteon gracilisetus*, attacks adult workers (Brown et al. 2011). This parasitoid is known only from the type locality in South Carolina (Brown et al. 2011) but could easily occur in Florida. *Pseudacteon* flies attack many species of ants and are often abundant, but they are microscopic and fast-moving, usually only seen by those who are specifically looking for them.

Name Derivation: From *arena* (Latin), meaning sand, and *vagus* (Latin), meaning "wandering," referring to the observation that this species wanders about on the sand.

Nylanderia bourbonica (Forel)— Robust Crazy Ant (Plate 85)

Taxonomy and Similar Species: This relatively large blackish species has a dense covering of fine hairs on the dorsal surface of the mesosoma, unlike any native Florida species. These hairs, which are blackish, can be seen best by viewing the mesosoma in profile so that the semidecumbent hairs stand out against a light background. *Nylanderia steinheili*, another introduced species, shares these mesosomal hairs, but is smaller, and the Florida population has conspicuous whitish coxae that contrast with the dark color of the body. The introduced species *pubens* and *fulva* are also hairy, but are pale brown rather than blackish, and the large hairs on the pronotum and mesonotum are tapering and brown, rather than blunt and blackish as in *bourbonica*.

Distribution: *Nylanderia bourbonica* probably originated in tropical Australasia, where there are several similar species that are probably closely related to *bourbonica*. The extent of its native range is not known exactly, but it has apparently been transported through commerce to various places, including Hawaii, Madagascar, Reunion (the type locality), and Florida (McGlynn 1999). Amazingly, there seem to be no records for the Greater and Lesser Antilles, South or Central America, or south coastal Texas. There are many Florida records from the southern two-thirds of the Peninsula; records from Mobile, Alabama (Trager 1984), and from Jacksonville, Florida, might represent populations in urban sites protected from cold weather. The first Florida record is 1924 (Deyrup et al. 2000).

Natural History: In Florida, *bourbonica* usually lives in wet, disturbed sites, including well-watered landscaped areas. It is listed as one of the top eight "urban pest ants" in peninsular Florida by Klotz et al. (1995), but often its chief crime seems to be that of traipsing about on the patio, where it seems to be a regular but not strikingly abundant visitor. In cool weather, however, it may move indoors (Trager 1984). I have never observed it overrunning premises in the manner of *N. fulva*. In natural habitats in Florida, *bourbonica*

is most often found in wet areas that are naturally disturbed, such as the edges of marshes and upper zones of beaches. In my experience, *bourbonica* is a ground-nesting species with superficial nests that are sometimes associated with a grass clump or with a pile of debris. I have not found arboreal colonies, but Simberloff and Wilson (1969, 1970) found *bourbonica* colonizing mangrove islands that have no dry land, or even mud flats, at high tide. Likewise, the spread of *bourbonica* as a "tramp species" in the Old World tropics suggests that it has an alternative nesting site, perhaps in dead wood.

On a larger scale, *bourbonica* shows a predilection for islands, although in the Caribbean, it is known only from the relatively small islands of the Bahamas (Deyrup et al. 1998). This suggests the possibility that *bourbonica*, which may well have an island origin, may not be a strong competitor when confronted by a well-developed tropical or subtropical ant fauna. Southern Florida, which has had little access to mainland subtropical ant faunas, could represent a virtual island to ants such as *bourbonica*.

Like other *Nylanderia*, *bourbonica* is a predator and scavenger of arthropods, and also collects honeydew. It appears unable to defend resources against *Solenopsis invicta*, a species that often occurs in habitats otherwise suitable for *bourbonica* (Trager 1984).

Name Derivation: Derived from the former name of the island of Reunion, which was called Bourbon for a lengthy period of its early history. Reunion is the type locality for *bourbonica*.

Nylanderia concinna (Trager)— Marsh Crazy Ant (Plate 86)

Taxonomy and Similar Species: Florida has a set of small, dark, shining *Nylanderia*, consisting of *concinna*, *faisonensis*, *vividula*, *parvula*, and *wojciki*. The last two species lack more than one or two large erect hairs on their antennal scapes. The most useful character for separating *concinna* from the remaining species is a patch of neatly aligned pubescence on the anterior part of the propodeum. This is best seen with strong, diffuse lighting, and shows up best on darker specimens. It helps to know the habitat where specimens were collected. If it was an open, wet area, such as the edge of a marsh, one may be more persistent in turning a specimen this way and that until the hairs are seen. If a specimen is from a dense, mesic woodland, it is probably *faisonensis*, and one would not expect to find any patch of propodeal hairs. The coxae of *faisonensis* are usually conspicuously paler than the tibiae, while those of concinna may be slightly paler, but not contrasting. *Nylanderia concinna* has closely spaced appressed hairs on the head, while those of *vividula* are widely spaced, so the head appears shinier.

Distribution: North Carolina south through peninsular Florida, west into north-central Alabama (Trager 1984). In Florida, there are no records from the western Panhandle, but targeted collecting in marshes would probably produce specimens.

Natural History: This species usually occurs in open, wet areas, such as the edges of marshes, seasonally flooded prairies, coastal marshes, low pastures, and open low flatwoods. Nests are often hollowed out of the root ball of grass tussocks, but may be in old pine logs in low flatwoods. *Nylanderia concinna* is able to tolerate considerable habitat disturbance, moving into wet pastures, the edges of ditches, and even lawns that have a high water table (Trager 1984). I have seldom found *concinna* in "improved" nonnative pasture in south Florida, a habitat that tends to be wall-to-wall *Solenopsis invicta*. Alates usually fly in summer but may appear at other times of year (Trager 1984).

Name Derivation: From *concinna* (Latin), meaning "well-arranged" or "well put together," referring to the orderly set of hairs on the propodeum (Trager 1984).

Nylanderia faisonensis (Forel)— Woodland Crazy Ant (Plate 86)

Taxonomy and Similar Species: *Nylanderia faisonensis* is a dark, shiny species that strongly resembles two other species that share the feature of prominent erect hairs on the antennal scapes (look for these against a dark background): *concinna* and *vividula*. *Nylanderia faisonensis* can be distinguished (with some difficulty) from *concinna* by the set of small hairs at the base of the propodeum of *concinna* and paler coxae that contrast with the tibiae and the mesosoma; *concinna* also tends to live in open wet areas, while *faisonensis* is usually a woodland species. *Nylanderia vividula* is less hairy and more shiny, a feature that Trager (1984) quantifies by stipulating that the small hairs on the upper part of the head should mostly be separated by spaces at least as wide as the length of the hairs themselves; *vividula* tends to live in open, disturbed sites.

Distribution: New Jersey south through Florida, west through the lower elevations of the Appalachians and into Arkansas (Trager 1984). In Florida, this mesic forest species is commonest in the northern part of the state and rarer in south Florida, where most forested areas are either wet flatwoods or swamp forest.

Natural History: This is a common leaf-litter inhabiting species of southern forests. It can live in wet sites, even sites that are seasonally flooded, by taking advantage of elevated rotten logs, tree buttresses, and fern tussocks. In north Florida, *faisonensis* thrives in ravines and other relict forests. It has apparently benefited from a reduction in fire frequency that has greatly expanded mesic forests into uplands. This opportunism has been documented in Maryland (Lynch 1981): *faisonensis* was scarce in an old field, but as abundant in a 35-year-old forest as in a forest at least 100 years old. There are a few records of *faisonensis* from tropical Florida, but these populations will probably be overrun and extinguished by the exploding population of *steinheili*. In the northern part of its range (north of Florida), *faisonensis* may face some competition from the exotic Japanese species *flavipes*, although Trager (1984) suggests that *faisonensis* may have prevented the spread of *flavipes*, especially in natural habitats.

The queen and much of the colony are usually in rotten wood, often a substantial hollow twig, buried in the leaf litter. Larvae, pupae, and callows, attended by some workers, are often in small "colony fragments" in cavities in leaf litter just below the surface. This may allow accelerated development at the higher temperatures near the surface. In Florida, this species forages on the ground or on vegetation, and recruits, although not very strongly, to various baits. Lynch et al. (1980) studied interactions between a northern population of *faisonensis*, *Prenolepis imparis*, and *Aphaenogaster rudis*. They discovered that *faisonensis* is "timid" at baits, does not recruit strongly, is active in warm weather, and prefers carbohydrate baits. The other two species, especially *P. imparis*, are relatively more aggressive at baits, active at cooler temperatures, and more strongly attracted to proteins and fats than *faisonensis*. While the measurement and statistical analyses of these trends are impressive, the ecological differences between these three ants were generally evident in advance, even at the genus level. As usual in studies of community or guild structure, it is easier to show that co-occurring species have bases for coexistence than it is to show that some species are being excluded by competition.

This species is the host of an undescribed parasitic species, discussed below as "Species B."

Name Derivation: Named by August Forel for the town of Faison, North Carolina, the source of the type series.

Nylanderia fulva (Mayr)—
Tawny Crazy Ant (Plate 85)

Taxonomy and Similar Species: There is an excess of taxonomic confusion associated with this ant, primarily because there are two species, fulva and pubens, whose workers are virtually identical. This problem of indistinguishable workers is also found among a few species of thief ants such as Solenopsis abdita, carolinensis, and texana, but the problem is less acute with those species as they are of interest to a relatively tiny number of people. In the case of fulva and pubens, the former species is a major pest with explosive outbreaks, whereas the latter species is apparently relatively innocuous. Nylanderia pubens was reported from Florida in 1984 in James Trager's revision of Nylanderia (under the name of Paratrechina); we know it was that species because this revision included a drawing of the hairy parameres of the male genitalia of pubens. The parameres are a pair of triangular appendages at the tip of the male gaster. They are usually visible without dissecting the ant. There is also a photo on AntWeb of a Florida male pubens collected by John Mangold. Some years later, it appeared that pubens had undergone some kind of adaptive shift that allowed it to produce dramatic outbreaks in various places. These included the West Indies, Florida, and Texas (Sharma et al. 2014). Genetic and morphological studies eventually revealed two similar species of Nylanderia, with fulva being the species that causes outbreaks (Gotzek et al. 2012). It was apparently introduced into North America sometime after 1984, as its massive outbreaks would have caught the attention of Trager while he was preparing his revision. The only reliable morphological feature separating the species is the structure of the male parameres (visible without dissection). Nylanderia fulva has slightly more pointed parameres, with relatively short curved hairs that are not obviously longer than the widest part of the parameres in lateral view; pubens has blunter parameres with much longer hairs that are strongly curved apically. Photos of parameres of both species are in the paper by Gotzek et al. (2012), Sharma et al. (2014), and in the Nylanderia revision by Kallal and LaPolla (2012). Workers of both species are distinguished from other Florida Nylanderia by their reddish brown color combined with abundant fine hair on the mesosoma. Any Nylanderia that fits this description and appears to be having a local population explosion is probably fulva.

Distribution: Nylanderia fulva is apparently native to Brazil, at least that is the origin of the type specimen (Gotzek et al. 2012). There are introduced populations in the West Indies (Wetterer and Keularts 2008), Florida, Mississippi, Louisiana, and Texas (Gotzek et al. 2012).

Natural History: This species has attained recent notoriety for explosive outbreaks in Florida and Texas. I have seen a highly localized but intense outbreak in Dominica, involving swarms of ants moving over the ground, up tree trunks, and pouring into a residence that lacked window screens. The species was absent 100 meters up or down the road from the site. The ants were blamed for injuring citrus trees, probably by fostering sap-sucking insects. Wetter and Keularts (2008) describe similar outbreaks at three sites on St. Croix. On this island, the dense populations of pubens reduced numbers of individuals and species of other ants on tree trunks. Population explosions are often associated with exotic, invasive species of ants. The mechanisms of some such explosions have been examined by Oliver et al. (2008). A species of exotic ant may occur at low levels until some factor, perhaps a year of especially favorable weather, allows the species to become abundant enough to displace native ants, especially those associated with honeydew-producing insects. A feedback loop ensues, with more individuals of the invasive species displacing more native species and fostering more honeydew producers. The importation of a new honeydew-producer could also, in theory, initiate a population explosion.

Nylanderia fulva collects honeydew from a wide variety of sap-sucking insects, including whiteflies, aphids, mealybugs, psyllids, lace bugs, and several families of scale insects (Sharma et al. 2013). The ants sometimes construct carton shelters over sessile honeydew producers (Sharma et al. 2013).

Wetterer and Keularts (2008) suggest that pubens may have population collapses as well, basing this hypothesis on historical records of "plague" ants, probably pubens, that beset Bermuda, followed by the decline and extinction of the species. A repeat visit to an outbreak area on St. Croix in the West Indies showed that local outbreaks may last only a few years (Wetterer et al. 2014). Anybody who has viewed one of these outbreaks will come away

ANTS OF FLORIDA *Deyrup*

with the impression that such a population density is unsustainable. The possible causes of such a collapse are most likely to be newly adapted or introduced diseases or enemies of the sap-sucking insects or of the ants themselves. Our rejoicing at such a fate for overpopulating ants might be tempered by the realization that a similar destiny might await our own exploding population.

Name Derivation: From *fulva* (Latin), meaning "tawny."

Nylanderia parvula (Mayr)—
Northern Crazy Ant (Plate 87)

Taxonomy and Similar Species: This small, black species is similar to *faisonensis* in the field, but lacks large hairs on the antennal scapes and is unlikely to occur in dense forest. Morphologically, it is most similar to *wojciki*, but *parvula* is much darker, not at all bicolored darker and lighter brown, as is the case in *wojciki*. Nests are often in soil rather than in leaf litter like most *wojciki* nests.

Distribution: The range of *parvula* extends farther north than that of other *Nylanderia* species, from northern Massachusetts west into North Dakota, south into Florida and Texas (Trager 1984). In Florida, *parvula* occurs in the Panhandle and northern half of the Peninsula, extending south into Hernando, Lake, and Brevard counties.

Natural History: This species usually occurs in semiopen areas such as the edges of fields and fire-maintained prairies and savannahs. While *parvula* is common in northern Florida, there are far fewer records than I would have expected from Alabama and Mississippi. The impression left by the distribution map (Trager 1984) and my own experience is that *parvula* has somewhat discontinuous eastern and Midwestern ranges, with more integration into southern ecosystems to the east of the Appalachians than to the west.

Alates are reared in the fall (in Florida) and fly the following spring (Trager 1984). In Massachusetts, *parvula* is host to a currently undescribed parasitic species of *Nylanderia* with flightless males.

Name Derivation: From *parvula* (Latin), meaning "littler," referring to the small size of this species.

Nylanderia phantasma (Trager)—
Ghostly Crazy Ant (Plate 84)

Taxonomy and Similar Species: *Nylanderia phantasma* is similar to *arenivaga*, but the enlarged hairs on the head and body are pale and there are only a few enlarged hairs on the antennal scapes. The tip of the gaster is not brownish as in *arenivaga*. It usually lives in open areas with well-drained white sand; members of populations in this habitat are an unusual whitish color. At the northern end of its range, *phantasma* may occur on open yellow sand, and there, its color may be pale yellow, more similar to that of *arenivaga* (Trager 1984). Trager (1984) found that crushed workers have (to him) a rose-like odor, while *arenivaga* has an acrid odor.

Distribution: *Nylanderia phantasma* is restricted to relict areas of scrub and open sandhill in the Southeastern Coastal Plain. It is most common in Florida scrub habitat on the Lake Wales Ridge, but is also known from scrub sites and a few sandhill sites scattered around the Peninsula, coastal Alabama and Mississippi (Kallal and LaPolla 2012), and interior sand dunes in Georgia east of the Ohoopee River (MacGown et al. 2009). There are no records from coastal scrub areas in the Florida Panhandle, but it probably occurs there.

Natural History: This nocturnal species is almost white, a rare coloration in ants. There are, however, many species of light yellow ants, most of which are subterranean (such as *Solenopsis tennesseensis*), or primarily nocturnal (such as *Dorymyrmex elegans*, *Temnothorax allardycei*, and *Nylanderia arenivaga*). The ultra-pale color of *phantasma* probably makes it less conspicuous to nocturnal predators in open areas of white sand. Diurnal ants in the same habitat tend to be black; for diurnal species, shielding from ultraviolet light may be more important than camouflage.

The normal habitat of *phantasma* is open Florida scrub, especially sites that are unusually well drained and dominated by the shrub scrub rosemary, *Ceratiola ericoides*. Rosemary Florida scrub covered much of the southern half of the Peninsula during parts of the Pleistocene, including a span from 40,000 to 5000 years before present (Delcourt and Delcourt 1981; Webb 1990). During this time, *phantasma* was probably a dominant species of *Nylanderia*, sharing its range with other open scrub species such as the lizard *Sceloporus woodi* and the tiger beetle *Cicindela scabrosa*. These species, like *phantasma*, now occupy disjunct patches of scrub habitat where rapid drainage on fossil dunes maintains the desert-like conditions of the late Pleistocene. *Nylanderia phantasma* and other species specialized for life in open rosemary scrub now face a new trial, the rapid conversion of their remaining habitat into housing and golf courses.

This species forages at night throughout the year, even during relatively cool nights (Trager 1984). Most colonies seem to have several nest entrances, but not all entrances are associated with concentrations of workers when I have attempted to excavate colonies. Alates appear in nests in winter and early spring.

Name Derivation: From *phantasma* (Greek), meaning "phantom," referring to the whitish color.

Nylanderia pubens (Forel)—
Hairy Crazy Ant (Plate 85)

Taxonomy and Similar Species: *Nylanderia pubens* is easily distinguished from most other Florida *Nylanderia* by its uniform medium brown color and by the dense pubescence covering the mesosoma, but *pubens* is apparently virtually identical to *fulva*, most easily distinguished by the hairiness of the male parameres, as discussed in the species account of *fulva*. There is, regrettably, no absolute reason why *pubens* could not be a more hairy-tailed form of *fulva*, perhaps with a different geographic origin. The genetic evidence (Gotzek et al. 2012) is not as clear cut as one might like, with a couple of outlier *fulva*, and genetic testing of only one sample of *pubens* associated with males.

Distribution: The distribution of *pubens* is unclear because of the apparently unstable taxonomy of the *fulva* complex. The only definite North American specimens are from the Miami area. Wetterer and Keularts (2008) suggested that *pubens* is not native in the Caribbean, but that suggestion actually referred to exploding populations of *fulva*, as Wetterer et al. explain in a later paper (2014). The types of *pubens* (with males) are from St. Vincent, and there is a male-associated record from Anguilla (Gotzek et al. 2012). These islands in the Lesser Antilles seem to share most species of ants with Trinidad and mainland South America, and lack almost all the Caribbean endemic species associated with the Greater Antilles. If *pubens* is native to the Lesser Antilles rather than being imported from some distant area, it should occur in Venezuela. In south Florida, *pubens* might have been displaced by *fulva* but there is no evidence of this.

Natural History: *Nylanderia pubens* in southern Florida appears to be restricted to disturbed areas. Most published information about *pubens* may refer to *fulva*. Any study of the natural history of *pubens* would need to start with definitive identification of the species observed, either by genetic analysis or by associating males and workers. Wetterer et al. (2014) consider population density to be a useful preliminary indicator to distinguish *pubens* from *fulva*, as *pubens* is not known to go through cycles of boom and bust. One reason for studying the natural history of *pubens* might be to compare it with that of *fulva*.

Name Derivation: From *pubens* (Latin), meaning "pubescent" or "hairy." The English name "hairy crazy ant" was suggested by Wetterer and Keularts (2008) to replace the sporadically applied "Caribbean crazy ant."

Nylanderia steinheili (Forel)—
West Indian Crazy Ant (Plate 85)

Taxonomy and Similar Species: *Nylanderia steinheili* is easily distinguished from other Florida *Nylanderia* by its contrasting whitish coxae, combined with abundant short pubescence on the mesosoma (best seen in lateral view against a dark background). This species appears under the name of *guatemalensis* in Trager's revision (1984). It is not clear whether *guatemalensis* ever occurred in Florida (Kallal and LaPolla 2012).

Distribution: West Indies and southern Florida. In Florida, *steinheili* was originally known only from Miami and Homestead (Trager 1984), but it now occurs sporadically as far north as Tampa (Hillsborough County), and appears to be rapidly expanding its range.

Natural History: In Florida, *steinheili* is usually found in shaded or mostly shaded sites. It may be abundant in the Florida Keys, where there is a prolonged dry season, but it also occurs farther north in the south-central Peninsula, where the dry season is less extreme. I have often found incipient colonies in hollow twigs buried in leaf litter. Larger colonies also tend to have their headquarters in a hollow twig, although parts of the nest may be spread out through adjacent leaf litter. This is exactly the nest site favored by *faisonensis*, and it seems likely that *steinheili* will replace *faisonensis* in southern Florida, where *faisonensis* is somewhat uncommon to begin with. Relatively cool winters with occasional freezes may exclude *steinheili* from northern Florida, but this is not certain. Several introduced tropical ants have ranges extending north into Alachua County or further. At first glance, this species, which nests outdoors and is unable to bite or sting, seems to have little potential as a pest.

Name Derivation: There is a sad story associated with the name of *Nylanderia steinheili*. Edouard

Steinheil was a German entomologist who had already returned from a collecting trip to northern South America when he met August Forel. In 1878, the two enthusiastic entomologists set off on a six-month collecting expedition to the West Indies and South America. Unfortunately, when they reached St. Thomas, Steinheil suddenly died of apparent heat stroke, and Forel immediately returned to Europe to inform the Steinheil family (Banani 2005). The name *steinheili* given by Forel to three species of Caribbean ants was undoubtedly intended as a tribute to Edouard Steinheil. This story has an upbeat sequel: some years later, Steinheil's daughter Emma married Forel; the couple had six children, with the oldest one named Edouard (Banani 2005).

Nylanderia vividula (Nylander)— Field Crazy Ant (Plate 86)

Taxonomy and Similar Species: *Nylanderia vividula* is a small, dark species that could be mistaken for *parvula*, *concinna*, or *faisonensis*. The large hairs on the antennal scapes of *vividula* separate it from *parvula*. Relative to *concinna* and *faisonensis*, the head of *vividula* is less hairy, with the hairs of the preoccipital area (posterior part of the head in front) usually separated at least by the length of one of the hairs. Although this distinction seems abstruse, the unusually shiny (less hairy) head of *vividula* can even be spotted with low magnification or in the field.

Distribution: In North America from North Carolina into Florida, west into Texas, Kansas, California, and Mexico (Trager 1984). It is probably native in Texas and Mexico, exotic in southeastern North America, Bermuda, Chile, southern Europe, and the Middle East; it has long been established in greenhouses in northern Europe (the types are from a greenhouse in Finland) (Trager 1984). Most of the relatively small numbers of Florida records are from northern Florida, but there are records from Highlands County.

Natural History: In the Southeast, *vividula* is usually found in fields and lawns, but may occur in other disturbed, open habitats. Nests may be subterranean or in cavities or debris close to the ground. It is more likely to be found in soils with significant amounts of clay or organic matter than in pure sand (Graham et al. 2008). It shares its habitat with *Solenopsis invicta*; increases in the populations of that species appear to be correlated with a decline in populations of *vividula* (Shawler

et al. 1989; Wojcik 1994). The future of *vividula* in Florida is uncertain, but perhaps *S. invicta* will decline in dominance, or there might be some ecological zone where the two do not seriously overlap. In any case, if *vividula* is exotic in Florida, its loss from the fauna would not be a blow to our myrmecodiversity.

Name Derivation: From *vividus* (Latin), meaning "lively." In his description of the species, Nylander (1846) notes its lively but timid disposition as it emerged from cracks in the cement to eagerly investigate plants in the Helsinki greenhouse where it was first found.

Nylanderia wojciki (Trager)— Pine Woods Crazy Ant (Plate 87)

Taxonomy and Similar Species: This species and *parvula* differ from other Florida *Nylanderia* in lacking a series of large setae on the antennal scapes, although there may be a few such setae on the scapes of *wojciki*. In Florida, *parvula* is very dark brown or blackish, while *wojciki* is pale brown or bicolored, and the mesosoma is lighter than the head and gaster. West of the Mississippi, *parvula* may be pale or bicolored (Trager 1984). This species is unusually small for the genus, under 2 mm in length (Trager 1984). I have seen specimens from Alabama pine woods that have numerous large setae on the antennal scapes, but otherwise appear to be *wojciki*.

Distribution: Throughout Florida, including the Keys; also known from Alabama. It has been collected on the eastern Florida–Georgia border, and must occur in southern Georgia.

Natural History: This species is usually found in open pinelands, both wet (pine flatwoods) and dry (high pine or sandhills). It might be a southeastern relative of *parvula* that has adapted to frequently burned landscapes. Nests are in grass clumps, rotten pine logs, thick leaf litter, and layers of loose bark at the bases of pine trees. Alates appear in the nest in fall and fly in spring (Trager 1984). At the Archbold Biological Station (Highlands County), *wojciki* is host to a workerless parasitic *Nylanderia* with flightless males, discussed below.

Name Derivation: James Trager (1984) named *wojciki* for a myrmecologist at the Medical-Veterinary Laboratory in Gainesville, Florida: "This species is named for Dr. Daniel P. Wojcik, who contributed the types and who has, through providing access to his literature files and collecting specimens for me, substantially aided this study."

Nylanderia, Undescribed Parasitic Species A (Plate 88)

Taxonomy and Similar Species: This appears to be a workerless parasite of *wojciki*. Males are flightless, unlike any other known Florida *Nylanderia*. They probably never leave the host nest. Females resemble miniature queens of *wojciki*, but have straight antennal scapes that are scarcely thickened apically, and the subapical rows of large, dark setae on tergites 2–5 on *wojciki* are replaced by short pale setae in parasite queens. Queens taken in flight traps do not have an enlarged gaster like those of dispersing queen *wojciki*. This species is similar to another undescribed parasitic species that has been collected from nests of *parvula* in Massachusetts. Both parasites have reduced setae relative to their hosts, but the reduction seems more extreme in the *wojciki* parasite. We have little information on variation in these parasites because they have been collected so rarely and from such widely separated sites. It is possible that a single species of parasite is associated with both *wojciki* and *parvula*, but for now, I am considering them separate species.

Distribution: Known from the Archbold Biological Station, and from two protected areas east of Sebring, Highlands County, Florida. It might have a wider distribution in Florida. The easiest way to get additional site records would be to examine flight trap catches from upland areas in fall.

Natural History: This species is known from Florida scrub habitat. Queens and the flightless males have been found in nests of the host. Queens are collected in flight traps in the fall. All specimens are from the Lake Wales Ridge, a sand ridge more than one million years old, the site of an unusual number of narrowly distributed endemic species (Turner et al. 2006). It is possible that *Nylanderia* species A is another endemic species confined to the Lake Wales Ridge.

It is reasonable to consider whether this supposed parasite might actually be an unusual form of *wojciki*. Some species of ants produce more than one form of reproductives. This is well known in the genera *Cardiocondyla* and *Hypoponera*. Moreover, ants generally have at least two female forms, a worker and a queen, and the developmental pathways occasionally go awry to produce intermediates. Queens of species A from flight traps at the Archbold Biological Station were examined by Trager when he was preparing his revision of U.S. *Nylanderia* (1984). Trager suggested (1984) that these specimens were microgynes, intermediate between workers and queens. He pointed out that these queens, even if they were functional, would have difficulty founding colonies because they fly late in fall, while males fly in spring. The subsequent discovery by Stefan Cover of wingless males in a nest of *wojciki* with small alate queens led to the hypothesis that these reproductives represented a workerless parasite. Furthermore, while the queens are intermediate in size between workers and queens of *wojciki*, they differ from both in structural features, especially the reduced setae on the gaster and the straight antennal scapes. The parasitic species B, discussed below, is more divergent from its host and provides a kind of precedent within *Nylanderia*. I suppose that it is still possible to hypothesize that some *Nylanderia* species have two reproductive forms, one of which is a small, low-budget, sib-mating form, but this would be more unlikely and novel than the occurrence of conventional workerless parasites.

Nylanderia, Undescribed Parasitic Species B (Plate 89)

Taxonomy and Similar Species: This undescribed species appears to be a workerless parasite of *faisonensis*, with a miniature queen and an almost normal-sized male. Both males and queens are much hairier than those of other native *Nylanderia*, with abundant large setae on the gastral tergites. The antennal scapes are straight, rather than curved, and the mandibles lack teeth. Males of some other *Nylanderia* species have negligible dentition, but the lack of teeth in the female appears to be a unique feature of this species.

Distribution: Known from a series of males and females taken in a nest of *faisonensis* in Hamilton County, Florida, and from a single female captured in a flight trap in Georgia.

Natural History: Several males and females were found in a nest of *faisonensis* in a rotten log in early July. The habitat was upland oak forest near a swamp thicket.

There is little doubt that this is a parasitic species rather than some form of *faisonensis*. The straight antennal scapes, toothless mandibles, and numerous, well-distributed, large hairs are all features lacking in alate *faisonensis* and in no way intermediate between workers and reproductives. The Georgia specimen also shows that this is not some one-time weird mutant. It seems unlikely that Species A evolved from Species B, or the reverse, because A lacks the derived structural features of B, while B has wings, which have been lost in A. No details are known of the biology of this species.

Genus *Paratrechina*

Longhorn Crazy Ant

The genus *Paratrechina* (pronounced **Par' ah trek ī" nuh**) consists of a single species, *longicornis*, originating in the Old World tropics. Until recently, numerous species now placed in the genus *Nylanderia* were included in *Paratrechina*. Phylogenetic analysis suggests that *Paratrechina longicornis* is only distantly related to species now deposited in *Nylanderia* (LaPolla et al. 2010). Distinguishing features of *P. longicornis* are listed below.

Name Derivation: From *para* (Greek), meaning "near" or "close," and *trechina* (Greek), meaning "little runner." Loosely interpreted, this could be expressed as "little one that runs all around." This genus was described in 1863 by Victor Ivanovitch von Motschulsky (1810–1871), an officer in the Russian Imperial Army. Motschulsky is better known, at least in the United States, as a describer of beetles. His descriptions, as usual for his times, are often insufficiently diagnostic to stand alone for recognition of species, but fortunately his collection survives. Like many early entomologists, Motschulsky did not routinely explain derivation of names he proposed for genera or species.

Paratrechina longicornis (Fabricius)— Longhorn Crazy Ant (Plate 87)

Taxonomy and Similar Species: A highly distinctive species with antennal scapes that are about twice the length of the head (including the jaws), proportionately long legs, a flat mesosomal profile and slender body, as well as the enlarged hairs also found in most *Nylanderia*. The somewhat similar *Nylanderia bourbonica* is more robust, with a convex mesosoma, proportionately shorter antennae, and a shining dark brown body.

Distribution: The following information on distribution is from an exhaustive study of the topic by Wetterer (2008a), who also suggested the English name of "longhorn crazy ant." This species occurs outdoors throughout the tropics and subtropics. It is not uniformly distributed through its range and is most likely to be found in coastal areas or cities. It has colonized greenhouses, zoos, and other buildings throughout the world. *Paratrechina longicornis* originated somewhere in the Old World tropics, probably Melanesia or Southeast Asia. It is a habitual stowaway in ships and vehicles, probably achieving much of its diaspora in the early days of world exploration, long before any surveys of ants. In Florida, *longicornis* occurs throughout the state, but is commonest in south and central Florida.

Natural History: This species is normally found in highly disturbed, man-modified habitats, but in Florida and elsewhere, it often occurs on beaches, which are naturally highly disturbed and productive, possibly the natural habitat of *longicornis* in its primeval range. Jaffe (1993) describes a dense population of colonies nesting in a wide, sandy intertidal zone in India. On a Venezuelan beach, Jaffe found *longicornis* foraging along the edge of the water, collecting small bits of animal detritus. When caught by a wavelet, the ants retracted their appendages and rode the surf up the beach until the water was absorbed by the sand, whereupon the ant unfolded and resumed foraging.

Paratrechina longicornis is a well-known pest species that infests the environs of buildings. Like other pest ants, it has ecological requirements that are fulfilled unintentionally but generously by humans. This species is especially common around buildings, even though the nests are usually outdoors; such species are sometimes called "peridomestic." It is normal to find a thriving population dominating the walkways and landscape plantings around a building, while in surrounding natural or landscaped areas a dozen meters away, there is not a *longicornis* to be found. The actual nest site is usually dry, but not as dry as the interior of most buildings, secure against predators, sheltered from flooding and cold, and spacious enough for a colony of at least a thousand ants. Nests are sometimes in the ground and enlarged by excavation but more frequently in piles of trash, in leaf bases of palm trees, or in deep mulch adjacent to a foundation.

Paratrechina longicornis seems to be programmed for a constant quest for an ideal nest site. Colonies, or large fragments of colonies, are often seen on the move, with dealate queens, males, and hundreds of workers toting larvae or pupae all racing in one direction, while an adjacent stream of workers runs back the other way for more larvae or pupae. It is easy to envisage longicornis colonies moving into containers and piles of supplies awaiting transport near the beach, and as quickly moving out again when unloaded at a distant port. Colonies frequently move into cars and trucks, and this may be a prime mode of dispersal today. On several occasions, I have used the home-upgrading instincts of longicornis to deal with colonies in a vehicle by placing a bucket of slightly moist leaf litter in the car for a day or two, removing the bucket after the ants have established residence. A tiny cricket, Myrmecophilus americanus, that lives with longicornis moves as readily as its host, and can be seen running along in columns of relocating ants. This cricket, which is almost as widespread as longicornis, solicits food from worker ants and feeds on food stored in the nest (Wetterer and Hugel 2014). Although longicornis relocates readily, there is no reason to believe that it has regular migrations or a nomadic lifestyle.

Myrmecologists fortunate enough to have a colony of longicornis patrolling the front walkway get to observe the remarkable foraging behavior of this species. Individual foragers are speedy and bold, traversing long distances, including vertical objects such as tree trunks or the legs of sitting humans. When a large resource is discovered, the forager hurries home, leaving a trail that is quickly followed by a band of recruits. Large dead or moribund arthropods, such as a large spider many times the weight of an ant, are quickly and efficiently carried back whole by a group of well-coordinated workers. Sizeable arthropods can even be hauled up vertical surfaces. The communication needed for these foraging feats has been studied by Witte et al. (2007). At least four pheromones are involved: a long-lasting but only mildly stimulating pheromone from the rectal gland, two less persistent and more stimulating compounds from the poison sac and Dufour gland, and a pheromone evoking defensive behavior from the mandibular gland. The rectal pheromone directs workers toward more stable resources such as honeydew-producing scale insects. The other three pheromones cause large groups of workers to approach, handle, and defend large, ephemeral resources. The secretion of the mandibular gland is responsible for the defense/repulsion reaction (Witte et al. 2007) that has given the whole genus the epithet of "crazy ants." Rapid, erratic movements appear to have two consequences: individuals are difficult to catch, and sooner or later they

encounter the threat, which can be mass-attacked if it is of appropriate size.

Efficient and preemptive resource use by longicornis would seem to give it a competitive advantage over many other ants, but a good variety of additional ants usually co-occur with longicornis in both natural and disturbed habitats. Greenhouses, with their limited fauna and stable environmental conditions, appear to be perfect for longicornis. Hordes of this species, subsisting largely on honeydew, dominated the insect fauna of Biosphere II, a supposedly sealed and self-supporting artificial ecosystem (Wetterer et al. 1999). We can thank longicornis for showing us the folly of imagining that we are able to create isolated, utopian ecosystems, and for saving us, at least temporarily, from more general production of such ultimate gated communities.

Although it neither bites nor stings, longicornis can be a genuine nuisance when it enters buildings in force. Colonies are usually located outdoors, and the infestation can often be controlled by sealing off access, or laying down a perimeter of insecticide around the outside of the opening that is admitting ants. Occasionally, longicornis relocates its colonies indoors, usually in a wall void; control of such infestations usually requires identification and treatment of nest sites (Hedges 1998). In Florida, longicornis is most likely to be a problem around outdoor eating areas, but in the tropics, the boundary between indoors and outdoors is often unclear, allowing ants to pass more freely in and out of living areas. This may be true even in relatively controlled buildings: in a survey of hospitals in Brazil, Bueno and Fowler (1994) found longicornis in 9 out of 15 hospitals; all these hospitals harbored ants, ranging from 10 to 20 species.

Colonies usually reproduce by budding or fission. Queens may have their wings removed while callow and mate with males just outside the nest entrance (Trager 1984). It is possible that some or all populations of this species are so dependent on large colony size for foraging and colony relocation that nest founding by individual queens has been eliminated by natural selection. Workers from well-separated colonies are not compatible, suggesting that there can be divergence in identifying nest odor, caused by either genetic or dietary factors (Lim et al. 2003). If mating normally occurs at the nest entrance (Trager 1984), there might be opportunities for outbreeding with males from other nests.

Name Derivation: From *longus* (Latin), meaning "elongate," and *cornu* (Latin), meaning "horn," referring to the excessively long antennae.

Genus *Polyergus*

Slave-Maker Ants

The genus *Polyergus* (pronounced **Pol' i ur" gus)** is a Northern Hemisphere genus of only 14 species, although this is a large increase from five species before James Trager's 2013 taxonomic revision of the genus. This revision also includes a helpful summary of the natural history of the genus. Since I had put off attempting to understand this genus for many years, it was a great relief to see Trager's excellent revision, and further evidence that procrastination sometimes pays off. At the generic level, these ants are easily recognized: they look at first glance like large species of *Formica*, but when their head is viewed face-on, as seen perhaps, by one of their victims, their jaws are long and sickle-shaped, with small teeth along the apical two-thirds of the inner edge. There is no other Florida species with similar jaws. The lower part of the face is broad, perhaps associated with the ability to spread the jaws widely, and the head somewhat elongated behind, perhaps to accommodate unusually strong jaw muscles. Almost all species are bright reddish brown with long appendages; Wheeler (1910a) calls a European species "one of the most beautiful of ants."

The slave-raiding behavior of *Polyergus* species, as well as their total dependence on kidnapped workers of other species for all the maintenance jobs of the colony, has intrigued myrmecologists for more than 200 years, more or less since the beginning of myrmecology. This early work was summarized by Wheeler (1910a) in his classic book *Ants*, in which there is a whole chapter devoted to *Polyergus*, which he called "amazons," after the mythical society of female warriors. (Wheeler was probably annoyed by the common propensity of the general public to assume that among social insects any strongly defensive or aggressive individuals must be males.) Wheeler, who made many of his own observations on *Polyergus* colonies, repeated earlier studies that showed that *Polyergus* workers are unable to excavate nests, care for their young, or find their own food. All these functions are carried out by *Formica* workers that are brought into the nest as pupae, integrating themselves into the colony after emerging. Wheeler was struck by the contrasting behavioral modes of *Polyergus* workers: "While in the home nest they sit about in stolid idleness or pass the long hours begging the slaves for food or cleaning themselves and burnishing their ruddy armor, but when outside the nest on one of their predatory expeditions they display a dazzling courage and capacity for concerted action...." Raids, which normally occur in the afternoon, begin with a sudden emergence of ants that assemble around the nest entrance and then move out in a rapidly moving column. Although these columns occasionally seem to have some difficulty finding a nest to raid, usually they move speedily and directly to a nest, pour into it, and quickly begin to emerge carrying worker pupae. Resisting host ant workers are easily incapacitated, as the strong curved jaws of *Polyergus* can easily pierce the head and mesosoma of *Formica* defenders. Wheeler found that species of *Formica* that put up little resistance are generally left almost completely unharmed.

Early myrmecologists such as Forel and Wheeler believed that the group raids followed discovery of *Formica* nests by scouts, and Wheeler made observations that confirmed this. Scouts were also reported by Talbot (1967), who noted that searching scouts wandered about, while scouts returning from a *Formica* nest came directly back along a path later followed by the raiding column. Topoff et al. (1987) discovered that a scout leaving its nest first moves out away from the nest in a linear path and then begins a circuitous searching phase, returning more or less directly to its nest once a suitable colony of *Formica* has been found. Scouts then lead the raiding column to the *Formica* nest. The column of outgoing *Polyergus* sometimes becomes stalled, whereupon an ant that had gone on ahead returns and recruits the raiders back in the right direction (Trager 2013). Until recently, it seemed reasonable to assume that after a scout finds a *Formica* colony, it would lay down a trail back to the *Polyergus* nest, analogous to the trail left by other, nonparasitic ants that have encountered a large food resource (Marlin 1969). This, however, does not seem to be the case. Studies of *P. topoffi* show that scouts use optical cues, especially polarized light, to return to the nest and to lead the raid back to the host nest, although the raiding column itself leaves a chemical trail (Topoff et al. 1984, 1985). It may be that depositing a chemical trail leading to a resource never evolved in the *Polyergus* lineage; it is also possible that since the success of *Polyergus* raids is strongly dependent on surprise, a chemical warning of a raid in the near future would be unadaptive and unlikely to evolve.

Colony founding in *Polyergus*, as assumed by Wheeler (1910a) and earlier myrmecologists, is completely dependent on taking over a colony of *Formica* whose workers will provide essential services. The exact mode of usurpation, however, seems to be variable, possibly according to the species of *Polyergus* or the species of host *Formica*. At least some species are able to form nests by "budding" (Marlin 1968). This occurs when one or more dealate inseminated queens remain in a raided nest with a few *Polyergus* workers and the remaining *Formica* workers. These queens have previously mated and returned to their nest, ready to go out on a raid (Topoff and Greenberg 1988). In at least one species, dealate queens in a raiding column use pheromones to attract males, so by the time the target colony has been reached and subdued, one or more *Polyergus* queens is ready to move in and replace the *Formica* queen; eventually, only one *Polyergus* queen is left in the newly formed colony (Topoff and Greenberg 1988). Solitary dispersing *Polyergus* queens can also take over a *Formica* colony, moving cautiously into the colony without fighting opposing workers until reaching the queen, which the *Polyergus* queen kills, remaining on her victim until she has absorbed the host scent and is accepted by the colony (Topoff and Zimmerli 1993). This usurpation may be facilitated, at least in the case of *P. topoffi*, by a pheromone from the Dufour's gland that appears to reduce the defensiveness of the host workers toward the invading queen (Topoff et al. 1988). If this "appeasement pheromone" is host specific, that might help explain why species of *Polyergus* often have a single host *Formica* and tend to have local host races. It appears that there may be another mode of colony usurpation when the *Polyergus* queen enters a recently founded *Formica* nest with a queen and a few workers. In this case, the invader queen allows the host queen to continue to produce offspring until the colony is larger and stronger, after which the host queen is killed (Trager 2013).

Polyergus species are scattered around the Northern Hemisphere, showing a distribution similar to that of *Formica*. Within this area, however, *Polyergus* species are often rare, and in some areas completely absent, for example, northern New England and most parts of the Appalachians. Through its wide geographic range, *Polyergus* species are so similar to each other that it has been difficult to distinguish species; there could still be undescribed species among the host races of *Polyergus*. This makes *Polyergus* an odd and intriguing lineage: its highly distinctive morphology and behavior suggests that the genus is very old, but all the species look as if they diverged from each other the day before yesterday. One possibility is that there are a few more basal and generalist lineages within the genus that are constantly producing host races that may solidify into highly specialized species, many of which eventually go extinct.

There has been some objection to using the traditional name "slave-makers" for these ants, based on the legitimate emotional baggage associated with the word *slave* (Trager 2013). Trager and some others prefer to use the word *dulotic*. Since this word means "slave-holder" in Greek, I am even less comfortable with this supposed compromise, as it retains the slave connotation, compounded with an elitist assumption that it won't cause offense because only a few highly educated people know any ancient Greek. As Trager (2013) points out, the analogy between the behavior of these ants and human slave-makers is imperfect. This is also true, however, of army ants, and weaver ants, to say nothing of all the general biological terms that we use divorced from their human social context, for example, mimicry, parasitism, dancing, singing, courtship, and parental care. "Slave-maker ants" has a useful functional meaning applied to a genus of ants that steals the young of another species to obtain workers that will do all the tasks of a normal nonparasitic ant colony. We should not be worrying about ethics being applied to insects, because ethical models are totally inapplicable to insects. We should be worrying about our own ethics, which are inconsistent, problematic, and evolving. The only ethical dilemma posed by the slave-maker ants of the genus *Polyergus* is that some of its species appear to be so rare that they should arouse our newfound ethical sensitivity about the fact that we are on the verge of annihilating much of the biodiversity of the planet Earth.

Name Derivation: From *polys* (Greek) meaning, in this case, "very," and *ergon* (Greek), meaning "work," supposedly meaning "hardworking." When these ants are out in public, so to speak, they are always dashing off on a raid or rushing back to their nest burdened with their prey, so they are certainly hardworking; otherwise, they hardly work. The work ethic, of course, does not apply to ants any more than other kinds of ethics.

Polyergus longicornis M. R. Smith—
Long-Horned Slave-Maker Ant (Plate 90)

Taxonomy and Similar Species: *Polyergus longicornis* is distinguished from other Florida species by the relatively abundant numerous, short, thick, black setae on the upper part of the head; there are at least 20 of these setae, and only a few in other Florida *Polyergus*. Sometimes, a number of these setae get knocked off, but the pits from which they originated are still obvious.

Distribution: North Carolina south into the Florida Panhandle, west into Mississippi (Trager 2013). In Florida, this species is known only from Leon, Walton, and Santa Rosa counties. In Walton County, *longicornis* was found in recently burned areas where their nests were largely exposed, and it was easy to trail host ants back to the nest. Searches for *longicornis* and other Florida *Polyergus* might be more likely to succeed if they concentrated on recently burned areas.

Natural History: *Polyergus longicornis* is closely associated with *Formica dolosa*, which is apparently its only host (King and Trager 2007; Trager 2013). As far as is known, the biology of *longicornis* resembles that of other *Polyergus*, except that it is behaviorally and probably chemically specialized to enslave *F. dolosa*. It seems probable that a population of *longicornis* can only persist where there is a dense population of *dolosa*. In the southeastern part of the range of *dolosa*, frequent fires are required to maintain the open sandhills and dry flatwoods where *dolosa* occurs. This means that *longicornis* may have an increased chance of local extirpation with a decrease in the amount of extensive, open, regularly burned natural habitat. In Florida and southern Georgia, efforts to protect the red-cockaded woodpecker, especially in well-drained habitats, might benefit *longicornis*.

Name Derivation: From *longius* (Latin), meaning "longer," and *cornu* (Latin), meaning "horn," referring to the relatively long antennal scapes.

Polyergus montivagus Wheeler—
Wandering Slave-Maker Ant (Plate 90)

Taxonomy and Similar Species: *Polyergus montivagus* is distinguished from the Florida species *longicornis* by the small number of conspicuous, black, stout setae on the upper part of the head; there are only a few such setae on the head of *montivagus*, and 20 or more on *longicornis*. It is much more difficult to distinguish between *montivagus* and *oligergus*. The queen of *montivagus* is larger than that of *oligergus*, and there are some small differences in average measurements of workers (Trager 2013). The easiest way by far to identify *montivagus* is by association with its host, *Formica pallidefulva*; *oligergus* is associated with *F. archboldi* (Trager 2013). This species was formerly combined with *lucidus*, and a series of biological studies of *montivagus* were published under the name *lucidus* (Trager 2013).

Distribution: Similar to that of its host, *F. pallidefulva*: southeastern Canada into Florida, west across the Great Plains and in the lower elevations of the Rocky Mountains from Wyoming into New Mexico (Trager 2013; Trager et al. 2007). Populations of *montivagus* are rare or scattered over this range and tend to be associated with areas with sandy soil (Trager 2013). In Florida, *montivagus* is only known from Leon County, even though *F. pallidefulva* is relatively common through much of the state. It is quite possible, however, that *montivagus* occurs in scattered localities elsewhere, for example, in the Osceola National Forest. *Polyergus* colonies are difficult to detect except when workers are out raiding, but these raids usually occur in late afternoon, around the time that entomologists are leaving the field, heading back for a shower, tick removal, and dinner.

Natural History: Unlike the other two Florida *Polyergus* species whose hosts require extensive natural habitats kept open by fire, *montivagus* is associated with *F. pallidefulva*, a more generalist species of open areas, including pastures, fields, and lawns. Talbot (1967), for example, studied a Michigan colony that she called the "Lawn Colony," which had persisted for at least a decade outside her laboratory at the George Reserve. Although rare in Florida, this species is probably of less conservation concern than some other *Polyergus* species because of its extensive range and adaptable host.

Raiding behavior of *montivagus* was studied by Marlin (1969) in Illinois; Marlin used the name *P. lucidus lucidus*, but his specimens were later determined as *montivagus* by Trager (2013). Marlin (1969) summarized his observations of several raids. Scouts leaving the *montivagus* nest wandered about until they found a foraging trail of the host, which they followed to the nest, returning back by a direct route to their own nest. A returning scout recruited additional scouts, which also visited the Formica nest; these apparently settled on a kind of consensus direct route between the two nests. Marlin was convinced that the scouts laid down a scent trail and the raiding colony eventually followed this trail. He did not, however, perform the kinds of experiments that led Topoff et al. (1984, 1985) to the conclusion that scouts of the western species topoffi use optical cues to determine the shortest route to the nest, and the raiding column follows these scouts rather than a chemical trail. Returning scouts seemed to excite "milling behavior" around the montivagus nest entrance, which usually leads to a mass exodus of workers that then form a coherent column. Such a column may lose its direction and form a traffic jam of milling workers that resumes marching only after scouts redirect the raiders. Marlin made the logical assumption that a raiding column becomes confused after it has lost contact with a scent trail, but in the light of studies of other species, the confusion may signal a loss of contact with scouts en route to the host nest. If a stalled raiding column receives no further direction, it eventually returns in disarray to the montivagus nest. On arriving at a pallidefulva nest, the raiders usually milled about a bit before pouring into the nest. Sometimes, the raiding column runs by the nest and begins to mill about beyond it. Marlin developed a plausible explanation for this: the scouts mark the nest entrance itself, but the trail to the nest dead-ends a little past the nest, ensuring that the ants in the column will pile up together for a mass attack on the nest. Considering the work by Topoff et al. (1984, 1985), it seems likely that montivagus scouts are using optical cues to lead the column to the host nest, as in topoffi, and the milling about may be because the scouts have reached their goal. Marlin writes, "Like most tactical maneuvers by P. lucidus, the surrounding of the enemy nest is a blind response to the scent trail." The instincts of insects may be blind, but they can still fool us from time to time. Marlin presents evidence that montivagus returning from a raid follow a scent trail that can be easily erased or interrupted, and only creative experimentalists like Howard Topoff and his collaborators would suspect that Polyergus raids involve one kind of orientation outbound and another when returning from the raid.

Flights of queens and males were repeatedly observed by Talbot (1968) and Marlin (1968, 1971). Usually, males departed first, followed sometime later by smaller numbers of queens. Marlin (1968) saw presumably mated dealate queens returning into the nest, and even found queens entering nests other than their own. Returning queens may have the opportunity to found a nest by budding (Marlin 1968).

Polyergus montivagus nests are often infested with the tiny thief ant Solenopsis molesta, probably preying on larvae or pupae of montivagus or F. pallidefulva (Marlin 1971).

Name Derivation: From mons (Latin), meaning "mountain" and vagus (Latin), meaning "wandering." The types were collected in canyons of the Rampart Range of the eastern Rocky Mountains (Wheeler 1915).

Polyergus oligergus Trager—
Florida Slave-Maker Ant (Plate 90)

Taxonomy and Similar Species: Polyergus oligergus is distinguished from the Florida species longicornis by the small number of conspicuous black setae on the top of its head: there are only a few in oligergus, whereas there are 20 or more in longicornis. It is much more difficult to distinguish between oligergus and montivagus. The queen oligergus is conspicuously smaller than that of montivagus, the workers are less noticeably smaller, and there are some small differences in average proportions of the workers (Trager 2013). The easiest way to separate these two species is by their hosts: oligergus enslaves Formica archboldi, and montivagus enslaves F. pallidefulva. Since Polyergus is always associated with a host ant, identification should not be a problem unless workers are collected from a raiding column.

Distribution: Polyergus oligergus is known from several counties in northern Florida and from Osceola County in central Florida (Trager 2013). This species appears to have the smallest geographic range of any North American species of Polyergus, and within that range, there must be relatively few sites where its host is abundant enough to support the population of a slave-maker. Fortunately, it occurs on several protected sites, including San Felasco State Park, the Ordway Preserve, The Nature Conservancy/Disney Wilderness Preserve, and the Apalachicola National Forest (Trager 2013). Within these sites, however, it may be rare: the colony studied by King and Trager (2007) was the only one found in two years of inspecting hundreds of Formica colonies through a huge swath of the Apalachicola National Forest.

Natural History: The biology of *oligergus* has been the subject of several studies (King and Trager 2007; Trager 2013; Trager and Johnson 1985). It is found in both sandhill and flatwoods habitat, usually associated with a more or less dense ground cover of herbaceous perennials, especially bunch grasses such as wiregrass (*Aristida beyrichiana*). Raids occur in June and early July, always in the afternoon. Worker *F. archboldi* flee when *oligergus* raiders enter the nest; in one case, a group of only four *oligergus* were able to successfully raid a colony of more than 100 *F. archboldi*. More normal raids are made by approximately 25–50 workers; this is a relatively small number for a *Polyergus* species and suggests that the total number of *oligergus* in a normal nest is also relatively small. Flights of alates have been observed in late morning on hot, sunny days in June and early July.

Name Derivation: "This species has the smallest worker populations of any *Polyergus* species. The name stems from Greek, *olig-* (few) plus *erg-* (work), roughly meaning few workers, and alliterating neatly with the genus name" (Trager 2013). While most alliterations, however, trip off the tongue, this one gurgles.

Genus *Prenolepis*

Fat-Belly Ants

The genus *Prenolepis* (usually pronounced **Pren' ō lep" iss**) is a small genus whose boundaries are currently unclear. It certainly includes, in addition to the European type species *nitens*, the North American species *imparis*, which strongly resembles *nitens*. *Prenolepis imparis* is the only known Nearctic species. This species stores fat in its distended gaster, hence the name "fat-belly ants."

Name Derivation: From *prenes* (Greek), meaning "prone" or "facing downward," and *lepis* (Greek), meaning "scale," probably referring to the small, shining, appressed hairs (chiefly on the head) of the type species.

Prenolepis imparis (Say)—Fat-Belly Ant (Plate 80)

Taxonomy and Similar Species: Distinguished from other small formicines such as *Lasius* and *Nylanderia* by the constricted "waist" between the anterior and posterior parts of the mesosoma. The legs and antennae are long and slender compared with those of *Lasius*, and the long, curved standing hairs, while positioned like those of *Nylanderia*, are not thickened as in *Nylanderia*. *Prenolepis imparis* shows some variation in size, color, and the slope of the propodeal declivity. This, combined with its anomalously extensive geographic range (almost all of temperate North America), has tempted various myrmecologists to describe a series of variants (Wheeler 1930). These were labeled "subspecies" before the geographic definition of subspecies became established, and most were emphatically rejected by Creighton (1950), often on the grounds of sympatry. It is still possible that there are cryptic species within *imparis*, as the situation has not been subjected to modern analysis.

Distribution: Ontario south into northern Florida, west into Oregon, south into the mountains of Mexico (Wheeler 1930). In Florida, *imparis* is known from Citrus and northern Orange counties northward, and west through the Panhandle. The historical biogeography of *imparis* was discussed by William Wheeler (1930), who was remarkable for the breadth of his intellectual curiosity, even (or perhaps especially) by today's standards. Wheeler was impressed by similarities between *imparis*, *nitens*, and the fossil species *henschi* found in Oligocene Baltic amber. He decided that each of these species represented little more than a "mere variety" of an ancient conservative lineage, a relict group from the temperate Boreal forest that was displaced by Pleistocene glaciation. Expanding on earlier observations of Emery, Wheeler was apparently the first myrmecologist to suggest that lineages of forest ants were eradicated in northern Europe by being pushed up against the Alps during episodes of glaciation, but survived in North America because of the north–south orientation of Nearctic mountain ranges. Although *imparis* and *nitens* seem to be obviously distinct species rather than "varieties," Wheeler's contention that the two represent a conservative, relict boreal lineage could be correct.

Natural History: The natural history of *imparis* has received considerable attention, partly, I suspect, because it is active in winter or very early spring when northern myrmecologists come prowling out in the throes of cabin fever. In addition to a long paper by Wheeler (1930), there are studies by Dennis (1941), Talbot (1943a, 1943b), Tarpley (1965), and Tschinkel (1987). Tschinkel's paper is the most applicable here, as the study was done in northern Florida. Tschinkel's study of the seasonality of *imparis* is summarized in the following paragraph.

The life history of *imparis* is unlike that of any other Florida ant. Foraging and nest excavation in are restricted to the coolest months, November through March or early April. In November, the colony consists of pupae (including those of both future workers and alates), recently developed callows, foragers, and multiple functional queens. Through the winter, foragers bring back food in liquid form, feeding it to the callows, which transform it into lipid, stored in an enormously expanded fat body. In older literature, these individuals were called "repletes," like the honey-pot ants of the genus *Myrmecocystus*, which store carbohydrates. Tschinkel points out that this is a completely different form of resource storage and uses the term *corpulents* for the fat-storing workers. Alates are produced in early spring, and soon afterward, the colony shuts down for the summer. At this point, most of the workers are corpulents, with approximately 10% of the older workforce remaining. Eggs and larvae are produced in late August through September, the brood apparently subsisting entirely on the resources stored by the corpulents, which slim down accordingly. By November, the former corpulents emerge as foragers of that winter. Each worker, therefore, lives for over a year, perhaps as long as two years. The period of estivation in Florida lasts seven to eight months, but only a few months in Tennessee and the Midwest. The abbreviated period of estivation in northern climates probably reflects a shorter period of hot temperatures but might also reflect the fact that the long spring and fall foraging period is subject to interruption by storms and extreme cold. In Florida, winters are usually cool and dry, probably ideal for almost continuous foraging.

Nests of *imparis* consist of a long, narrow vertical shaft with low-ceiling chambers opening directly off the shaft (Dennis 1941; Talbot 1943b; Tschinkel 1987). This restricts nest sites to deep soils, usually those with good drainage. The association that Wheeler made between oak forests and *imparis* probably reflects the soil preferred by some species of oaks; *imparis* often inhabits sites without oak trees. Florida nests are unusually deep, 2.5 to 3.6 meters, with the uppermost chambers at least 50 cm below the surface (Tschinkel 1987). A major benefit of the deep nest is that it provides excellent protection from extreme temperatures (Tschinkel 1987). Conversely, the brood chambers are always cool, in contrast to those of most species of northern ants, which bring their larvae up to the warmest part of the nest. Wheeler (1930) claimed that the strong preference of *imparis* for cool temperatures, which he termed *cryophily*, was unique among ants. We now know of a few other ants active in the cold, especially species of *Stenamma*, and possibly *Lasius*. Even if there were more winter-active ants, however, "cryophily" would be an unnecessary bit of jargon. At first glance, it seems that *imparis* has made an evolutionary breakthrough in adaptations for foraging in cold weather and in protecting the colony from extreme temperatures. This adaptive suite, however, did not lead to any phylogenetic radiation. If, as Wheeler says, *imparis* is a representative of a conservative, archaic lineage, it may be appropriate to consider it a specialized relict species with a clumsy but effective series of physiological and behavioral adaptations. These are basically less efficient than those found in other co-occurring ants, which display a pattern of inactivity during hot and freezing temperatures, with colony activity focused on the period of peak biological activity in the surrounding environment.

Prenolepis imparis has a liquid diet consisting of honeydew, the juice of fallen fruits, and the juices of dead insects or annelids (Talbot 1943a; Wheeler 1930). Workers from adjacent colonies mingle freely in an extended common feeding arena (Talbot 1943a). When a worker encounters a large resource, such as a piece of rotting fruit, a foraging trail is established and the resource is defended against workers from other colonies (Talbot 1943a).

Males and alate queens emerge in early spring (Talbot 1943b; Tarpley 1965; Wheeler 1930). From the accounts of these three authors, it appears that mating occurs on low vegetation near the nest. Males fly about actively, and probably seek out females that have just emerged from their nest entrance. If the weather suddenly becomes unsuitable, queens and at least some males can retreat back into the nest (Talbot 1943b). According to Tarpley (1965), who observed an aggregation on a yellow bedspread hung out to dry near a nest, females sometimes mate more than once. Although some pairs were seen flying away while still attached, males and females did not meet and mate on the wing.

Name Derivation: From *impar* (Latin), meaning "unequal," referring to the great disparity in size between males and females.

NW: introduced from New World tropics or subtropics

OW: introduced from Old World tropics or subtropics

AMBLYOPONINAE

Stigmatomma pallipes

Prionopelta antillana, NW

PROCERATIINAE

Discothyrea testacea

Proceratium chickasaw

Proceratium crassicorne

Proceratium creek

Proceratium croceum

Proceratium pergandei

Proceratium silaceum

PONERINAE

Anochetus mayri, NW

Cryptopone gilva

Gnamptogenys triangularis, NW

Hypoponera inexorata

Hypoponera opaciceps

Hypoponera opacior

Hypoponera punctatissima, OW

Leptogenys manni

Odontomachus brunneus

Odontomachus haematodus, NW

Odontomachus relictus

Odontomachus ruginodis, NW

Pachycondyla stigma

Platythyrea punctata

Ponera exotica

Ponera pennsylvanica

ECITONINAE

Neivamyrmex carolinensis

Neivamyrmex nigrescens

Neivamyrmex opacithorax

Neivamyrmex texanus

PSEUDOMYRMECINAE

Pseudomyrmex cubaensis

Pseudomyrmex ejectus

Pseudomyrmex elongatus

Pseudomyrmex gracilis, NW

Pseudomyrmex leptosus

Pseudomyrmex pallidus

Pseudomyrmex seminole

Pseudomyrmex simplex

MYRMICINAE

Aphaenogaster ashmeadi

Aphaenogaster carolinensis

Aphaenogaster flemingi

Aphaenogaster floridana

Aphaenogaster fulva

Aphaenogaster lamellidens

Aphaenogaster mariae

Aphaenogaster miamiana

Aphaenogaster tennesseensis

Aphaenogaster treatae

Aphaenogaster umphreyi

Cardiocondyla emeryi, OW

Cardiocondyla mauritanica, OW

Cardiocondyla minutior, OW

Cardiocondyla obscurior, OW

Cardiocondyla venustula, OW

Cardiocondyla wroughtonii, OW

Cephalotes varians

Crematogaster ashmeadi

Crematogaster atkinsoni

Crematogaster cerasi

Crematogaster lineolata

Crematogaster minutissima

Crematogaster missuriensis

Crematogaster obscurata, NW

Crematogaster pilosa

Crematogaster pinicola

Crematogaster vermiculata

Cyphomyrmex minutus

Cyphomyrmex rimosus, NW

Eurhopalothrix floridana

Monomorium destructor, OW

Monomorium ebeninum, NW

Monomorium floricola, OW

Monomorium pharaonis, OW

Monomorium subcoecum, OW

Monomorium trageri

Monomorium viridum

Myrmecina americana

Myrmecina cooperi

Myrmecina Species A

Myrmica punctiventris

Pheidole adrianoi

Pheidole bicarinata

Pheidole carrolli

Pheidole crassicornis

Pheidole dentata

Pheidole dentigula

Pheidole diversipilosa

Pheidole flavens, NW

Pheidole floridana

Pheidole lamia

Pheidole littoralis

Pheidole megacephala, OW

Pheidole metallescens

Pheidole moerens, NW

Pheidole morrisi

Pheidole obscurithorax, NW

Pheidole tysoni

Pogonomyrmex badius

Solenopsis abdita

Solenopsis carolinensis/texana

Solenopsis geminata

Solenopsis globularia

Solenopsis invicta, NW

Solenopsis nickersoni

Solenopsis pergandei

Solenopsis phoretica

Solenopsis picta

Solenopsis tennesseensis

Solenopsis terricola, NW

Solenopsis tonsa

Solenopsis validiuscula

Solenopsis xyloni

Solenopsis zeteki, NW

Solenopsis Species A, NW

Solenopsis Species B, NW

Stenamma foveolocephalum

Strumigenys abdita

Strumigenys angulata

Strumigenys apalachicolensis

Strumigenys archboldi

Strumigenys boltoni

Strumigenys bunki

Strumigenys carolinensis

Strumigenys cloydi

Strumigenys clypeata

Strumigenys creightoni

Strumigenys dietrichi

Strumigenys eggersi, NW

Strumigenys emmae, OW

Strumigenys epinotalis, NW

Strumigenys gundlachi, NW

Strumigenys hexamera, OW

Strumigenys hyalina

Strumigenys inopina

Strumigenys laevinasis

Strumigenys lanuginosa, NW

Strumigenys louisianae

Strumigenys margaritae, NW

Strumigenys membranifera, OW

Strumigenys metazytes

Strumigenys missouriensis

Strumigenys nigrescens

Strumigenys ohioensis

Strumigenys ornata

Strumigenys pilinasis

Strumigenys pulchella

Strumigenys reflexa

Strumigenys rogeri, OW

Strumigenys rostrata

Strumigenys silvestrii, NW

Strumigenys talpa

Strumigenys wrayi

Temnothorax allardycei

Temnothorax bradleyi

Temnothorax curvispinosus

Temnothorax palustris

Temnothorax pergandei

Temnothorax schaumii

Temnothorax smithi

Temnothorax texanus

Temnothorax torrei

Temnothorax tuscaloosae

Tetramorium bicarinatum, OW

Tetramorium caldarium, OW

Tetramorium lanuginosum, OW

Tetramorium simillimum, OW

Trachymyrmex jamaicensis

Trachymyrmex septentrionalis

Wasmannia auropunctata, NW

Xenomyrmex floridanus

DOLICHODERINAE

Dolichoderus mariae

Dolichoderus pustulatus

Dorymyrmex bossutus

Dorymyrmex bureni

Dorymyrmex elegans

Dorymyrmex flavopectus

Dorymyrmex grandula

Dorymyrmex medeis

Dorymyrmex reginicula

Forelius pruinosus

Forelius Species A

Forelius Species B

Linepithema humile, NW

Ochetellus glaber, OW

Tapinoma litorale

Tapinoma melanocephalum, OW

Tapinoma sessile

Technomyrmex difficilis, OW

FORMICINAE

Brachymyrmex depilis

Brachymyrmex minutus

Brachymyrmex obscurior

Brachymyrmex patagonicus, NW

Brachymyrmex Species A

Brachymyrmex Species B

Brachymyrmex Species C

Brachymyrmex Species D

Camponotus americanus

Camponotus caryae

Camponotus castaneus

Camponotus decipiens

Camponotus discolor

Camponotus floridanus

Camponotus impressus

Camponotus inaequalis

Camponotus mississippiensis

Camponotus nearcticus

Camponotus novogranadensis, NW

Camponotus obliquus

Camponotus pennsylvanicus

Camponotus planatus, NW

Camponotus riehlii

Camponotus sexguttatus, NW

Camponotus snellingi

Camponotus socius

Formica archboldi

Formica biophilica

Formica dolosa

Formica pallidefulva

Formica subsericea

Lasius alienus

Lasius claviger

Lasius interjectus

Lasius neoniger

Lasius umbratus

Myrmelachista ramulorum, NW

Nylanderia arenivaga

Nylanderia bourbonica, OW

Nylanderia concinna

Nylanderia faisonensis

Nylanderia fulva, NW

Nylanderia parvula

Nylanderia phantasma

Nylanderia pubens, NW

Nylanderia steinheili, NW

Nylanderia vividula

Nylanderia wojciki

Nylanderia Species A

Nylanderia Species B

Paratrechina longicornis, OW

Polyergus longicornis

Polyergus montivagus

Polyergus oligergus

Prenolepis imparis

LITERATURE CITED

Adams, R. M. M., U. G. Mueller, A. K. Holloway, A. M. Green, and J. Narozniak. 2000. Garden sharing and garden stealing in fungus-growing ants. *Naturwissenschaften* 87: 491–493.

Aguayo, C. G. 1931. New ants of the genus *Macromischa*. *Psyche* 38: 175–183.

Akre, R. D., and L. D. Hansen. 1990. Management of carpenter ants. pp. 693–700 in Vander Meer, R. K., K. Jaffe and A. Cedeno, Eds. *Applied Myrmecology, a World Perspective*. Westview Press, San Francisco, California. 741 pp.

Alayo, P. D., and L. Z. Montero. 1977. Estudios sobre los himenopteros de Cuba. VII. Dos nuevas especies para la fauna mirmecologica Cubana. *Poeyana* 174: 1–5.

Amador-Vargas, S, J. Martínez, P. Giraldo-Beltrán, R. González, S. Rifkin, and V. Gamarra-Toledo. 2011. Ant body posture: Gaster curling increases ant speed. *Ecological Entomology* 36: 663–666.

Amoore, J. E., G. Palmieri, E. Wanke, and M. S. Blum. 1969. Ant alarm pheromone activity: Correlation with molecular shape by scanning computer. *Science* 165: 1266–1269.

Andersen, A. N. 2002. Common names for Australian ants (Hymenoptera: Formicidae). *Australian Journal of Entomology* 41: 285–293.

AntWeb. Available from https://www.antweb.org /browse.do?subfamily=myrmecinae&genus=Myr mecina&rank=genus&project=allantwebants. Accessed August 4, 2015.

Aron, S., J. M. Pasteels, S. Gross, and J. L. Deneubourg. 1990. Self-organizing spatial patterns in the Argentine ant *Iridomyrmex humilis* (Mayr). pp. 438–451 in R. K. Vander Meer, K. Jaffe, and A. Cedeno, Eds. *Applied Myrmecology, a World Perspective*. Westview Press, Boulder, Colorado. 715 pp.

Ashmead, W. H. 1905. A skeleton of a new arrangement of the families, subfamilies, tribes and genera of ants, or the superfamily Formicoidea. *Canadian Entomologist* 37: 381–384.

Astruc, C., C. Malosse, and C. Errard. 2001. Lack of intraspecific aggression in the ant *Tetramorium bicarinatum*: A chemical hypothesis. *Journal of Chemical Ecology* 27: 1229–1248.

Axelrod, D. L. 1983. Biogeography of oaks in the Arcto-Tertiary Province. *Annals of the Missouri Botanical Garden* 70: 629–657.

Ayre, G. L. 1962. *Pseudometagea schwarzii* (Ashm.) (Echaritidae: Hymenoptera), a parasite of *Lasius neoniger* Emery (Formicidae: Hymenoptera). *Canadian Journal of Zoology* 40: 157–164.

Bahls, P., and M. Deyrup. 1988. A habitual lurking predator of the Florida harvester ant. pp. 547–551 in J. Trager, Ed. *Advances in Myrmecology*. E. J. Brill, New York. 551 pp.

Baldacci, J., and W. R. Tschinkel. 1999. An experimental study of colony-founding in pine saplings by queens of the arboreal ant, *Crematogaster ashmeadi*. *Insectes Sociaux* 46: 41–44.

Baldridge, R. S., C. W. Rettenmeyer, and J. F. Watkins II. 1980. Seasonal, nocturnal and diurnal flight periodicities of Nearctic army ant males (Hymenoptera: Formicidae). *Journal of the Kansas Entomological Society* 53: 189–204.

Banani, S. 2005. The life and times of August Forel. *Lights of Irfán* 6: 1–20.

Barber, E. R. 1916. The Argentine ant: Distribution and control in the United States. *United States Department of Agriculture Bulletin* 337: 1–23.

Barber, E. R. 1920. The Argentine ant as a household pest. *United States Department of Agriculture Farmer's Bulletin* 1101: 1–11.

Baroni Urbani, C. 1978. Materiali per una revision dei *Leptothorax* neotropicali appartementi al sottogenere *Macromischa* Roger, n. comb. (Hymenoptera: Formicidae). *Entomologica Basiliensia* 3: 395–618.

Baroni Urbani, C., and M. L. De Andrade. 2003. The ant genus *Proceratium* in the extant and fossil record. *Museo Regionale di Scienze Naturale Torino* 36: 1–492.

Baroni Urbani, C., and M. L. De Andrade. 2007. The ant tribe Dacetini: Limits and constituent genera, with descriptions of new species (Hymenoptera, Formicidae). *Annali Del Museo Civico Di Storia Naturale "Giacomo Doria"* 99: 1–191.

Bartels, P. J. 1985. Field observations of multiple matings in *Lasius alienus* Foerster (Hymenoptera: Formicidae). *American Midland Naturalist* 113: 190–192.

Barth, R. 1960. Ueber den Bewegungsmechanismus der Mandibelm von *Odontomachus chelifer* Latr. (Hymenopt. Formicidae). *Anais da Academia Brazileira de Ciências* 32: 379–384.

Bates, H. W. 1864. *The Naturalist on the River Amazons*. John Murray, London. Reprinted facsimile, 1962, Foreword by R. L. Usinger, University of California Press, Berkeley, California. 469 pp.

Beattie, A. J. 1985. *The Evolutionary Ecology of Ant–Plant Mutualisms*. Cambridge University Press, New York. 182 pp.

Bequaert, J. 1923. Ants accidentally introduced into New York and New Jersey; and a correction. *Bulletin of the Brooklyn Entomological Society* 18: 165.

Bernstein, R. A. 1978. Slavery in the subfamily Dolichoderinae (F. Formicidae) and its ecological consequences. *Experientia* 34: 1281–1282.

Blum, M. S. 1981. *Chemical Defenses of Arthropods*. Academic Press, New York. 562 pp.

Blum, M. S., S. L. Warter, and J. G. Traynham. 1966. Chemical releasers of social behavior. VI. The relation of structure to activity of ketones as releasers of alarm for *Iridomyrmex pruinosus* (Roger). *Journal of Insect Physiology* 12: 419–427.

Blum, M. S., and E. O. Wilson. 1964. The anatomical source of trail substances in formicine ants. *Psyche* 71: 28–31.

Boer, P, and B. Vierbergen. 2008. Exotic ants in The Netherlands. *Entomologische Bericheten* 68: 121–129.

Bolton, B. 1975. A revision of the ant genus *Leptogenys* Roger (Hymenoptera: Formicidae) in the Ethiopian Region, with a review of the Malagasy species. *Bulletin of the British Museum (Natural History) Entomology* 31: 237–305.

Bolton, B. 1976. The ant tribe Tetramoriini (Hymenoptera: Formicidae): Constituent genera, review of smaller genera and revision of *Triglyphothrix* Forel. *Bulletin of the British Museum (Natural History) Entomology* 34: 283–379.

Bolton, B. 1977. The ant tribe Tetramoriini (Hymenoptera: Formicidae). The genus *Tetramorium* Mayr in the Oriental and Indo-Australian regions, and in Australia. *Bulletin of the British Museum (Natural History) Entomology* 36: 67–151.

Bolton, B. 1979. The ant tribe Tetramoriini. The genus *Tetramorium* Mayr in the Malagasy region and in the New World. *Bulletin of the British Museum (Natural History) (Entomology)* 38: 129–181.

Bolton, B. 1985. The ant genus *Triglyphothrix* Forel a synonym of *Tetramorium* Mayr. (Hymenoptera: Formicidae). *Journal of Natural History* 19: 243–248.

Bolton, B. 1995. *A New General Catalogue of the Ants of the World*. Harvard University Press, Cambridge, Massachusetts. 504 pp.

Bolton, B. 1999. Ant genera of the tribe Dacetonini. *Journal of Natural History* 33: 1639–1689.

Bolton, B. 2000. The ant tribe Dacetini. *Memoirs of the American Entomological Institute* 65: 1–1028.

Bolton, B. 2003. Synopsis and classification of Formicidae. *Memoirs of the American Entomological Institute* 71: 1–370.

Bolton, B. 2007. Taxonomy of the dolichoderine ant genus *Technomyrmex* Mayr (Hymenoptera: Formicidae) based on the worker caste. *Contributions of the American Entomological Institute* 35: 1–150.

Brandão, C. R. F. 1983. Sequential ethograms along colony development of *Odontomachus affinis* Guerin (Hymenoptera, Formicidae, Ponerinae). *Insectes Sociaux* 30: 193–203.

Brandão, C. R. F., and A. J. Mayhé-Nunes. 2007. A phylogenetic hypothesis for the *Trachymyrmex* species groups, and the transition from fungus-growing to leaf-cutting in the Attini. pp. 72–88 in Snelling, R. R., B. L. Fisher, and P. S. Ward, Eds. Advances in ant systematics (Hymenoptera: Formicidae): Homage to E. O. Wilson—50 years of contributions. *Memoirs of the American Entomological Institute* 80.

LITERATURE CITED

Brown, B. V., S. A. Schneider, and J. S. LaPolla. 2011. A new North American species of *Pseudacteon* (Diptera: Phoridae), parasitic on *Nylanderia arenivaga* (Hymenoptera: Formicidae). *Annals of the Entomological Society of America* 104: 37–38.

Brown, J. J., and J. F. A. Traniello. 1998. Regulation of brood-care behavior in the dimorphic castes of the ant *Pheidole morrisi* (Hymenoptera: Formicidae): Effects of caste ratio, colony size, and colony needs. *Journal of Insect Behavior* 11: 209–219.

Brown, W. L., Jr. 1943. A new metallic ant from the pine barrens of New Jersey. *Entomological News* 54: 243–248.

Brown, W. L., Jr. 1948. A preliminary generic revision of the higher Dacetini. *Transactions of the American Entomological Society* 74: 101–129.

Brown, W. L., Jr. 1949a. A new American *Amblyopone*, with notes on the genus (Hymenoptera: Formicidae). *Psyche* 56: 81–88.

Brown, W. L., Jr. 1949b. Revision of the ant tribe Dacetini: 3. *Epitritus* Emery and *Quadristruma* new genus. (Hymenoptera: Formicidae). *Transactions of the American Entomological Society* 75: 43–51.

Brown, W. L., Jr. 1949c. Revision of the ant tribe Dacetini. 1. Fauna of Japan, China and Taiwan. *Mushi* 20: 1–25

Brown, W. L., Jr. 1949d. Synonymic and other notes on Formicidae (Hymenoptera). *Psyche* 56: 41–49.

Brown, W. L., Jr. 1950. Preliminary descriptions of seven new species of the dacetine ant genus *Smithistruma* Brown. *Transactions of the American Entomological Society* 76: 37–45.

Brown, W. L., Jr. 1951. New synonymy of a few genera and species of ants. *Bulletin of the Brooklyn Entomological Society* 46: 101–106.

Brown, W. L., Jr. 1953. Revisionary studies in the ant tribe Dacetini. *American Midland Naturalist* 50: 1–137.

Brown, W. L., Jr. 1958a. Contributions toward a reclassification of the Formicidae. II. Tribe Ectatommini (Hymenoptera). *Bulletin of the Museum of Comparative Zoology at Harvard* 118: 171–362.

Brown, W. L., Jr. 1958b. Predation of arthropod eggs by the genera *Proceratium* and *Discothyrea*. *Psyche* 64: 115.

Brown, W. L., Jr. 1960a. Contributions toward a reclassification on the Formicidae. II. Tribe Amblyoponini (Hymenoptera). *Bulletin of the Museum of Comparative Zoology at Harvard* 122: 145–230.

Brown, W. L., Jr. 1960b. The Neotropical species of the ant genus *Strumigenys* Fr. Smith: Group of *gundlachi* (Roger). *Psyche* 66: 37–52.

Brown, W. L., Jr. 1963. Characters and synonymies among the genera of ants. Part III. Some members of the tribe Ponerini (Ponerinae, Formicidae). *Breviora* 190: 1–10.

Brown, W. L., Jr. 1964. The ant genus *Smithistruma*: A first supplement to the World revision (Hymenoptera: Formicidae). *Transactions of the American Entomological Society* 89: 183–200.

Brown, W. L., Jr. 1967. Studies on North American Ants. II. *Myrmecina*. *Entomological News* 9: 233–240.

Brown, W. L., Jr. 1973. A comparison of the Hylean and Congo-West African Rain Forest ant faunas. pp. 161–185 in Meggers, B. J., E. S. Ayensu and W. D. Duckworth, Eds. *Tropical Forest Ecosystems in Africa and South America: A Comparative Review*. Smithsonian Institution Press, Washington, D.C. 350 pp.

Brown, W. L., Jr. 1975. Contributions toward a reclassification of the Formicidae. V. Ponerinae, tribes Platythyreini, Cerapachyini, Cylindromyrmecini, Acanthostichini, and Aenictogitini. *Search Agriculture Cornell University Agricultural Experiment Station Paper* 15: 1–115.

Brown, W. L., Jr. 1976. Contributions toward a reclassification of the Formicidae. Part VI. Ponerinae, tribe Ponerini, subtribe Odontomachiti. Section A. Introduction, subtribal characters, genus *Odontomachus*. *Studia Entomologica* 19: 67–171.

Brown, W. L., Jr. 1978. Contributions toward a reclassification of the Formicidae. Part 6. Ponerinae, tribe Ponerini, subtribe Odontomachiti. Section B. Genus *Anochetus* and bibliography. *Studia Entomologica* 20: 549–652.

Brown, W. L., Jr. 1979. A remarkable new species of *Proceratium*, with dietary and other notes on the genus (Hymenoptera: Formicidae). *Psyche* 86(4): 337–346.

Brown, W. L., Jr., and W. W. Kempf. 1960. A world revision of the ant tribe Basicerotini. *Studia Entomologica* 3: 161–250.

Brown, W. L., Jr., and E. O. Wilson. 1959. The evolution of the dacetine ants. *Quarterly Review of Biology* 34: 278–294.

Buckley, S. B. 1866. Descriptions of new species of North American Formicidae. *Proceedings of the Entomological Society of Philadelphia* 6: 152–172.

Buczkowski, G., and G. W. Bennett. 2006. Dispersed central-place foraging in the polydomous odorous house ant, *Tapinoma sessile* as revealed by a protein marker. *Insectes Sociaux* 53: 282–290.

Buczkowski, G., and G. Bennett. 2008. Seasonal polydomy in a polygynous supercolony of the odorous house ant, *Tapinoma sessile*. *Ecological Entomology* 33: 780–788.

Buczkowski, G., and P. Krushlycky. 2012. The odorous house ant, *Tapinoma sessile* (Hymenoptera: Formicidae), as a new temperate-origin invader. *Myrmecological News* 16: 61–66.

Bueno, O. C., and H. G. Fowler. 1994. Exotic ants and native ant fauna of Brazilian hospitals. pp. 191–198 in D. F. Williams, Ed. *Exotic Ants: Biology, Impact and Control of Introduced Species*. Westview Press, San Francisco. 332 pp.

Buren, W. F. 1944. A list of Iowa ants. *Iowa State College Journal of Science* 18: 277–312.

Buren, W. F. 1968. A review of the species of *Crematogaster*, sensu stricto, in North America (Hymenoptera: Formicidae). Part II. Descriptions of new species. *Journal of the Georgia Entomological Society* 3: 91–121.

Buren, W. F. 1972. Revisionary studies on the taxonomy of the imported fire ants. *Journal of the Georgia Entomological Society* 7: 1–26.

Buren, W. F., M. A. Naves, and T. C. Carlysle. 1977. False phragmosis and apparent specialization for subterranean warfare in *Pheidole lamia* Wheeler (Hymenoptera: Formicidae). *Journal of the Georgia Entomological Society* 12: 100–108.

Buren, W. F., J. C. Nickerson, and C. R. Thompson. 1975. Mixed nests of *Conomyrma insana* and C. *flavopecta*—Evidence of parasitism (Hymenoptera: Formicidae). *Psyche* 82: 306–314.

Burges, R. J. 1979. A rare fly and its parasitic behavior toward an ant (Diptera: Phoridae, Hymenoptera: Formicidae). *Florida Entomologist* 62: 413–414.

Buschinger, A. 2003. Mating behavior in the ant, *Myrmecina graminicola* (Myrmicinae). *Insectes Sociaux* 50: 295–296.

Buschinger, A., B. Schlick-Steiner, F. M. Steiner and X. Espadaler. 2003. On the geographic distribution of queen polymorphism in *Myrmecina graminicola* (Hymenoptera: Formicidae). *Myrmecological News* 5: 37–41.

Calabi, P. 1988. Behavioral flexibility in Hymenoptera: A re-examination of the concept of caste. pp. 237–253 in J. C. Trager, Ed. *Advances in Myrmecology*. E. J. Brill Press, Leiden. 551 pp.

Calabi, P., and J. F. A. Traniello. 1989. Social organization in the ant *Pheidole dentata*: Physical and temporal caste ratios lack ecological correlates. *Behavioral and Ecological Sociobiology* 24: 69–78.

Calcaterra, L. A., J. A. Briano, and D. F. Williams. 2000. New host for the parasitic ant *Solenopsis daguerrei* (Hymenoptera: Formicidae) in Argentina. *Florida Entomologist* 83: 363–365.

Cannon, C. A. 1998. *Nutritional ecology of the carpenter ant Camponotus pennsylvanicus (De Geer): Macronutrient preference and particle consumption*. PhD dissertation, Virginia Polytechnic Institute and State University. 147 + xi pp.

Capinera, J. L., C. W. Scherer, and J. M. Squitier. 2001. *Grasshoppers of Florida*. University Press of Florida, Gainesville, Florida. 143 pp.

Carlin, N. F., and D. S. Gladstein. 1989. The "bouncer" defense of *Odontomachus ruginodis* and other Odontomachine ants (Hymenoptera: Formicidae). *Psyche* 96: 1–19.

Carlin, N. F., and B. Hölldobler. 1986. The kin recognition system of carpenter ants (*Camponotus* spp.). I. Hierarchical cues in small colonies. *Behavioral Ecology and Sociobiology* 19: 123–134.

Carlin, N. F., and B. Hölldobler. 1987. The kin recognition system of carpenter ants (*Camponotus* spp.). II. Larger colonies. *Behavioral Ecology and Sociobiology* 20: 209–217.

Carlin, N. F., and B. Hölldobler. 1988. Influence of virgin queens on kin recognition in the carpenter ant *Camponotus floridanus* (Hymenoptera: Formicidae). *Insectes Sociaux* 35: 191–197.

Carlin, N. F., H. K. Reeve, and S. P. Cover. 1993. Kin discrimination and division of labor among matrilines in the polygynous carpenter ant, *Camponotus planatus*. pp. 362–401 in L. Keller, Ed. *Queen Number and Sociality in Insects*. Oxford University Press, Oxford. 439 pp.

Carlton, R. G., and C. R. Goldman. 1984. Effects of a massive swarm of ants on ammonia concentrations in a subalpine lake. *Hydrobiologia* 111: 113–117.

Carpenter, F. M. 1930. The fossil ants of North America. *Bulletin of the Museum of Comparative Zoology at Harvard College* 70: 1–66.

Carroll, C. R., and S. J. Risch. 1984. The dynamics of seed harvesting in early successional communities by a tropical ant, *Solenopsis geminata*. *Oecologia* 61: 388–392.

Carroll, J. F. 1975. *Biology and ecology of ants of the genus Aphaenogaster in Florida*. Unpublished PhD Thesis, University of Florida, Gainesville. 177 pp.

Carter, W. G. 1962. Ant distribution in North Carolina. *Journal of the Elisha Mitchell Science Society* 78: 150–204.

Cerquera, L. M., and W. R. Tschinkel. 2010. The nest architecture of the ant *Odontomachus brunneus*. *Journal of Insect Science* 10: 64.

Chapela, I. H., S. A. Rehner, T. R. Schultz, and U. G. Mueller. 1994. Evolutionary history of the symbiosis between fungus-growing ants and their fungi. *Science* 266: 1691–1694.

Chen, X., J. A. MacGown, B. J. Adams, K. A. Parys, R. M. Strecker, and L. Hooper-Bui. 2012. First record of *Pyramica epinotalis* (Hymenoptera: Formicidae) for the United States. *Psyche*: 2012, Article ID 850893, 7 pp.

Clark, D. B., C. Guayasamín, O. Pazmiño, C. Donoso, and Y. Páez de Villacís. 1982. The tramp ant *Wasmannia auropunctata*: Autecology and effects on ant diversity and distribution on Santa Cruz Island, Galapagos. *Biotropica* 14: 196–207.

Cokendolpher, J. C. 1990. The ants (Hymenoptera, Formicidae) of western Texas. Part III. Additions and corrections. *Special Publication of the Museum of Texas Technical University* 31: 5–19.

Cole, A. C. 1938. Descriptions of new ants from the western United States. *American Midland Naturalist* 20: 368–373.

Cole, A. C. 1940. A guide to the ants of the Great Smoky Mountains National Park, Tennessee. *American Midland Naturalist* 24: 1–88.

Cole, A. C. 1952. A new *Pheidole* (Hymenoptera: Formicidae) from Florida. *Annals of the Entomological Society of America* 45: 443–444.

Cole, A. C. 1968. *Pogonomyrmex harvester ants. A study of the genus in North America.* The University of Tennessee Press, Knoxville. 222 pp.

Cole, B. J. 1980. Repertoire convergence in two mangrove ants, *Zacryptocerus varians* and *Camponotus* (*Colobopsis*) sp. *Insectes Sociaux* 27: 265–275.

Cole, B. J. 1981. Dominance hierarchies in *Leptothorax* ants. *Science* 212: 83–84.

Cole, B. J. 1982. The guild of sawgrass-inhabiting ants in the Florida Keys. *Psyche* 89: 351–356.

Cole, B. J. 1983a. Assembly of mangrove ant communities: Patterns of geographical distribution *Journal of Animal Ecology* 52: 339–347.

Cole, B. J. 1983b. Assembly of mangrove ant communities: Colonization abilities. *Journal of Animal Ecology* 52: 349–355.

Cole, B. J. 1986. The social behavior of *Leptothorax allardycei* (Hymenoptera: Formicidae): Time budgets and the evolution of worker reproduction. *Behavioral Ecology and Sociobiology* 18: 165–173.

Cole, B. J. 1992. Short-term activity cycles in ants: Age-related changes in tempo and colony synchrony. *Behavioral Ecology and Sociobiology* 31: 181–187.

Cole, B. J., J. McDowell and D. Cheshire. 1994. Demography of the worker caste of *Leptothorax allardycei* (Hymenoptera: Formicidae). *Annals of the Entomological Society of America* 87: 562–565.

Coleman, D. C., D. A. Crossley, Jr., and P. F. Hendrix. 2004. *Fundamentals of Soil Ecology*. Second edition. Elsevier Press, New York. 386 pp.

Coody, C. J., and J. F. Watkins. 1986. The correlation of eye size with circadian flight periodicity of Nearctic army ant males of the genus *Neivamyrmex*. (Hymenoptera: Formicidae: Ecitoninae). *Texas Journal of Science* 38: 3–7.

Coovert, G. A. 2005. The ants of Ohio (Hymenoptera: Formicidae). *Bulletin of the Ohio Biological Survey New Series* 2: 1–196.

Cornelius, M. L., and J. K. Grace. 1996. Effect of two ant species (Hymenoptera: Formicidae) on the foraging behavior and survival of the Formosan subterranean termite (Isoptera: Rhinotermitidae). *Environmental Entomology* 25: 85–89.

Cornelius, M. L., and J. K. Grace. 1997. Influence of brood on the nutritional preferences of the tropical ant species, *Pheidole megacephala* (F.) and *Ochetellus glaber* (Mayr). *Journal of Entomological Science* 32: 421–429.

Covell, C. V. 1984. *A Field Guide to the Moths of Eastern North America*. Houghton Mifflin Company, Boston. 496 pp.

Cowan, F. 1865. *Curious Facts in the History of Insects; Including Spiders and Scorpions*. J. B. Lippincott and Co., Philadelphia. 396 pp.

Creighton, W. S. 1939. A new subspecies of *Crematogaster minutissima* with revisionary notes concerning that species (Hymenoptera: Formicidae). *Psyche* 46: 137–140.

Creighton, W. S. 1950. The ants of North America. *Bulletin of the Museum of Comparative Zoology at Harvard* 104: 1–585

Creighton, W. S. 1957. A study of the genus *Xenomyrmex* (Hymenoptera, Formicidae). *American Museum Novitates* 1843: 1–14.

Creighton, W. S., and R. R. Snelling. 1974. Notes on the behavior of three species of *Cardiocondyla* in the United States. (Hymenoptera: Formicidae). *Journal of the New York Entomological Society* 82: 82–92.

Crewe, R. M., M. S. Blum and C. A. Collingwood. 1972. Comparative analysis of alarm pheromones in the ant genus *Crematogaster*. *Comparative Biochemistry and Physiology* 43: 703–716.

Cuezzo, F. 2000. Revision del genero *Forelius* (Hymenoptera: Formicidae: Dolichoderinae). *Sociobiology* 35: 197–277.

Culver, D. C., and A. J. Beattie. 1978. Myrmecochory in *Viola*: Dynamics of seed-ant interactions in some West Virginia species. *Journal of Ecology* 66: 53–72.

Currie, C. R., U. G. Mueller, and D. Mallock. 1999. The agricultural pathology of fungus ant gardens. *Proceedings of the National Academy of Sciences USA* 96: 7998–8002.

Currie, C. R., M. Poulsen, J. Mendenhall, J. J. Boomsma, and J. Billen. 2006. Coevolved crypts and exocrine glands support mutualistic bacteria in fungus-growing ants. *Science* 311: 81–83.

Damon, A., and M. A. Perez-Soriano. 2005. Interactions between ants and orchids in the Soconusco region, Chiapas, Mexico. *Entomotropica* 20: 59–65.

Davidson, D. 1977. Species diversity and community organization in desert seed-eating ants. *Ecology* 58: 711–724.

Davis, L. R., and M. Deyrup. 2006. *Solenopsis phoretica*, a new species of apparently parasitic ant from Florida (Hymenoptera: Formicidae). *Florida Entomologist* 89: 141–143.

Davis, L. R., Jr., and D. P. Jouvenaz. 1990. *Obeza floridana*, a parasitoid of *Camponotus abdominalis floridanus* from Florida (Hymenoptera: Eucharitidae, Formicidae). *Florida Entomologist* 73: 335–337.

De Andrade, M. L., and C. Barone Urbani. 1999. Diversity and adaptation in the ant genus *Cephalotes*, past and present. *Stuttgarter Bieträge zur Naturkunde Serie B (Geologie und Paläontologie* 271: 1–889.

De Biseau, J. C., M. Schuiten, J. M. Pasteels, and J. L. Deneubourg. 1994. Respective contributions of leader and trail during recruitment to food in *Tetramorium bicarinatum* (Hymenoptera: Formicidae). *Insectes Sociaux* 41: 241–254.

DeGeer, C. 1773. *Memoires pour Servir à l'Histoire des Insectes* 3: 696 pp.

Dejean, A. 1985. Etude eco-ethologique de la predation chez les fourmis du genre *Smithistruma* (Formicidae-Myrmicinae-Dacetini) II. Attraction des prois principales (Collemboles). *Insectes Sociaux* 32: 158–172.

Dejean, A., and A. Dejean. 1998. How a ponerine ant acquired the most evolved mode of colony foundation. *Insectes Sociaux* 45: 343–346.

Del Toro, I., M. Vásquez, W. P. Mackay, P. Rojas, and R. Zapata-Mata. 2009. Hormigas (Hymenoptera: Formicidae) de Tabasco: Explorando la diversidad de la mirmecofauna en las selvas tropicales de baja altitude. *Dugesiana* 16: 1–14.

Delabie, J. H. C., and F. Bland. 2002. The tramp ant *Hypoponera punctatissima* (Roger) (Hymenoptera: Formicidae): New records from the Southern Hemisphere. *Neotropical Entomology* 31: 1–3.

Delcourt, H. R., and P. A. Delcourt. 1984. Ice age haven for hardwoods. *Natural History* 93: 22–28.

Delcourt, P. A., and H. R. Delcourt. 1981. Vegetation maps for eastern North America: 40,000 years BP to the present. pp.123–165 in R. Romans, Ed. *Geobotany* II. Plenum, New York.

Dennis, C. A. 1938. The distribution of ant species in Tennessee, with reference to ecological factors. *Annals of the Entomological Society of America* 31: 267–308.

Dennis, C. A. 1941. Some notes on the nest of the ant *Prenolepis imparis* Say. *Annals of the Entomological Society of America* 34: 82–86.

Deyrup, M. 1975. The insect community of dead and dying Douglas-fir: I. The Hymenoptera. *Coniferous Forest Ecosystem Analysis Studies Bulletin* 6: 1–104.

Deyrup, M. 1989. Arthropods endemic to Florida Scrub. *Florida Scientist* 52: 254–270.

Deyrup, M. 1990. Arthropod footprints in the sands of time. *Florida Entomologist* 73: 529–538.

Deyrup, M. 1991a. Exotic ants of the Florida Keys (Hymenoptera: Formicidae). *Proceedings of the Fourth Symposium on the Natural History of the Bahamas:* 15–22.

Deyrup, M. 1991b. *Technomyrmex albipes*, a new exotic ant in Florida (Hymenoptera: Formicidae). *Florida Entomologist* 74: 147–148.

Deyrup, M. 2002. The exotic ant *Anochetus mayri* in Florida (Hymenoptera: Formicidae). *Florida Entomologist* 85: 658–659.

Deyrup, M. 2003. An updated list of Florida ants (Hymenoptera: Formicidae). *Florida Entomologist* 86: 43–48.

Deyrup, M. 2006. *Pyramica boltoni*, a new species of leaf-litter inhabiting ant from Florida (Hymenoptera: Formicidae: Dacetini). *Florida Entomologist* 89: 1–5.

Deyrup, M. 2015. A new species of *Myrmecina* (Hymenoptera: Formicidae) from southeastern North America. *Florida Entomologist* 98: 1203–1205.

Deyrup, M., and R. A. Belmont. 2013. First record of a Florida population of the Neotropical carpenter ant *Camponotus novogranadensis* (Hymenoptera: Formicidae). *Florida Entomologist* 96: 283–285.

Deyrup, M., N. Carlin, J. Trager, and G. Umphrey. 1988. A review of the ants of the Florida Keys. *Florida Entomologist* 71: 163–176.

Deyrup, M., and S. Cover. 1998. Two new species of *Smithistruma* Brown (Hymenoptera: Formicidae) from Florida. *Proceedings of the Entomological Society of Washington* 100: 214–221.

Deyrup, M., and S. Cover. 2004a. A new species of the ant genus *Leptothorax* from Florida, with a key to the *Leptothorax* of the Southeast (Hymenoptera: Formicidae). *Florida Entomologist* 87: 51–59.

Deyrup, M., and S. Cover. 2004b. A new species of *Odontomachus* ant (Hymenoptera: Formicidae) from inland ridges of Florida, with a key to *Odontomachus* of the United States. *Florida Entomologist* 87: 136–144.

Deyrup, M., and S. Cover. 2007. A new species of *Crematogaster* from the pinelands of the southeastern United States. *Memoirs of the American Entomological Institute* 80: 100–112.

Deyrup, M., and S. Cover. 2009. Dacetine ants in Southeastern North America (Hymenoptera: Formicidae). *Southeastern Naturalist* 8: 191–212.

Deyrup, M., and L. Davis. 1998. A new species of *Aphaenogaster* (Hymenoptera: Formicidae) from upland habitats in Florida. *Entomological News* 109: 88–94.

Deyrup, M., L. Davis, and S. Buckner. 1998. Composition of the ant fauna of three Bahamian islands. *Proceedings of the Seventh Symposium on the Natural History of the Bahamas*: 23–32.

Deyrup, M., L. Davis, and S. Cover. 2000. Exotic ants in Florida. *Transactions of the American Entomological Society* 126: 293–326.

Deyrup, M., L. Deyrup, and J. Carrel. 2013. Ant species in the diet of a Florida population of eastern narrow-mouthed toads, *Gastrophryne carolinensis*. *Southeastern Naturalist* 12: 367–378.

Deyrup, M., and S. Deyrup. 1999. Notes on the introduced ant *Quadristruma emmae* (Hymenoptera: Formicidae) in Florida. *Entomological News* 110: 13–21.

Deyrup, M., and F. Fisk. 1984. A myrmecophilous cockroach new to the United States (Blattaria: Polyphagidae). *Entomological News* 95: 183–185.

Deyrup, M., C. Johnson, and L. Davis. 1997. Notes on the ant *Eurhopalothrix floridana*, with a description of the male (Hymenoptera: Formicidae). *Entomological News* 108: 183–189.

Deyrup, M., C. Johnson, G. C. Wheeler, and J. Wheeler. 1989. A preliminary list of the ants of Florida. *Florida Entomologist* 72: 91–101.

Deyrup, M., and D. Lubertazzi. 2001. A new species of ant (Hymenoptera: Formicidae) from north Florida. *Entomological News* 112: 15–21.

Deyrup, M., and Z. A. Prusak. 2008. *Solenopsis enigmatica*, a new species of inquilines ant from the island of Dominica, West Indies (Hymenoptera: Formicidae). *Florida Entomologist* 91: 70–74.

Deyrup, M., and J. Trager. 1984. *Strumigenys rogeri*, an African dacetine ant new to the U.S. (Hymenoptera: Formicidae). *Florida Entomologist* 67: 512–516.

Deyrup, M., J. Trager, and N. Carlin. 1985. The genus *Odontomachus* in the southeastern United States (Hymenoptera: Formicidae). *Entomological News* 96: 188–195.

Donisthorpe, H. 1915. *British Ants, Their Life History and Classification*. William Brendon and Son, Plymouth. 379 pp.

Dreisig, H. 2000. Defense by exploitation in the Florida carpenter ant, *Camponotus floridanus*, at an extrafloral nectar resource. *Behavioral Ecology and Sociobiology* 47: 274–279.

DuBois, M. B. 1985. Distribution of Kansas Ants. *Sociobiology* 11: 153–188.

DuBois, M. B. 1986. A revision of the native New World species of the ant genus *Monomorium* (minimum group) (Hymenoptera: Formicidae). *University of Kansas Science Bulletin* 53: 65–119.

DuBois, M. B. 1988. Distribution of army ants (Hymenoptera: Formicidae) in Illinois. *Entomological News* 99: 157–160.

DuBois, M. B. 1998. A revision of the ant genus *Stenamma* in the Palearctic and Oriental Regions. *Sociobiology* 32: 193–403.

DuBois, M. B., and L. R. Davis, Jr. 1998. *Stenamma foveolocephalum* (= *S. carolinensis*) rediscovered (Hymenoptera: Formicidae: Myrmicinae). *Sociobiology* 32: 125–138.

Duffield, R. M., J. W. Wheeler, and M. S. Blum. 1980. Methyl anthranilate: Identification and possible function in *Aphaenogaster fulva* and *Xenomyrmex floridanus*. *Florida Entomologist* 63: 203–206.

Dunkle, S. W. 1989. *Dragonflies of the Florida Peninsula, Bermuda, and the Bahamas*. Scientific Publishers, Gainesville. 155 pp.

Dunkle, S. W. 1990. *Damselflies of Florida, Bermuda and the Bahamas*. Scientific Publishers, Gainesville. 148 pp.

Dunning, J. W. 1879. Biographical notices no. III. Frederick Smith. *The Entomologist* 12: 89–92.

Ehmer, B., and B. Hölldobler. 1995. Foraging behavior of *Odontomachus bauri* on Barro Colorado Island, Panama. *Psyche* 102: 215–224.

Eichler, Wd. 1990. Health aspects and control of *Monomorium pharaonis*. pp. 671–675 in Vander Meer, R. K., K. Jaffe and A. Cedeno, Eds. *Applied Myrmecology, a World Perspective*. Westview Press, Boulder, Colorado. 741 pp.

Eisner, T. 1972. Chemical ecology: On arthropods and how they live as chemists. *Verhandlungsbericht der Deutschen Zoologischen Gesellschaft* 65: 123–137.

Eisner, T. 1991. The insect as chemist. *Proceedings of the Centennial Symposium of the Entomological Society of America*, Washington, D.C.: 5–15.

Eisner, T. 2003. *For Love of Insects*. Belknap Press of Harvard University Press, Cambridge, Massachusetts. 448 pp.

Eisner, T., I. T. Baldwin, and J. Conner. 1993. Circumvention of prey defense by a predator: Ant lion vs. ant. *Proceedings of the National Academy of Sciences USA* 90: 6716–6720.

Eisner, T., and M. Eisner. 1992. Operation and defensive role of "gin traps" in a coccinellid pupa (*Cycloneda sanguinea*). *Psyche* 99: 265–274.

Eisner, T., M. Eisner, and M. Deyrup. 1996. Millipede defense: Use of detachable bristles to entangle ants. *Proceedings of the National Academy of Sciences USA* 93: 10848–10851.

Eisner, T., M. Eisner, and M. Seigler. 2005. *Secret Weapons: Defenses of Insects, Spiders, Scorpions, and Other Many-Legged Creatures*. Belknap Press of Harvard University Press, Cambridge, Massachusetts. 372 pp.

Eisner, T., P. Jutro, D. J. Aneshansley, and R. Niedhauk. 1972. Defense against ants in a caterpillar that feeds on ant-guarded scale insects. *Annals of the Entomological Society of America* 65: 987–988.

Ellison, A. M., N. J. Gotelli, E. J. Farnsworth, and G. D. Alpert. 2012. *A Field Guide to the Ants of New England*. Yale University Press, New Haven. 398 pp.

Emery, C. 1888. Über den sogenannten Kaumagen einiger Ameisen. *Zeitschrift für Wissenschaftliche Zoologie* 46: 378–412.

Emery, C. 1890. Studi sulle formiche della fauna Neotropica. *Bullettino della Società Entomologica Italiana* 22: 38–30.

Emery, C. 1893. Beiträge zur Kenntniss der nordamerikanischen Ameisenfauna. *Zoologische Jahrbücher. Abtheilung für Systematik, Geographie und Biologie der Tiere* 7: 633–682.

Emery, C. 1894. Studi sulle formiche della fauna Neotropica. *Bullettino della Società Entomologica Italiana* 26: 137–241.

Emery, C. 1895. Beiträge zur Kenntniss der nordamerikanischen Ameisenfauna (Schluss). *Zoologische Jahrbücher. Abtheilung für Systematik, Geographie und Biologie der Tiere* 8: 257–360.

Emery, C. 1906. Note sur *Prenolepis vividula* Nyl. et sur le classification des espèces du genre *Prenolepis*. *Annales de la Société Entomologique de Belgique* 50: 130–134.

Erickson, J. M. 1971. The displacement of native ant species by the introduced Argentine ant *Iridomyrmex humilis* Mayr. *Psyche* 78: 257–266.

Errard, C., J. Delabie, H. Jourdan, and A. Hefetz. 2005. Intercontinental chemical variation in the invasive and *Wasmannia auropunctata* (Roger) (Hymenoptera Formicidae): A key to the invasive success of a tramp species. *Naturwissenschaften* 92: 319–323.

Faeth, S. H. 1980. Invertebrate predation of leaf-miners at low densities. *Ecological Entomology* 5: 111–114.

Fairchild, D. 1939. *The World Was My Garden: Travels of a Plant Explorer*. Charles Scribner's Sons, New York. 494 pp.

Feener, D. H., Jr. 1981a. Notes on the biology of *Pheidole lamia* (Hymenoptera: Formicidae) at its type locality (Austin, Texas). *Journal of the Kansas Entomological Society* 54: 269–277.

Feener, D. H., Jr. 1981b. Competition between ant species: Outcome controlled by parasitic flies. *Science* 214: 815–817.

Feener, D. H., Jr. 1987. Response of *Pheidole morrisi* to two species of enemy ants, and a general model of defense behavior in *Pheidole*. *Journal of the Kansas Entomological Society* 60: 569–575.

Fellers, J. H. 1987. Interference and exploitation in a guild of woodland ants. *Ecology* 68: 1466–1478.

Fellers, J. H. 1989. Daily and seasonal activity in woodland ants. *Oecologia* 78: 69–76.

Fellers, J. H., and G. M. Fellers. 1976. Tool use in a social insect and its implications for competitive interactions. *Science* 192: 70–72.

Fellers, J. H., and G. M. Fellers. 1982. Status and distribution of ants in the Crater District of Haleakala National Park. *Pacific Science* 36: 427–437.

Fernández, F., and R. J. Guerrero. 2008. *Technomyrmex* (Formicidae: Dolichoderinae) in the New World: Synopsis and description of a new species. *Revista Colombiana de Entomología* 34: 110–115.

Fernández-Marín, H., G. Bruner, E. B. Gomez, D. R. Nash, J. J. Boomsma, and W. T. Wcislo. 2013. Dynamic disease management in *Trachymyrmex* fungus-growing ants (Attini: Formicidae). *The American Naturalist* 181: 571–582.

Fernández-Marín, H., J. K. Zimmerman, and W. T. Wcislo. 2006. *Acanthopria* and *Mimopriella* parasitoid wasps (Diapriidae) attack *Cyphomyrmex* fungus-growing ants (Formicidae, Attini). *Naturwissenschaften* 93: 17–21.

Ferster, B., M. R. Pie, and J. F. A. Traniello. 2006. Morphometric variation in North American *Pogonomyrmex* and *Solenopsis* ants: Caste evolution through ecological release or dietary change? *Ethology Ecology and Evolution* 18: 19–32.

Ferster, B., and J. F. A. Traniello. 1995. Polymorphism and foraging behavior in *Pogonomyrmex badius* (Hymenoptera: Formicidae); worker size, foraging distance, and load size associations. *Environmental Entomology* 24: 673–678.

Fisher, B. L., and S. P. Cover. 2007. *Ants of North America: A Guide to the Genera*. University of California Press, Berkeley. 194 pp.

Fontenla, J. L. 1994. Biogeografia de *Macromischa* (Hymenoptera: Formicidae) en Cuba. *Avicenna* 1: 19–29.

Fontenla, J. L. 1997. Lista preliminar de las hormigas de Cuba (Hymenoptera: Formicidae). *Cucuyo* 6: 18–21.

Forbes, S. A. 1908. Habits and behavior of the corn-field ant, *Lasius niger americanus*. Bulletin of the University of Illinois Agricultural Experiment Station 131: 30–45.

Forel, A. 1874. Les fourmis de la Suisse. Systématique. Notices anatomiques et physiologiqes. Architecture. Distribution géographique. Nouvelles experiences et observations de moeurs. *Neue Denkschriften der allgemeinen Schweizerischen Gesellschaft für die gesammten Naturwissenschaften* 26: 1–447.

Forel, A. 1885. Études myrmécologiques en 1884; avec une description des organs sensoriels des antennes. Bulletin de la Société Vaudoise des Sciences Naturelles 20: 316–380.

Forel, A. 1886. Espèces nouvelles de fourmis Américaines. *Annales de la Société Entomolgique de Belgique. Comptes-rendus* 30: 37–49.

Forel, A. 1892. Le mâle des *Cardiocondyla*, et la reproduction consanguine perpétuée. *Annales de la Société Entomologique de Belgique* 36: 458–462.

Forel, A. 1893. Formicides de l'Antille St. Vincent. Récoltées par Mons. H. H. Smith. *Transactions of the Entomological Society of London* 1893: 333–418.

Forel, A. 1901. Varietes myrmecologiques. *Annales de la Société Entomologique de Belgique* 45: 334–382.

Fournier, D., A. Estoup, J. Orivel, J. Foucaud, H. Jourdan, J. Le Breton, and L. Keller. 2005. Clonal reproduction by males and females in the little fire ant. *Nature* 235: 1230–1234.

Fowler, H. G. 1990. Carpenter ants (*Camponotus* spp.): Pest status and human perception. pp. 525–532 in Vander Meer, R., K. Jaffe and A. Cedeno, Eds. *Applied Myrmecology, a World Perspective*. Westview Press, San Francisco, California. 741 pp.

Fowler, H. G., and R. B. Roberts. 1980. Foraging behavior of the carpenter ant, *Camponotus pennsylvanicus*, (Hymenoptera: Formicidae) in New Jersey. *Journal of the Kansas Entomological Society* 53: 295–304.

Fowler, H. G., and R. B. Roberts. 1982. Carpenter ant (Hymenoptera: Formicidae) induced wind breakage in New Jersey shade trees. *Canadian Entomologist* 114: 649–650.

Francoeur, A. 1973. Révision taxonomique des espèces néarctiques du groupe fusca, genre *Formica* (Formicidae, Hymenoptera). *Mémoires de la Société Entomologique du Québec* 3: 1–316.

Frank, J. H., and M. C. Thomas. 1981. Myrmedoniini (Coleoptera, Staphylinidae, Aleocharinae) associated with army ants (Hymenoptera, Formicidae, Ecitoninae) in Florida. *Florida Entomologist* 64: 138–146.

Frost, C. B. 1993. Four centuries of changing landscape patterns in the longleaf pine ecosystem. *Proceedings of the Tall Timbers Fire Ecology Conference* 18: 17–43.

Gadau, J., J. Heinze, B. Hölldobler, and M. Schmid. 1996. Population and colony structure of the carpenter ant *Camponotus floridanus*. *Molecular Ecology* 5: 785–792.

Garrison, R. W. 1995. New agricultural pests for southern California: Two new ants, *Pheidole fervens* and *Pheidole moerens*. *California Plant Pest and Disease Report* Oct.–Dec. 1995: 69–70.

Gentry, J. B. 1974. Response to predation by colonies of the Florida harvester ant, *Pogonomyrmex badius*. *Ecology* 55: 1328–1338.

Gordon, D. M. 1984. The harvester ant (*Pogonomyrmex badius*) midden: Refuse or boundary. *Ecological Entomology* 9: 403–412.

Gotwald, W. H., Jr. 1995. *Army Ants: The Biology of Social Predation*. Cornell University Press, Ithaca, New York. 302 pp.

Gotwald, W. H., and J. Levieux. 1972. Taxonomy and biology of a new West African ant belonging to the genus *Amblyopone*. *Annals of the Entomological Society of America* 65: 383–396.

Gotzek, D., H. J. Axen, A. V. Suarez, S. H. Cahan, and D. Shoemaker. 2015. Global invasion history of the tropical fire ant: A stowaway on the first global trade routes. *Molecular Ecology* 24: 374–388.

Gotzek, D., S. G. Brady, R. J. Kallal, and J. S. LaPolla. 2012. The importance of using multiple approaches for identifying emerging invasive species: The case of the rasberry crazy ant in the United States. *PLoS ONE* 7: e45314. doi: 1-.1371 /journal.pone.0045314.

Graham, J. H., A. J. Krzysik, D. A. Kovacic, J. J. Duda, D. C. Freeman, J. M. Emlen, J. C. Zak, W. R. Long, M. P. Wallace, C. Chamberlin-Grahamm, J. P. Nutter, and H. E. Balbach. 2008. Ant community composition across a gradient of disturbed military landscapes at Fort Benning, Georgia. *Southeastern Naturalist* 7: 429–448.

Grangier, J., J. Le Breton, A. Dejean, and J. Orivel. 2007. Coexistence between *Cyphomyrmex* ants and dominant populations of *Wasmannia auropunctata*. *Behavioral Processes* 74: 93–96.

Gregg, R. E. 1956. An extension of range for the ant, *Pheidole lamia* Wheeler (Hymenoptera: Formicidae). *Entomological News* 67: 37–39.

Gregg, R. E. 1958. Key to the species of *Pheidole* (Hymenoptera: Formicidae) in the United States. *Journal of the New York Entomological Society* 66: 7–48

Gregg, R. E. 1963. *The Ants of Colorado, with Reference to Their Ecology, Taxonomy, and Geographic Distribution.* University of Colorado Press, Boulder. 792 pp.

Gronenberg, W. 1996. The trap-jaw mechanism in the dacetine ants *Daceton armigerum* and *Strumigenys* sp. *Journal of Experimental Biology* 199: 2021–2033.

Gronenberg, W., J. Tautz, and B. Hölldobler. 1993. Fast trap jaws and giant neurons in the ant *Odontomachus. Science* 262: 561–563.

Guénard, B., K. A. McCaffrey, A. Lucky, and R. R. Dunn. 2012. Ants of North Carolina: An updated list. *Zootaxa* 3552: 1–36.

Gulmahamad, H. 1997. Ecological studies on *Cardiocondyla ectopia* Snelling (Hymenoptera: Formicidae) in Southern California. *Pan-Pacific Entomologist* 73: 21–27.

Haak, U., B. Hölldobler, H. J. Bestman, and F. Kern. 1996. Species-specificity in trail pheromones and Dufour's gland contents of *Camponotus atriceps* and *C. floridanus* (Hymenoptera: Formicidae). *Chemoecology* 7: 85–93.

Hahn, D. A., and W. R. Tschinkel. 1997. Settlement and distribution of colony-founding queens of the arboreal ant, *Crematogaster ashmeadi*, in a longleaf pine forest. *Insectes Sociaux* 44: 323–336.

Hamilton, W. D. 1979. Wingless and fighting males in fig wasps and other insects. pp. 167–220 in *Sexual Selection and Reproductive Competition in Insects.* M. S. Blum and N. A. Blum, Eds. Academic Press, New York.

Hamm, C. A. 2010. Multivariate discrimination and description of a new species of *Tapinoma* from the western United States. *Annals of the Entomological Society of America* 103: 20–29.

Handel, S. N. 1978. New ant-dispersed species in the genera *Carex, Luzula*, and *Claytonia. Canadian Journal of Botany* 56: 2925–2927.

Handel, S. N., and A. J. Beattie. 1990. Seed dispersal by ants. *Scientific American*: 76–83.

Hansen, L. D., and J. H. Klotz. 2005. *Carpenter Ants of the United States and Canada.* Comstock Publishing Associates, Ithaca, New York. 204 pp.

Hare, J. F., and T. Eisner. 1993. Pyrrolizidine alkaloid deters ant predators of *Utetheisa ornatrix* eggs: Effects of alkaloid concentration, oxidation state, and prior exposure of ants to alkaloid-laden prey. *Oecologica* 96: 9–18.

Harmon, G. 1993. Mating in *Pogonomyrmex badius* (Hymenoptera: Formicidae). *Florida Entomologist* 76: 524–526.

Hart, L. M., and W. R. Tschinkel. 2012. A seasonal natural history of the ant, *Odontomachus brunneus. Insectes Sociaux* 59: 45–54.

Hartmann, A., J. Wantia, and J. Heinze. 2005. Facultative sexual reproduction in the parthenogenetic ant *Platythyrea punctata. Insectes Sociaux* 52: 155–162.

Hartwick, E. B., W. G. Friend, and C. E. Atwood. 1977. Trail-laying behavior of the carpenter ant *Camponotus pennsylvanicus* (Hymenoptera: Formicidae). *Canadian Entomologist* 109: 129–136.

Hashimoto, Y., K. Yamauchi, and E. Hasegawa. 1995. Unique habits of stomodeal trophallaxis in the ponerine ant *Hypoponera* sp. *Insectes Sociaux* 42: 137–144.

Hashmi, A. A. 1973. A revision of the Neotropical ant subgenus *Myrmothrix* of the genus *Camponotus. Studia Entomologica (N.S.)* 16: 1–140

Haskins, C. P. 1930. Preliminary notes on certain phases of the behavior and habits of *Proceratium croceum* Roger. *Journal of the New York Entomological Society* 38: 121–126.

Haskins, C. P. 1931. Notes on the biology and social life of *Euponera gilva* Roger var. *harnedi* M. R. Smith. *Journal of the New York Entomological Society* 39: 507–521.

Haskins, C. P., and E. V. Enzmann. 1938. Studies of certain sociological and physiological features in the Formicidae. Part II. Types of colony-initiation in the Ponerinae and degeneration of the wing-musculature in the queen. *Annals of the New York Academy of Science* 37: 147–161.

Hassell, M. P., and H. C. J. Godfray. 1992. The population biology of insect parasitoids. In M. J. Crawley, Ed. *Natural Enemies: The Population Biology of Predators, Parasites and Diseases.* Blackwell Scientific Publications, Oxford, U.K.

Haug, G. W. 1932. Description of the male of *Strumigenys louisianae* subsp. *laticephala* M.R. Smith. *Annals of the Entomological Society of America* 25: 170–172.

Hayes, W. P. 1920. *Solenopsis molesta* Say (Hym.): A biological study. *Kansas State Agricultural Experiment Station Department of Entomology Contribution* 55: 1–55.

Headley, A. E. 1943. Population studies of two species of ants, *Leptothorax longispinosus* Roger and *Leptothorax curvispinosus* Mayr. *Annals of the Entomological Society of America* 36: 743–753.

Hedges, S. A. 1998. *Field Guide for the Management of Structure-Infesting Ants, Second Edition.* G.I.E. Publishers, Cleveland, Ohio. 304 pp.

Hedrick, U. P. 1919. Sturtevant's notes on edible plants. *Report of the New York Agricultural Experiment Station* 27, Vol. II: 1–686.

Heinze, J., S. P. Cover and B. Hölldobler. 1995. Neither worker nor queen: An ant caste specialized in the production of unfertilized ants. *Psyche* 102: 173–185.

Heinze, J., S. Cremer, N. Eckl, and A. Schrempf. 2006. Stealthy invaders: The biology of *Cardiocondyla* tramp ants. *Insectes Sociaux* 53: 1–7.

Heinze, J., J. Gadeau, B. Hölldobler, I. Nanda, M. Schmid, and K. Scheller. 1994. Genetic variability in the ant *Camponotus floridanus* detected by multilocus DNA fingerprinting. *Naturwissenschaften* 81: 34–36.

Heinze, J., and B. Hölldobler. 1993. Fighting for a harem of queens: Physiology of reproduction in *Cardiocondyla* male ants. *Proceedings of the National Academy of Science USA* 90: 8412–8414.

Heinze, J., and B. Hölldobler. 1995. Thelytokous parthenogenesis and dominance hierarchies in the ponerine ant, *Platythyrea punctata. Naturwissenschaften* 82: 40–41.

Heinze, J., B. Hölldobler and S. Trenkle. 1995. Reproductive behavior of the ant *Leptothorax* (*Dichothorax*) *pergandei. Insectes Sociaux* 42: 309–305.

Heinze, J., and S. Trenkle. 1997. Male polymorphism and gynandromorphs in the ant *Cardiocondyla emeryi. Naturwissenschaften* 84: 129–131.

Helfer, J. H. 1953. *How to Know the Grasshoppers, Cockroaches and Their Allies.* Wm. C. Brown Company Publishers, Dubuque. 353 pp.

Hespenheide, H. A. 1986. Mimicry of ants of the genus *Zacryptocerus* (Hymenoptera: Formicidae). *Journal of the New York Entomological Society* 94: 394–408.

Hess, C. A., and F. C. James. 1998. Diet of the red-cockaded woodpecker in the Apalachicola National Forest. *Journal of Wildlife Management* 62: 509–517.

Higashi, S., S. Tsuyuzaki, M. Ohara and F. Ito. 1989. Adaptive advantages of ant-dispersed seeds in the myrmecochorus plant *Trillium tschonoskii* (Liliaceae). *Oikos* 54: 389–394.

Hill, J. G., and B. V. Brown. 2006. New records of the rarely collected ant-decapitating fly *Apocephalus tenuipes* Borgmeier (Diptera: Phoridae). *Southeastern Naturalist* 5: 367–368.

Holland, W. J. 1903. *The Moth Book. A Popular Guide to a Knowledge of the Moths of North America.* Doubleday, Page and Company, New York. 479 pp.

Hölldobler, B. 1971a. Sex pheromone in the ant *Xenomyrmex floridanus. Journal of Insect Physiology* 17: 1497–1499.

Hölldobler, B. 1971b. Recruitment behavior in *Camponotus socius* (Hym. Formicidae). *Zeitschrift für Vergleichende Physiologie* 75: 123–142.

Hölldobler, B. 1982. Interference strategy of *Iridomyrmex pruinosus* (Hymenoptera: Formicidae) during foraging. *Oecologia* 52: 208–213.

Hölldobler, B., E. Janssen, H. J. Bestmann, I. R. Leal, P. S. Oliveira, F. Kern, and W. A. König. 1996. Communication in the migratory termite-hunting ant *Pachycondyla* (= *Termitopone*) *marginata* (Formicidae, Ponerinae). *Journal of Comparative Physiology A* 178: 47–53.

Hölldobler, B., M. Obermayer, and E. O. Wilson. 1992. Communication in the primitive cryptobiotic ant *Prionopelta amabilis* (Hymenoptera: Formicidae). *Journal of Comparative Physiology A* 170: 9–16.

Hölldobler, B., and E. O. Wilson. 1986a. Ecology and behavior of the primitive cryptobiotic ant *Prionopelta amabilis* (Hymenoptera: Formicidae). *Insectes Sociaux* 33: 45–58.

Hölldobler, B., and E. O. Wilson. 1986b. Soil-binding pilosity and camouflage in ants of the tribes Basicerotini and Stegomyrmecini (Hymenoptera, Formicidae). *Zoomorphology* 106: 12–20.

Hölldobler, B., and E. O. Wilson. 1990. *The Ants.* Belknap Press of Harvard University Press, Cambridge, Massachusetts. 732 pp.

Hooker, C. W. 1913. Report of the entomologist. *Journal of the Annual Report of the Porto Rico Agricultural Experiment Station for 1912:* 34–38.

Hornung, E. 2011. Evolutionary adaptation of oniscidean isopods to terrestrial life: Structure, physiology and behavior. *Terrestrial Arthropod Reviews* 4: 95–130.

Howard, D. F., M. S. Blum, T. H. Jones, and M. D. Tomalski. 1982. Behavioral responses to an alkylpyrazine from the mandibular gland of the ant *Wasmannia auropunctata. Insectes Sociaux* 29: 369–374.

Huddleston, E. W., and S. S. Fluker. 1968. Distribution of ant species of Hawaii. *Proceedings of the Hawaiian Entomological Society* 20: 45–69.

Human, K. G., and D. M. Gordon. 1999. Behavioral interactions of the invasive Argentine ant with native ant species. *Insectes Sociaux* 46: 159–163.

Hunt, J. H., and R. R. Snelling. 1975. A checklist of the ants of Arizona. *Arizona Academy of Sciences* 10: 20–23.

Hunt, R. 1993. The role of vibrational signals in mating behavior of *Spissistilus festinus* (Homoptera: Membracidae). *Annals of the Entomological Society of America* 86: 356–361.

Imae, H. 2003. Ant image database. http://ant.edb.miyakyo-u.ae.jp/E/GUIDE/MAEGAKI.HTM.

Ipser, R. M., M. A. Brinkman, W. A. Gardner and H. B. Peeler. 2004. A Survey of ground-dwelling ants (Hymenoptera: Formicidae) in Georgia. *Florida Entomologist* 87: 253–260.

Ishak, H. D., J. L. Miller, R. Sen, S. E. Dowd, E. Meyer, and U. G. Mueller. 2011. Microbiomes of ant castes implicate new microbial roles in the fungus-growing ant *Trachymyrmex septentrionalis*. *Scientific Reports* 204: 1–12.

Ito, F., Y. Touyama, A. Gotoh, S. Kitahiro, and J. Billen. 2010. Thelytokous parthenogenesis by queens in the dacetine ant *Pyramica membranifera*. *Naturwissenschaften* 97: 725–728.

Jackson, J. F., and B. A. Drummond, III. 1974. A batesian ant–mimicry complex from the mountain pine ridge of British Honduras, with an example of transformational mimicry. *American Midland Naturalist* 91: 248–251.

Jaffe, K. 1993. Surfing ants. *Florida Entomologist* 76: 182–183.

Janda, M., D. Folkova, and J. Zrzavy. 2004. Phylogeny of *Lasius* ants based on mitochondrial DNA and morphology, and the evolution of social parasitism in the Lasiini (Hymenoptera: Formicidae). *Molecular Phylogeny and Evolution* 33: 595–614.

Jansen, D. H. 1966. Coevolution of mutualism between ants and acacias in Central America. *Evolution* 20: 249–275.

Jeanne, R. L. 1970. Chemical defense of a brood by a social wasp. *Science* 168: 1465–1466.

Jerdon, T. C. 1851. A catalogue of species of ants found in southern India. *Madras Journal of Literature and Science* 17: 103–127.

Johnson, A. B., and E. O. Wilson. 1985. Correlates of variation in the major/minor ratio of the ant *Pheidole dentata* (Hymenoptera: Formicidae). *Annals of the Entomological Society of America* 78: 8–11.

Johnson, C. 1987. Biogeography and habitats of *Ponera exotica* (Hymenoptera: Formicidae). *Journal of Entomological Science* 22: 358–361.

Johnson, C. 1988a. Species identification in the eastern *Crematogaster* (Hymenoptera: Formicidae). *Journal of Entomological Science* 23: 314–332.

Johnson, C. 1988b. Colony structure and behavioral observations in *Pheidole morrisi* (Hymenoptera: Formicidae). pp. 371–383 in J. C. Trager, Ed. *Advances in Myrmecology*. E. J. Brill, New York. 551 pp.

Johnson, C. 1989a. Identification and nesting sites of North American species of *Dolichoderus* Lund (Hymenoptera: Formicidae). *Insecta Mundi* 3: 1–9.

Johnson, C. 1989b. Taxonomy and diagnosis of *Conomyrma insana* (Buckley) and *C. flava* (McCook) (Formicidae). *Insecta Mundi* 3: 179–194.

Jones, T. H., M. S. Blum, H. M. Fales, and C. R. Thompson. 1980. (5Z, 8E)-3-heptyl-5-methyl pyrrolizidine from a thief ant. *Journal of Organic Chemistry* 45: 4778–4780.

Jourdan, H., R. A. Sadlier, and A. M. Bauer. 2001. Little fire ant invasion (*Wasmannia auropunctata*) as a threat to New Caledonian lizards: Evidences from a sclerophyll forest (Hymenoptera: Formicidae). *Sociobiology* 38: 283–301.

Kallal, R. J., and J. S. LaPolla. 2012. Monograph of *Nylanderia* (Hymenoptera: Formicidae) of the World, part II: *Nylanderia* in the Nearctic. *Zootaxa* 3508: 1–64.

Kaspari, M., D. Donoso, J. A. Lucas, T. Zumbusch, and A. D. Kay. 2012. Using nutritional ecology to predict community structure: A field test in the Neotropics. *Ecosphere* 3: 93. http://dx.doi .org/10.1890/ES12-00136.1.

Kaufmann, S., D. B. McKey, M. Hossaert-McKey, and C. C. Horvitz. 1991. Adaptations for a two-phase seed dispersal system involving vertebrates and ants in a hemiepiphitic fig (*Ficus microcarpa*: Moraceae). *American Journal of Botany* 78: 971–977.

Keller, L., L. Passera, and J. P. Suzzoni. 1989. Queen execution in the Argentine ant *Iridomyrmex humilis* (Mayr). *Physiological Entomology* 14: 157–163.

Kempf, W. W. 1972. Catalago abreviado das formigas da Regiao Neotropical (Hymenoptera: Formicidae). *Studia Entomologica* 15: 3–344.

Kennedy, C. H., and M. M. Schramm. 1933. A new *Strumigenys* with notes on Ohio species (Formicidae: Hymenoptera). *Annals of the Entomological Society of America* 26: 95–104.

Kettle, D. S. 1984. *Medical and Veterinary Entomology*. John Wiley and Sons, New York. 658 pp.

King, J. R., and S. D. Porter. 2007. Body size, colony size, abundance, and ecological impact of exotic ants in Florida's upland ecosystems. *Evolutionary and Ecological Research* 9: 757–774.

King, J. R., and J. C. Trager. 2007. Natural history of the slave making ant, *Polyergus lucidus, sensu lato* in northern Florida and its three *Formica pallidefulva* group hosts. 14 pp. *Journal of Insect Science*, available online: insectscience.org/7.42.

King, J. R., and W. R. Tschinkel. 2006. Experimental evidence that the introduced fire ant, *Solenopsis invicta*, does not competitively suppress co-occurring ants in a disturbed habitat. *Journal of Animal Ecology* 75: 1370–1378.

LITERATURE CITED

King, J. R., and W. R. Tschinkel. 2007. Range extension and local population increase of the exotic ant *Pheidole obscurithorax*, in the southeastern United States (Hymenoptera: Formicidae). *Florida Entomologist* 90: 435–439.

King, J. R., and W. R. Tschinkel. 2008. Experimental evidence that human impacts drive fire ant invasions and ecological change. *Proceedings of the National Academy of Science* 105: 20339–20343.

Kitahiro, S., K. Yamamoto, Y. Touyama, and F. Ito. 2014. Habitat preferences of *Strumigenys* ants in western Japan (Hymenoptera: Formicidae). *Asian Myrmecology* 6: 91–94.

Klotz, J. H., J. R. Mangold, K. M. Vail, L. R. Davis, Jr., and R. S. Patterson. 1995. A survey of the urban pest ants (Hymenoptera: Formicidae) of peninsular Florida. *Florida Entomologist* 78: 109–118.

Klotz, J. H., and B. L. Reid. 1993. Nocturnal orientation in the black carpenter ant *Camponotus pennsylvanicus* (DeGeer) (Hymenoptera: Formicidae). *Insectes Sociaux* 40: 95–106.

Kohl, E., B. Hölldobler, and H. J. Bestman. 2001. Trail and recruitment pheromones in *Camponotus socius*. *Chemoecology* 11: 67–73.

Koptur, S. 1992. Plants with extrafloral nectaries and ants in Everglades habitats. 1992. *Florida Entomologist* 75: 38–50.

Kugler, C. 1978. A comparative study of the myrmicine sting apparatus (Hymenoptera, Formicidae). *Studia Entomologica* 20: 413–548.

Kugler, C. 1979. Evolution of the sting apparatus in myrmicine ants. *Evolution* 33: 117–130.

Kugler, J. 1983. The males of *Cardiocondyla* Emery (Hymenoptera: Formicidae) with the description of the winged male of *Cardiocondyla wroughtonii* (Forel). *Israel Journal of Entomology* 17: 1–21.

Kusnezov, N. 1951. El género *Wasmannia* en la Argentina (Hymenoptera, Formicidae). *Acta Zoologica Lilloana* 10: 173–182.

LaPolla, J. S., S. G. Brady, and S. O. Shattuck. 2010. Phylogeny and taxonomy of the *Prenolepis* genus-group of ants (Hymenoptera: Formicidae). *Systematic Entomology* 35: 118–131.

LaPolla, J. S., S. G. Brady, and S. O. Shattuck. 2011. Monograph of *Nylanderia* (Hymenoptera: Formicidae) of the World: An introduction to the systematics and biology of the genus. *Zootaxa* 3110: 1–9.

LaPolla, J. S., U. G. Mueller, M. Seid, and S. P. Cover. 2002. Predation by the army ant *Neivamyrmex rugulosus* on the fungus-growing ant *Trachymyrmex arizonensis*. *Insectes Sociaux* 49: 251–256.

Laskis, K. O., and W. R. Tschinkel. 2008. The seasonal natural history of the ant, *Dolichoderus mariae*, in northern Florida. *Journal of Insect Science* 9: 26 pp.

Latreille, P. A. 1802. *Histoire Naturelle des Formis*. Paris, 445 pp.

Lattke, J. E. 1990. Revision del genero *Gnamptogenys* Mayr in Venezuela (Hymenoptera: Formicidae). *Acta Terramaris* 2: 1–46.

Lattke, J. E. 1995. Revision of the ant genus *Gnamptogenys* in the New World (Hymenoptera: Formicidae). *Journal of Hymenoptera Research* 4: 139–193.

Lattke, J. E. 2011. Revision of the New World species of the genus *Leptogenys* Roger (Insects: Hymenoptera: Formicidae: Ponerinae). *Journal of Arthropod Systematics and Phylogeny* 69: 127–264.

Lavigne, R. J. 1977. Notes on the ants of Luquillo Forest, Puerto Rico (Hymenoptera: Formicidae). *Proceedings of the Entomological Society of Washington* 79: 216–228.

Le Breton, J., J. Chazeau, and H. Jourdan. 2003. Immediate impacts of invasion by *Wasmannia auropunctata* (Hymenoptera: Formicidae) on native litter ant fauna in a New Caledonian rainforest. *Austral Ecology* 28: 204–209.

Letourneau, D. K., and J. C. Choe. 1987. Homopteran attendance bay wasps and ants: The stochastic nature of interactions. *Psyche* 94: 81–91.

Leuthold, R. H. 1968. A tibial gland scent-trail and trail-laying behavior in the ant *Crematogaster ashmeadi* Mayr. *Psyche* 75: 233–248.

Lim, S., S. A. Chong, and C. Lee. 2003. Nestmate recognition and intercolonial aggression in the crazy ant, *Paratrechina longicornis* (Hymenoptera: Formicidae). *Sociobiology* 41: 295–305.

Longino, J. T. 2002. *Camponotus novogranadensis* Mayr. 2002. *Ants of Costa Rica*. Accessed August 1, 2012.

Longino, J. T. 2003. The *Crematogaster* (Hymenoptera, Formicidae, Myrmicinae) of Costa Rica. *Zootaxa* 151: 1–150.

Longino, J. T. 2004. Ants of Costa Rica. *Cardiocondyla* of Costa Rica. http://www.evergreen.edu/ants/genera/cardiocondyl/home.html.

Longino, J. T. 2005. Ants of Costa Rica. *Solenopsis* of Costa Rica. http://www.evergreen.edu/ants/genera/cardiocondyl/home.html.

Longino, J. T. 2006. A taxonomic review of the genus *Myrmelachista* (Hymenoptera: Formicidae) in Costa Rica. *Zootaxa* 1141: 1–54.

Longino, J. T., and F. Fernández. 2007. Taxonomic review of the genus *Wasmannia*. *Memoirs of the American Entomological Institute* 80: 271–289.

Lubin, Y. D. 1984. Changes in the native fauna of the Galapagos Islands following invasion by the little red fire ant, *Wasmannia auropunctata*. *Biological Journal of the Linnean Society* 21: 229–242.

Lynch, J. F. 1981. Seasonal, successional, and vertical segregation in a Maryland ant community. *Oikos* 37: 183–198.

Lynch, J. F., C. Balinsky, and S. G. Vail. 1980. Foraging patterns in three sympatric ant species, *Prenolepis imparis*, *Paratrechina melanderi* and *Aphaenogaster rudis* (Hymenoptera: Formicidae). *Ecological Entomology* 5: 353–371.

MacGown, J. A., and R. L. Brown. 2006. Survey of ants (Hymenoptera: Formicidae) of the Tombigbee National Forest in Mississippi. *Journal of the Kansas Entomological Society* 79: 325–340.

MacGown, J. A., and J. A. Forster. 2005. A preliminary list of the ants (Hymenoptera: Formicidae) of Alabama, U.S.A. *Entomological News* 116: 61–74.

MacGown, J. A., J. G. Hill, and R. L. Brown. 2010. Dispersal of the exotic *Brachymyrmex patagonicus* (Hymenoptera: Formicidae) in the United States. *Proceedings of the 2010 Imported Fire Ant Conference* 24–26.

MacGown, J. A., B. Boudinot, M. Deyrup, and D. M. Sorger. 2014. A review of the Nearctic *Odontomachus* (Hymenoptera: Formicidae: Ponerinae) with a treatment of the males. *Zootaxa* 3802: 515–552.

MacGown, J. A., R. L. Brown, J. G. Hill and B. Layton. 2007a. Carpenter ants of Mississippi. *Mississippi Agricultural and Forestry Experiment Station Bulletin* 1158: 1–28.

MacGown, J. A., J. G. Hill, and M. A. Deyrup. 2007b. *Brachymyrmex patagonicus* (Hymenoptera: Formicidae), an emerging pest species in the southeastern United States. *Florida Entomologist* 90: 457–464.

MacGown, J. A., J. G. Hill, R. L. Brown, and T. L. Schieffer. 2009a. Ant diversity at Noxubee National Wildlife Refuge in Oktibbeha, Noxubee, and Winston Counties. *Mississippi Entomological Museum Report* 2009-01.

MacGown, J. A., J. G. Hill, and M. A. Deyrup. 2009b. Ants (Hymenoptera: Formicidae) of the Little Ohoopee River Dunes, Emanuel County, Georgia. *Journal of Entomological Science* 44: 193–197.

MacGown, J. A., and J. K. Wetterer. 2012. Geographic spread of *Pyramica hexamera* (Hymenoptera: Formicidae: Dacetini) in the southeastern USA. *Terrestrial Arthropod Reviews* 5: 3–14.

MacGown, J. A., J. K, Wetterer, and J. G. Hill. 2012. Geographic spread of *Strumigenys silvestrii* (Hymenoptera: Formicidae: Dacetini). *Terrestrial Arthropod Reviews* 5: 10 pp.

Mackay, W. P. 1993a. The status of the ant *Leptothorax pergandei* Emery (Hymenoptera: Formicidae. *Sociobiology* 21: 287–297.

Mackay, W. P. 1993b. A review of the New World ants of the genus *Dolichoderus* (Hymenoptera: Formicidae). *Sociobiology* 22: 1–148.

Mackay, W. P. 1995. Range extensions for the ant *Leptothorax pergandei* (Hymenoptera: Formicidae): A mesic forest species discovered in the Chihuahuan Desert. *Proceedings of the Entomological Society of Washington* 97: 888.

Mackay, W. P. 2000. A review of the New World ants of the subgenus *Myrafant* (genus *Leptothorax*). Hymenoptera: Formicidae. *Sociobiology* 36: 265–444.

Mackay, W. P., and R. S. Anderson. 1991. New distributional records of the ant genus *Ponera* (Hymenoptera: Formicidae) in North America. *Journal of the New York Entomological Society* 99: 696–699.

Mackay, W. P., and E. Mackay. 2002. *The Ants of New Mexico* (Hymenoptera: Formicidae). The Edwin Mellen Press, Lewiston, New York. 400 pp.

Maehr, D. S. 1997. The comparative ecology of bobcat, black bear, and Florida panther in south Florida. *Bulletin of the Florida Museum of Natural History* 40: 1–176.

Mallis, A. 1971. *American Entomologists*. Rutgers University Press, New Brunswick, New Jersey. 549 pp.

Mann, W. M. 1920. Additions to the ant fauna of the West Indies and Central America. *Bulletin of the American Museum of Natural History* 42: 403–439.

Mann, W. M. 1948. *Ant Hill Odyssey*. Little, Brown and Company, Boston. 338 pp.

Markin, G. P. 1970. The seasonal life cycle of the Argentine ant, *Iridomyrmex humilis* (Hymenoptera: Formicidae), in southern California. *Annals of the Entomological Society of America* 63: 1238–1242.

Marlin, J. C. 1968. Notes on a new method of colony formation employed by *Polyergus lucidus lucidus* Mayr (Hymenoptera: Formicidae). *Transactions of the Illinois Academy of Science* 61: 207–209.

Marlin, J. C. 1969. The raiding behavior of *Polyergus lucidus lucidus* in central Illinois (Hymenoptera: Formicidae). *Journal of the Kansas Entomological Society* 42: 108–115.

Marlin, J. C. 1971. The mating, nesting and ant enemies of *Polyergus lucidus* Mayr (Hymenoptera: Formicidae). *American Midland Naturalist* 86: 181–189.

Martinez, M. J. 1995. The first record of mixed nests of *Conomyrma bicolor* (Wheeler) and *Conomyrma insana* (Buckley) (Hymenoptera: Formicidae). *Pan-Pacific Entomologist* 71: 252.

LITERATURE CITED

Martinez, M. J., W. J. Wrenn, A. Tilger, and R. F. Cummings. 2011. New records for the exotic ants *Brachymyrmex patagonicus* Mayr and *Pheidole moerens* Wheeler (Hymenoptera: Formicidae) in California. *Pan-Pacific Entomologist* 87: 47–50.

Maruyama, M., F. M. Steiner, C. Stauffer, T. Akino, R. H. Crozier, and B. C. Schlick-Steiner. 2008. A DNA and morphology-based phylogenetic framework of the ant genus *Lasius* with hypotheses for the evolution of social parasitism and fungiculture. *BMC Evolutionary Biology* 8: 237.

Maschwitz, U., and P. Schönegge. 1983. Forage communication, nest moving recruitment, and prey specialization in the Oriental ponerine *Leptogenys chinensis. Oecologia* 57: 175–182.

Masner, L., and J. L. García. 2002. The genera of Diapriinae (Hymenoptera: Diapriidae) in the New World. *Bulletin of the American Museum of Natural History* 268: 1–138.

Masuko, K. 1985. Studies on the predatory behavior of Oriental dacetine ants (Hymenoptera: Formicidae). I. Some Japanese species of *Strumigenys, Pentastruma,* and *Epitritus,* and a Malaysian *Labidogenys,* with special reference to hunting tactics in short-mandibulate forms. *Insectes Sociaux* 31: 429–451.

Masuko, K. 1986. Larval hemolymph feeding: A non-destructive parental cannibalism in the primitive ant *Amblyopone silvestrii* Wheeler (Hymenoptera: Formicidae) *Behavioral and Ecological Sociobiology* 19: 249–255.

Masuko, K. 1994. Specialized predation on oribatid mites by two species of the ant genus *Myrmecina* (Hymenoptera: Formicidae). *Psyche* 101: 159–173.

Masuko, K. 2013. Thelytokous parthenogenesis in the ant *Strumigenys hexamera* (Hymenoptera: Formicidae). *Annals of the Entomological Society of America* 106: 479–484.

Mayhé-Nunes, A. J., and C. R. Brandão. 2007. Revisionary studies on the attine genus *Trachymyrmex* Forel. Part 3. The Jamaicensis group (Hymenoptera: Formicidae). *Zootaxa* 1444: 1–21.

Mayr, E. 1942. *Systematics and the Origin of Species.* Columbia University Press, New York. 334 pp.

Mayr, G. 1866a. Myrmecologische Beiträge. *Sitzungberichte der k. Akademie der Wissenhaften. Mathematisch-Natuwissenscaftliche Classe* 53: 484–517.

Mayr, G. 1866b. Diagnosen neuer und wenig gekannter Formiciden. *Verhandlungen der k.k. Zoologische-Botanischen Gesellschaft in Wien* 16: 885–908.

Mayr, G. 1870. Neue Formiciden. *Verhadlungen der k.k. Zoologische-Botanischen Gesellschaft in Wein* 20: 939–996.

Mayr, G. 1872. Formicidae Borneensis collectae a J. Doria et O. Beccari in territorio Sarawak annis 1865–1867. *Annali de Museo Civico di Storia Naturale di Genova* 2: 133–135.

Mayr, G. 1886. Die Formiciden der Vereinigten Staaten von Nordamerika. *Verhandlungen Zoologisch-Botanischen Gesellschaft Wein* 36: 419–464.

McCook, H. C. 1879. Family Formicidae. pp. 182–189 in J. H. Comstock, *Report upon Cotton Insects.* Department of Agriculture, Washington, D.C. 511 pp.

McGlynn, T. P. 1999. The worldwide transfer of ants: Geographical distribution and ecological invasions. *Journal of Biogeography* 26: 535–548.

McGurk, D. J., J. Frost, G. R. Waller, E. J. Eishenbraun, K. Vick, W. A. Drew, and J. Young. 1968. Iridoidal isomer variation in Dolichoderine ants. *Journal of Insect Physiology* 14: 841–845.

McInnes, D. A., and W. R. Tschinkel. 1996. Mermithid nematode parasitism of *Solenopsis* ants (Hymenoptera: Formicidae) of northern Florida. *Annals of the Entomological Society of America* 89: 231–237.

Medeiros, F. N. S., L. E. Lopes, P. R. S. Moutinho, P. S. Oliveira and B. Hölldobler. 1992. Functional polygyny, agonistic interactions and reproductive dominance in the Neotropical ant *Odontomachus chelifer* (Hymenoptera: Formicidae, Ponerinae). *Ethology* 91: 134–146.

Mehdiabadi, N. J., and T. R. Schultz. 2009. Natural history and phylogeny of the fungus-farming ants (Hymenoptera: Formicidae: Myrmecinae: Attini). *Myrmecological News* 13: 37–55.

Mikheyev, A. S., S. Bresson, and P. Conant. 2009. Single-queen introductions characterize regional and local invasions by the facultatively clonal little fire ant *Wasmannia auropunctata. Molecular Ecology* 2009: 1–10.

Mikheyev, A. S., and U. Mueller. 2007. Genetic relationships between native and introduced populations of the little fire ant *Wasmannia auropunctata. Diversity and Distributions* 13: 573–579.

Mikheyev, A. S., U. G. Mueller, and P. Abbot. 2010. Comparative dating of attine ant and lepiotaceous cultivar phylogenies reveals coevolutionary synchrony and discord. *The American Naturalist* 175: E126–E133.

Miller, S. E. 1994. Dispersal of plant pests into the Virgin Islands. *Florida Entomologist* 77: 520–521.

Moore, W. S. 1995. Northern Flicker, *Colaptes auratus. Birds of North America* 166: 7–25.

Morawetz, W., M. Henzl, and B. Wallnöfer. 1992. Tree killing by herbicide producing ants for the establishment of pure *Tococa occidentalis* populations in the Peruvian Amazon. *Biodiversity and Conservation* 1: 19–33.

Moreau, C. S. 2008. Unraveling the evolutionary history of the hyperdiverse ant genus *Pheidole* (Hymenoptera: Formicidae). *Molecular Phylogenetics and Evolution* 48: 224–239.

Moreau, C. S., M. A. Deyrup, and L. R. Davis, Jr. 2014. Ants of the Florida Keys: Species accounts, biogeography, and conservation (Hymenoptera: Formicidae). *Journal of Insect Science* 14: 1–8.

Morrison, L. W. 2002. Island biogeography and metapopulation dynamics of Bahamian ants. *Journal of Biogeography* 29: 387–394.

Morrison, L. W. 2006. The ants of small Bahamian cays. *Bahamas Naturalist and Journal of Science* September 2006: 27–32.

Mueller, U. G., N. M. Gerardo, D. K. AAnen, D. L. Six, and T. R. Schultz. 2005. The evolution of agriculture in insects. *Annual Review of Evolution and Systematics* 36: 563–595.

Mueller, U. G., T. R. Schultz, C. R. Currie, R. M. M. Adams, and D. Malloch. 2001. The origin of the attine ant-fungus mutualism. *The Quarterly Review of Biology* 76: 169–197.

Murakami, T., K. Ohkawara and S. Higashi. 2002. Morphology and developmental plasticity of reproductive females in *Myrmecina nipponica* (Hymenoptera: Formicidae). *Annals of the Entomological Society of America* 95: 577–582.

Murphree, L. C. 1947. *Alabama ants: Description, distribution and biology, with notes on the control of the most important household species.* Unpublished Master's thesis, Department of Entomology, Mississippi State College. 144 pp.

Myers, R. L. 1990. Scrub and high pine. pp. 150–193 in R. L. Myers and J. J. Ewel, Eds. *Ecosystems of Florida*. University of Central Florida Press, Orlando, Florida. 765 pp.

Navarrete, B., H. McAuslane, M. Deyrup, and J. E. Peña. 2013. Ants (Hymenoptera: Formicidae) associated with *Diaphorina citri* (Hemiptera: Liviidae) and their role in its biological control. *Florida Entomologist* 96: 590–597.

Naves, M. A. 1985. A monograph of the genus *Pheidole* in Florida (Hymenoptera: Formicidae). *Insecta Mundi* 1: 53–90.

Newell, W. 1909. The life history of the Argentine ant *Iridomyrmex humilis* Mayr. *Journal of Economic Entomology* 2: 174–192.

Nickerson, J. C., and D. L. Harris. 1985. The Florida carpenter ant, *Camponotus abdominalis floridanus* (Buckley) (Hymenoptera: Formicidae). *Florida Department of Agriculture and Consumer Services Division of Plant Industry Entomology Circular* 269: 2 pp.

Nickerson, J. C., and W. H. Whitcomb. 1988. Sources of carbohydrates utilized by *Conomyrma medeis*. pp. 465–472 in J. C. Trager, Ed. *Advances in Myrmecology*. E. J. Brill, New York. 551 pp.

Nickerson, J. C., H. L. Cromroy, W. H. Whitcomb, and J. A. Cornell. 1975a. Colony organization and queen numbers in two species of *Conomyrma*. *Annals of the Entomological Society of America* 68: 1083–1085.

Nickerson, J. C., W. H. Whitcomb, A. P. Bhatkar and M. A. Naves. 1975b. Predation on founding queens of *Solenopsis invicta* by workers of *Conomyrma insana*. *Florida Entomologist* 58: 75–82.

Nielsen, M. G., and T. F. Jensen. 1975. Økologiske studier over *Lasius alienus* (Först.) (Hymenoptera, Formicidae). *Entomogiske Meddelelser* 43: 5–16.

Nylander, W. 1846. Adnotationes in monographiam formicarum borealium Europae. *Acta Societatis Scientiarum Fennicae* 2: 875–944.

Oliveira, P. S., and B. Hölldobler. 1989. Orientation and communication in the Neotropical ant *Odontomachus bauri* Emery (Hymenoptera, Formicidae, Ponerinae). *Ethology* 83: 154–166.

Oliveira, P. S., M. Obermayer, and B. Hölldobler. 1998. Division of labor in the Neotropical ant, *Pachycondyla stigma* (Ponerinae), with special reference to mutual antennal rubbing between nestmates (Hymenoptera). *Sociobiology* 31: 9–24.

Oliver, T. H., T. Pettitt, S. R. Leather, and J. M. Cook. 2008. Numerical abundance of invasive ants and monopolization of exudate-producing resources—A chicken and egg situation. *Insect Conservation and Diversity* 1: 208–214.

Olubajo, O., R. M. Duffield and J. W. Wheeler. 1980. 4-Heptanone in the mandibular gland secretin of the Nearctic ant, *Zacryptocerus varians* (Hymenoptera: Formicidae). *Annals of the Entomological Society of America* 73: 93–94.

Osborne, L. S., J. E. Peña, and D. H. Oi. 1995. Predation by *Tapinoma melanocephalum* (Hymenoptera: Formicidae) on twospotted spider mites (Acari: Tetranychidae) in Florida Greenhouses. *Florida Entomologist* 78: 565–570.

Pacheco, J. A. 2007. *The New World thief ants of the genus Solenopsis.* Unpublished PhD Dissertation, University of Texas at El Paso. 543 pp.

Pacheco, J. A., and W. P. Mackay. 2013. *The Systematics and Biology of the New World Thief Ants of the Genus Solenopsis* (Hymenoptera: Formicidae). Edwin Mellen Press, Lampeter, U.K.

Page, R. E. 1982. Polyandry in *Brachymyrmex depilis* Emery (Hymenoptera: Formicidae). *Pan-Pacific Entomologist* 58: 258.

Passera, L. 1994. Characteristics of tramp species. pp. 23–43 in Williams, D. F., Ed. *Exotic Ants, Biology, Impact, and Control of Introduced Species.* Westview Press, San Francisco. 332 pp.

Patel, A. D. 1990. An unusually broad behavioral repertory for a major worker in a dimorphic ant species: *Pheidole morrisi* (Hymenoptera: Formicidae). *Psyche* 97: 181–191

Patton, W. H. 1894. Habits of the leaping-ant of southern Georgia. *American Naturalist* 28: 618–619.

Peters, C., and B. Hölldobler. 1992. Notes on the morphology of the sticky "doorknobs" of larvae in an Australian *Hypoponera* sp. (Formicidae: Ponerinae). *Psyche* 99: 23–30.

Petralia, R. S., and S. B. Vinson. 1979. Comparative anatomy of the ventral region of ant larvae, and its relation to feeding behavior. *Psyche* 86: 375–394.

Pfitzer, D. W. 1951. A new species of *Smithistruma* from Tennessee (Hymenoptera: Formicidae). *Journal of the Tennessee Academy of Sciences* 26: 198–200.

Piso, M. A., and P. S. Oliveira. 1998. Interaction between ants and seeds of a nonmyrmecochorus Neotropical tree, *Cabralea canjerana* (Meliaceae), in the Atlantic forest of southeast Brazil. *American Journal of Botany* 85: 669–674.

Porter, S. D., and C. D. Jorgensen. 1990. Psammophores: Do harvester ants (Hymenoptera: Formicidae) use these pouches to transport seeds? *Journal of the Kansas Entomological Society* 62: 138–149.

Powell, S., and W. R. Tschinkel. 1999. Ritualized conflict in *Odontomachus brunneus* and the generation of interaction-based task allocation: A new organizational mechanism in ants. *Animal Behavior* 58: 965–972.

Pratt, S. C., N. F. Carlin and P. Calabi. 1994. Division of labor in *Ponera pennsylvanica* (Formicidae: Ponerinae). *Insectes Sociaux* 41: 43–61.

Pricer, J. L. 1908. The life history of the carpenter ant. *Biologists Bulletin* 14: 177–218.

Queller, D. 2005. Males from Mars. *Nature* 435: 1167–1168.

Quirán, E. M., J. J. Martínez, and A. O. Bachmann. 2004. The Neotropical genus *Brachymyrmex* Mayr, 1868 (Hymenoptera: Formicidae) in Argentina.

Redescription of the type species, *B. patagonicus* Mayr, 1868, *B. bruchi* Forel, 1912 and *B. oculatus* Santschi, 1919. *Acta Zoológica Mexicana (n.s.)* 20: 273–285.

Rabeling, C., S. P. Cover, R. A. Johnson, and U. G. Mueller. 2007. A review of the North American species of the fungus-gardening ant genus *Trachymyrmex*. *Zootaxa* 1664: 1–53.

Raczkowski, J. M., and G. M. Luque. 2011. Colony founding and social parasitism in *Lasius* (*Acanthomyops*). *Insectes Sociaux* 58: 227–244.

Raven, P. H., and D. I. Axelrod. 1978. Origin and relationships of the California flora. *University of California Publications in Botany* 72: 1–134.

Regnier, F. E., and E. O. Wilson. 1969. The alarm-defense system of the ant *Lasius alienus*. *Journal of Insect Physiology* 15: 893–898.

Reimer, N. J. 1994. Distribution and impact of alien ants in vulnerable Hawaiian ecosystems. pp. 11–22 in D. F. Williams, Ed. *Exotic Ants: Biology, Impact, and Control of Introduced Species.* Westview Press, San Francisco. 332 pp.

Rettenmeyer, C. W. 1963. Behavioral studies of army ants. *University of Kansas Science Bulletin* 44: 281–465.

Rettenmeyer, C. W., and J. F. Watkins, II. 1978. Polygyny and monogyny in army ants (Hymenoptera: Formicidae). *Journal of the Kansas Entomological Society* 51: 581–591.

Reuter, M., F. Balloux, L. Lehmann, and L. Keller. 2001. Kin structure and queen execution in the Argentine ant *Linepithema humile*. *Journal of Evolutionary Biology* 14: 954–958.

Rheindt, F. E., J. Gadau, C.-P. Strehl, and B. Hölldobler. 2004. Extremely high mating frequency in the Florida harvester ant (*Pogonomyrmex badius*). *Behavioral Ecology and Sociobiology* 56: 472–481.

Ribas, C. R., R. B. F. Campos, F. A. Schmidt, and R. R. C. Solar. 2012. Ants as indicators in Brazil: A review with suggestions to improve the use of ant in environmental monitoring programs. *Psyche* 2012: 1–23.

Rico-Gray, V., J. T. Barber, L. B. Thien, E. G. Ellgaard, and J. J. Toney. 1989. An unusual animal–plant interaction: Feeding of *Schomburgkia tibicinis* (Orchidaceae) by ants. *American Journal of Botany* 76: 603–608.

Rico-Gray, V., and L. S. L. Sternberg. 1991. Carbon isotopic evidence for seasonal change in feeding habits of *Camponotus planatus* Roger (Formicidae) in Yucatan, Mexico. *Biotropica* 23: 93–95.

Rico-Gray, V., and L. B. Thien. 1989. Effect of different ant species on reproductive fitness of *Schomburkia tibicinis* (Orchidaceae). *Oecologia* 81: 487–489.

Rifflet, A, S. Gavalda, T. Téné, J. Orivel, F. Leprince, L. Guilhaudis, E. Génin, A. Vétillard, and M. Treilhou. 2012. Identification and characterization of a novel antimicrobial peptide from the venom of the ant *Tetramorium bicarinatum*. *Peptides* 38: 363–370

Risch, S. J., and C. R. Carroll. 1982a. The ecological role of ants in two Mexican agroecosystems. *Oecologia* 55: 114–119.

Risch, S. J., and C. R. Carroll. 1982b. Effect of a keystone predaceous ant, *Solenopsis geminata*, on arthropods in a tropical agroecosystem. *Ecology* 63: 1979–1983.

Risch, S. J., and C. R. Carroll. 1986. Effects of seed production by a tropical ant on competition among weeds. *Ecology* 67: 1319–1327.

Roger, J. 1857. Einiges über Ameisen. *Berliner Entomologische Zeitschrift* 1: 10–20.

Roger, J. 1863. Die neu aufgeführten Gattungen und Arten meines Formiciden-Verzeichnesses, nebst Ergänzung einiger früher gegeben Beschreibungen. *Berliner Entomologische Zeitschrift* 7: 131–214.

Ross, M. S., J. O. O'Brien, R. G. Ford, K. Zhang, and A. Morkill. 2009. Disturbance and the rising tide: The challenge of biodiversity management on low-island ecosystems. *Frontiers in Ecology and Environment* 7: 471–478

Saarinen, E. V., and J. C. Daniels. 2006. Miami blue butterfly larvae (Lepidoptera: Lycaenidae) and ants (Hymenoptera: Formicidae): New information on the symbionts of an endangered taxon. *Florida Entomologist* 89: 69–74.

Santschi, F. 1921. Ponerinae, Dorylinae et quelques autres formicides neotropiques. *Bulletin de la Société Vaudoise des Sciences Naturelles* 54: 81–103.

Santschi, F. 1923. Revue des fourmis du genre "Brachymyrmex" Mayr. *Anales del Museo Nacional de Historia Natural de Buenos Aires* 31: 650–678.

Saporito, R. A., M. A. Donnelly, R. A. Norton, J. M. Garraffo, T. F. Spande, and J. W. Daly. 2007. Oribatid mites as a major dietary source for alkaloids in poison frogs. *Proceedings of the National Academy of Sciences USA* 104: 8885–8890.

Saporito, R. A., H. M. Garraffo, M. A. Donnelly, A. L. Edwards, J. T. Longino and J. W. Daly. 2004. Formicine ants: An arthropod source for the pumiliotoxin alkaloids of dendrobatid poison frogs. *Proceedings of the National Academy of Sciences USA* 101: 8045–8050.

Sauer, C., D. Dudaczek, B. Hölldobler, and R. Gross. 2002. Tissue location of the endosymbiotic bacterium "*Candidatus* Blochmannia floridanus" in adults and larvae of the carpenter ant *Camponotus floridanus*. *Applied and Experimental Microbiology* 68: 4187–4193.

Say, T. 1836. Descriptions of new species of North American Hymenoptera, and observations on some already described. *Boston Journal of Natural History* 1: 209–305.

Schauff, M. E., and Z. Bouček. 1987. *Alachua floridensis*, a new genus and species of Entedoninae (Hymenoptera: Eulophidae) parasitic on the Florida carpenter ant, *Camponotus abdominalis* (Formicidae). *Proceedings of the Entomological Society of Washington* 89: 660–664.

Schemske, D. W. 1980. The evolutionary significance of extrafloral nectar production by *Costus woodsonii* (Zingibeaceae): An experimental analysis of ant protection. *Journal of Ecology* 69: 959–967.

Schilder, K., J. Heinze, and B. Hölldobler. 1999. Colony structure and reproduction in the thelytokous parthenogenetic ant *Platythyrea punctata* (F. Smith) (Hymenoptera, Formicidae). *Insectes Sociaux* 46: 150–158.

Schmidt, J. O., and M. S. Blum. 1978. A harvester ant venom: Chemistry and pharmacology. *Science* 200: 1064–1066.

Schneirla, T. C. 1944. Results of the Archbold Expeditions. No. 51. Behavior and ecological notes on some ants from south-central Florida. *American Museum Novitates* 1261: 1–5.

Schneirla, T. C. 1958. The behavior and biology of certain Nearctic army ants. Last part of the functional season, southeastern Arizona. *Insectes Sociaux* 5: 215–255.

Schultz, G. W. 1982. Soil-dropping behavior of the pavement ant, *Tetramorium caespitum* (L.) (Hymenoptera: Formicidae) against the alkali bee (Hymenoptera: Halictidae). *Journal of the Kansas Entomological Society* 55: 277–282.

Sciaky, R., and F. Rigato. 1987. Studies on the behavior of the parasitic queens of the genus *Lasius* (Hymenoptera Formicidae). *Pubblicazioni dell' Instituto di Entomologica dell'Universita di Pavia* 36: 39–41.

Seal, J. W., and W. R. Tschinkel. 2006a. Colony productivity of the fungus-gardening ant *Trachymyrmex septentrionalis* (Hymenoptera: Formicidae) in a Florida pine forest. *Annals of the Entomological Society of America* 99: 673–681.

Seal, J. W., and W. R. Tschinkel. 2006b. Energetics of newly-mated queens and colony founding in the fungus-gardening ants *Cyphomyrmex rimosus* and *Trachymyrmex septentrionalis* (Hymenoptera: Formicidae). *Physiological Entomology* 2006: 1–8.

Seal, J. W., and W. R. Tschinkel. 2008. Food limitation in the fungus-gardening ant, *Trachymyrmex septentrionalis*. *Ecological Entomology* 33: 597–607.

Seifert, B. 2003. The ant genus *Cardiocondyla* (Insecta: Hymenoptera: Formicidae)—A taxonomic revision of the *C. elegans*, *C. bulgarica*, *C. batesii*, *C. nuda*, *C. shuckardi*, *C. stambulofii*, *C. wroughtonii*, *C. emeryi*, and *C. minutior* species groups. *Annalen des Naturhistorischen Museums in Wien. Serie B, Für Botanik und Zoologie* 104B: 203–338.

Sharma, S., D. H. Oi, and E. A. Buss. 2013. Honeydew-producing hemipterans in Florida associated with *Nylanderia fulva* (Hymenoptera: Formicidae), an invasive crazy ant. *Florida Entomologist* 96: 538–547.

Sharma, S., J. Warner, and R. H. Scheffrahn. 2014. Tawny crazy ant (previously known as Caribbean crazy ant) *Nylanderia* (formerly *Paratrechina*) *fulva* (Mayr) (Insecta: Hymenoptera: Formicidae: Formicinae). *Entomology and Nematology Department University of Florida/Institute of Food and Agricultural Services Extension Document* EENY610: 1–5.

Shattuck, S. O. 1992. Generic revision of the ant subfamily Dolichoderinae. *Sociobiology* 21: 1–181.

Shattuck, S. O. 2009. A revision of the Australian species of the ant genus *Myrmecina* (Hymenoptera: Formicidae). *Zootaxa* 2146: 1–29.

Shattuck, S. O., S. D. Porter, and D. P. Wojcik. 1999. *Solenopsis invicta* Buren, 1972 (Insecta, Hymenoptera): Proposed conservation of the specific name. *Bulletin of Zoological Nomenclature* 56: 27–30.

Shawler, A. T., R. M. Kraus, and T. E. Reagan. 1989. Foraging territoriality of the imported fire ant, *Solenopsis invicta* Buren, in sugarcane as determined by neutron activation analysis. *Insectes Sociaux* 36: 235–239.

Simberloff, D., and E. O. Wilson. 1969. Experimental zoogeography of islands: The colonization of empty islands. *Ecology* 50: 278–296.

Simberloff, D., and E. O. Wilson. 1970. Experimental zoogeography of islands: A two-year record of colonization. *Ecology* 51: 934–937.

Sivinski, J. M., J. E. Lloyd, S. N. Beshers, L. R. Davis, R. G. Sivinski, S. R. Wing, R. T. Sullivan, P. E. Cushing, and E. Petersson. 1998. A natural history of *Pleotomodes needhami* Green (Coleoptera: Lampyridae): A firefly symbiont of ants. *The Coleopterists Bulletin* 52: 23–30.

Skelley, P. E., and R. E. Woodruff. 1991. Five new species of *Aphodius* (Coleoptera: Scarabaeidae) from Florida pocket gopher burrows. *Florida Entomologist* 74: 517–536.

Smedley, S. R., E. Ehrhardt, and T. Eisner. 1993. Defensive regurgitation by a noctuid moth larva (*Litoprosopa futilis*). *Psyche* 100: 209–221.

Smith, C. R., and W. R. Tschinkel. 2006. The sociometry and sociogenesis of reproduction in the Florida harvester ant, *Pogonomyrmex badius*. *Journal of Insect Science* 6: 11 pp.

Smith, C. R., and W. R. Tschinkel. 2007. The adaptive nature of non-food collection for the harvester ant, *Pogonomyrmex badius*. *Ecological Entomology* 32: 105–112.

Smith, D. R. 1979. Formicoidea. pp. 1323–1467 in Krombein, K. V., P. D. Hurd, Jr., D. R. Smith, and B. D. Burks, Eds. *Catalog of Hymenoptera in America north of Mexico*. Smithsonian Institution Press, Washington, D.C.: 2735 pp.

Smith, M. R. 1923. Two new Mississippi ants of the subgenus *Colobopsis*. *Psyche* 30: 82–83.

Smith, M. R. 1928. An additional annotated list of the ants of Mississippi. *Entomological News* 39: 275–279.

Smith, M. R. 1929a. Descriptions of five new North American ants, with biological notes. *Annals of the Entomological Society of America* 22: 543–551.

Smith, M. R. 1929b. Two introduced ants not previously known to occur in the United States. *Journal of Economic Entomology* 22: 241–243.

Smith, M. R. 1930a. A list of Florida ants. *Florida Entomologist* 14: 1–6.

Smith, M. R. 1930b. Descriptions of three new North American ants, with biological notes. *Annals of the Entomological Society of America* 23: 564–568.

Smith, M. R. 1931a. A revision of the genus *Strumigenys* or America, north of Mexico, based on a study of the workers. (Hymen: Formicidae). *Annals of the Entomological Society of America* 24: 686–710.

Smith, M. R. 1931b. An additional annotated list of the ants of Mississippi. *Entomological News* 42: 16–24.

Smith, M. R. 1933. Additional species of Florida ants, with remarks. *Florida Entomologist* 17: 21–26.

Smith, M. R. 1934. Ponerine ants of the genus *Euponera* in the United States. *Annals of the Entomological Society of America* 27: 557–564.

Smith, M. R. 1936. Ants of the genus *Ponera* in America, north of Mexico. *Annals of the Entomological Society of America* 29: 420–430.

Smith, M. R. 1937. The ants of Puerto Rico. *Journal of Agriculture of the University of Puerto Rico* 20: 819–875.

Smith, M. R. 1943a. A generic and subgeneric synopsis of the male ants of the United States. *American Midland Naturalist* 30: 273–321.

Smith, M. R. 1943b. A new North American *Solenopsis* (*Diplorhoptrum*). *Proceedings of the Entomological Society of Washington* 44: 209–211.

Smith, M. R. 1944. Additional ants recorded from Florida, with descriptions of two new subspecies. *Florida Entomologist* 27: 14–17.

Smith, M. R. 1948. A new species of *Myrmecina* from California (Hymenoptera: Formicidae). *Proceedings of the Entomological Society of Washington* 50: 238–240.

Smith, M. R. 1951. Family Formicidae. pp. 778–875 in C. F. W. Muesebeck, K. V. Krombein, and H. K. Townes, Eds. *Hymenoptera of North America North of Mexico. Synoptic Catalog.* U.S. Department of Agriculture Monograph 2: 1420 pp.

Smith, M. R. 1954. Ants of the Bimini Island group, Bahamas, British West Indies (Hymenoptera, Formicidae). *American Museum Novitates* 1617: 1–16.

Smith, M. R. 1957. Revision of the genus *Stenamma* Westwood in America north of Mexico (Hymenoptera, Formicidae). *American Midland Naturalist* 57: 133–174.

Smith, M. R. 1958. Family Formicidae. pp. 108–162 in K. V. Krombein, Ed. *Hymenoptera of North America North of Mexico. Synoptic Catalog.* U.S. Department of Agriculture Monograph 2, First Supplement: 305 pp.

Smith, M. R. 1961. A study of New Guinea ants of the genus *Aphaenogaster* Mayr (Hymenoptera: Formicidae). *Acta Hymenopterologica* 1: 213–237.

Smith, M. R. 1962. A new species of exotic *Ponera* from North Carolina. *Acta Hymenopterologica* 1: 377–382.

Smith, M. R. 1965. House-infesting ants of the eastern United States: Their recognition, biology, and economic importance. *U.S. Department of Agriculture Technical Bulletin* 1326: 1–105

Smith, M. R., and M. W. Wing. 1954. Redescription of *Discothyrea testacea* Roger, a little-known North American ant, with notes on the genus (Hymenoptera: Formicidae). *Contributions to Science of the Los Angeles County Museum* 124: 1–10.

Smith, S. Q., and T. H. Jones. 2004. Tracking the cryptic pumiliotoxins. *Proceedings of the National Academy of Sciences U.S.A.* 101: 7841–7842.

Snelling, R. R. 1965. Studies on California ants. 2. *Myrmecina californica* M. R. Smith (Hymenoptera: Formicidae). *Bulletin of the Southern California Academy of Science* 64: 101–105.

Snelling, R. R. 1973. Studies on California ants. 7. The genus *Stenamma* (Hymenoptera: Formicidae). *Contributions to Science of the Los Angeles County Museum* 245: 1–38.

Snelling, R. R. 1974. Studies on California ants. 8. A new species of *Cardiocondyla* (Hymenoptera: Formicidae). *Journal of the New York Entomological Society* 82: 76–81.

Snelling, R. R. 1988. Taxonomic notes on Nearctic species of *Camponotus*, subgenus *Myrmentoma* (Hymenoptera: Formicidae). pp. 55–78 in Trager, J. C., Ed. *Advances in Myrmecology.* E. J. Brill, New York. 551 pp.

Snelling, R. R. 1995. Systematics of Nearctic ants of the genus *Dorymyrmex* (Hymenoptera: Formicidae). *Contributions to Science of the Los Angeles County Museum* 454: 1–14.

Snelling, R. R. 2001. Two new species of thief ants (*Solenopsis*) from Puerto Rico (Hymenoptera: Formicidae. *Sociobiology* 37: 511–525.

Snelling, R. R., and J. T. Longino. 1992. Revisionary notes on the fungus-growing ants of the genus *Cyphomyrmex, rimosus* group (Hymenoptera: Formicidae: Attini). pp. 479–494 in Quintero, D., and A. Aeillo, Eds. *Insects of Panama and Mesoamerica: Selected Studies.* Oxford University Press, Oxford. 692 pp.

Spencer, H. 1941. The small fire ant *Wasmannia* in citrus groves—A preliminary report. *Florida Entomologist* 24: 6–14.

Starr, F., and K. Starr. 2013. New insect records from Maui. *Bishop Museum Occasional Papers* 114: 69.

Steghaus-Kovak, S., and U. Maschwitz. 1993. Predation on earwigs: A novel diet specialization within the genus *Leptogenys. Insectes Sociaux* 40: 337–340.

Steiner, F. M., B. C. Schlick-Steiner, H. Konrad, T. A. Linksvayer, S.-P. Quek, E. Christian, C. Stauffer and A. Buschinger. 2006. Phylogeny and evolutionary history of queen polymorphic *Myrmecina* ants (Hymenoptera: Formicidae). *European Journal of Entomology* 103: 619–626.

Storz, S. R., and W. R. Tschinkel. 2004. Distribution, spread, and ecological associations of the introduced ant *Pheidole obscurithorax* in the southeastern United States. *Journal of Insect Science* 4: 12, available online: insectscience.org/4.12.

Stuart, R. J. 1987. Individual workers produce colony-specific nestmate recognition cues in the ant, *Leptothorax curvispinosus. Animal Behavior* 35: 1062–1069.

Stuart, R. J., A. Francoeur and R. Loiselle. 1987. Lethal fighting among dimorphic males of the ant *Cardiocondyla wroughtonii. Naturwissenschaften* 74: 548–549.

Suarez, A. V., D. T. Bolger, and T. J. Case. 1998. Effects of fragmentation and invasion on native ant communities in coastal southern California. *Ecology* 79: 2041–2056.

Suarez, A. V., D. A. Holway, and T. J. Case. 2001. Patterns of spread in biological invasions dominated by long-distance jump dispersal: Insights from Argentine ants. *Proceedings of the National Academy of Sciences* 98: 1095–1100.

Sutherland, D. W. S. 1978. Common names of insects and related organisms (1978 revision). *Entomological Society of America Special Publications* 78-1: 1–132.

Swanton, J. R. 1946. Indians of the Southeastern United States. *Smithsonian Institution Bureau of American Ethnology* 137: 943 pp.

Sykes, W. H. 1835. Descriptions of new species of Indian ants. *Transactions of the Entomological Society of London* 1: 99–107.

Talbot, M. 1943a. Population studies of the ant, *Prenolepis imparis* Say. *Ecology* 24: 31–44.

Talbot, M. 1943b. Response of the ant *Prenolepis imparis* Say to temperature and humidity changes. *Ecology* 24: 345–352.

Talbot, M. 1954. Populations of the ant *Aphaenogaster* (*Attomyrma*) *treatae* Forel on abandoned fields on the Edwin S. George Preserve. *University of Michigan Contributions of the Laboratory of Invertebrate Biology* 69: 1–9.

Talbot, M. 1956. Flight activities of the ant *Dolichoderus* (*Hypoclinea*) *mariae* Forel. *Psyche* 63: 134–139.

Talbot, M. 1965. Populations of ants in a low field. *Insectes Sociaux* 12: 19–47.

Talbot, M. 1966. Flights of the ant *Aphaenogaster treatae*. *Journal of the Kansas Entomological Society* 39: 67–77.

Talbot, M. 1967. Slave-raids of the ant *Polyergus lucidus* Mayr. *Psyche* 74: 299–313.

Talbot, M. 1968. Flights of the ant *Polyergus lucidus*. *Psyche* 75: 46–52.

Tarpley, W. A. 1965. Nuptial flight of *Prenolepis imparis* (Say) (Hymenoptera: Formicidae). *Journal of the New York Entomological Society* 73: 6–12.

Taylor, R. W. 1967. A monographic revision of the ant genus *Ponera* Latreille (Hymenoptera: Formicidae). *Pacific Insects Monographs* 13: 1–112.

Taylor, R. W. 1968. Nomenclature and synonymy of the North American ants of the genera *Ponera* and *Hypoponera*. *Entomological News* 79: 63–66.

Thomas, M. L., C. M. Payne-Makrisâ, A. V. Suarez, N. D. Tsutsui, and D. A. Holway. 2007. Contact between supercolonies elevates aggression in Argentine ants. *Insectes Sociaux* 54: 225–233.

Thompson, C. R. 1980. *Monograph of the Solenopsis* (*Diplorhoptrum*) *of Florida*. PhD. Dissertation, University of Florida. 115 pp.

Thompson, C. R. 1982. A new *Solenopsis* (*Diplorhoptrum*) species from Florida (Hym. Formicidae). *Journal of the Kansas Entomological Society* 55: 485–488.

Thompson, C. R. 1989. The thief ants, *Solenopsis molesta* group of Florida (Hymenoptera: Formicidae). *Florida Entomologist* 72: 268–283.

Thompson, C. R., and C. Johnson. 1989. Rediscovered species and revised key to the Florida thief ants (Hymenoptera: Formicidae). *Florida Entomologist* 72: 697–698.

Thompson, J. N. 1981. Elaiosomes and fleshy fruits: Phenology and selection pressures for ant-dispersed seeds. *American Naturalist* 117: 104–108.

Topoff, H., D. Bodoni, P. Sherman, and L. Goodloe. 1987. The role of scouting in slave raids by *Polyergus breviceps* (Hymenoptera: Formicidae). *Psyche* 94: 261–270.

Topoff, H., S. Cover, L. Greenberg, L. Goodloe, and P. Sherman. 1988. Colony founding by queens of the obligatory slave-making ant, *Polyergus breviceps*: The role of the Dufour's gland. *Ethology* 78: 209–218.

Topoff, H., and L. Greenberg. 1988. Mating behavior of the socially-parasitic ant *Polyergus breviceps*: The role of the mandibular glands. *Psyche* 81: 81–87.

Topoff, H., B. LaMon, L. Goodloe, and M. Goldstein. 1984. Social and orientation behavior of *Polyergus breviceps* during slave-making raids. *Behavioral Ecology and Sociobiology* 15: 273–279.

Topoff, H., and J. Mirenda. 1980. Army ants do not eat and run: Influence of food supply on emigration behavior in *Neivamyrmex nigrescens*. *Animal Behavior* 28: 1040–1045.

Topoff, H., M. Pagani, M. Goldstein, and L. Mack. 1985. Orientation behavior of the slave-making ant *Polyergus breviceps* in an oak-woodland habitat. *Journal of the New York Entomological Society* 93: 1041–1046.

Topoff, H., and E. Zimmerli. 1993. Colony takeover by a socially parasitic ant, *Polyergus breviceps*: The role of chemicals obtained during host-queen killing. *Animal Behavior* 46: 479–486.

Torres, J. A. 1984a. Niches and coexistence of ant communities in Puerto Rico: Repeated patterns. *Biotropica* 16: 284–295.

Torres, J. A. 1984b. Diversity and distribution of ant communities in Puerto Rico. *Biotropica* 16: 296–303.

LITERATURE CITED

Torres, J. A., M. Santiago, and M. Salgado. 1999. The effects of the fungus-growing ant, *Trachymyrmex jamaicensis*, on soil fertility and seed germination in a subtropical dry forest. *Tropical Ecology* 40: 237–245.

Torres, J. A., and R. R. Snelling. 1997. Biogeography of Puerto Rican ants: A non-equilibrium case? *Biodiversity and Conservation* 6: 1103–1121.

Torres, J. A., R. R. Snelling and T. H. Jones. 2000a. Distribution, ecology and behavior of *Anochetus kempfi* (Hymenoptera: Formicidae) and description of the sexual forms. *Sociobiology* 36: 505–516.

Torres, J. A., R. Thomas, M. Leal, and T. Gush. 2000b. Ant and termite predation by the tropical blind-snake *Typhlops platycephalus*. *Insectes Sociaux* 47: 1–6.

Trager, J. C. 1984. A revision of the genus *Paratrechina* (Hymenoptera: Formicidae) of the Continental United States. *Sociobiology* 9: 49–162.

Trager, J. C. 1988. A revision of *Conomyrma* (Hymenoptera: Formicidae) from the southeastern United States, especially Florida, with keys to the species. *Florida Entomologist* 71: 11–29.

Trager, J. C. 1991. A revision of the fire ants, *Solenopsis geminata* group (Hymenoptera: Formicidae: Myrmicinae). *Journal of the New York Entomological Society* 99: 141–198.

Trager, J. C. 1997. A preliminary list of ants of the St. Louis Region. On line at: http://research.amnh .org/entomology/social_insects/invtrager .html.

Trager, J. C. 2013. Global revision of the dulotic ant genus *Polyergus* (Hymenoptera: Formicidae, Formicinae, Formicini). *Zootaxa* 3722: 501–548.

Trager, J. C., and C. Johnson. 1985. A slave making ant in Florida: *Polyergus lucidus*, with observations on the natural history of its host, *Formica archboldi* (Hymenoptera: Formicidae). *Florida Entomologist* 68: 261–266.

Trager, J. C., and C. Johnson. 1988. The ant genus *Leptogenys* (Hymenoptera: Formicidae, Ponerinae) in the United States. pp. 29–34 in *Advances in Myrmecology*, J. C. Trager, Ed. E. J. Brill, N.Y. 551 + xxvii pp.

Trager, J. C., J. A. MacGown and M. D. Trager. 2007. Revision of the Nearctic endemic *Formica pallidefulva* group. *Memoirs of the American Entomological Institute* 80: 610–636.

Traniello, J. F. A. 1977. Recruitment behavior, orientation, and the organization of foraging in the carpenter ant *Camponotus pennsylvanicus* DeGeer (Hymenoptera: Formicidae). *Behavioral Ecology and Sociobiology* 2: 61–79.

Traniello, J. F. A. 1980. Colony specificity in the trail pheromone of an ant. *Naturwissenschaften* 67: 361.

Traniello, J. F. A. 1982. Population structure and social organization in the primitive ant *Amblyopone pallipes* (Hymenoptera: Formicidae). *Psyche* 89: 65–80.

Traniello, J. F. A. 1983. Social organization and foraging success in *Lasius neoniger* (Hymenoptera: Formicidae): Behavioral and ecological aspects of recruitment communication. *Oecologia* (Berlin) 59: 94–100.

Traniello, J. F. A., and S. N. Beshers. 1991. Polymorphism and size-pairing in the harvester ant *Pogonomyrmex badius*: A test of the ecological release hypothesis. *Insectes Sociaux* 38: 121–127.

Traniello, J. F. A., and S. C. Levings. 1986. Intra- and intercolony patterns of nest dispersion in the ant *Lasius neoniger*: Correlations with territoriality and foraging ecology. *Oecologia* (Berlin) 69: 413–419.

Treat, M. 1879. A chapter in the history of ants. *Harper's New Monthly Magazine* 58: 176–184.

Tschinkel, W. R. 1987. Seasonal life history and nest architecture of a winter-active ant, *Prenolepis imparis*. *Insectes Sociaux* 34: 143–164.

Tschinkel, W. R. 1988. Distribution of fire ants *Solenopsis invicta* and *S. geminata* in north Florida in relation to habitat and disturbance. *Annals of the Entomological Society of America* 81: 76–81.

Tschinkel, W. R. 1999. Sociometry and sociogenesis of colonies of the harvester ant, *Pogonomyrmex badius*: Distribution of workers, brood and seeds within the nest in relation to colony size and season. *Ecological Entomology* 24: 222–237.

Tschinkel, W. R. 2002. The natural history of the arboreal ant, *Crematogaster ashmeadi*. *Journal of Insect Science* 2: 1–15.

Tschinkel, W. R. 2004. The nest architecture of the Florida harvester ant, *Pogonomyrmex badius*. *Journal of Insect Science* 4: 19 pp.

Tschinkel, W. R. 2005. The nest architecture of the ant, *Camponotus socius*. *Journal of Insect Science* 5: 18 pp.

Tschinkel, W. R. 2006. *The Fire Ants*. The Belknap Press of Harvard University Press, Cambridge, Massachusetts. 723 pp.

Tschinkel, W. R. 2014. Nest relocation and excavation in the Florida harvester ant, *Pogonomyrmex badius*. *PLoS ONE* 9: 18 pp.

Tschinkel, W. R., and A. Bhatkar. 1974. Oriented mound building in the ant, *Trachymyrmex septentrionalis*. *Environmental Entomology* 3: 667–673.

Tschinkel, W. R., and C. A. Hess. 1999. Arboreal ant community of a pine forest in northern Florida. *Annals of the Entomological Society of America* 92: 63–70.

Tsutsui, N. D., and T. J. Case. 2001. Population genetics and colony structure of the Argentine ant (*Linepithema humile*) in its native and introduced ranges. *Evolution* 55: 976–985.

Turner, W. R., D. D. Wilcove, and H. M. Swain. 2006. *State of the Scrub: Conservation, Progress, Management Responsibilities, and Land Acquisition Priorities for Imperiled Species of Florida's Lake Wales Ridge.* Archbold Biological Station, Lake Placid, Florida. 160 pp.

Tynes, J. S. 1964. *Biological and ecological studies of ants of the subgenus Colobopsis in Mississippi.* Unpublished doctoral dissertation, Mississippi State University. 73 pp.

Umphrey, G. J. 1996. Morphometric discrimination among sibling species in the fulva-rudis-texana complex of the ant genus *Aphaenogaster* (Hymenoptera: Formicidae). *Canadian Journal of Zoology* 74: 528–559.

Van Pelt, A. F., Jr. 1950. *Orasema* in nests of *Pheidole dentata* Mayr (Hymenoptera: Formicidae). *Entomological News* 61: 49–163.

Van Pelt, A. F., Jr. 1953. Notes on the above-ground activity and a mating flight of *Pogonomyrmex badius* (Latr.). *Journal of the Tennessee Academy of Science* 28: 164–168.

Van Pelt, A. F., Jr. 1958. The ecology of the ants of the Welaka Reserve, Florida. (Hymenoptera: Formicidae). II. Annotated list. *American Midland Naturalist* 59: 1–57

Van Pelt, A. F., Jr., and J. B. Gentry. 1985. *The Ants (Hymenoptera: Formicidae) of the Savannah River Plant, South Carolina.* Publication of the Savannah River Plant, Natural Environmental Research Park Program SRO-NERP-14. 56 pp.

Vasconcelos, H. L. 1999. Effects of forest disturbance on the structure of ground-foraging ant communities in central Amazonia. *Biodiversity Cons.* 8: 409–420.

Vazquez, R. J., P. G. Koehler, and R. Pereira. 2008. Black pyramid ants emerging as another serious pest in Florida. *Florida Pest Pro* November–December 2008: 10–13.

Vonshak, M., T. Dayan, and A. Hefetz. 2012. Interspecific displacement mechanisms by the invasive little fire ant *Wasmannia auropunctata*. *Biological Invasions.* 14: 851–861.

Walker, J. M. 1984. *The soldier caste and colony structure in the ant Camponotus (Colobopsis) impressus.* PhD dissertation, University of California, Davis. 141 pp.

Walker, J. M., and J. Stamps. 1986. A test of optimal caste ratio theory using the ant *Camponotus (Colobopsis) impressus. Ecology* 67: 1052–1062.

Wang, Y., U. G. Mueller, and J. Clardy. 1999. Antifungal diketopiperazines from symbiotic fungus of fungus-growing ant *Cyphomyrmex minutus. Journal of Chemical Ecology* 25: 935–941.

Ward, P. S. 1985. The Nearctic species of the genus *Pseudomyrmex* (Hymenoptera: Formicidae). *Quaestiones Entomologicae* 21: 209–246.

Ward, P. S. 1987. Distribution of the introduced Argentine ant (*Iridomyrmex humilis*) in natural habitats of the Lower Sacramento Valley and its effects on the indigenous ant fauna. *Hilgardia* 55: 1–16.

Ward, P. S. 1988. Mesic elements in the western Nearctic ant fauna: Taxonomic and biological notes on *Amblyopone, Proceratium,* and *Smithistruma* (Hymenoptera: Formicidae). *Journal of the Kansas Entomological Society* 61: 102–124.

Ward, P. S. 1989. Systematic studies on pseudomyrmecine ants: Revision of the *Pseudomyrmex oculatus* and *P. subtillissimus* species groups, with taxonomic comments on other species. *Quaestiones Entomologicae* 25: 393–468.

Ward, P. S. 1992. Ants of the genus *Pseudomyrmex* (Hymenoptera: Formicidae) from Dominican amber, with a synopsis of the extant Antillean species. *Psyche* 99: 55–85.

Ward, P. S. 1993. Systematic studies on *Pseudomyrmex* acacia-ants. *Journal of Hymenoptera Research* 2: 117–168.

Ward, P. S. 1999. Deceptive similarity in army ants of the genus *Neivamyrmex* (Hymenoptera: Formicidae). Taxonomy, distribution and biology of *N. californicus* (Mayr) and *N. nigrescens* (Cresson). *Journal of Hymenoptera Research* 8: 74–97.

Ward, P. S. 2005. Synoptic review of the ants of California. *Zootaxa* 936: 1–68.

Ward, P. S. 2007. Phylogeny, classification, and species-level taxonomy of ants (Hymenoptera: Formicidae). *Zootaxa* 1668: 549–563.

Warner, J., R. H. Scheffrahn, and B. Cabrera. 2013. White-footed ant, *Technomyrmex difficilis* (= *albipes*) Forel (Insecta: Hymenoptera: Formicidae: Dolichoderinae). *University of Florida IFAS Extension Document EENY-273*: 1–5.

Watkins, J. F., II. 1972. The taxonomy of *Neivamyrmex texanus,* n. sp. *N. nigrescens* and *N. californicus* (Formicidae: Dorylinae), with distribution maps and keys to the species of *Neivamyrmex* of the United States. *Journal of the Kansas Entomological Society* 45: 347–372.

Watkins, J. F., II. 1982. The army ants of Mexico. *Journal of the Kansas Entomological Society* 55: 197–247.

Watkins, J. F., II. 1985. The identification and distribution of the army ants of the United States of America (Hymenoptera: Formicidae: Ecitoninae). *Journal of the Kansas Entomological Society* 58: 479–502.

Webb, S. D. 1990. Historical biogeography. pp. 70–100 in R. L. Myers and J. J. Ewel, Eds. *Ecosystems of Florida.* University of Central Florida Press, Orlando. 765 pp.

Weber, N. A. 1934a. A new *Strumigenys* from Illinois (Hymenoptera: Formicidae). *Psyche* 41: 63–65.

Weber, N. A. 1934b. Notes on Neotropical ants, including the descriptions of new forms. *Revista de Entomología* 4: 22–59.

Weber, N. A. 1955. Fungus-growing ants and their fungi: *Cyphomyrmex rimosus minutus* Mayr. *Journal of the Washington Academy of Sciences* 45: 275–281.

Weber, N. A. 1967. The fungus-growing ant, *Trachymyrmex jamaicensis*, on Bimini Island, Bahamas (Hymenoptera: Formicidae). *Entomological News* 78: 107–109.

Weems, H. V., Jr., F. C. Thompson, G. Rotheray, and M. A. Deyrup. 2003. The genus *Rhopalosyrphus* (Diptera: Syrphidae). *Florida Entomologist* 86: 186–193.

Weseloh, R. M. 1994. Spatial distribution of the ants *Formica subsericea*, *F. neogagates*, and *Aphaenogaster fulva* (Hymenoptera: Formicidae) in Connecticut. *Environmental Entomology* 23: 1165–1170.

Wesson, L. G. Jr. 1936. Contributions toward the biology of *Strumigenys pergandei*: A new food relationship among ants (Hym: Formicidae). *Entomological News* 47: 171–174.

Wesson, L. G., Jr. 1939. Contribution to the natural history of *Harpagoxenus americanus* (Hymenoptera: Formicidae). *Transactions of the American Entomological Society* 65: 97–122.

Wesson, L. G., Jr., and R. G. Wesson. 1939. Notes on *Strumigenys* from southern Ohio, with descriptions of six new species. *Psyche* 46: 91–112.

Wesson, L. G., Jr., and R. G. Wesson. 1940. A collection of ants from southcentral Ohio. *American Midland Naturalist* 21: 89–103.

Westoby, M., L. Hughes, and B. L. Rice. 1991. Seed dispersal by ants: Comparing infertile with fertile soils. pp. 434–447 in Huxley, C. R. and D. F. Cutler, Eds. *Ant–plant interactions.* Oxford University Press, New York. 601 pp.

Westwood, J. O. 1838. *The Entomologist's Text Book.* Wm. S. Orr and Co., London. 432 pp.

Westwood, J. O. 1839. *An Introduction to the Modern Classification of Insects; Founded on the Natural Habits and Corresponding Organization of the Different Families* 2 (part 11). London. pp. 193–224.

Westwood, J. O. 1840. Observations on the genus *Typhlopone*, with descriptions of several exotic species of ants. *Annals of the Magazine of Natural History* 6: 81–89.

Wetterer, J. K. 2008a. Worldwide spread of the longhorn crazy ant, *Paratrechina longicornis*. *Myrmecological News* 11: 137–149.

Wetterer, J. K. 2008b. *Technomyrmex difficilis* (Hymenoptera: Formicidae) in the West Indies. *Florida Entomologist* 91: 428–430.

Wetterer, J. K. 2009a. Worldwide spread of the penny ant, *Tetramorium bicarinatum* (Hymenoptera: Formicidae). *Sociobiology* 54: 811–830.

Wetterer, J. K. 2009b. Worldwide spread of the ghost ant, *Tapinoma melanocephalum* (Hymenoptera: Formicidae). *Myrmecological News* 12: 23–33.

Wetterer, J. K. 2009c. Worldwide spread of the flower ant, *Monomorium floricola* (Hymenoptera: Formicidae). *Myrmecological News* 13: 19–27.

Wetterer, J. K. 2010. Worldwide spread of the wooly ant, *Tetramorium lanuginosum* (Hymenoptera: Formicidae). *Myrmecological News* 13: 81–88.

Wetterer, J. K. 2011a. Worldwide spread of the tropical fire ant, *Solenopsis geminata* (Hymenoptera: Formicidae). *Myrmecological News* 14: 21–45.

Wetterer J. K. 2011b. Worldwide spread of the membraniferous dacetine ant, *Strumigenys membranifera* (Hymenoptera: Formicidae). *Myrmecological News* 14: 129–135.

Wetterer, J. K. 2012a. Worldwide spread of the African big-headed ant, *Pheidole megacephala* (Hymenoptera: Formicidae). *Myrmecological News* 17: 51–62.

Wetterer, J. K. 2012b. Worldwide spread of Emma's dacetine ant, *Strumigenys emmae* (Hymenoptera: Formicidae). *Myrmecological News* 16: 69–74.

Wetterer, J. K. 2012c. Worldwide spread of Roger's dacetine ant, *Strumigenys rogeri* (Hymenoptera: Formicidae). *Myrmecological News* 16: 1–6.

Wetterer, J. K. 2013a. Worldwide spread of the difficult white-footed ant, *Technomyrmex difficilis* (Hymenoptera: Formicidae). *Myrmecological News* 18: 93–97.

Wetterer, J. K. 2013b. World-wide spread of the little fire ant, *Wasmannia auropunctata* (Hymenoptera: Formicidae). *Terrestrial Arthropod Reviews* 6: 173–184.

Wetterer, J. K., O. Davis, and J. R. Williamson. 2014. Boom and bust of the tawny crazy ant, *Nylanderia fulva* (Hymenoptera: Formicidae), on St. Croix, U.S. Virgin Islands. *Florida Entomologist* 97: 1099–1103.

Wetterer, J. K., and S. Hugel. 2014. First North American records of the Old World ant cricket *Myrmecophilus americanus* (Orthoptera: Myrmecophilidae). *Florida Entomologist* 97: 126–129.

LITERATURE CITED

Wetterer, J. K., and J. L. W. Keularts. 2008. Population explosion of the hairy crazy ant, *Paratrechina pubens* (Hymenoptera: Formicidae), on St. Croix, U.S. Virgin Islands. *Florida Entomologist* 91: 423–427.

Wetterer, J. K., S. E. Miller, D. E. Wheeler, C. A. Olson, D. A. Polhemus, M. Pitts, I. W. Ashton, A. G. Himler, M. M. Yospin, K. R. Helms, E. L. Harken, J. Gallaher, M. Nelson, J. Litsinger, A. Southern, and T. L. Burgess. 1999. Ecological dominance by *Paratrechina longicornis* (Hymenoptera: Formicidae), an invasive tramp ant, in Biosphere 2. *Florida Entomologist* 82: 381–388.

Wetterer, J. K., and S. D. Porter. 2003. The little fire ant, *Wasmannia auropunctata*: Distribution, impact and control. *Sociobiology* 41: 1–41.

Wetterer, J. K., A. L. Wild, A. V. Suarez, N. Roura-Pascual, and X. Espadalar. 2009. Worldwide spread of the Argentine ant, *Linepithema humile* (Hymenoptera: Formicidae). *Myrmecological News* 12: 187–194.

Wheeler, D. E., and B. Hölldobler. 1985. Cryptic phragmosis: The structural modification. *Psyche* 92: 337–353.

Wheeler, D. E., and H. F. Nijhout. 1981. Soldier determination in ants: A new role for juvenile hormone. *Science* 213: 361–363.

Wheeler, G. C., and J. Wheeler. 1956. The ant larvae of the subfamily Pseudomyrmecinae (Hymenoptera: Formicidae). *Annals of the Entomological Society of America* 49: 374–398.

Wheeler, G. C., and J. Wheeler. 1963. *The Ants of North Dakota*. University of North Dakota Press, Grand Forks. 326 pp.

Wheeler, G. C., and J. Wheeler. 1978. *Brachymyrmex musculus*, a new ant in the United States. *Entomological News* 89: 189–190.

Wheeler, G. C., and J. Wheeler. 1986. *The Ants of Nevada*. Los Angeles County Museum of Natural History, Los Angeles. 138 pp.

Wheeler, G. C., J. Wheeler, and P. B. Kannowski. 1994. Checklist of the ants of Michigan (Hymenoptera: Formicidae). *Great Lakes Entomologist* 26: 297–310.

Wheeler, J. W., O. Olubajo, C. B. Storm and R. M. Duffield. 1981. Anabaseine: Venom alkaloid of *Aphaenogaster* ants. *Science* 211: 1051–1052.

Wheeler, W. M. 1900. The habits of *Ponera* and *Stigmatomma*. *Biologists Bulletin* 2: 43–69.

Wheeler, W. M. 1901. The compound and mixed nests of American ants. *American Naturalist* 35: 513–539.

Wheeler, W. M. 1902. A consideration of S. B. Buckley's "North American Formicidae." *Transactions of the Texas Academy of Sciences* 4: 1–15.

Wheeler, W. M. 1903. A revision of the North American ants of the genus *Leptothorax*. *Proceedings of the Philadelphia Academy of Natural Science* 55: 215–260.

Wheeler, W. M. 1904. The American ants of the subgenus *Colobopsis*. *Bulletin of the American Museum of Natural History* 20: 139–158.

Wheeler, W. M. 1905a. The ants of the Bahamas, with a list of the known West Indian species. *Bulletin of the American Museum of Natural History* 21: 79–135.

Wheeler, W. M. 1905b. The North American ants of the genus *Dolichoderus*. *Bulletin of the American Museum of Natural History* 21: 305–319.

Wheeler, W. M. 1905c. An annotated list of the ants of New Jersey. *Bulletin of the American Museum of Natural History* 21: 371–403.

Wheeler, W. M. 1907. The fungus-growing ants of North America. *Bulletin of the American Museum of Natural History* 23: 669–807.

Wheeler, W. M. 1908. The ants of Texas, new Mexico and Arizona (Part I). *Bulletin of the American Museum of Natural History* 24: 399–485.

Wheeler, W. M. 1910a. *Ants. Their Structure, Development and Behavior*. The Columbia University Press, New York. 663 pp.

Wheeler, W. M. 1910b. The North American ants of the genus *Camponotus* Mayr. *Annals of the New York Academy of Sciences* 20: 295–354.

Wheeler, W. M. 1911. Additions to the ant fauna of Jamaica. *Bulletin of the American Museum of Natural History* 30: 21–29.

Wheeler, W. M. 1912. New names for some ants of the genus *Formica*. *Psyche* 19: 90.

Wheeler, W. M. 1913a. Ants collected in Georgia by Dr. J. C. Bradley and Mr. W. T. Davis. *Psyche* 20: 112–117.

Wheeler, W. M. 1913b. A revision of the ants of the genus *Formica* (Linné) Mayr. *Bulletin of the Harvard Museum of Comparative Zoology* 53: 379–565.

Wheeler, W. M. 1913c. The ants of Cuba. *Bulletin of the Harvard Museum of Comparative Zoology* 54: 477–505.

Wheeler, W. M. 1915. Some additions to the North American ant-fauna. *Bulletin of the American Museum of Natural History* 34: 389–421.

Wheeler, W. M. 1916. An Indian ant introduced into the United States. *Journal of Economic Entomology* 9: 566–569.

Wheeler, W. M. 1919. A new paper-making *Crematogaster* from the southeastern United States. *Psyche* 26: 107–112.

Wheeler, W. M. 1923. The occurrence of winged females in the ant genus *Leptogenys* Roger, with descriptions of new species. *American Museum Novitates* 90: 1–16.

Wheeler, W. M. 1927. The occurrence of the pavement ant (*Tetramorium caespitum* L.) in Boston. *Psyche* 34: 164–165.

Wheeler, W. M. 1929. Two Neotropical ants established in the United States. *Psyche* 22: 89–90.

Wheeler, W. M. 1930. The ant *Prenolepis imparis* Say. *Annals of the Entomological Society of America* 23: 1–26.

Wheeler, W. M. 1931a. New and little known ant of the genera *Macromischa, Croesomyrmex*, and *Antillaemyrmex*. *Bulletin of the Museum of Comparative Zoology at Harvard University* 72: 1–34.

Wheeler, W. M. 1931b. Neotropical ants of the genus *Xenomyrmex*. *Revista de Entomología* 1: 129–139.

Wheeler, W. M. 1932. A list of the ants of Florida with descriptions of new forms. *Journal of the New York Entomological Society* 40: 1–17.

Wheeler, W. M. 1933. A second parasitic *Crematogaster*. *Psyche* 40: 83–86.

Wheeler, W. M. 1942. Studies of Neotropical ant-plants and their ants. *Bulletin of the Museum of Comparative Zoology at Harvard* 90: 1–262.

Wheeler, W. M., and I. W. Bailey. 1920. The feeding habits of pseudomyrmecine and other ants. *Transactions of the American Philosophical Society (N.S.)* 22: 237–279.

Whitcomb, W. H., H. A. Denmark, A. P. Bhatkar and G. L. Greene. 1972a. Preliminary studies on the ants of Florida soybean fields. *Florida Entomologist* 55: 129–142.

Whitcomb, W. H., H. A. Denmark, W. F. Buren, and J. F. Carroll. 1972b. Habits and present distribution in Florida of the exotic ant, *Pseudomyrmex mexicanus*. *Florida Entomologist* 55: 31–33.

Whitney, E., D. B. Means, and A. Rudloe. 2004. *Priceless Florida: Natural Ecosystems and Native Species*. Pineapple Press, Sarasota, Florida. 423 pp.

Wild, A. 2004. Taxonomy and distribution of the Argentine ant, *Linepithema humile* (Hymenoptera: Formicidae). *Annals of the Entomological Society of America* 97: 1204–1215.

Wild, A. 2007. Taxonomic revision of the ant genus *Linepithema* (Hymenoptera: Formicidae). *University of California Publications in Entomology* 126: 1–151.

Willson, M. F., B. L. Rice, and M. Westoby. 1990. Seed dispersal: A comparison of temperate plant communities. *Journal of Vegetation Science* 1: 547–562.

Wilson, E. O. 1950a. Notes on the food habits of *Strumigenys louisianae* Roger (Hymenoptera: Formicidae). *Bulletin of the Brooklyn Entomological Society* 45: 85–86.

Wilson, E. O. 1950b. A new *Leptothorax* from Alabama. *Psyche* 57: 128–130.

Wilson, E. O. 1952. Notes on *Leptothorax bradleyi* Wheeler and L. *wheeleri* M. R. Smith (Hymenoptera: Formicidae). *Entomological News* 63: 67–71.

Wilson, E. O. 1953. The ecology of some North American dacetine ants. *Annals of the Entomological Society of America* 46: 479–495.

Wilson, E. O. 1955. A monographic revision of the ant genus *Lasius*. *Bulletin of the Harvard Museum of Comparative Zoology* 113: 1–199.

Wilson, E. O. 1958. Studies on the ant fauna of Melanesia. IV. The tribe Ponerini. *Bulletin of the Harvard Museum of Comparative Zoology Harvard* 119: 320–371.

Wilson, E. O. 1959. Communication by tandem running in the ant genus *Cardiocondyla*. *Psyche* 66: 29–34.

Wilson, E. O. 1964. The ants of the Florida Keys. *Breviora* 210: 1–14.

Wilson, E. O. 1974a. Aversive behavior and competition within colonies of the ant *Leptothorax curvispinosus*. *Annals of the Entomological Society of America* 67: 777–780.

Wilson, E. O. 1974b. The soldier of the ant, *Camponotus* (*Colobopsis*) *fraxinicola*, as a trophic caste. *Psyche* 81: 182–188.

Wilson, E. O. 1975. Enemy specification in the alarm-recruitment system of an ant. *Science* 190: 798–800.

Wilson, E. O. 1976a. A social ethogram of the Neotropical arboreal ant *Zacryptocerus varians* (Fr. Smith). *Animal Behavior* 24: 354–363.

Wilson, E. O. 1976b. The organization of colony defense in the ant *Pheidole dentata* Mayr (Hymenoptera: Formicidae). *Behavioral and Ecological Sociobiology* 1: 63–81.

Wilson, E. O. 1978. Division of labor in fire ants based on physical castes (Hymenoptera, Formicidae, *Solenopsis*). *Journal of the Kansas Entomological Society* 51: 615–636.

Wilson, E. O. 1988. The biogeography of the West Indian ants (Hymenoptera: Formicidae). pp. 214–230 in J. K. Liebherr, Ed. *Zoogeography of Caribbean Insects*. Cornell University Press, Ithaca, New York. 285 pp.

Wilson, E. O. 2003. Pheidole in the New World: A dominant, hyperdiverse genus. Harvard University Press, Cambridge Massachusetts. 794 pp.

Wilson, E. O. 2005. Early ant plagues in the New World. *Nature* 433: 32.

Wilson, E. O., and W. L. Brown, Jr. 1953. The subspecies concept and its taxonomic application. *Systematic Zoology* 2: 97–111.

Wilson, E. O., and W. L. Brown, Jr. 1984. Behavior of the cryptobiotic predaceous ant *Eurhopalothrix heliscata*, n. sp. (Hymenoptera: Formicidae: Basicerotini). *Insectes Sociaux* 31: 408–428.

Wilson, E. O., and A. Francoeur. 1974. Ants of the Formica fusca group in Florida. *Florida Entomologist* 57: 115–116.

Wilson, E. O., and R. W. Taylor. 1967. The ants of Polynesia. *Pacific Insects Monographs* 14: 1–109.

Wing, M. W. 1968. Taxonomic revision of the Nearctic genus *Acanthomyops* (Hymenoptera: Formicidae). *Memoirs of the Cornell University Agricultural Experiment Station* 405: 1–173.

Witte, V., A. B. Attygalle, and J. Meinwald. 2007. Complex chemical communication in the crazy ant *Paratrechina longicornis* Latreille (Hymenoptera: Formicidae). *Chemoecology* 17: 57–62.

Wojcik, D. P. 1994. Impact of the native red imported fire ant on native ant species in Florida. pp. 269–281 in Williams, D. T., Ed. *Exotic Ants: Biology, Impact, and Control of Introduced Species.* Westview Press, San Francisco. 332 pp.

Wojcik, D. P., W. A. Banks, and W. F. Buren. 1975. First report of *Pheidole moerens* in Florida (Hymenoptera: Formicidae). *U.S. Department of Agriculture Cooperative Economic Insect Report* 25: 906.

Wolcott, G. N. 1948. Formicidae: ANTS. pp. 810–839 in The Insects of Puerto Rico—Hymenoptera. *Journal of the Agricultural University of Puerto Rico* 32: 749–975.

Woodruff, R. E., and B. M. Beck. 1989. The scarab beetles of Florida (Coleoptera: Scarabaeidae) Part II. The May or June beetles (Genus *Phyllophaga*). *Arthropods of Florida and Neighboring Land Areas* 13: 226 pp.

Wray, D. L. 1938. Notes on the southern harvester ant (*Pogonomyrmex badius* Latr.) in North Carolina. *Annals of the Entomological Society of America* 31: 196–201.

Yamauchi, K., T. Furukawa, K. Kinomura, H. Takamine, and K. Tsuji. 1991. Secondary polygyny by inbred wingless sexuals in the dolichoderine ant *Technomyrmex albipes. Behavioral Ecology and Sociobiology* 29: 313–319.

Yamauchi, K., Y. Kimura, B. Corbara, K. Kinomura and K. Tsuji. 1996. Dimorphic ergatoid males and their reproductive behavior in the ponerine ant *Hypoponera bondroiti. Insectes Sociaux* 43: 119–130.

Yamauchi, K., S. Oguchi, Y. Nakamura, H. Suetake, N. Kawada and K. Kinomura. 2001. Mating behavior of dimorphic reproductives of the ponerine ant, *Hypoponera nubatama. Insectes Sociaux* 48: 83–87.

Yang, A. S. 2006. Seasonality, division of labor, and dynamics of colony-level nutrient storage in the ant *Pheidole morrisi. Insectes Sociaux* 53: 456–452.

Yang, A. S., C. H. Martin, and H. F. Nijhout. 2004. Geographic variation of caste structure among ant populations. *Current Biology* 14: 514–519.

Yanoviak, S. P., R. Dudley, and M. Kaspari. 2005. Directed aerial descent in canopy ants. *Nature* 433: 624–626.

Yoshimura, M., and B. L. Fisher. 2012. A revision of male ants of the Malagasy Amblyoponinae (Hymenoptera: Formicidae) with resurrections of the genera *Stigmatomma* and *Xymmer. PLoS ONE* 7(3): e33325. doi:10.1371/journal.pone.0033325.

Zayas, F. de. 1981. *Entomofauna Cubana. Tomo VIII. Seccion Oligneoptera. Ordenes Hymenoptera y Strepsiptera.* Editorial Cientifica-Technica, Havana. 111 pp.

PLATES

Morphology (*Hypoponera*) (p. 9)

Plate 1

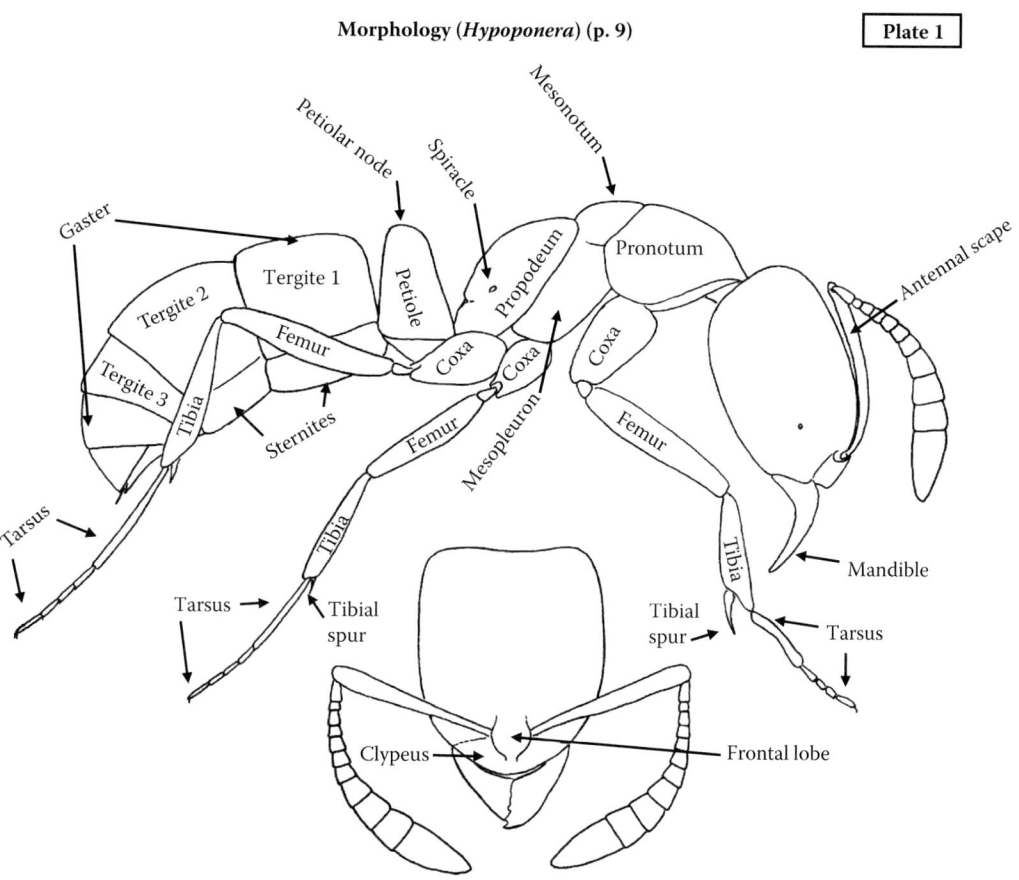

Morphology (*Cardiocondyla*) (p. 9)

Plate 2

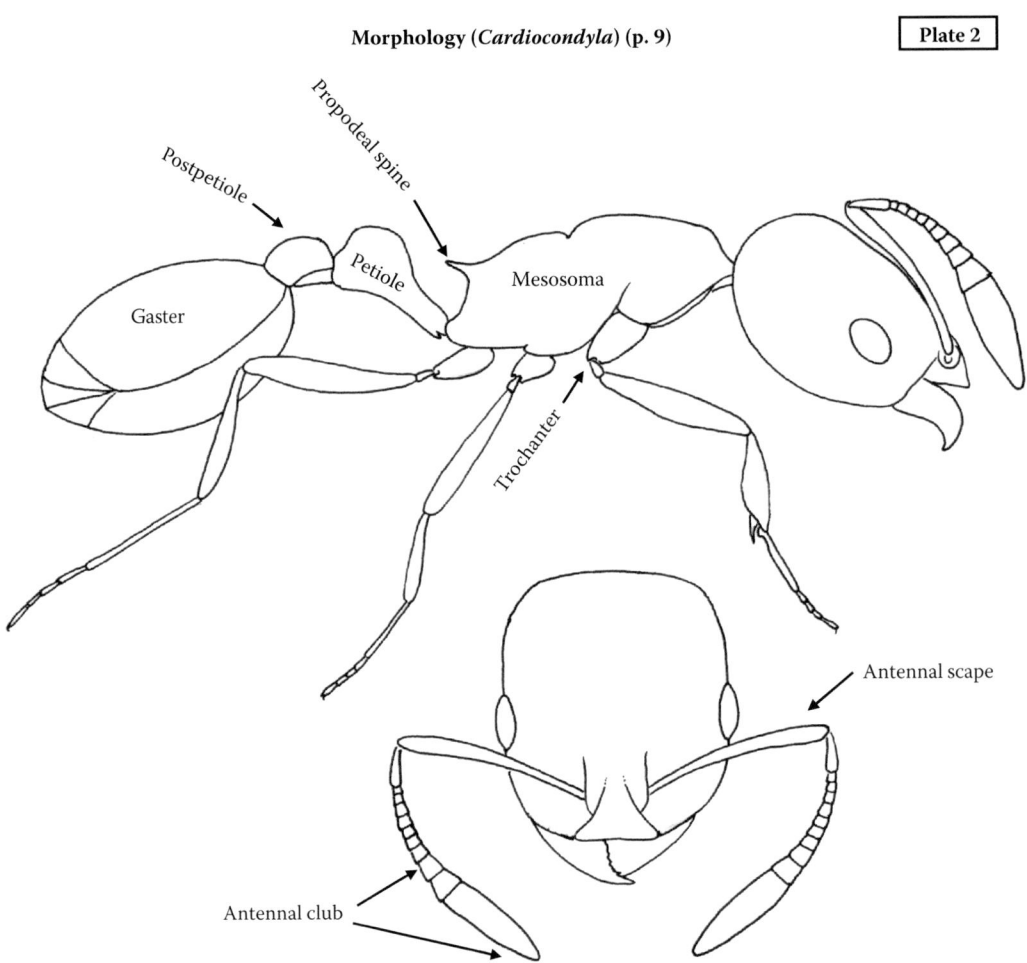

Plate 3

Florida *Stigmatomma* (saw-toothed ants) and
***Prionopelta* (minute three-toothed hunter ants)**

Tiny eyes

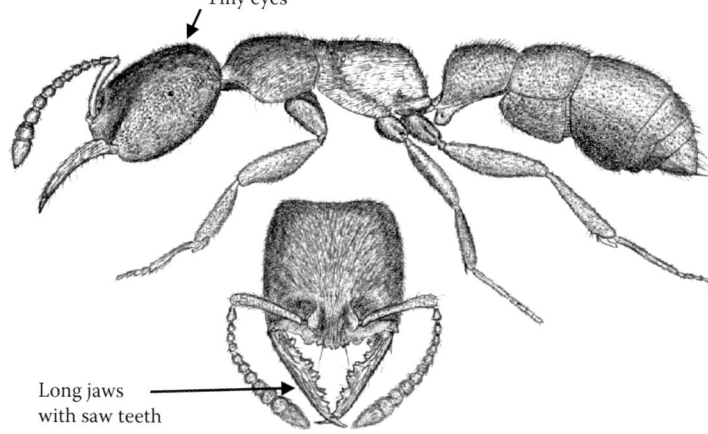

• Found in leaf litter

• Usually dark brown,
  may be reddish brown

• Widespread in Florida

• More than 5 mm in length

Long jaws
with saw teeth

***Stigmatomma pallipes* (p. 13)**

• Male *Stigmatomma*
  common in flight traps

***Stigmatomma pallipes* male**

Tiny eyes

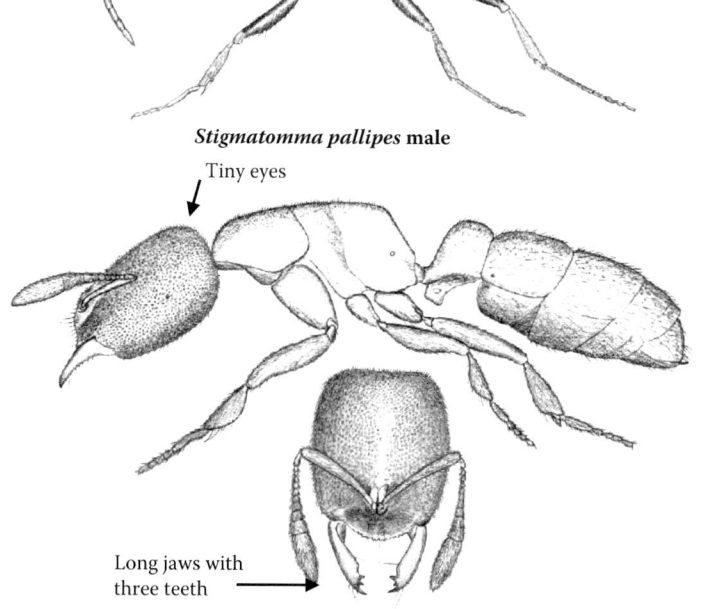

• Found in leaf litter

• In Florida, known from
  Marion and Sumter
  counties

• Small, pale yellow, about
  2.5 mm long

• Introduced from West
  Indies

Long jaws with
three teeth

***Prionopelta antillana* (p. 14)**

**Florida *Discothyrea* (pygmy egg-eating ants)
and *Gnamptogenys* (grooved ants)**

Plate 4

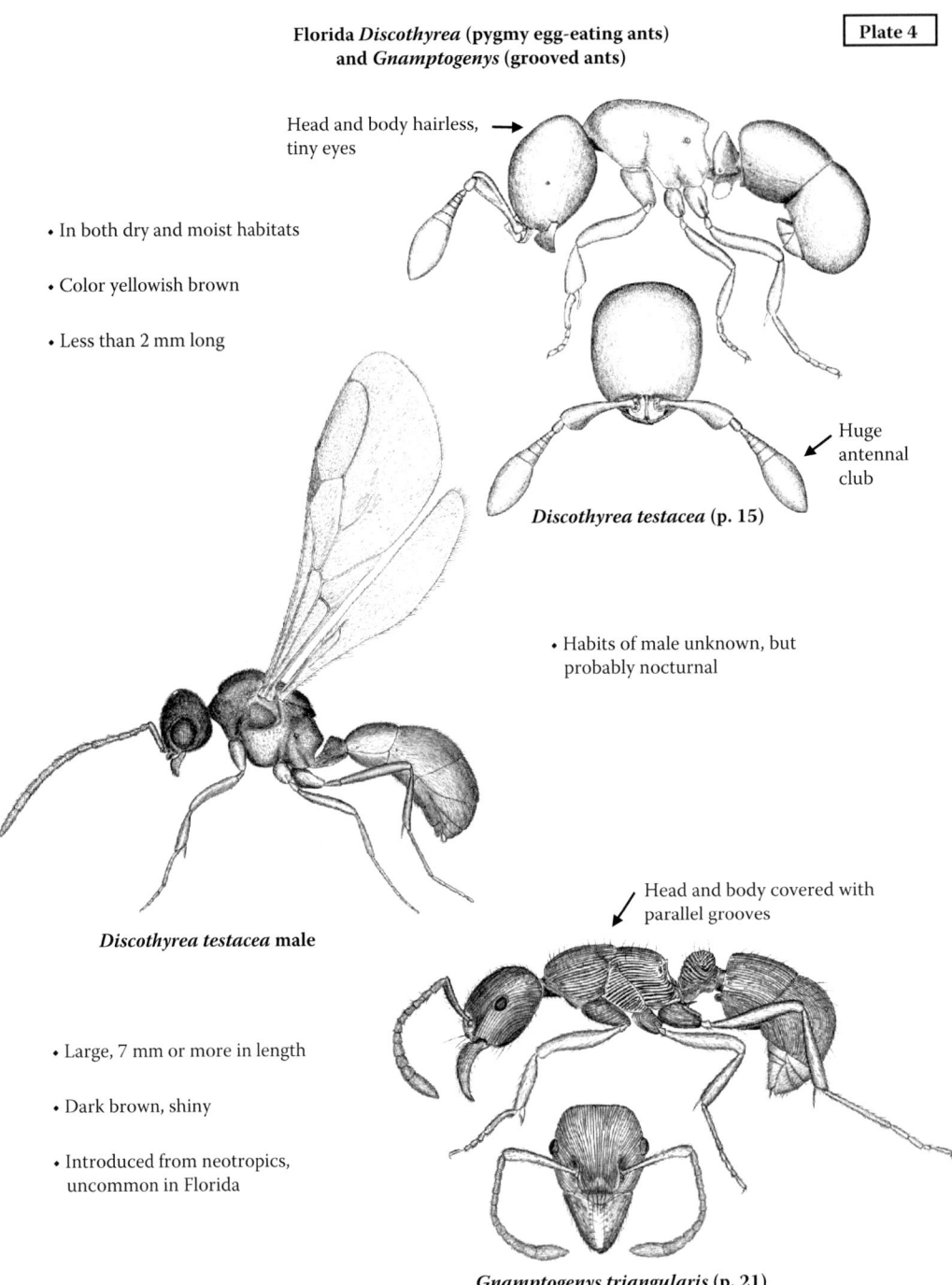

Head and body hairless,
tiny eyes →

- In both dry and moist habitats

- Color yellowish brown

- Less than 2 mm long

Huge
antennal
club

Discothyrea testacea (p. 15)

- Habits of male unknown, but
  probably nocturnal

Discothyrea testacea male

- Large, 7 mm or more in length

- Dark brown, shiny

- Introduced from neotropics,
  uncommon in Florida

Head and body covered with
parallel grooves

Gnamptogenys triangularis (p. 21)

Florida *Proceratium* (egg-eating ants)

Plate 5

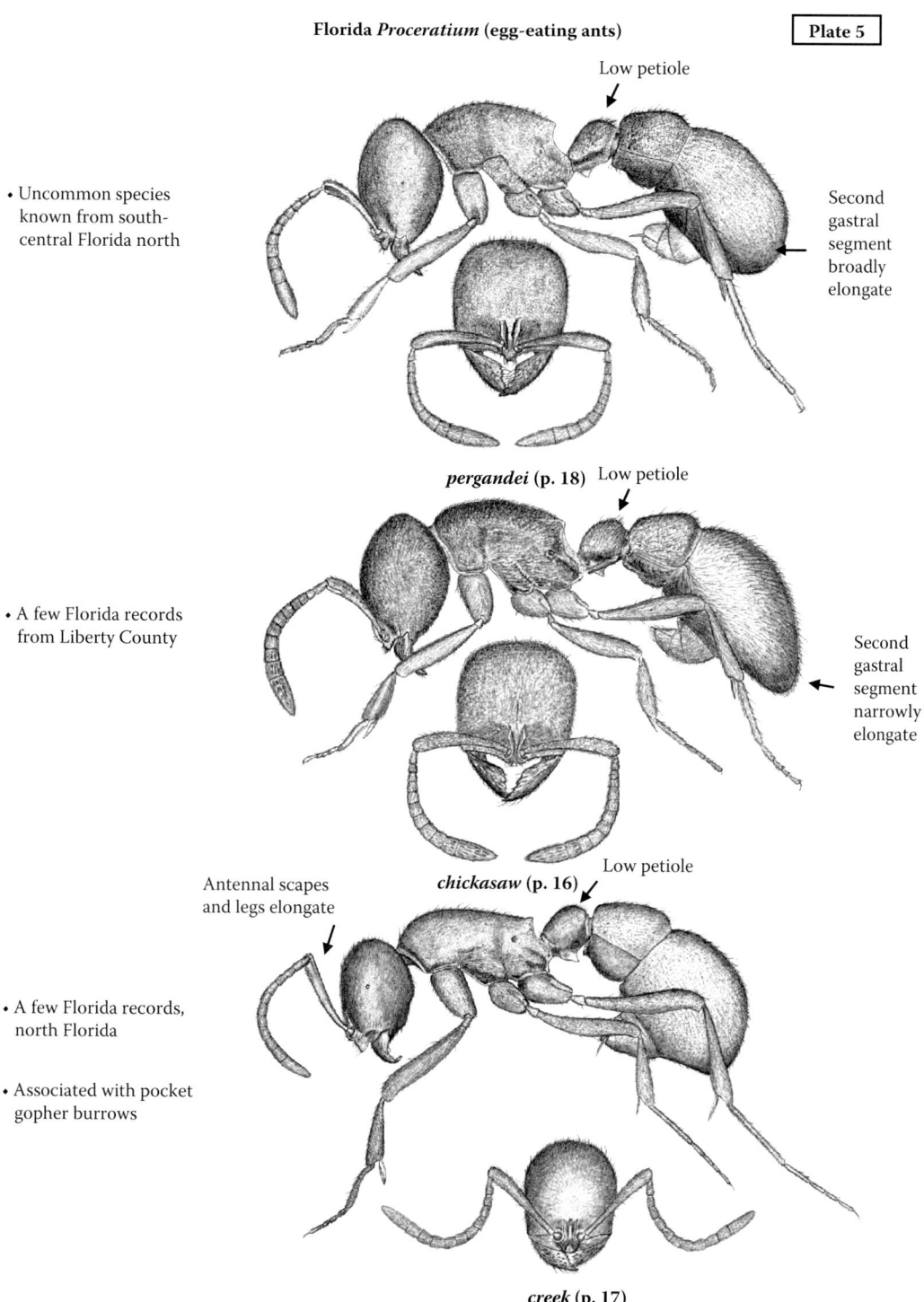

Low petiole

Second gastral segment broadly elongate

- Uncommon species known from south-central Florida north

pergandei (p. 18)

Low petiole

Second gastral segment narrowly elongate

- A few Florida records from Liberty County

chickasaw (p. 16)

Low petiole

Antennal scapes and legs elongate

- A few Florida records, north Florida

- Associated with pocket gopher burrows

creek (p. 17)

Florida *Proceratium* (egg-eating ants)

Plate 6

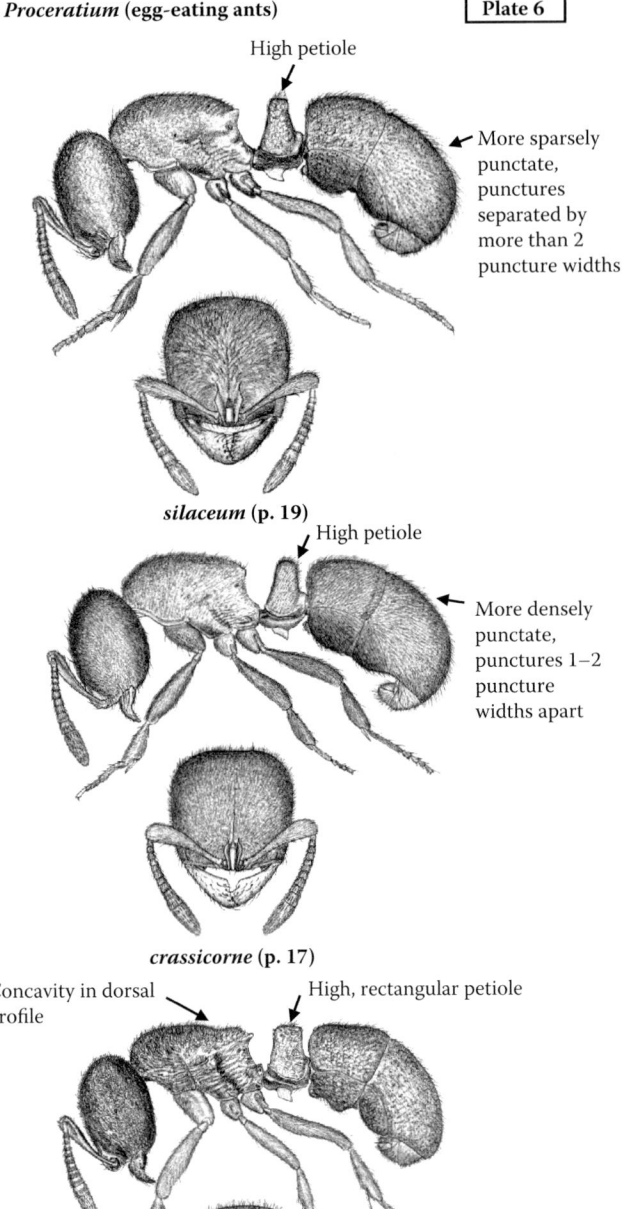

- Nests in rotten wood

- From south-central Florida north

- Length about 2.5 mm long, smaller than *croceum*

High petiole

More sparsely punctate, punctures separated by more than 2 puncture widths

silaceum (p. 19)

- Nests in rotten wood

- A northern species, in Florida collected twice

- Length about 2.5 mm long, smaller than *croceum*

High petiole

More densely punctate, punctures 1–2 puncture widths apart

crassicorne (p. 17)

- Nests in rotten wood

- North Florida

- Larger (about 3 mm long) than *silaceum* or *croceum*

Concavity in dorsal profile

High, rectangular petiole

croceum (p. 18)

266

PLATES

**Florida *Odontomachus* (snapping ants)
and *Anochetus* (lesser snapping ants)**

Plate 7

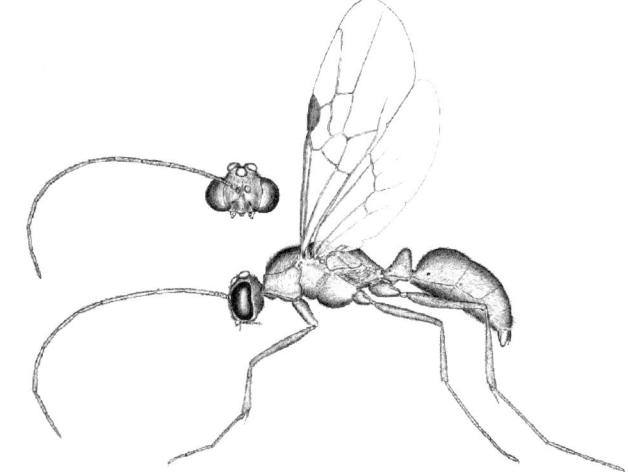

Gaster with
sparse hairs

• In natural scrub and
sandhill areas on the Lake
Wales Ridge and
Brooksville Ridge

• Dark brown with red legs

• Usually nocturnal

No ridges on
back side of
petiole

Odontomachus relictus (p. 30)

• Common in flight traps
and at lights within its
restricted range

Odontomachus relictus male

• Small (about 4 mm), in
contrast to *Odontomachus*
species (about 9 mm)

• In Florida in southern
tropical counties

• Introduced, probably from
the West Indies

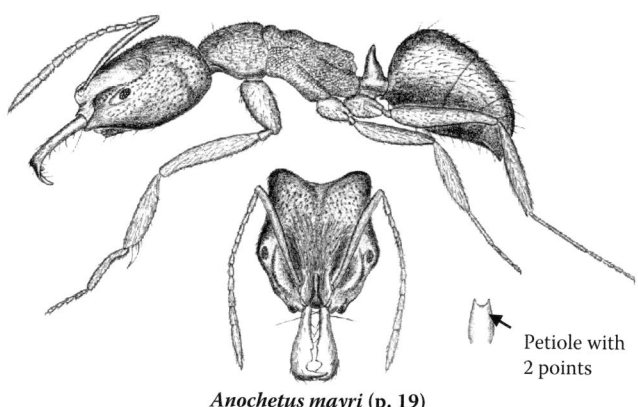

Petiole with
2 points

Anochetus mayri (p. 19)

Plate 8

Florida *Odontomachus* (snapping ants)

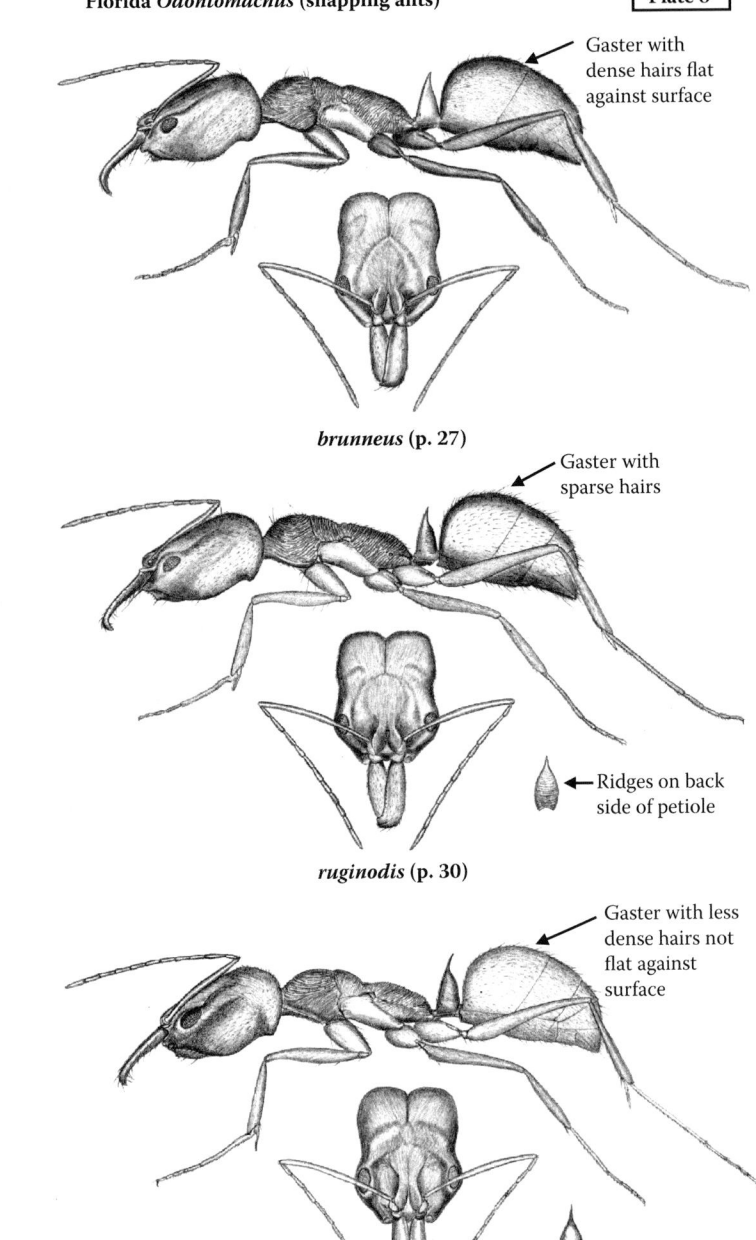

Gaster with dense hairs flat against surface

- In both dry and wet habitats

- Color dark brown

- Usually nocturnal

brunneus (p. 27)

Gaster with sparse hairs

- Often in disturbed habitats

- Usually dark brown with red legs

- Usually nocturnal

- Probably introduced from West Indies

Ridges on back side of petiole

ruginodis (p. 30)

Gaster with less dense hairs not flat against surface

- Usually in mesic habitats, sometimes in wet habitats

- Color dark brown

- Usually nocturnal

- Introduced from southern South America

No ridges on back side of petiole

haematodus (p. 29)

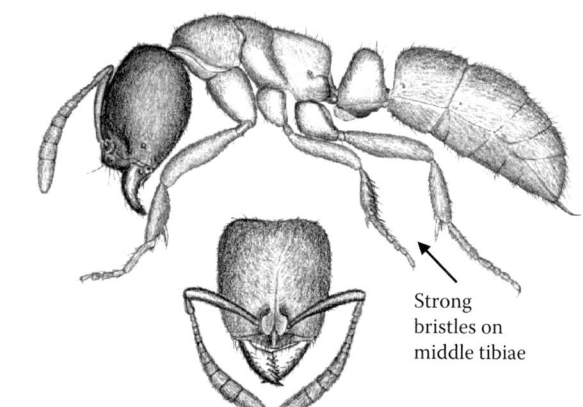

Cryptopone gilva (p. 20)

- Lives in rotten wood in wooded areas

- Reddish brown

- Larger than *Hypoponera* or *Ponera*, smaller than *Pachycondyla*

- Uncommon in Florida

Strong bristles on middle tibiae

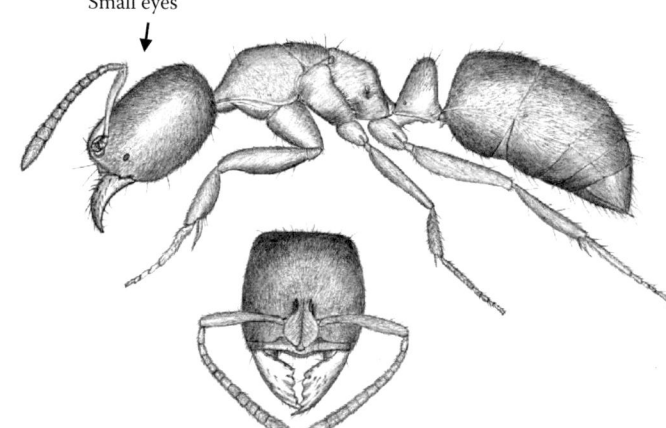

Small eyes

Pachycondyla stigma (p. 32)

- Usually associated with dead wood

- Relatively large (4–5 mm long)

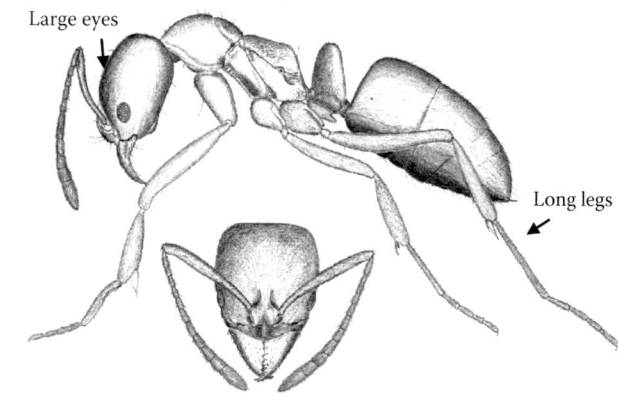

Large eyes

Long legs

Pachycondyla chinensis

- Not known from Florida, but expanding south in other southeastern states

- Introduced from Asia

Plate 10

Petiole tapers strongly above

• Lives in dry sandy areas, including coastal hammocks, scrub and sandhills

• Orange brown color

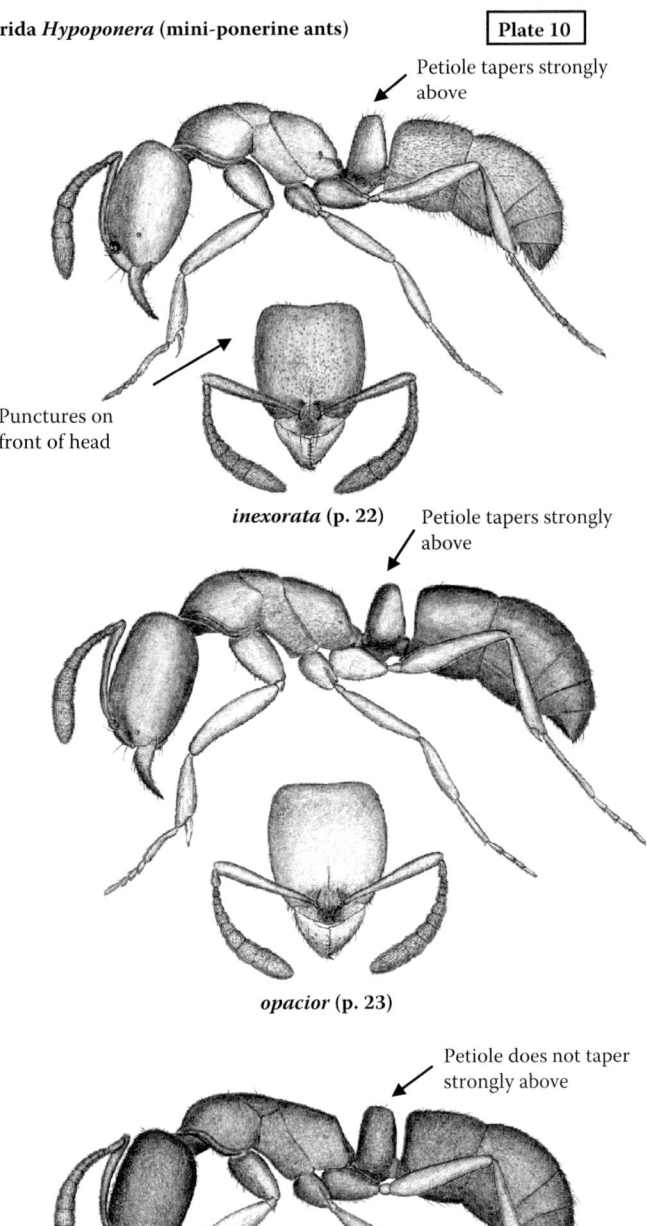

Punctures on front of head

inexorata (p. 22)

Petiole tapers strongly above

• One of the commonest ants in leaf litter in many habitats

• Color dark brown (reddish when immature)

opacior (p. 23)

Petiole does not taper strongly above

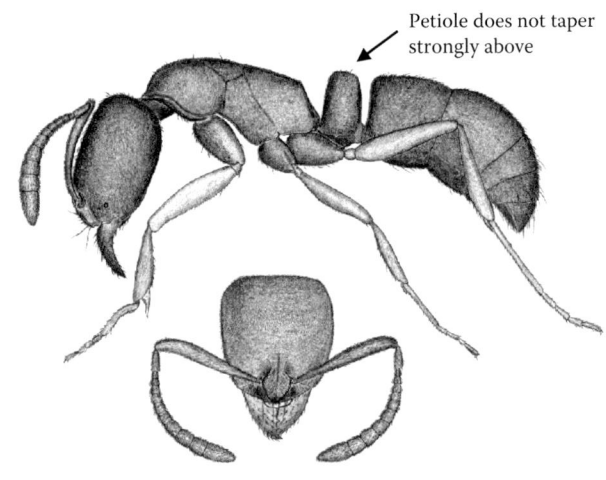

• Common in wet areas, including edges of marshes, ditches, and seaweed just above high tide level

• Color dark brown (reddish when immature)

opaciceps (p. 23)

Plate 11

**Florida *Hypoponera* (mini-ponerine ants)
and *Ponera* (little porthole ants)**

Petiole does not taper
strongly above

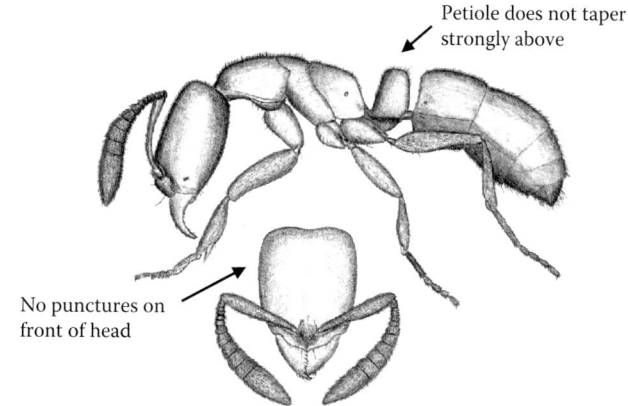

• Usually in moist disturbed
areas

• Color reddish brown

• Introduced from Old World
Tropics

No punctures on
front of head

***Hypoponera punctatissima* (p. 24)**

Dense
punctures
on head

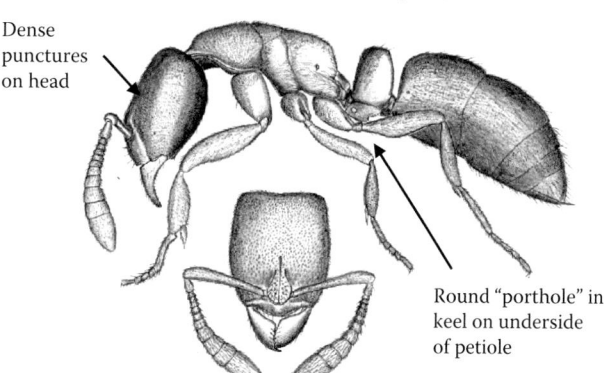

• In moist areas, in leaf litter or
rotten wood

• Color dark brown (reddish
when immature)

• In Florida in northern counties

Round "porthole" in
keel on underside
of petiole

***Ponera pennsylvanica* (p. 34)**

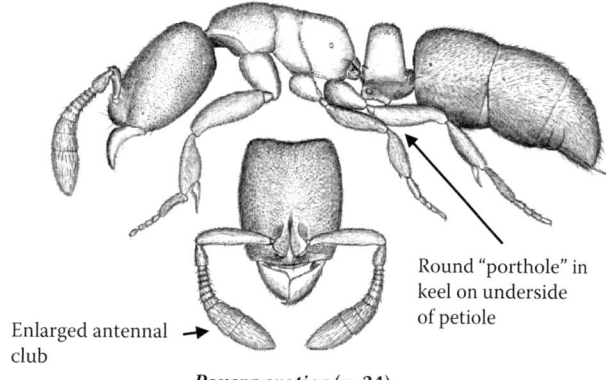

• Widespread but uncommon

• Color reddish brown

• Small: about 2 mm long

• Probably not exotic, as name
implies, but native to Florida

Round "porthole" in
keel on underside
of petiole

Enlarged antennal
club

***Ponera exotica* (p. 34)**

**Florida *Leptogenys* (beaked hunter ants)
and *Platythyrea* (silvery hunter ants)**

Plate 12

- Uncommon, generally
  nocturnal species, usually
  in woodlands

- Run very fast

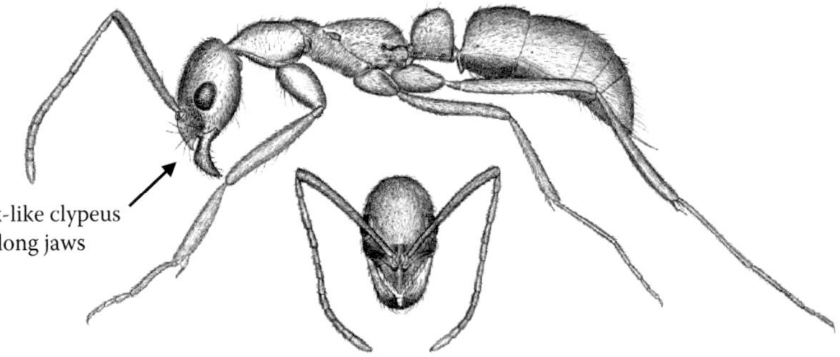

Beak-like clypeus
and long jaws

Leptogenys manni (p. 26)

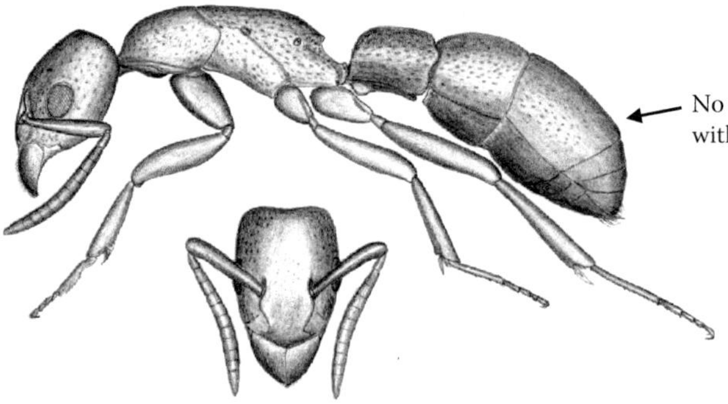

No erect hairs, charcoal gray
with silvery surface hairs

Platythyrea punctata (p. 32)

- Tropical and subtropical
  species

- Nests in dead wood

Platythyrea punctata male

PLATES

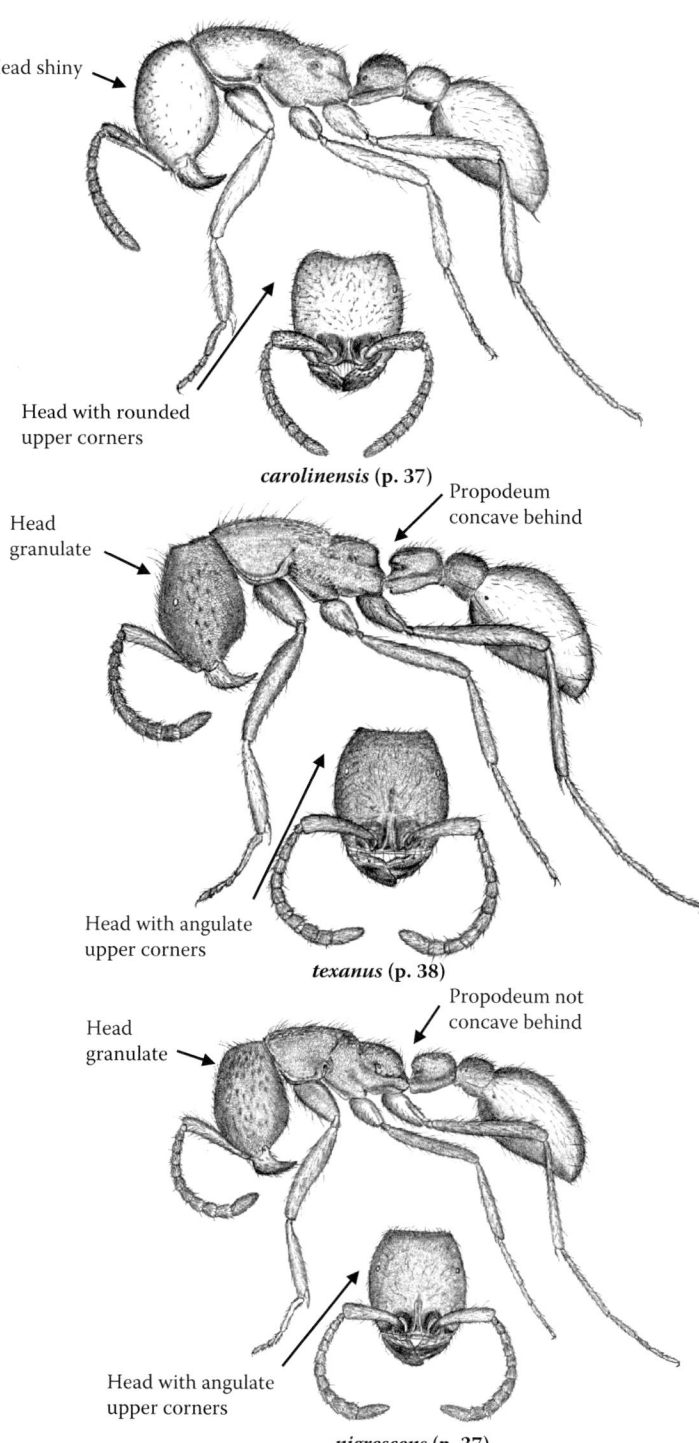

• Usually subterranean, rarely seen

• Yellowish color

• Compare with *opacithorax*

Head shiny

Head with rounded upper corners

***carolinensis* (p. 37)**

Head granulate

Propodeum concave behind

• Usually subterranean, sometimes seen during day

• Uncommon in Florida, judging by rarity of males in flight traps

• Compare with *nigrescens*

Head with angulate upper corners

***texanus* (p. 38)**

Head granulate

Propodeum not concave behind

• Not known from Florida, but occurs nearby in coastal Alabama

• Compare with *texanus*

Head with angulate upper corners

***nigrescens* (p. 37)**

Florida *Neivamyrmex* (army ants)

Plate 14

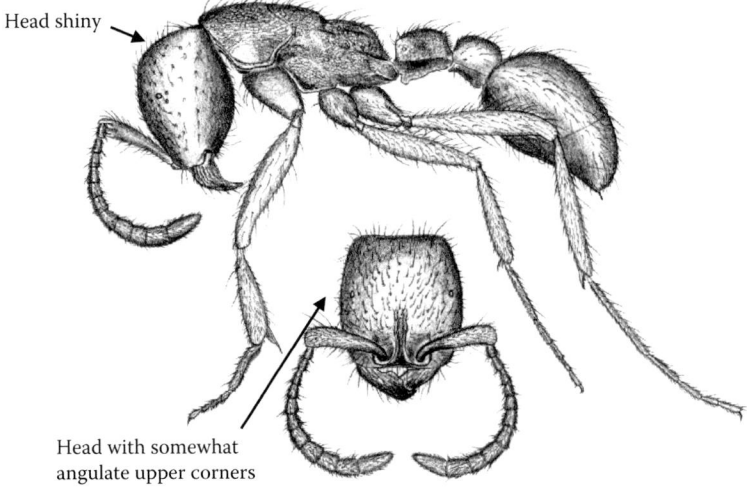

Head shiny

• Usually subterranean or nocturnal

• Reddish brown color

• Commonest Florida *Neivamyrmex*

Head with somewhat angulate upper corners

opacithorax (p. 38)

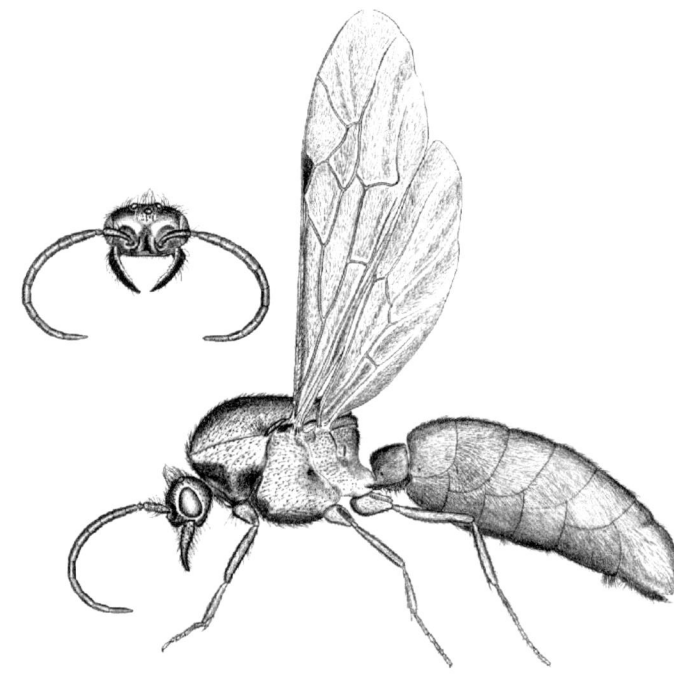

• Not usually seen in colonies

• Mimics wasps that are black with red gaster

• Often found in flight traps, a useful indicator of cryptic army ants

opacithorax **male**

Florida *Pseudomyrmex* (slender twig ants)

Plate 15

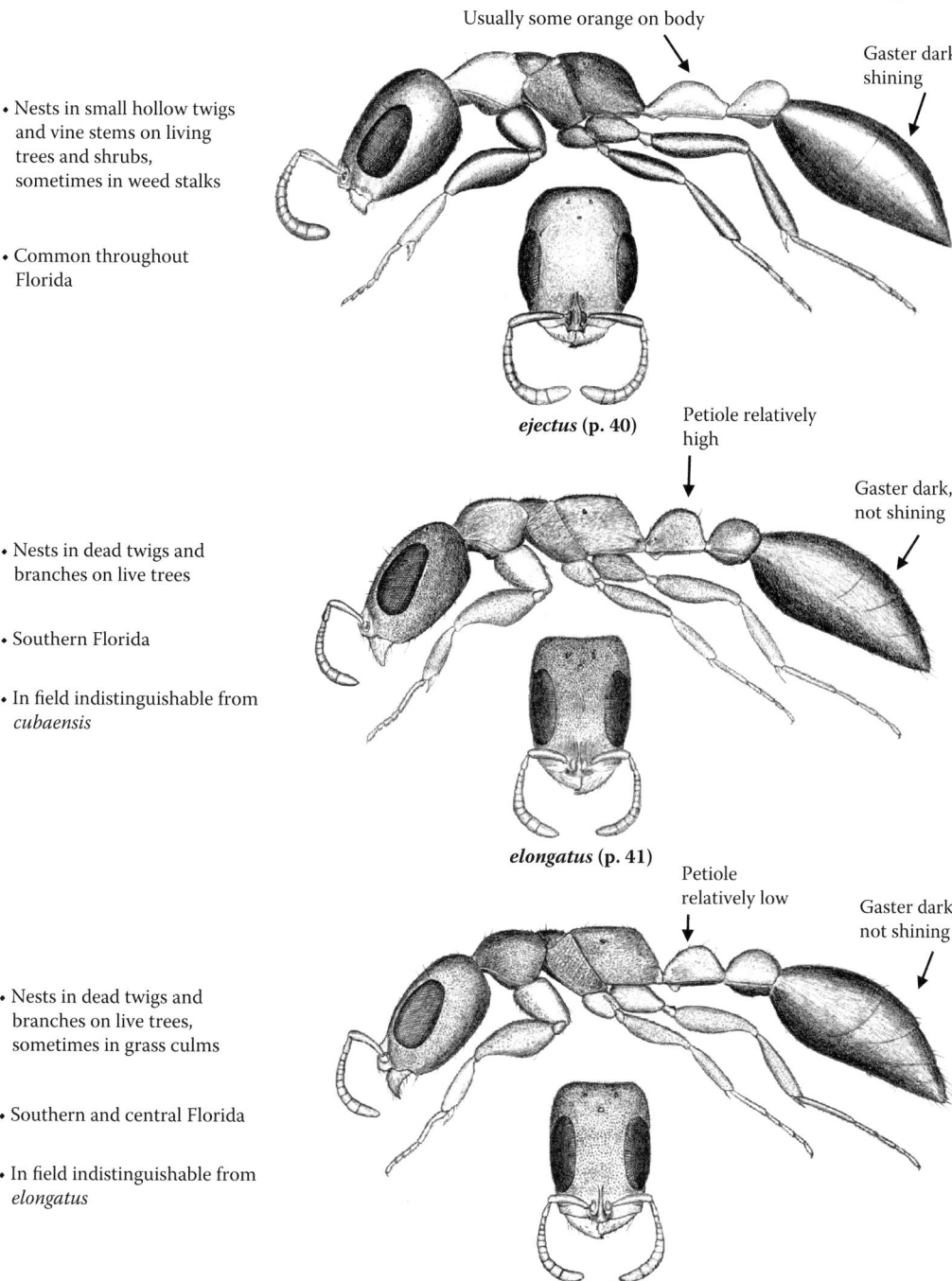

Usually some orange on body

Gaster dark, shining

• Nests in small hollow twigs and vine stems on living trees and shrubs, sometimes in weed stalks

• Common throughout Florida

ejectus (p. 40)

Petiole relatively high

Gaster dark, not shining

• Nests in dead twigs and branches on live trees

• Southern Florida

• In field indistinguishable from *cubaensis*

elongatus (p. 41)

Petiole relatively low

Gaster dark, not shining

• Nests in dead twigs and branches on live trees, sometimes in grass culms

• Southern and central Florida

• In field indistinguishable from *elongatus*

cubaensis (p. 40)

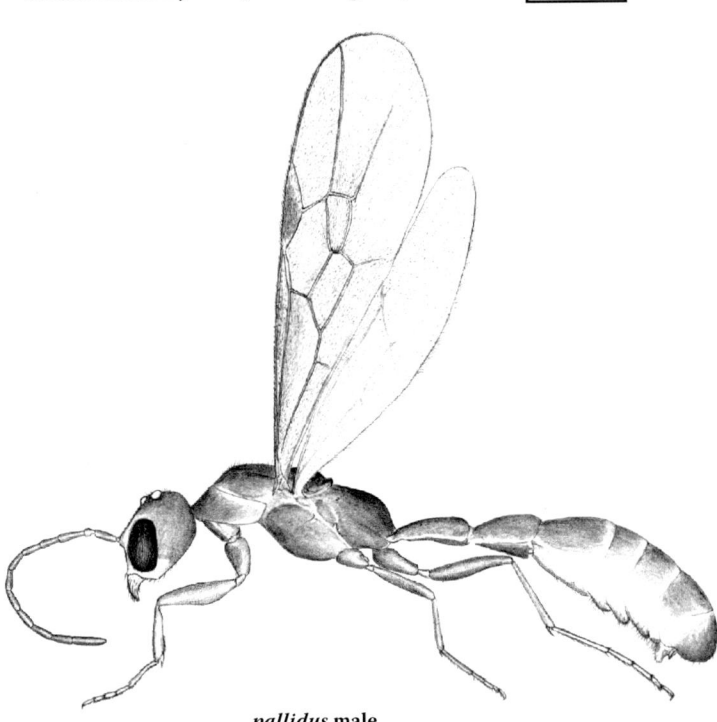

- Male *Pseudomyrmex* are often easy to find in nests; they have additional features for distinguishing between species

pallidus **male**

Only species with erect hairs

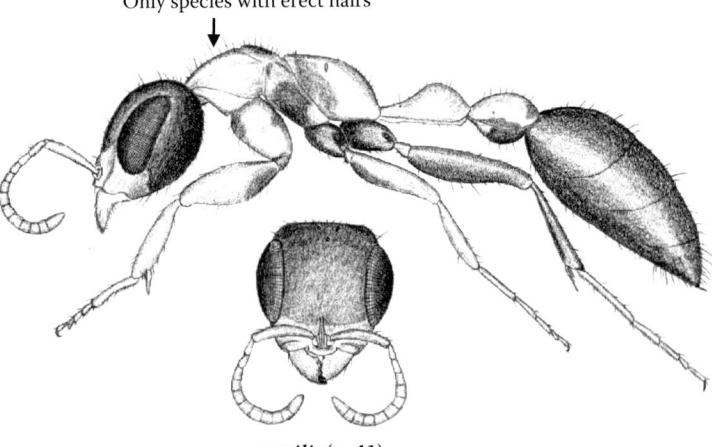

- Common in Florida

- Color orange and black

- Large: 7–8 mm

- Nests in hollow twigs, dead branches, sometimes in large weed stems

- Introduced from southwestern North America or neotropics

gracilis **(p. 41)**

Plate 17

Black spot on gaster, very shiny,
no small punctures on gaster

- Arboreal, in twigs and vines, not usually in stems of grass or weeds

- South and central Florida

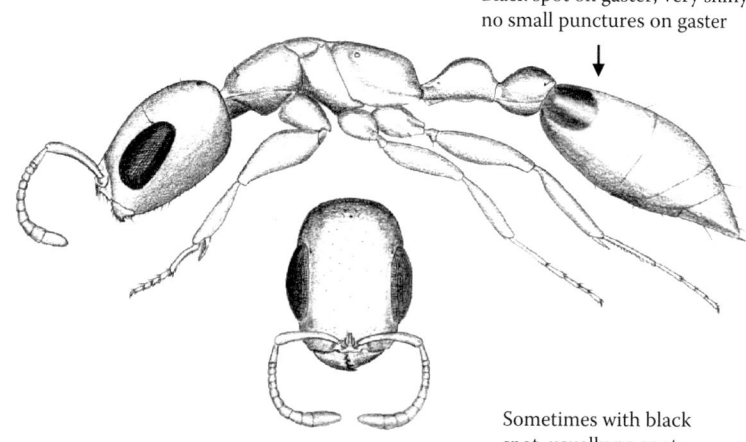

simplex (p. 43)

Sometimes with black spot, usually no spot. Small punctures on gaster

- Arboreal in dead twigs and vines, also in stems of grass and weeds

- A little smaller and usually paler orange than *seminole*

- Throughout Florida

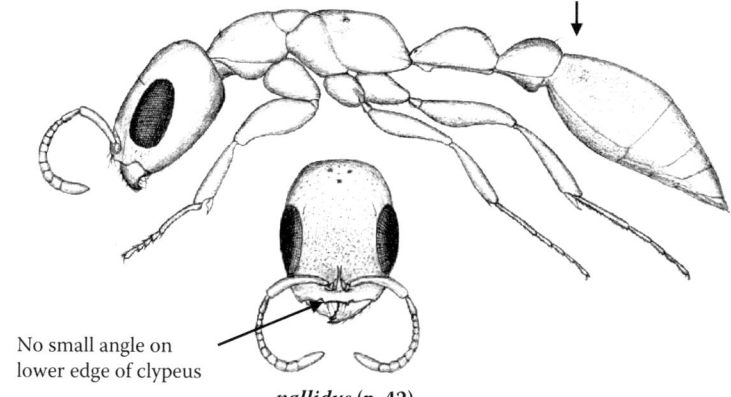

No small angle on lower edge of clypeus

pallidus (p. 42)

No black spot, with small punctures on gaster

- Usually in grass or sedge stems, sometimes weed stems, not arboreal

- A little larger and usually deeper orange than *pallidus*

- Throughout Florida in marshes and along coasts

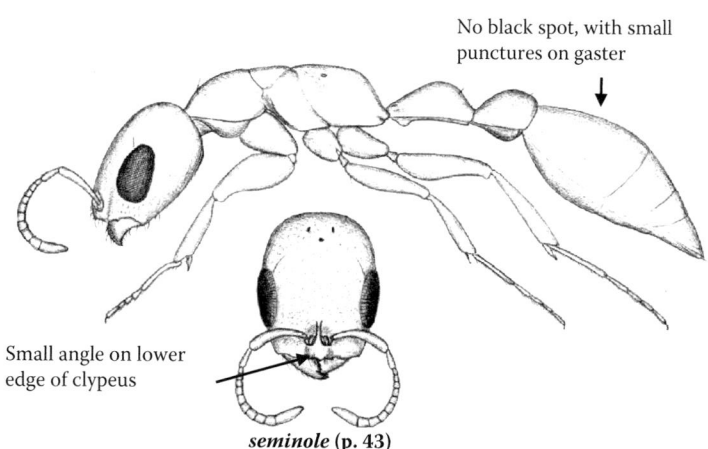

Small angle on lower edge of clypeus

seminole (p. 43)

Plate 18

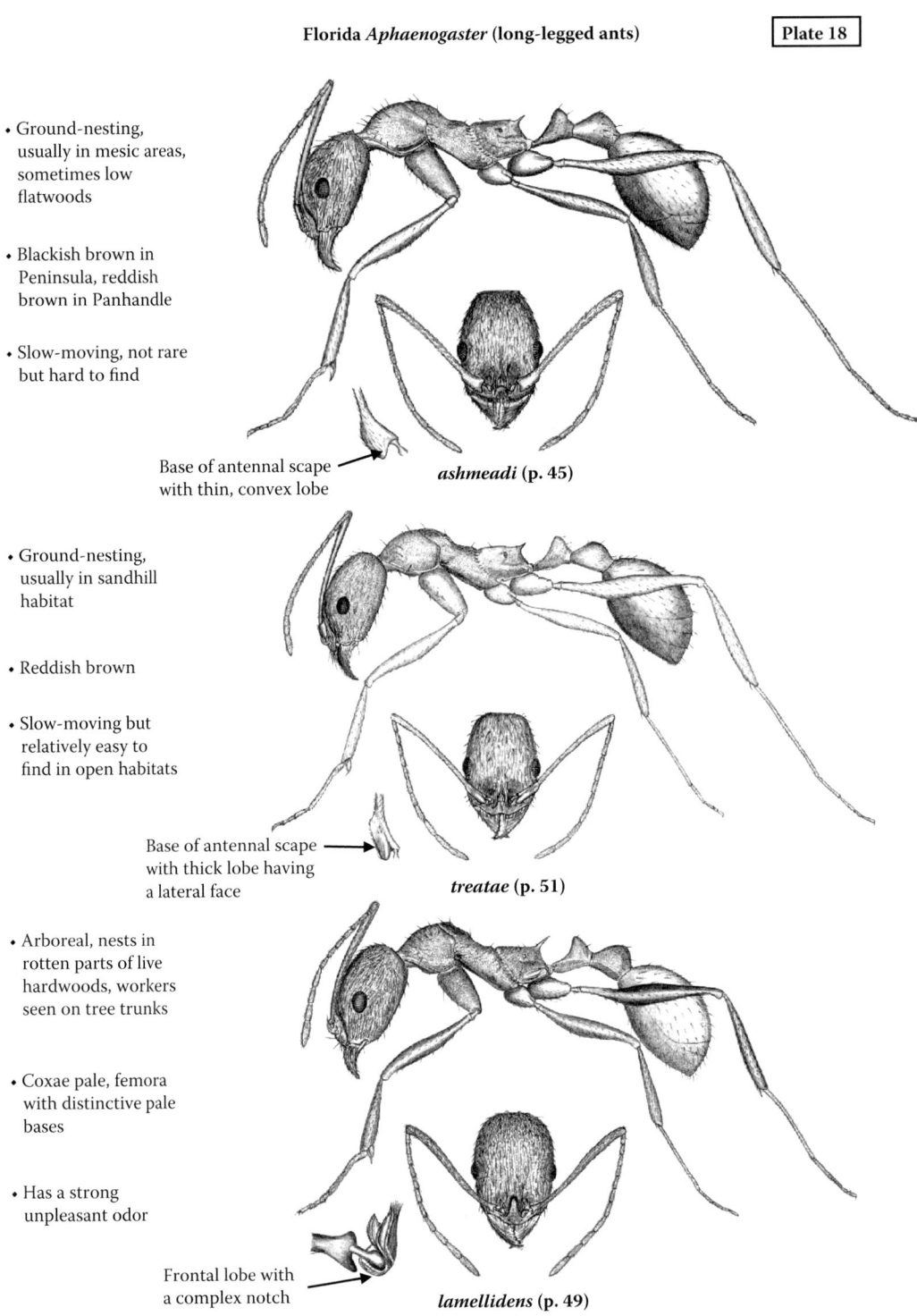

- Ground-nesting, usually in mesic areas, sometimes low flatwoods

- Blackish brown in Peninsula, reddish brown in Panhandle

- Slow-moving, not rare but hard to find

Base of antennal scape with thin, convex lobe

ashmeadi (p. 45)

- Ground-nesting, usually in sandhill habitat

- Reddish brown

- Slow-moving but relatively easy to find in open habitats

Base of antennal scape with thick lobe having a lateral face

treatae (p. 51)

- Arboreal, nests in rotten parts of live hardwoods, workers seen on tree trunks

- Coxae pale, femora with distinctive pale bases

- Has a strong unpleasant odor

Frontal lobe with a complex notch

lamellidens (p. 49)

Florida *Aphaenogaster* (long-legged ants)

Plate 19

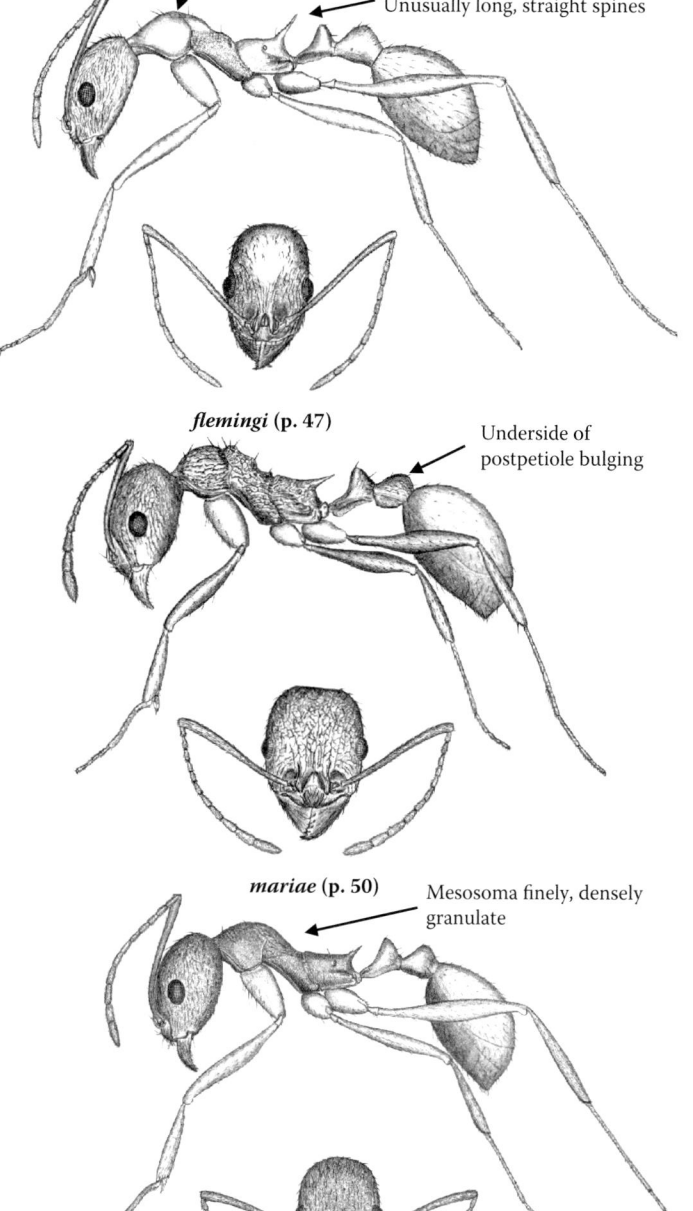

Shiny pronotum

Unusually long, straight spines

• Wide range of habitats, wet to dry, including salt marsh

• Ground nesting

• Sometimes nocturnal

flemingi (p. 47)

Underside of postpetiole bulging

• A rare arboreal species collected twice in Florida

• Possibly a temporary nest parasite of other *Aphaenogaster*

mariae (p. 50)

Mesosoma finely, densely granulate

• Nests in mesic shaded areas, often in dense leaf litter or in rotten wood

• Two species differ genetically

• South Florida specimens probably *miamiana*, in north Florida could be either *miamiana* or *carolinensis*

miamiana (p. 50) and *carolinensis* (p. 46)
(similar species)

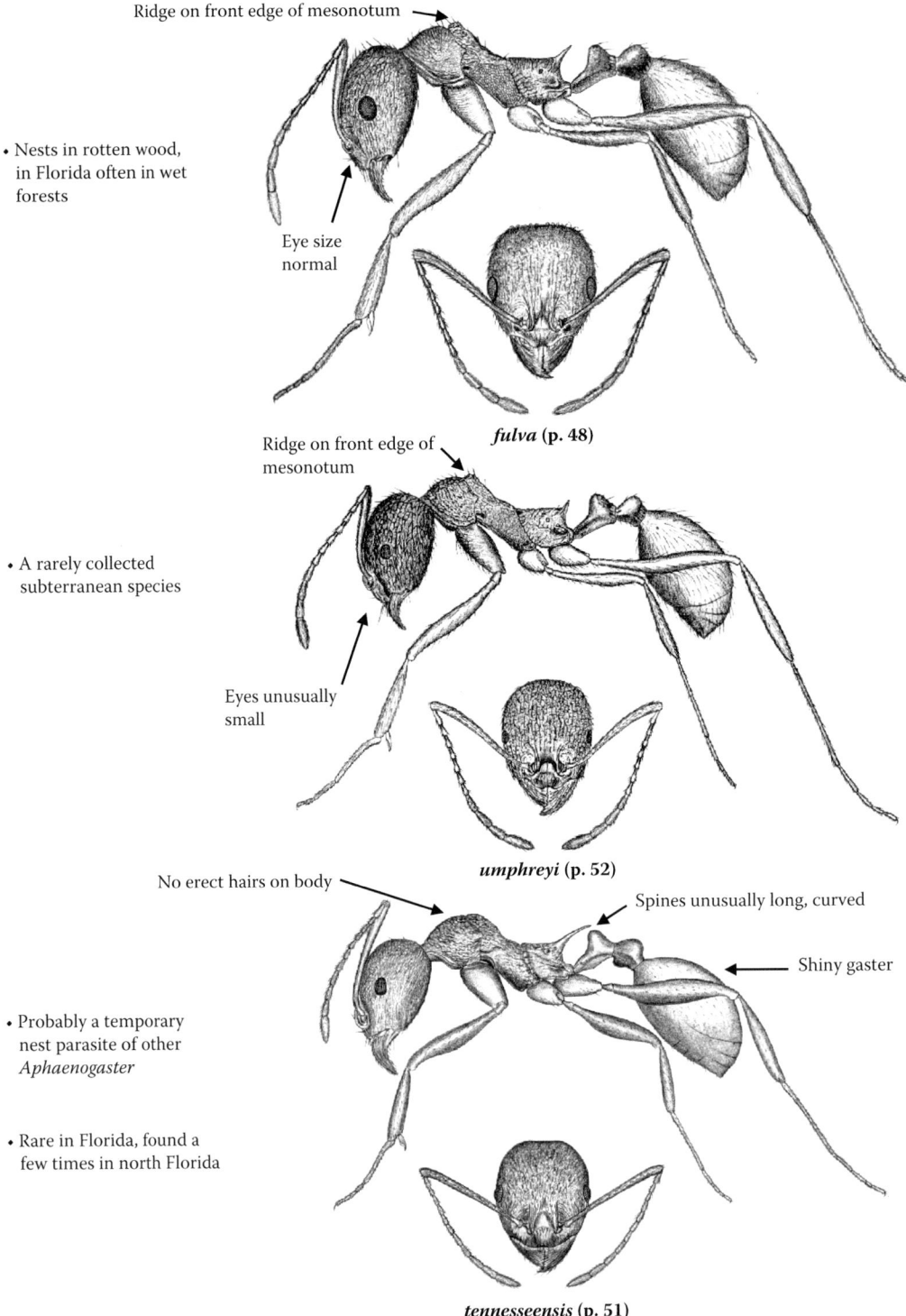

Ridge on front edge of mesonotum

• Nests in rotten wood, in Florida often in wet forests

Eye size normal

fulva (p. 48)

Ridge on front edge of mesonotum

• A rarely collected subterranean species

Eyes unusually small

umphreyi (p. 52)

No erect hairs on body

Spines unusually long, curved

Shiny gaster

• Probably a temporary nest parasite of other *Aphaenogaster*

• Rare in Florida, found a few times in north Florida

tennesseensis (p. 51)

Florida *Aphaenogaster* (long-legged ants), Florida *Myrmica* (furrowed ants), and Florida *Stenamma* (retiring ground ants)

Plate 21

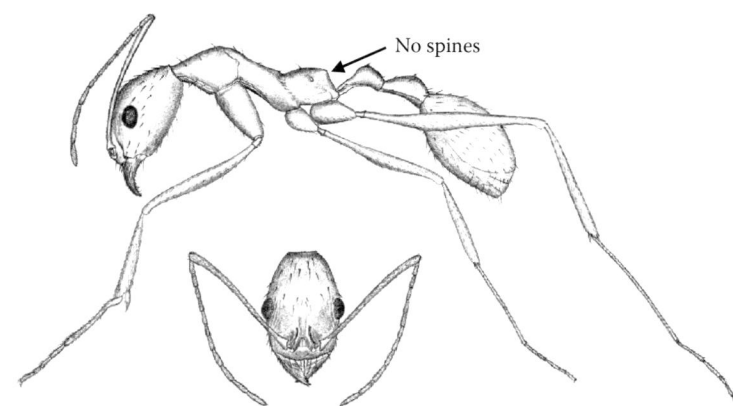

No spines

- Nests in open sandy areas, nest often with a low turret at entrance

- Yellow

- Usually nocturnal

Aphaenogaster floridana (p. 47)

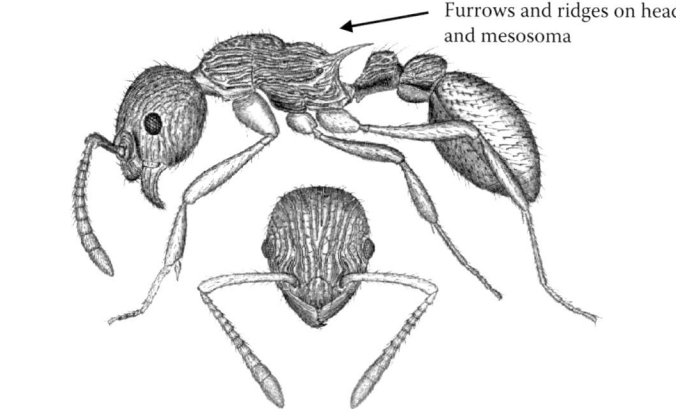

Furrows and ridges on head and mesosoma

- North Florida, in hardwood forests

- Uncommon in Florida, more common farther north

Myrmica punctiventris (p. 80)

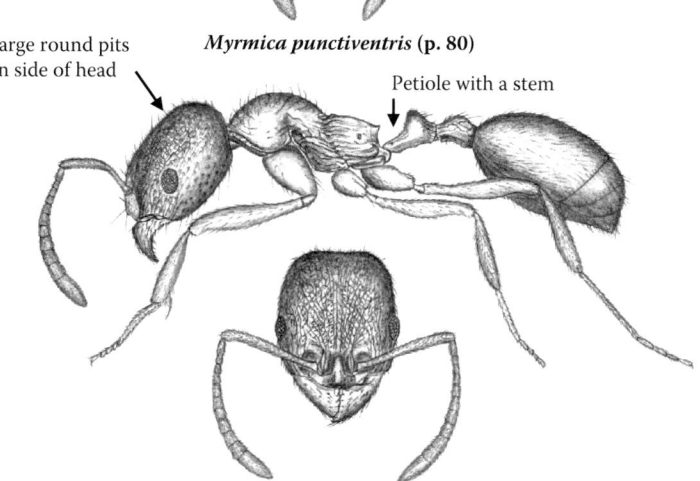

Large round pits on side of head

Petiole with a stem

- Nests in open sandy areas

- Active in winter only

- North Florida, a southeastern species rare everywhere

- Resembles small fire ant when seen in field

Stenamma foveolocephalum (p. 118)

Florida *Cardiocondyla* (sneaking ants)

Plate 22

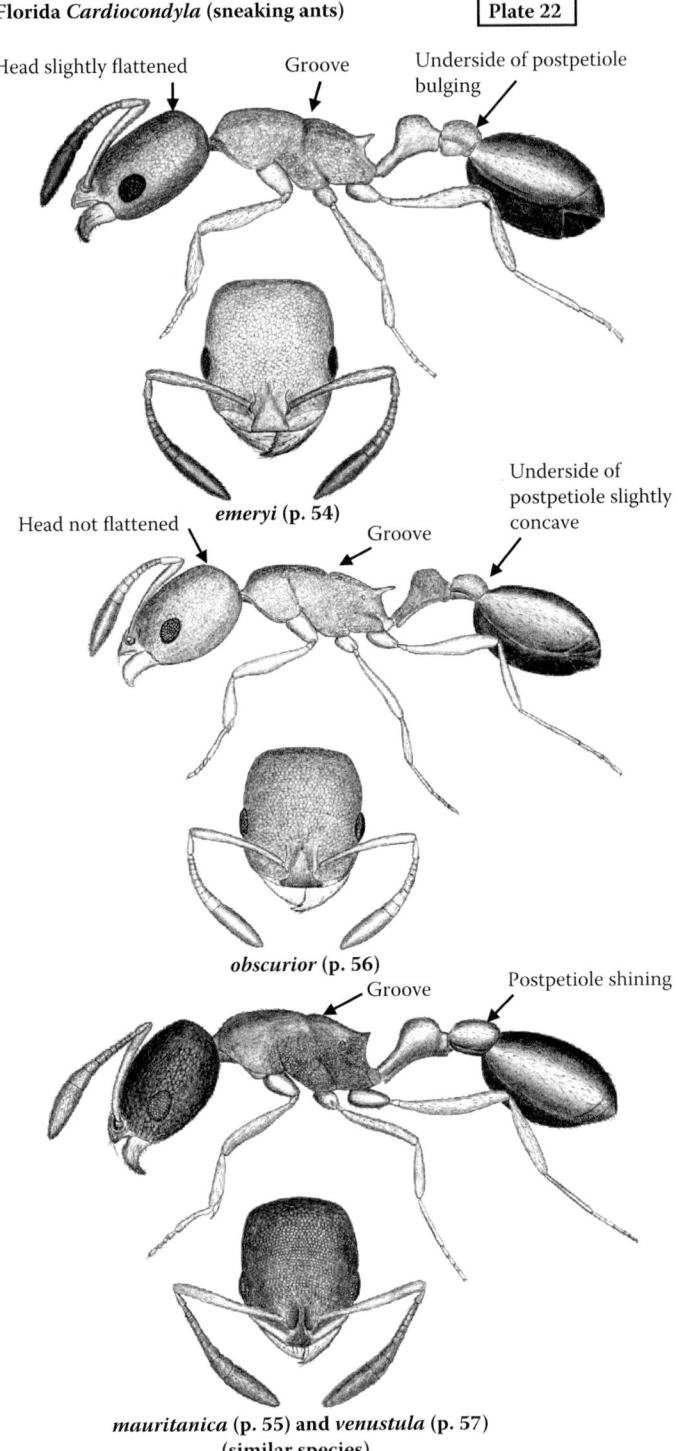

Head slightly flattened · Groove · Underside of postpetiole bulging

emeryi (p. 54)

Head not flattened · Groove · Underside of postpetiole slightly concave

obscurior (p. 56)

Groove · Postpetiole shining

mauritanica (p. 55) and *venustula* (p. 57)
(similar species)

- Nest in ground in open areas

- Reddish brown with black gaster

- Usually in disturbed sites, including urban

- Introduced from Old World Tropics

- Nest in trees or shrubs, sometimes in dead weed stalks

- Reddish brown with black gaster

- Introduced from Old World Tropics

- Nests in ground

- Color of *mauritanica* slightly lighter, size smaller than *venustula*

- Possibly *mauritanica* and *venustula* inbred variants of one species

- Introduced from Old World Tropics

282

PLATES

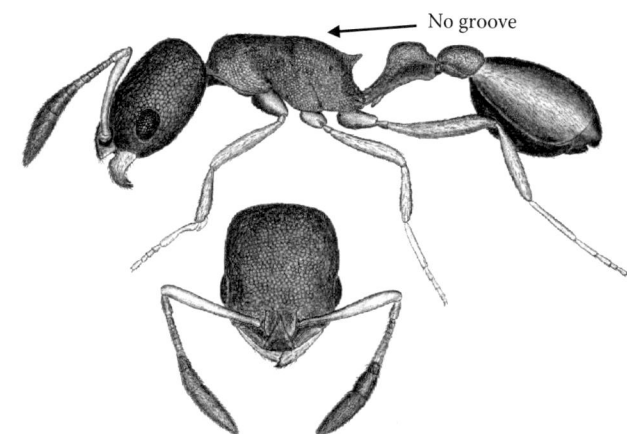

No groove

• Nests in ground in open, disturbed sites

• Color blackish to pale reddish brown

• Introduced from Old World Tropics

Cardiocondyla minutior (p. 56)

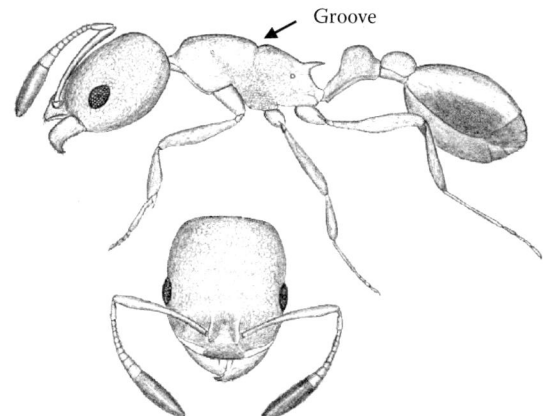

Groove

• Nests in hollow stems on ground

• Yellow color distinctive

• Introduced from Old World Tropics

Cardiocondyla wroughtonii (p. 57)

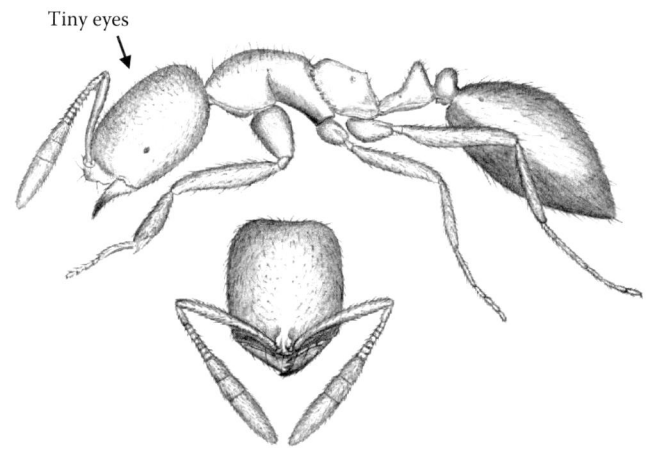

Tiny eyes

• In Florida known from one site in Miami

• Distinctive pale yellow subterranean species

• Introduced from Old World Tropics

Monomorium subcoecum (p. 76)

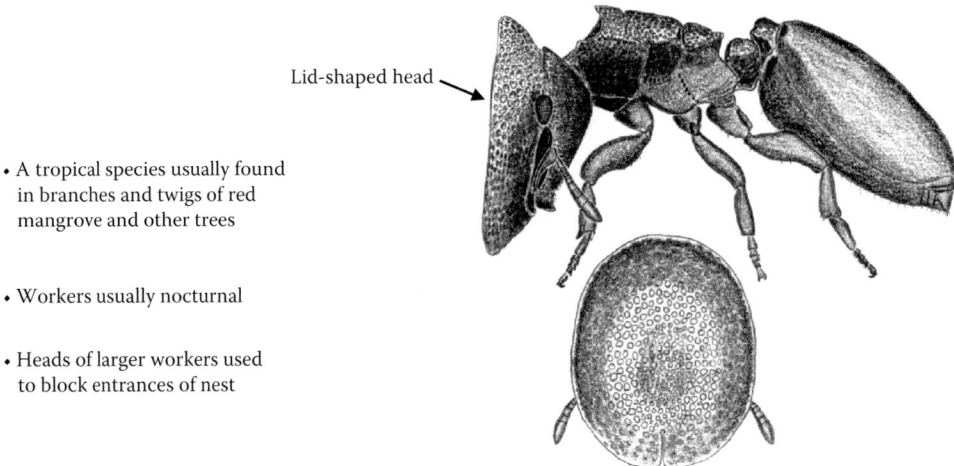

Lid-shaped head

- A tropical species usually found
  in branches and twigs of red
  mangrove and other trees

- Workers usually nocturnal

- Heads of larger workers used
  to block entrances of nest

Cephalotes varians (p. 58)

Club-shaped hairs

Broad antennal
scapes

Eurhopalothrix floridana (p. 71)

Club-shaped hairs

- Lives in leaf litter

- Difficult to see, and also
  plays dead

Eurhopalothrix floridana male

Florida *Crematogaster* (acrobat ants)

Plate 25

No sculpture on sides of pronotum

Short spines

- *C. ashmeadi* is all black

- Nests in dead parts of trees, shrubs and vines

- *C. pinicola* is dark red with black gaster

- Nests in dead branches and twigs of pines in open

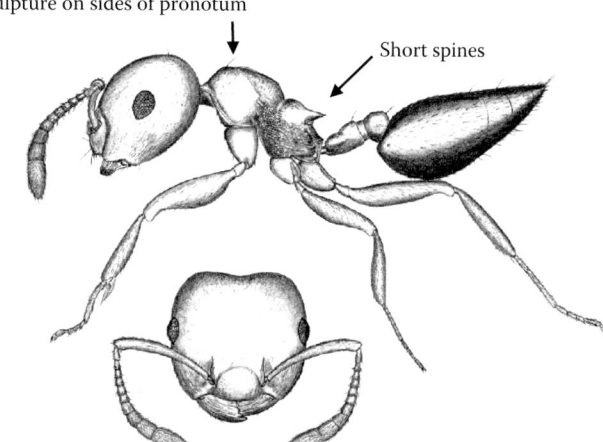

ashmeadi (p. 60) and *pinicola* (p. 66)
(similar species)

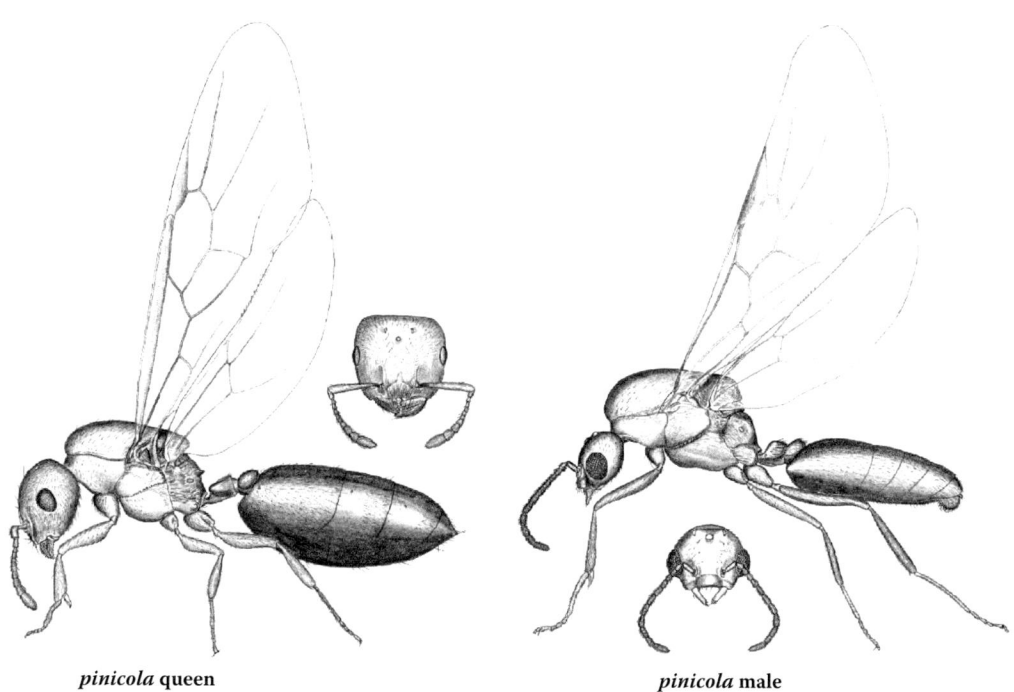

pinicola queen

pinicola male

Sculpture on mesosoma, few erect hairs

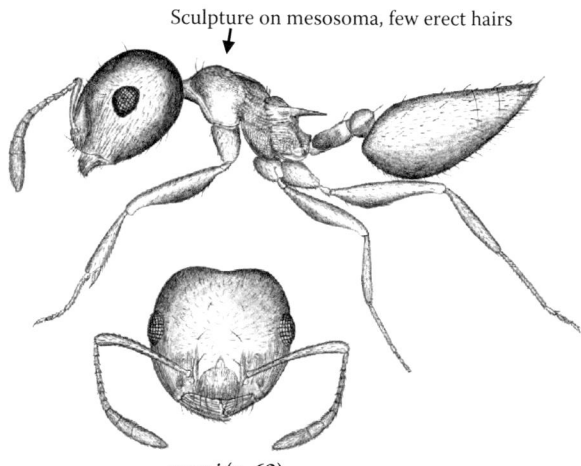

cerasi (p. 62)

- In Florida, a woodland species living in rotten wood and leaf litter

- Most common in north Florida

Sculpture on mesosoma, many erect hairs

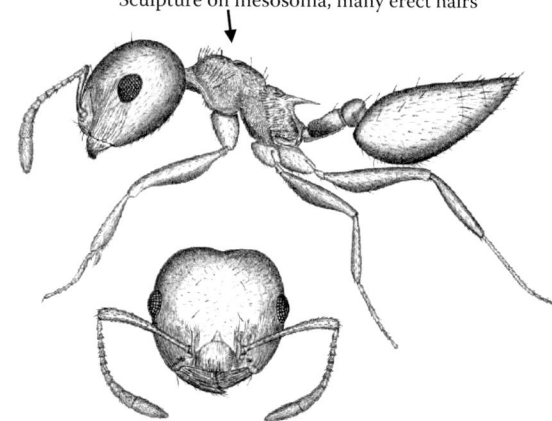

lineolata (p. 63)

- In Florida, a woodland species living in rotten wood and leaf litter

- North Florida

Irregular ridges on mesosoma

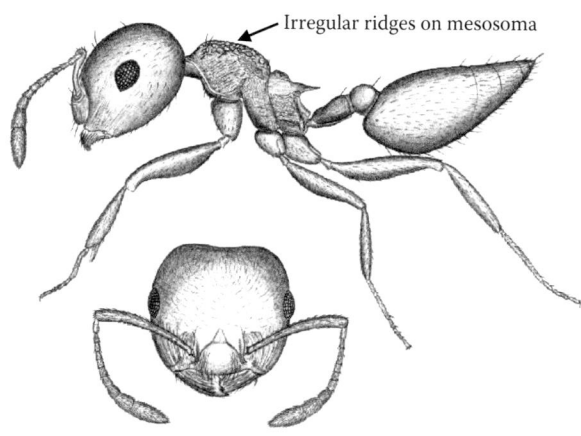

vermiculata (p. 67)

- A small black species in dead twigs and branches in swamp forests

- North and central Florida

- Nests in ground, north Florida

- Yellow color

- Probably does not have multiple queens

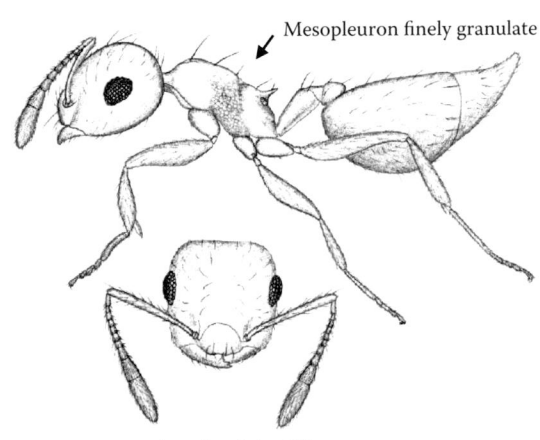

Mesopleuron finely granulate

***missuriensis* (p. 64)**

- Nests in rotten wood or leaf litter, throughout Florida

- Yellow color

- Multiple queens easy to find

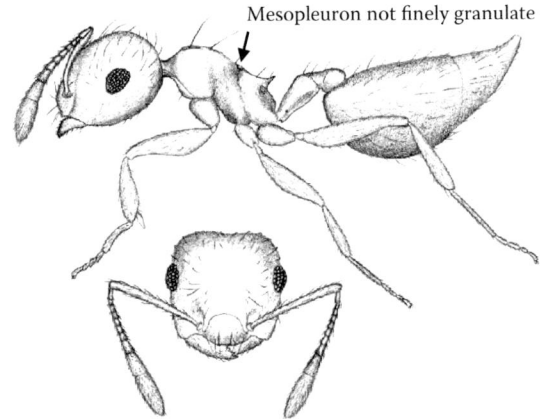

Mesopleuron not finely granulate

***minutissima* (p. 63)**

- Nests in hollow twigs and branches, Bahamas, not known from Florida

- Yellow color

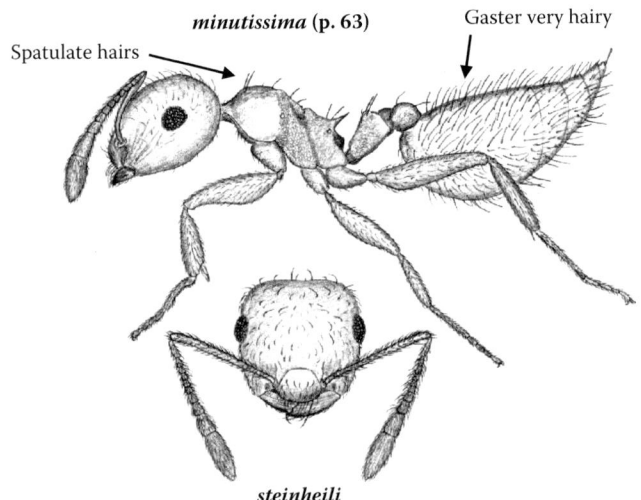

Spatulate hairs

Gaster very hairy

steinheili

Weakly developed
dorsal tubercles

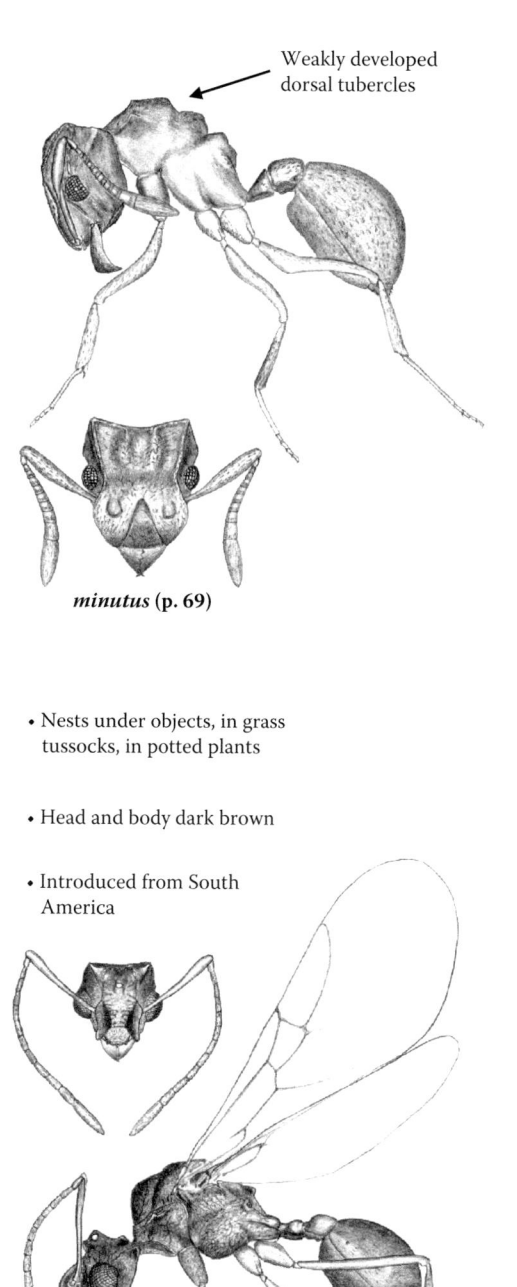

minutus (p. 69)

• Nests in rotten wood, under
objects, in grass tussocks

• Mesosoma pale brown,
slightly contrasting with head
and gaster

• Native to south and central Florida

Strongly developed
dorsal tubercles

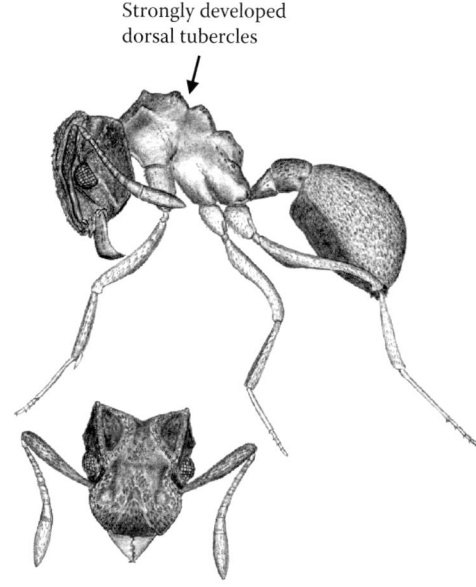

rimosus (p. 70)

• Nests under objects, in grass
tussocks, in potted plants

• Head and body dark brown

• Introduced from South
America

rimosus **male**

• Flies early in morning, numerous
at lights

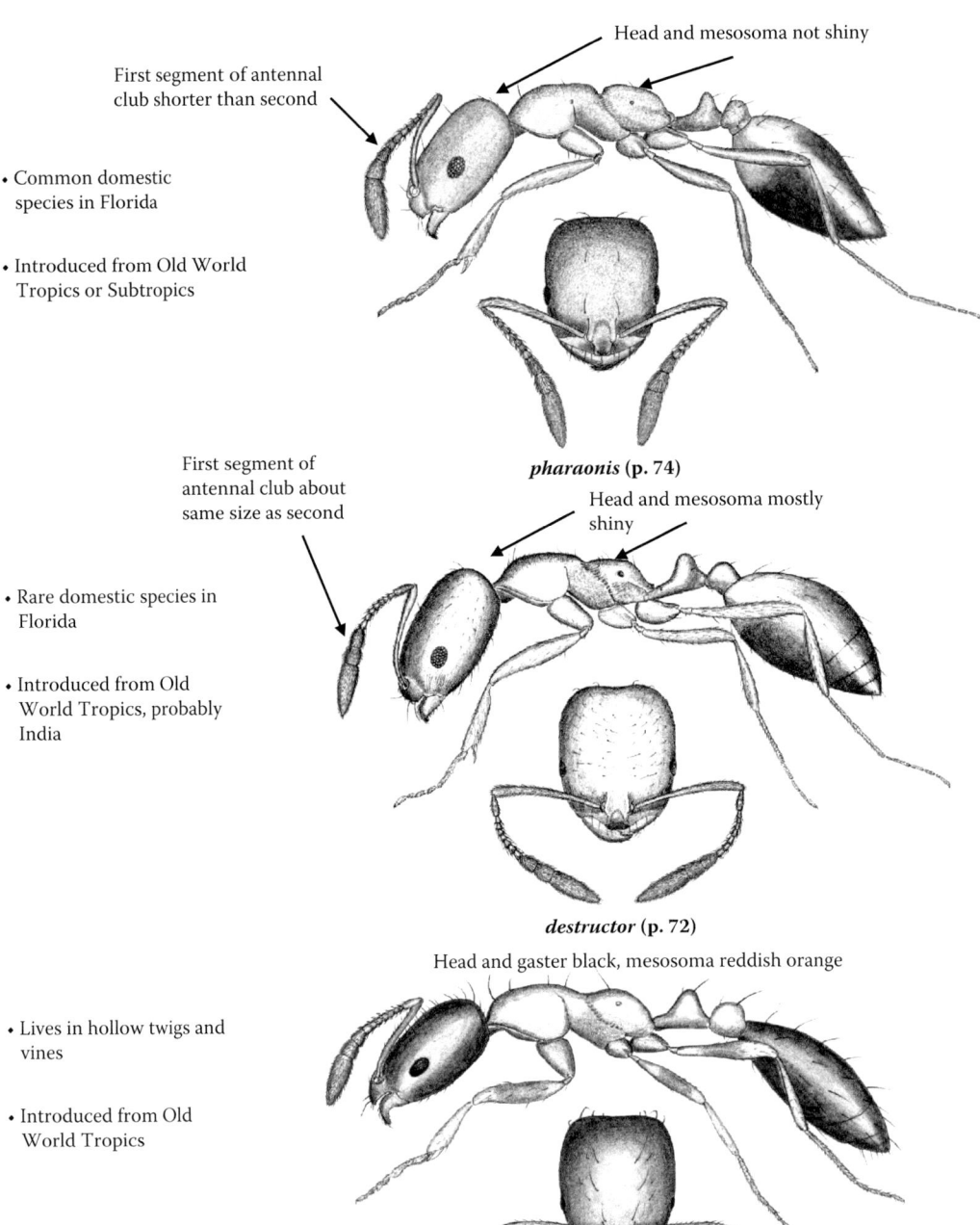

First segment of antennal club shorter than second

Head and mesosoma not shiny

- Common domestic species in Florida

- Introduced from Old World Tropics or Subtropics

pharaonis (p. 74)

First segment of antennal club about same size as second

Head and mesosoma mostly shiny

- Rare domestic species in Florida

- Introduced from Old World Tropics, probably India

destructor (p. 72)

Head and gaster black, mesosoma reddish orange

- Lives in hollow twigs and vines

- Introduced from Old World Tropics

floricola (p. 74)

Florida *Monomorium* (trailing ants)   Plate 31

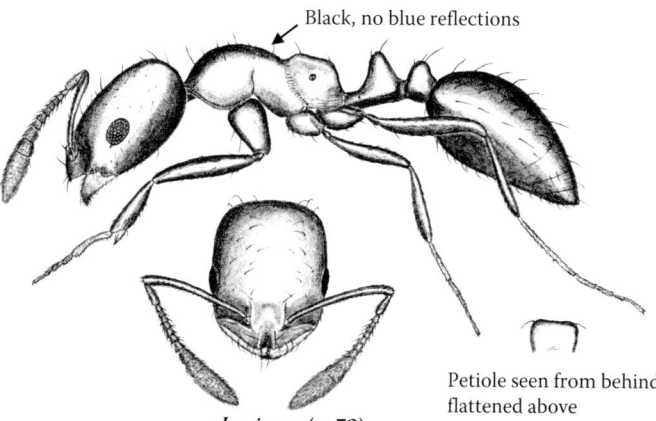

Black, no blue reflections

• Nests in ground or in dead branches or in leaves of bromeliads

• In Florida Keys

• Introduced from Caribbean

• Queens flightless

Petiole seen from behind flattened above

ebeninum (p. 73)

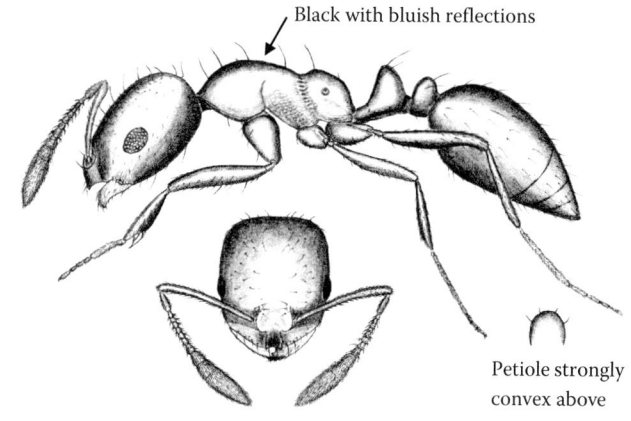

Black with bluish reflections

• Nests in ground in open sandy areas

• Widespread in uplands, not in Florida Keys

• Young queens have wings for dispersal

Petiole strongly convex above

viridum (p. 77)

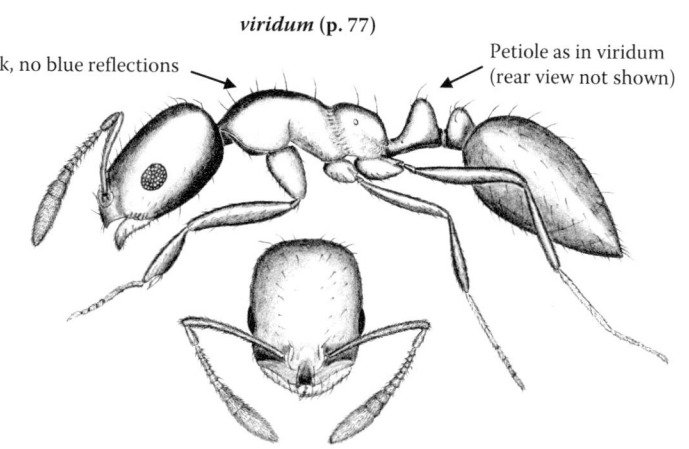

Black, no blue reflections

Petiole as in viridum (rear view not shown)

• Nests in trees

• Widespread in Florida

• Queens usually flightless

trageri (p. 76)

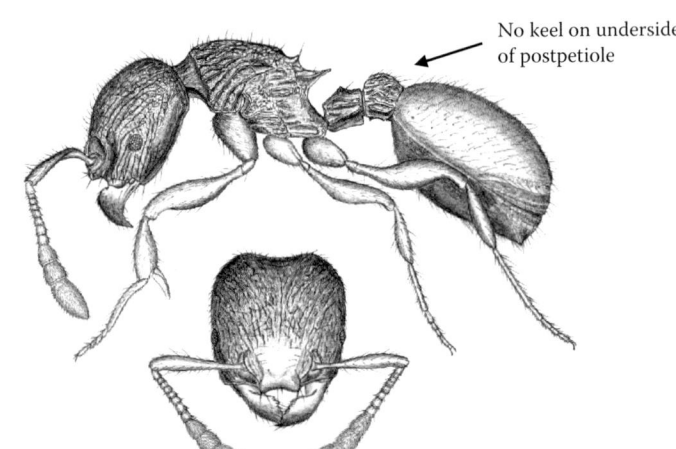

No keel on underside
of postpetiole

- Woodland species
  widespread in Florida

- Mature color dark brown

americana (p. 78)

Head and mesosoma
without strong sculpture

Keel on underside of postpetiole

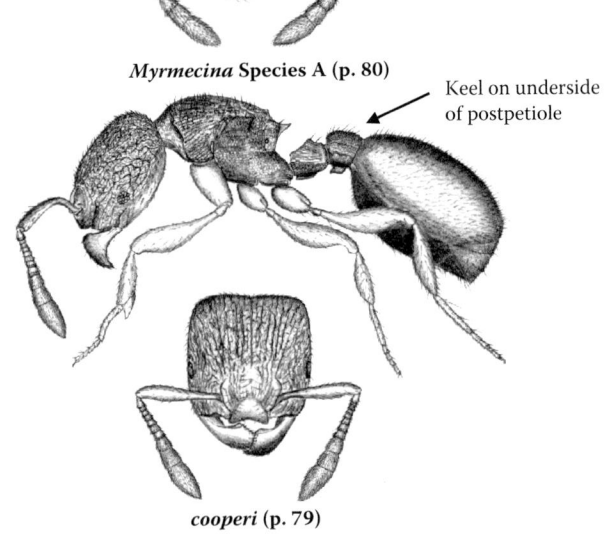

- Widespread where host
  occurs, found once in
  Florida

- Mature color red

- Parasitic on *americana*

Myrmecina Species A (p. 80)

Keel on underside
of postpetiole

- Rare woodland species
  known from few specimens

- Mature color dark brown

- Northwest Florida

- Very small, about 1.8 mm long

cooperi (p. 79)

Shiny black head

• Lives in open sandy areas, primarily Florida scrub

• Black, head of major not reddish

• Small, minor less than 2 mm long

Shiny black head

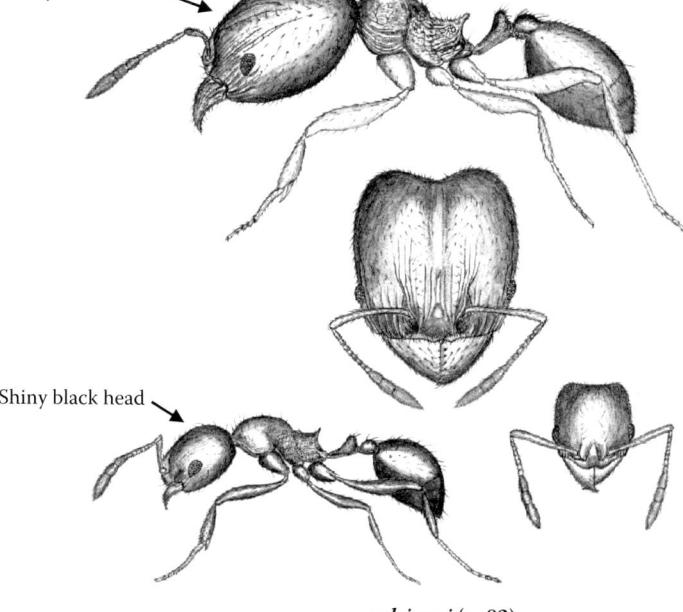

adrianoi (p. 82)

Heavily sculptured dark red head

• Usually found in lightly shaded sandy areas

• Head of major usually reddish, majors and minors with strong metallic reflections

• Small, minor less than 2 mm long

Heavily sculptured head

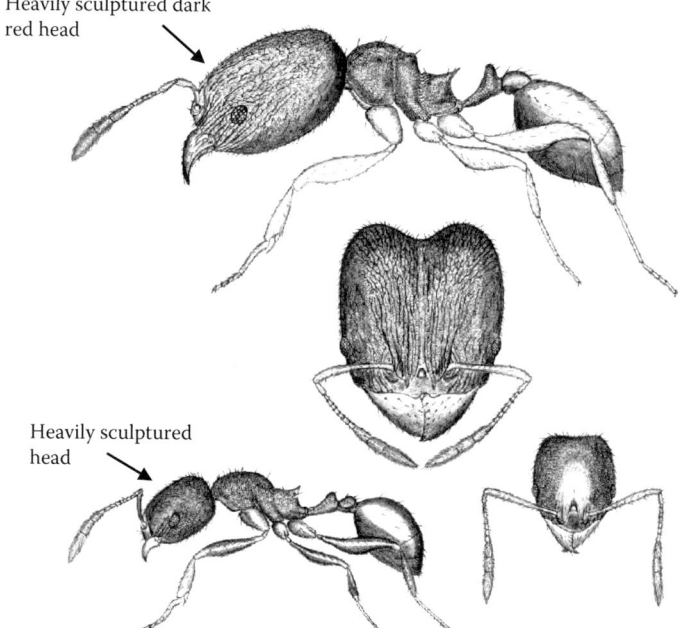

metallescens (p. 91)

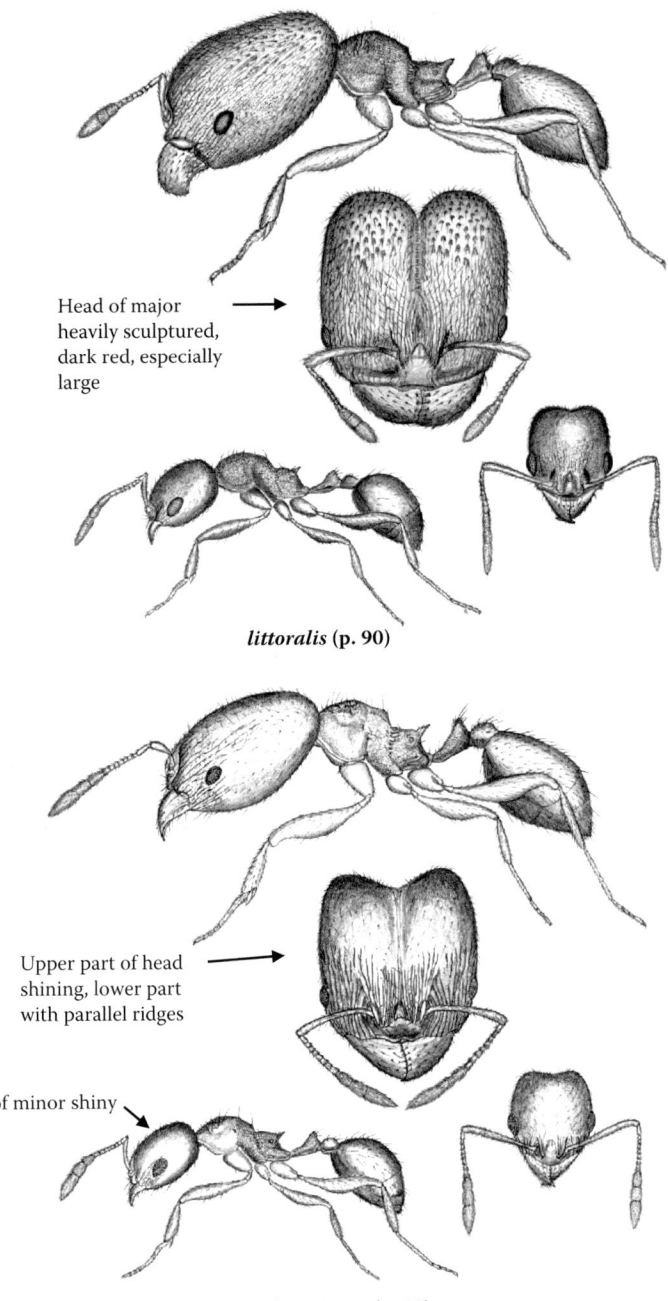

- Lives in very open sandy areas such as beaches and open scrub

- Black, except for head of major

- Small species, minor about 2 mm long

Head of major heavily sculptured, dark red, especially large

littoralis (p. 90)

- In north Florida, usually in open disturbed sites

- Brownish, minor looks dark in field

- Small species, minor less than 2 mm long

Upper part of head shining, lower part with parallel ridges

Head of minor shiny

bicarinata (p. 83)

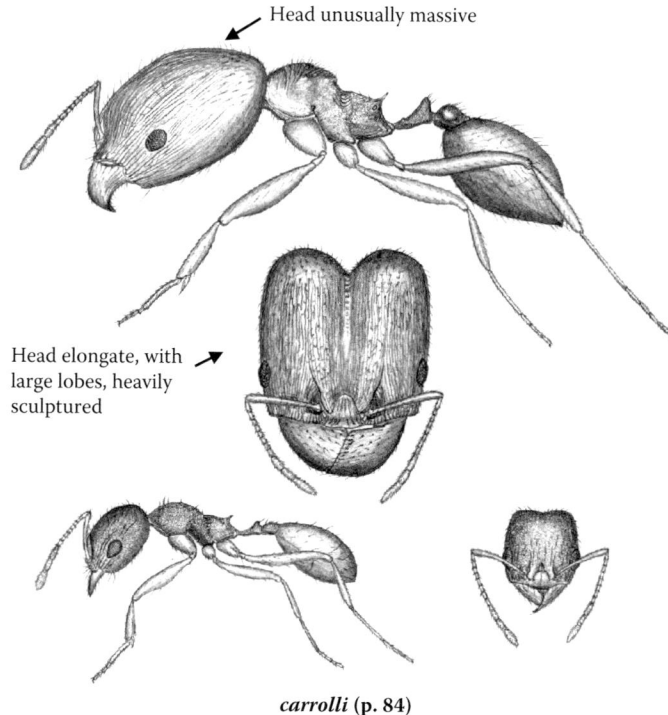

Head unusually massive

- Lives in sandhill habitat, north Florida, not common

- Reddish brown

- Mid-sized species, minor more than 2 mm long

Head elongate, with large lobes, heavily sculptured

carrolli (p. 84)

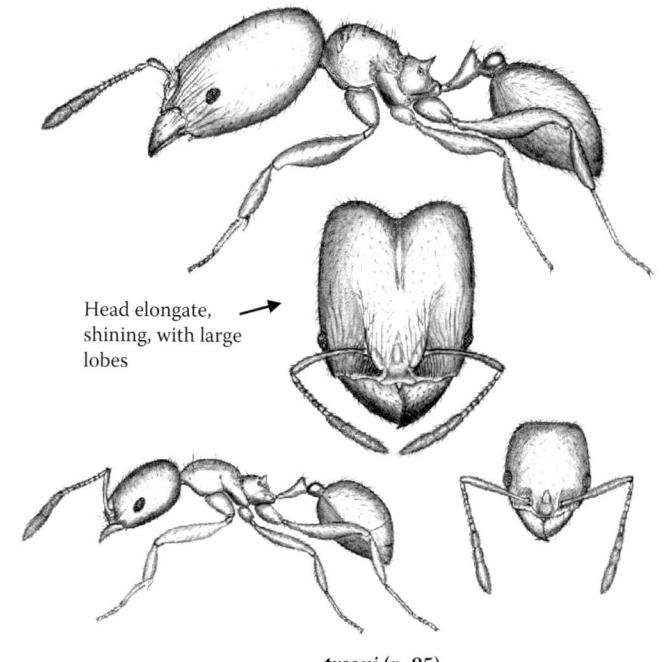

- Lives in lawns and other open grassy habitats, north Florida

- Yellow color

- Small species, minor 2 mm long or less

Head elongate, shining, with large lobes

tysoni (p. 95)

Florida *Pheidole* (big-headed ants)

Plate 36

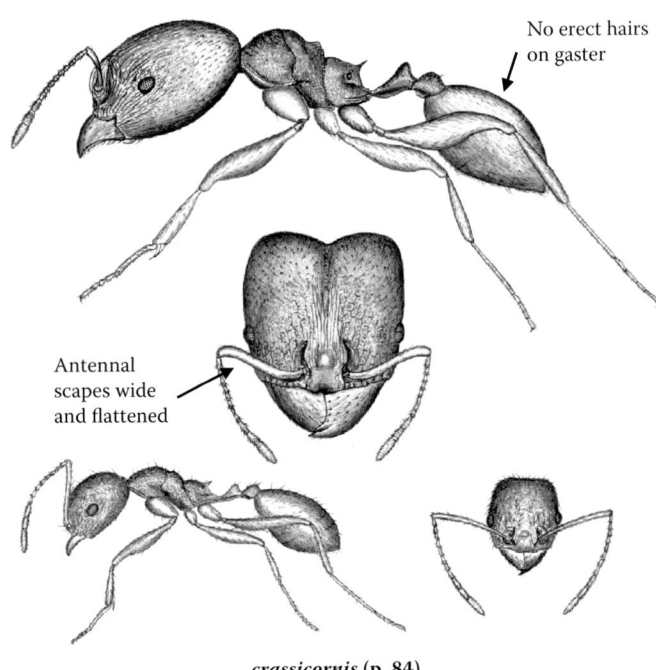

No erect hairs
on gaster

• Uncommon species
usually found in open
forest in north Florida

• Reddish brown color

• Mid-sized species,
minors longer than 2 mm

Antennal
scapes wide
and flattened

crassicornis (p. 84)

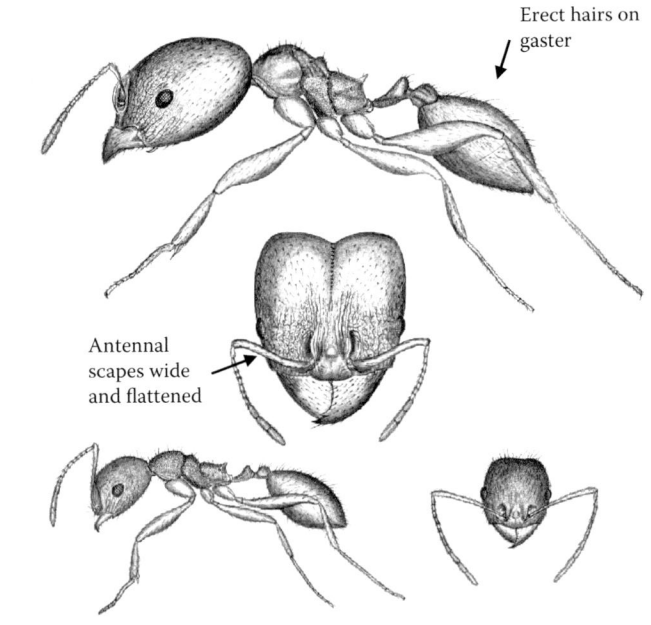

Erect hairs on
gaster

• Uncommon species
usually found in open
forest in north Florida

• Reddish brown color

• Mid-sized species,
minor longer than 2 mm

Antennal
scapes wide
and flattened

diversipilosa (p. 87)

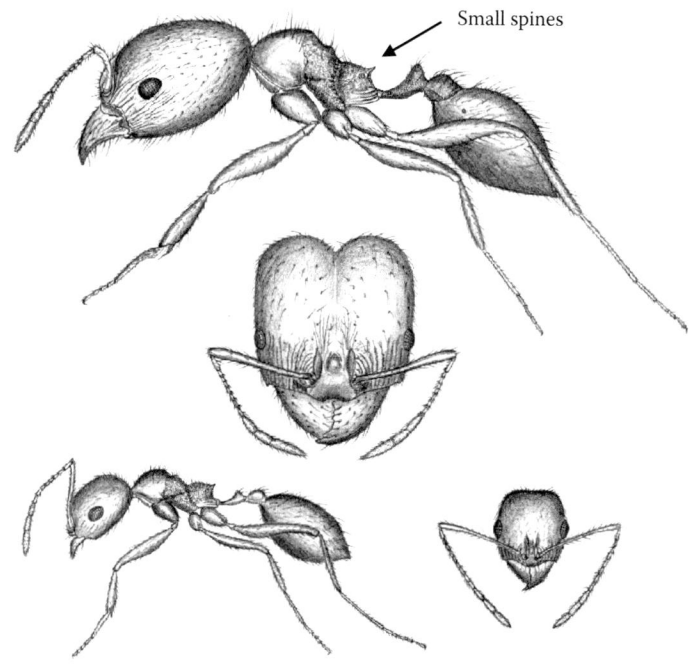

Small spines

- Many different habitats throughout Florida, very common

- Variable color from blackish to pale brown

- Compare with *morrisi* and *megacephala*

- Mid-sized species, minor longer than 2 mm

dentata (p. 85)

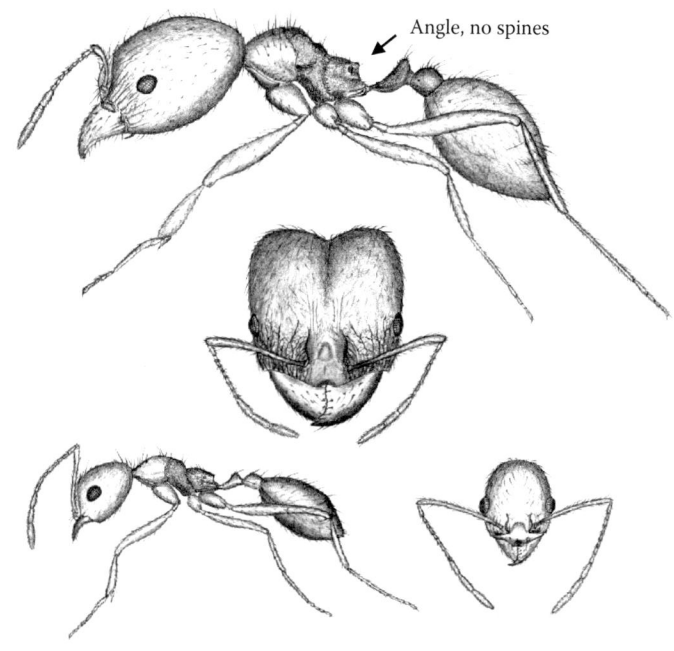

Angle, no spines

- Open sandy areas, widespread

- Yellow color

- Compare with *dentata*

- Mid-sized species, minor longer than 2 mm

morrisi (p. 92)

Plate 38

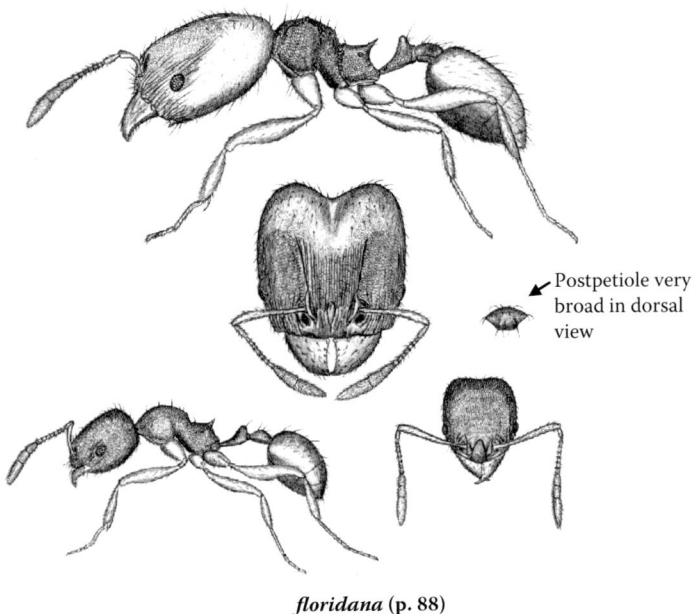

- Lives in many types of habitats, often in rotten wood or grass tussocks

- Reddish brown color

- Small species, minor less than 2 mm long

- Compare with *moerens*, often in similar habitats

Postpetiole very broad in dorsal view

floridana (p. 88)

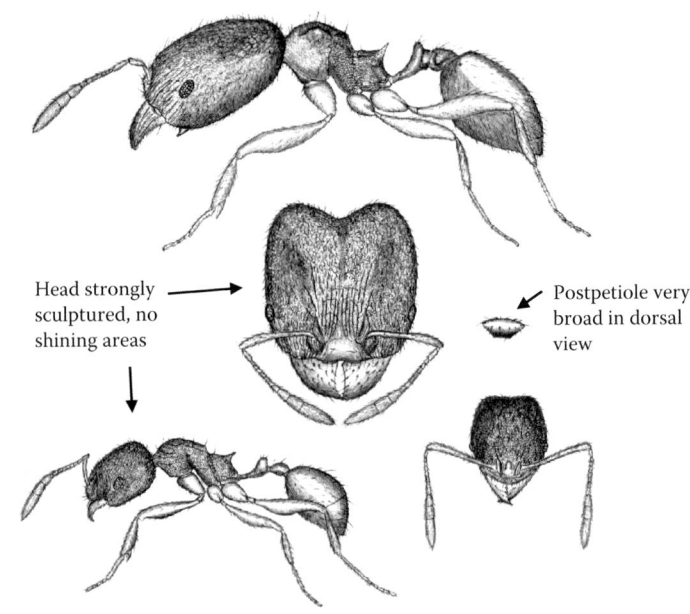

- Usually in moist areas, nests usually in rotten wood

- Reddish brown color

- Small species, minor less than 2 mm long

Head strongly sculptured, no shining areas

Postpetiole very broad in dorsal view

dentigula (p. 86)

Plate 39

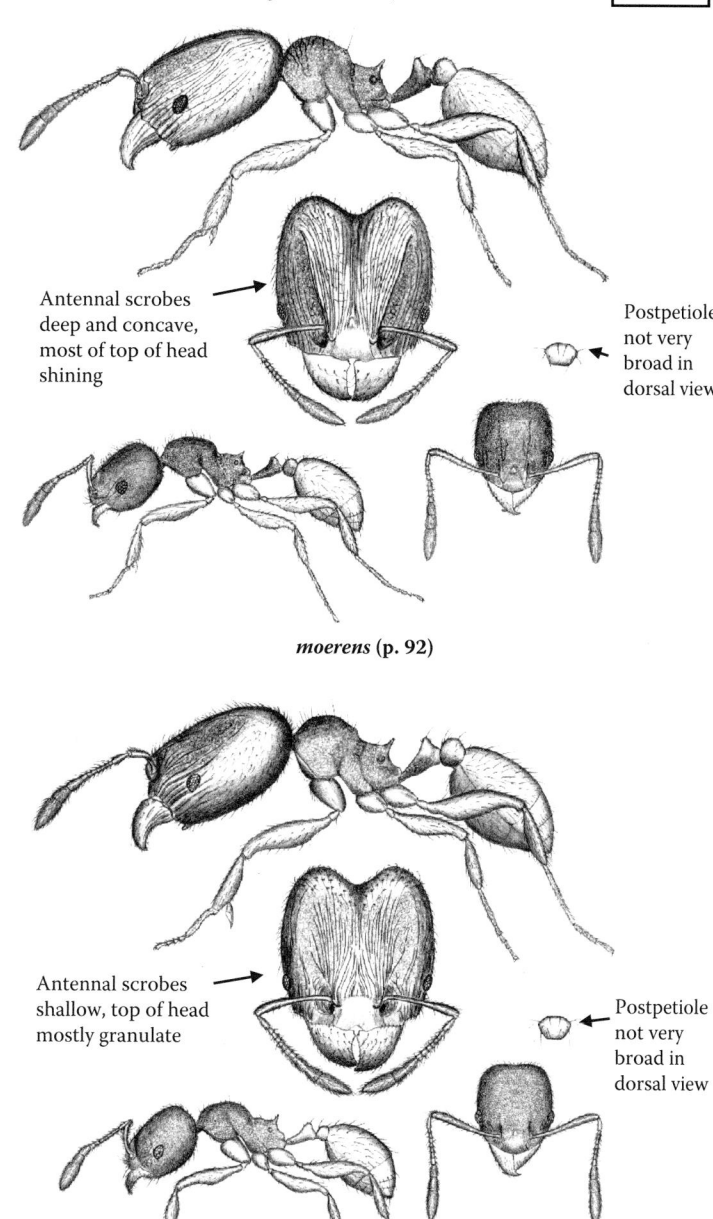

- Natural and disturbed areas, very common throughout Florida

- Small species, minor less than 2 mm long

- Compare with *floridana, flavens*

- Introduced from neotropics

Antennal scrobes deep and concave, most of top of head shining

Postpetiole not very broad in dorsal view

moerens (p. 92)

- Usually in tropical hammocks, tropical Florida only

- Small species, minor less than 2 mm long

- Compare with *floridana, moerens*

- Introduced from neotropics

Antennal scrobes shallow, top of head mostly granulate

Postpetiole not very broad in dorsal view

flavens (p. 88)

Florida *Pheidole* (big-headed ants) and
Pogonomyrmex (harvester ants)

Plate 40

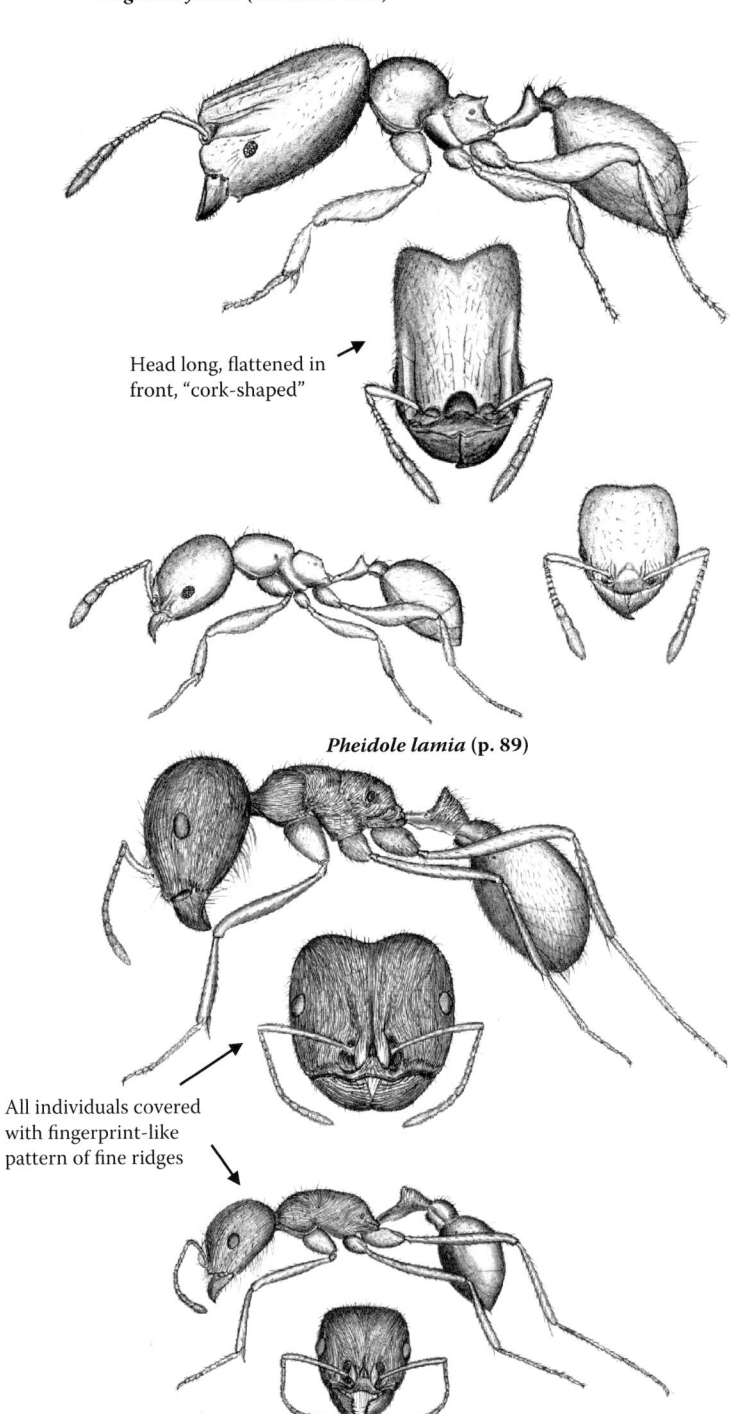

- In Florida, known from the Red Hills area near Tallahassee

- Color yellow, shining

- Small species, minors about 2 mm long

Head long, flattened in front, "cork-shaped"

Pheidole lamia (p. 89)

- Lives in well-drained sandy areas with patches of bare sand, common in Florida

- Dark reddish brown

- Nest area marked with bits of charcoal and plant debris

All individuals covered with fingerprint-like pattern of fine ridges

- Workers occur in varying sizes, the largest with outsized heads

- Large species, smaller workers more than 5 mm long

Pogonomyrmex badius (p. 96)

Florida *Pheidole* (big-headed ants)

Plate 41

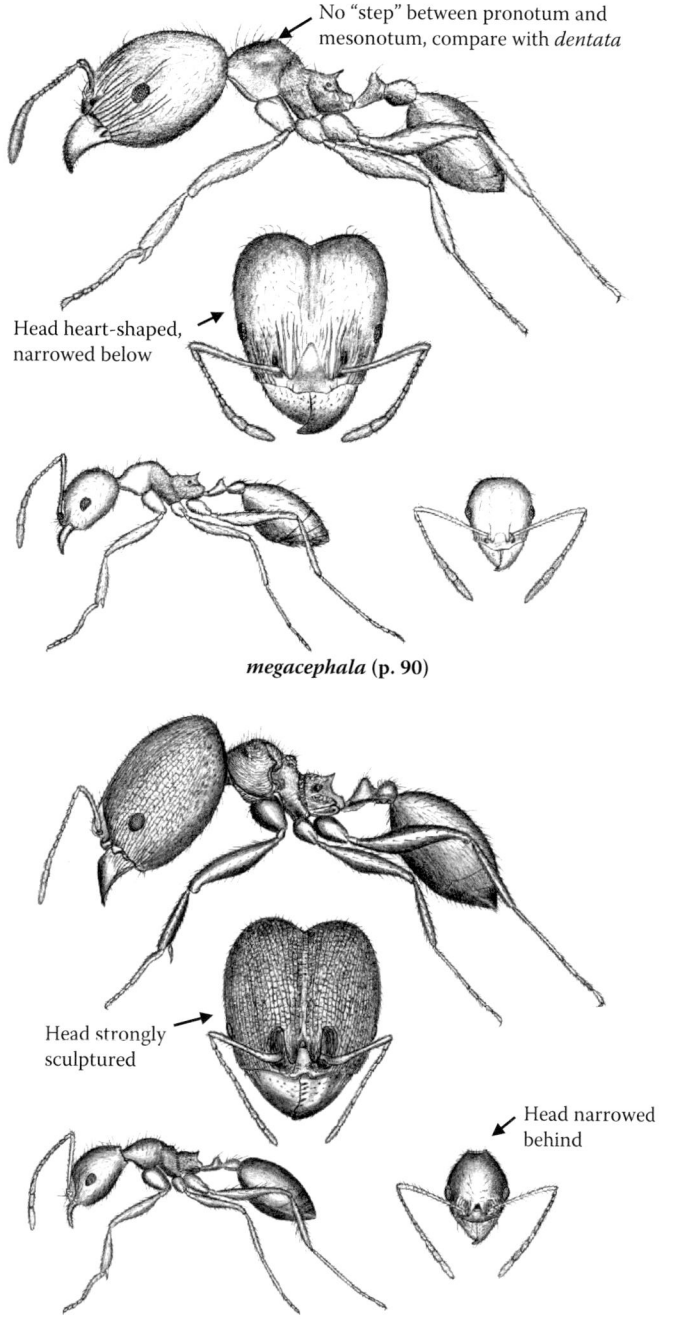

No "step" between pronotum and mesonotum, compare with *dentata*

Head heart-shaped, narrowed below

- Usually in disturbed areas, including urban areas

- Color blackish brown

- Majors relatively rare in Florida colonies

- Mid-sized species, minors just more than 2 mm long

- Introduced from Africa

megacephala (p. 90)

- Often but not always in disturbed areas

- Black, head blackish red

- Largest Florida *Pheidole*, minors more than 3 mm long

- Strong repellent odor

- Introduced from southern South America

Head strongly sculptured

Head narrowed behind

obscurithorax (p. 94)

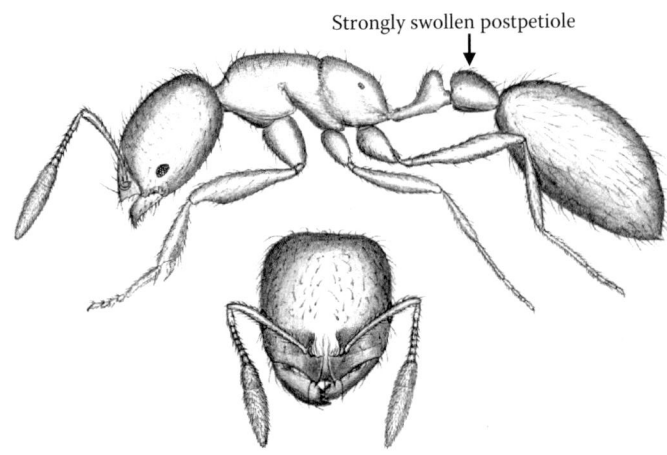

Strongly swollen postpetiole

- Beaches and open areas, throughout Florida

- Color usually orange with blackish gaster

- Nests usually in ground

Solenopsis globularia (p. 105)

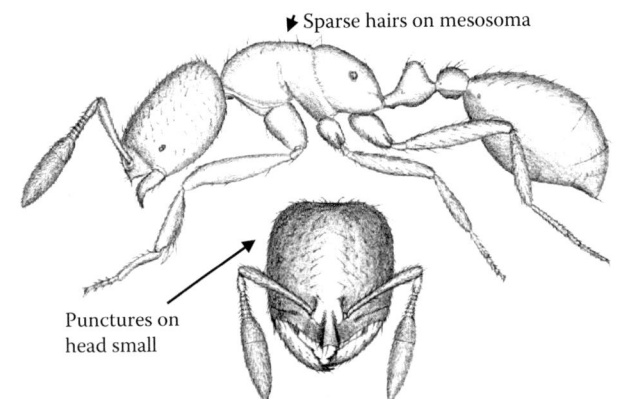

Sparse hairs on mesosoma

- Yellow species found in leaf litter

- *S. abdita* distinguished by black queens and males, those of *S. carolinensis* complex yellow

- Workers of this group among the most abundant Florida ants

Punctures on head small

Solenopsis abdita (p. 101) and *carolinensis* (p. 102) complex
(similar species)

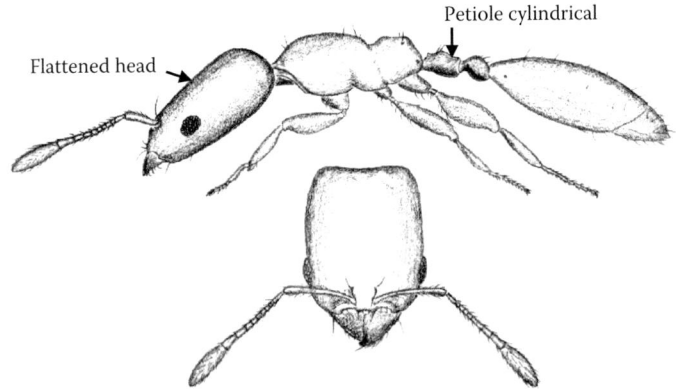

Petiole cylindrical

Flattened head

- Arboreal, often in red mangroves, inland in wet areas

- Color dark browns to yellow

- The flattest Florida ant

Xenomyrmex floridanus (p. 161)

Larger workers with
disproportionately large
head

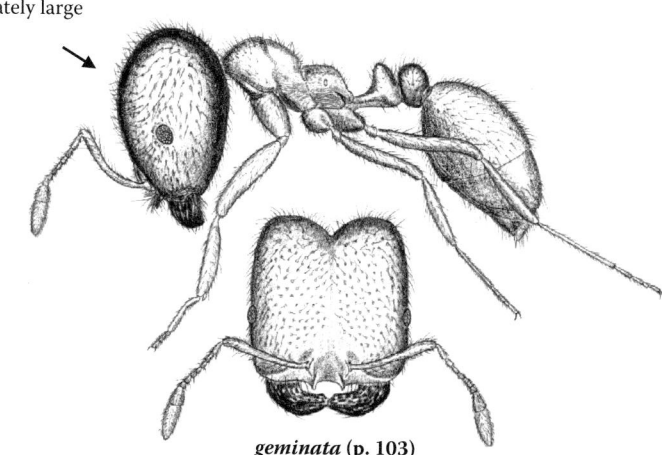

• Throughout Florida, in open
sites, both natural and
disturbed

• In many sites apparently
displaced by *invicta*

geminata (p. 103)

Head of larger workers
proportional

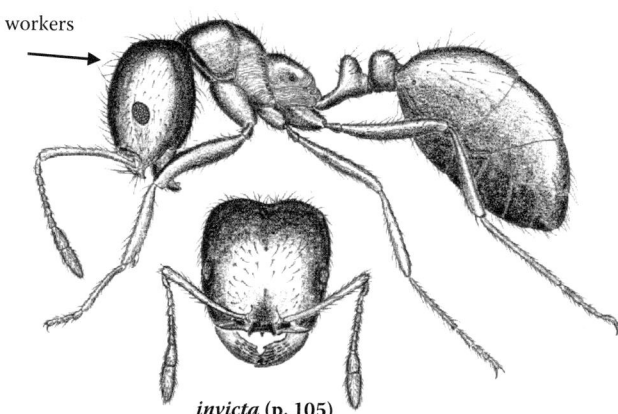

• Throughout Florida in open
sites, much commoner in
disturbed habitat

invicta (p. 105)

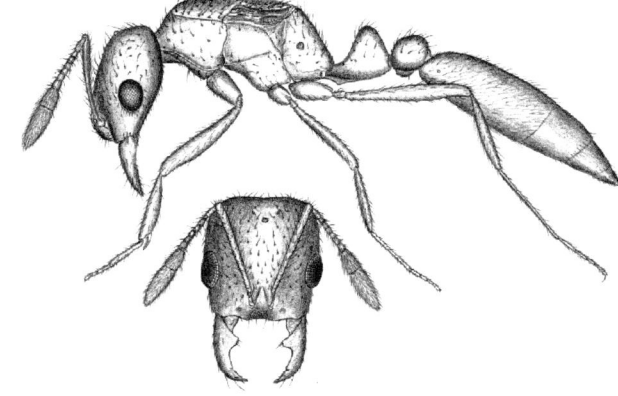

• Only a single specimen
known of this species, found
in Gilchrist County

• Found clinging to back of
queen *Pheidole dentata*

phoretica (p. 111)

Florida *Solenopsis* (thief ants)

Plate 44

Dark brown with
yellowish white
appendages

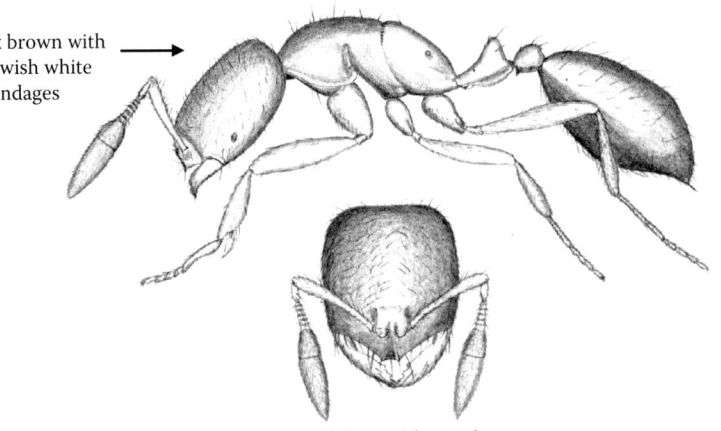

- Most easily identified by color

- Open sandy habitats, often
found among grass roots

- Widespread in peninsular
Florida

nickersoni (p. 109)

Petiole narrowly rounded above

Dark or yellowish
brown, including
appendages

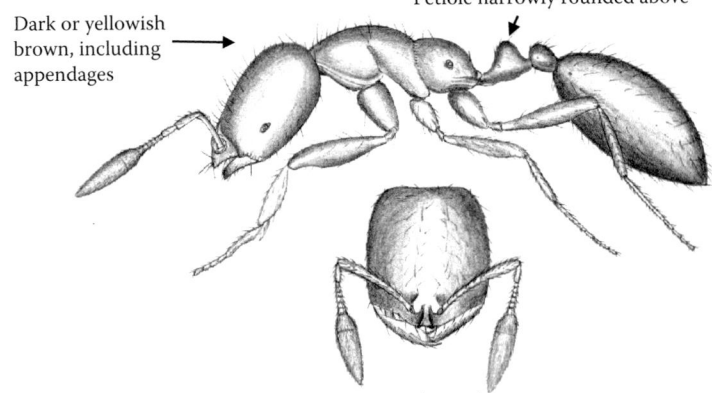

- Only arboreal brown thief
ant in Florida

- Lives in dead twigs on live
trees

- Widespread in Florida

picta (p. 111)

Dark brown
including
appendages

Petiole broadly rounded above

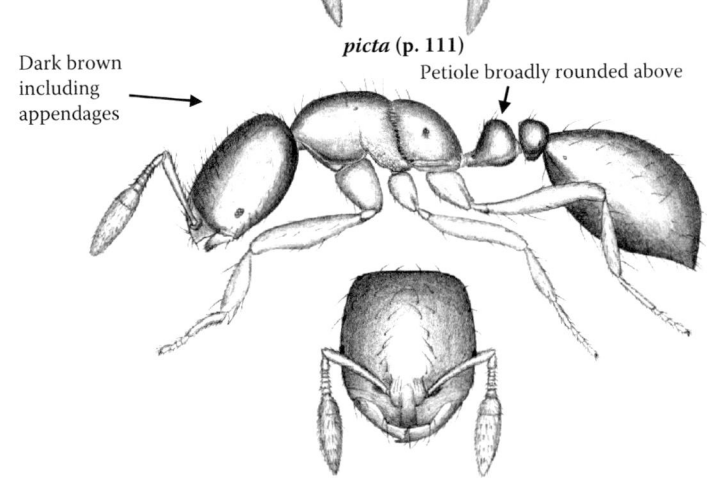

- In Florida, known only from
Matheson Hammock in
south Miami

- Nests in rotten wood on
ground

- Introduced from neotropics

terricola (p. 113)

Hairs on pronotum and mesonotum
irregular, some curved

• Length about 2 mm

• Pale yellow

• Usually in open sandy areas

• Subterranean, but
conspicuous mounds in early
morning when queens and
males fly

• Throughout Florida

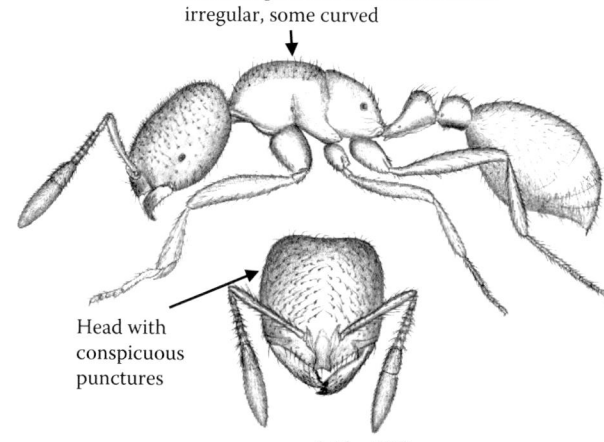

Head with
conspicuous
punctures

pergandei (p. 110)

Hairs on pronotum and mesonotum short, even

• Length about 1.5 mm

• Pale yellow

• Seldom seen subterranean
species from north Florida
sandhills

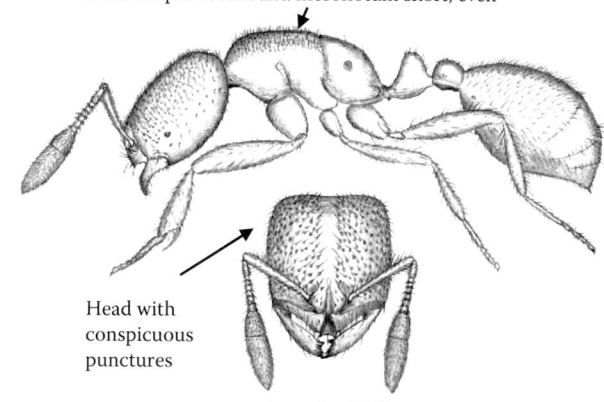

Head with
conspicuous
punctures

tonsa (p. 114)

Head flattened

• Small, 1.2 mm long,
smaller than other
Solenopsis (except species A)

• Pale yellow

• Common and widespread,
often extracted from leaf
litter

Head with
conspicuous
punctures

tennesseensis (p. 112)

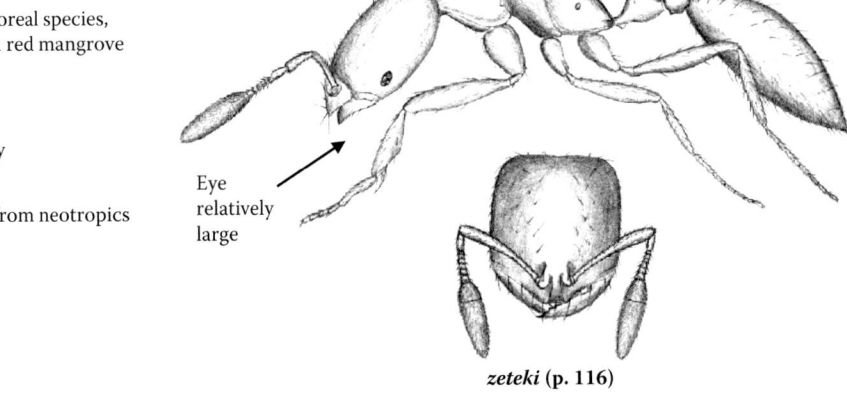

Long sparse hairs

- Tropical arboreal species, often lives in red mangrove twigs

- Yellow, shiny

- Introduced from neotropics

Eye relatively large

zeteki (p. 116)

- Found in leaf litter, known from Matheson Hammock

- Yellow, shiny

- Small, length 1.2 mm

- Compare with *tennesseensis*

- Probably introduced from neotropics

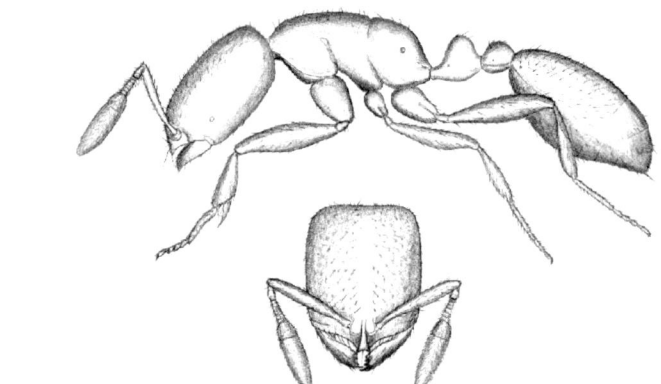

Solenopsis Species A (p. 116)

Pyramidal petiole

- Nests on ground in hollow twigs

- Orange yellow, shiny

- In Florida known from three widely separated sites

- Introduced from neotropics

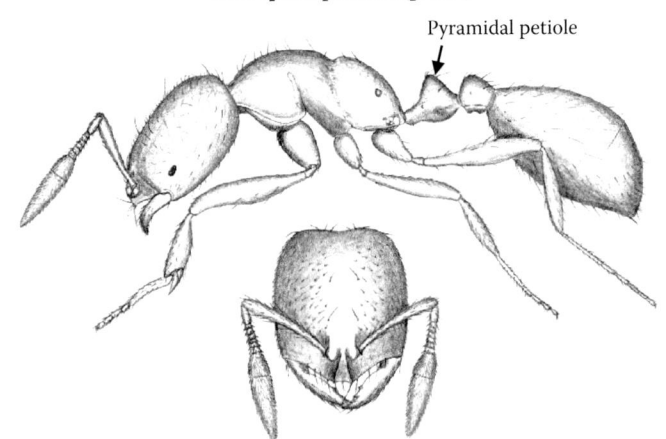

Solenopsis Species B (p. 117)

No spoon-shaped hairs on mesosoma

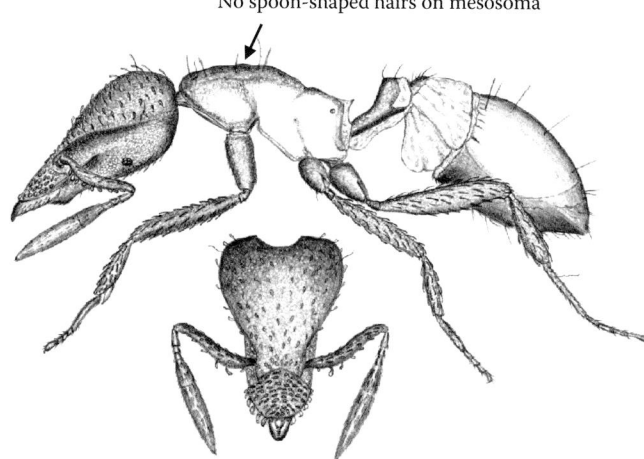

• In Florida rare, known
  from western Panhandle,
  two specimens

• Compare with *bunki,
  carolinensis, creightoni*

hyalina (p. 130)

• Rare in Florida

• Possibly subterranean

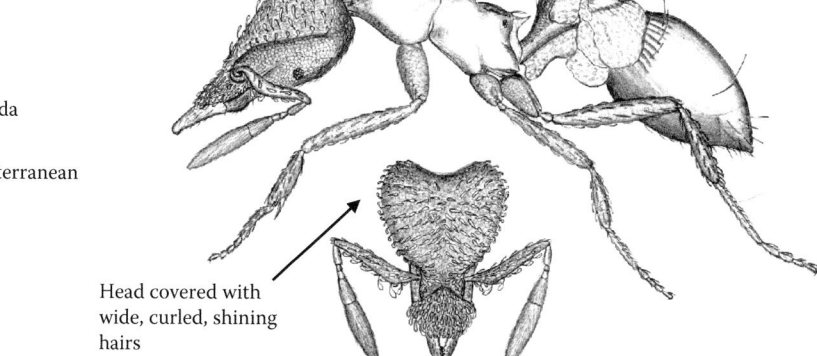

Head covered with
wide, curled, shining
hairs

abdita (p. 122)

Sides of mesosoma granulate, not shining

Relatively large eyes

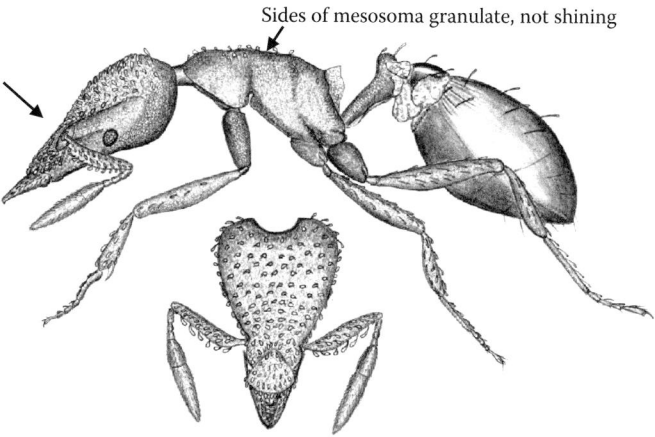

• Only arboreal Florida
  Strumigenys

• Introduced from
  neotropics

epinotalis (p. 128)

No membranous structures on petiole or
postpetiole

Gaster not
shining

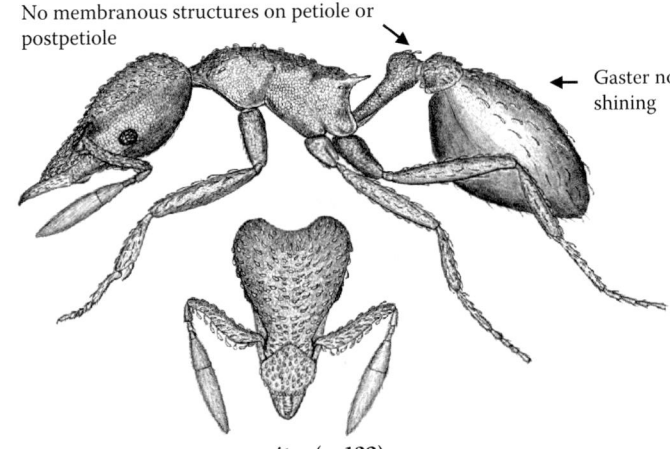

• Uncommon species in
  Florida

• Introduced from
  neotropics

margaritae (p. 132)

• Rare species in Florida

• Nests usually in rotten
  wood

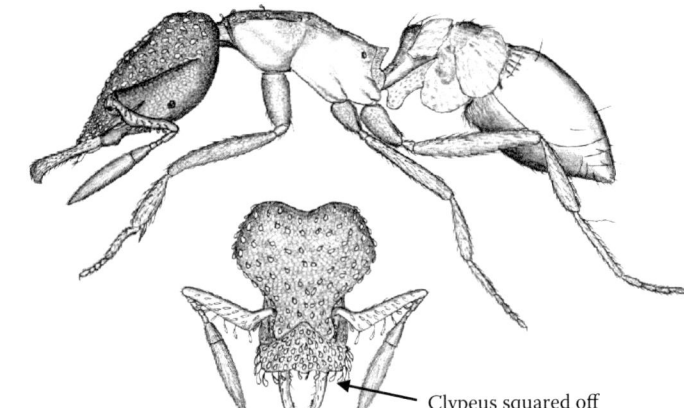

Clypeus squared off

angulata (p. 122)

Two small hairs on
head otherwise
hairless

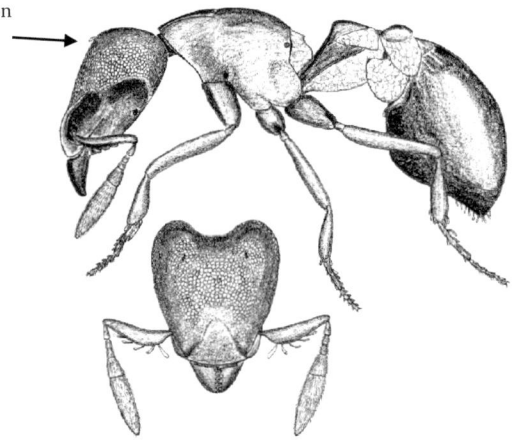

• Found in both disturbed
  and natural areas

• Introduced from Old
  World Tropics and
  Subtropics

membranifera (p. 133)

Florida *Strumigenys* (mustache ants)

Plate 49

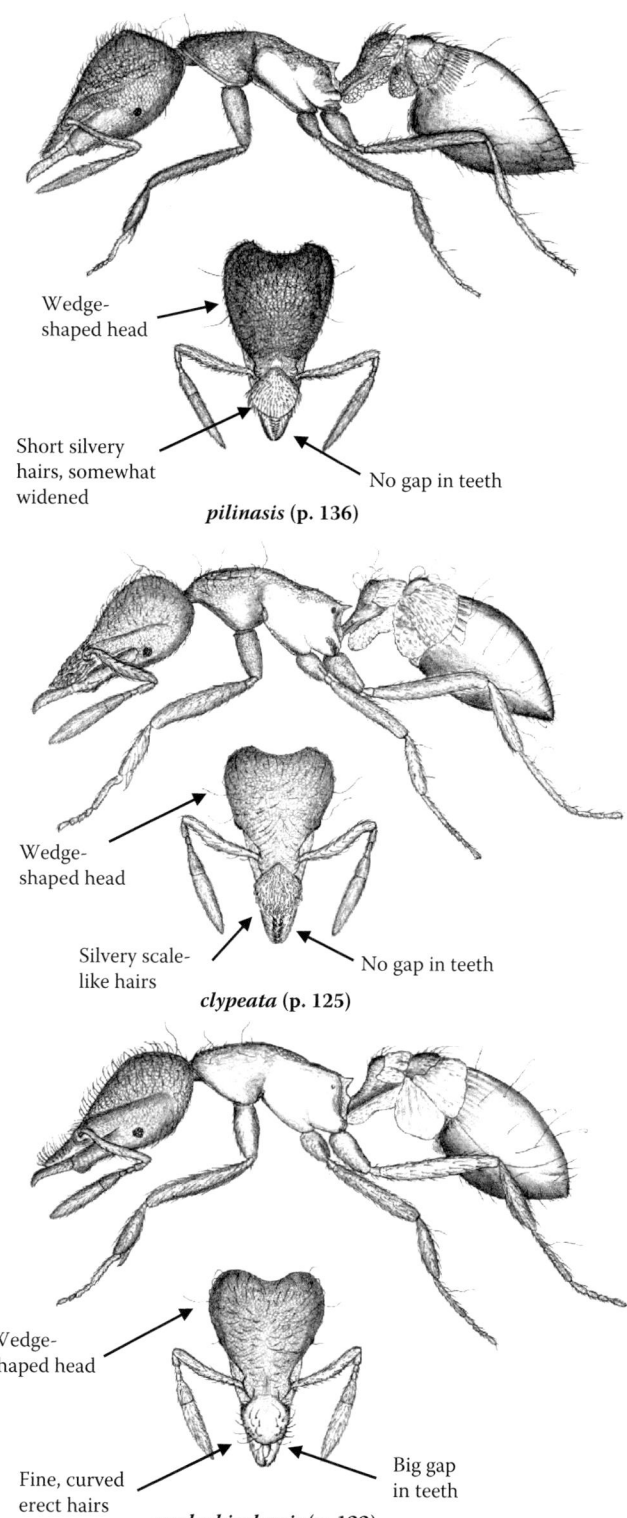

• A rarely collected
  woodland species

• Compare with *laevinasis*,
  which has much finer
  clypeal hairs

• Compare with *clypeata*,
  which has scale-like hairs
  flat on surface of clypeus

Wedge-
shaped head

Short silvery
hairs, somewhat
widened

No gap in teeth

pilinasis (p. 136)

• A relatively common
  woodland species

• Compare with *pilinasis*

Wedge-
shaped head

Silvery scale-
like hairs

No gap in teeth

clypeata (p. 125)

• Found twice in Florida
  flatwoods habitat, once
  in South Carolina

Wedge-
shaped head

Fine, curved
erect hairs

Big gap
in teeth

apalachicolensis (p. 122)

PLATES   309

- Commonest native *Strumigenys* in Florida

- In wide variety of forest habitats

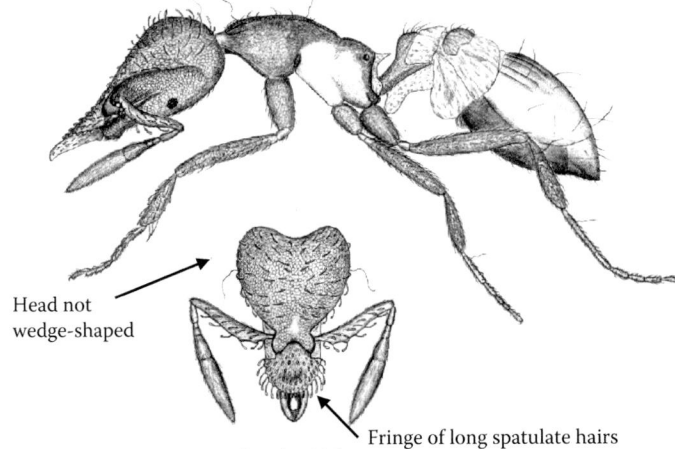

Head not wedge-shaped

Fringe of long spatulate hairs

talpa (p. 140)

- Relatively uncommon forest species

- Found in northern peninsular Florida and adjacent Georgia

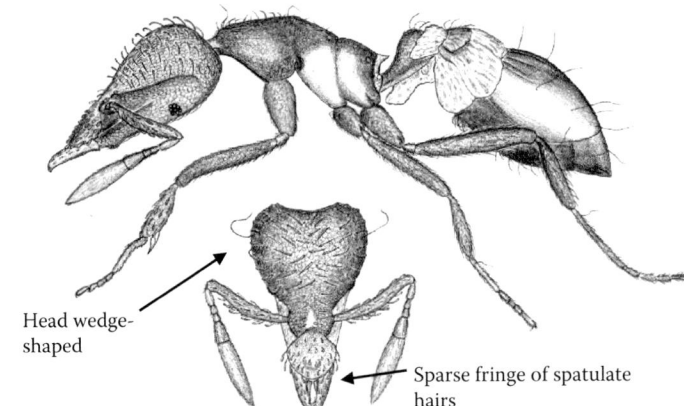
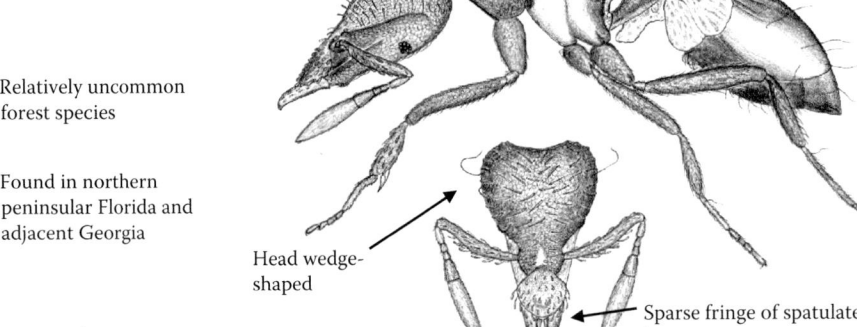

Head wedge-shaped

Sparse fringe of spatulate hairs

archboldi (p. 123)

- Rare species, a few specimens from north Florida

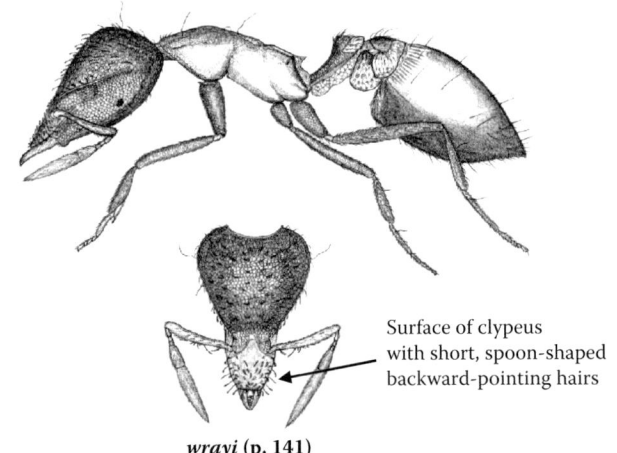

Surface of clypeus with short, spoon-shaped backward-pointing hairs

wrayi (p. 141)

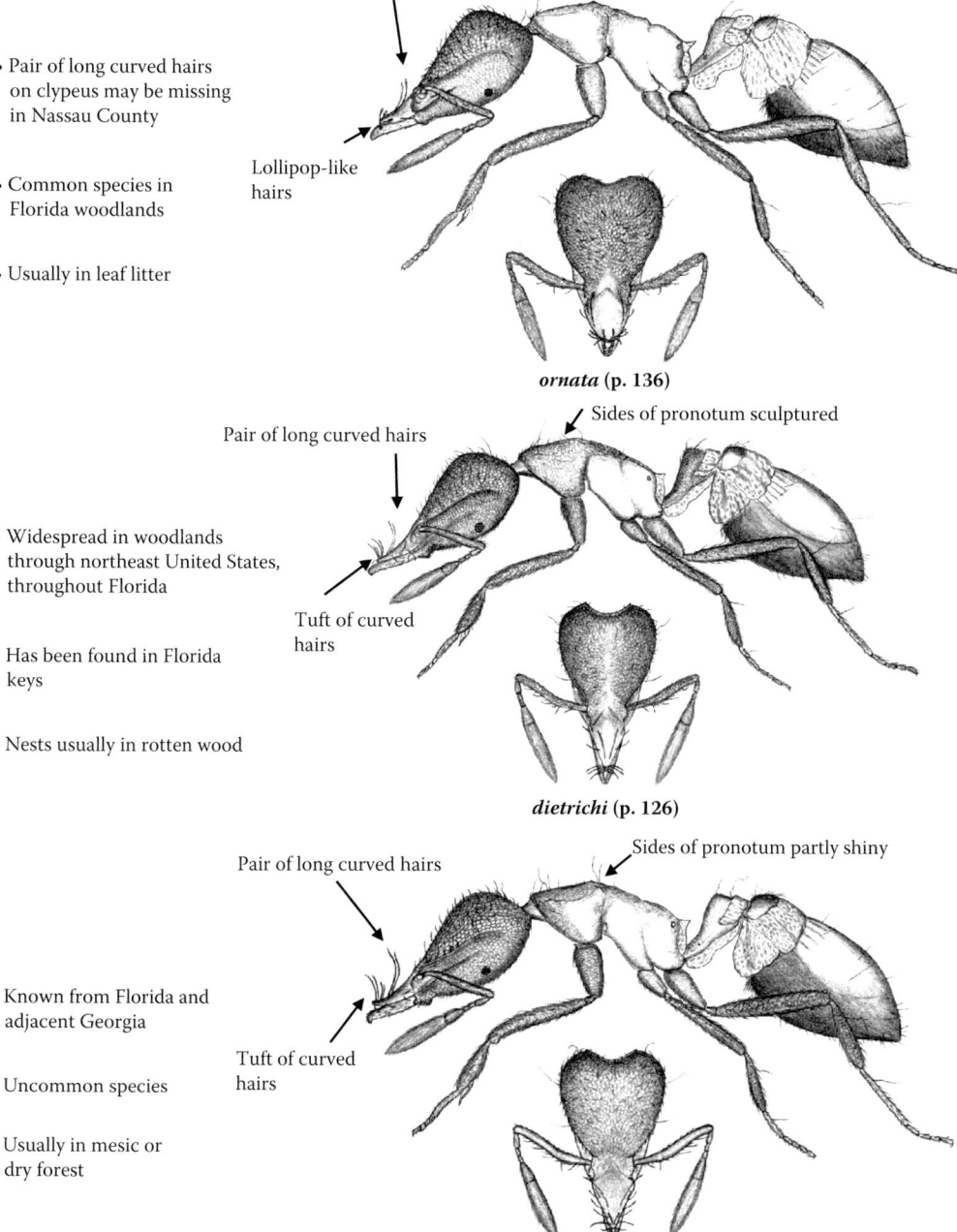

Pair of long curved hairs

- Pair of long curved hairs on clypeus may be missing in Nassau County

- Common species in Florida woodlands

- Usually in leaf litter

Lollipop-like hairs

ornata (p. 136)

Pair of long curved hairs

Sides of pronotum sculptured

- Widespread in woodlands through northeast United States, throughout Florida

- Has been found in Florida keys

- Nests usually in rotten wood

Tuft of curved hairs

dietrichi (p. 126)

Pair of long curved hairs

Sides of pronotum partly shiny

- Known from Florida and adjacent Georgia

- Uncommon species

- Usually in mesic or dry forest

Tuft of curved hairs

boltoni (p. 123)

Plate 52

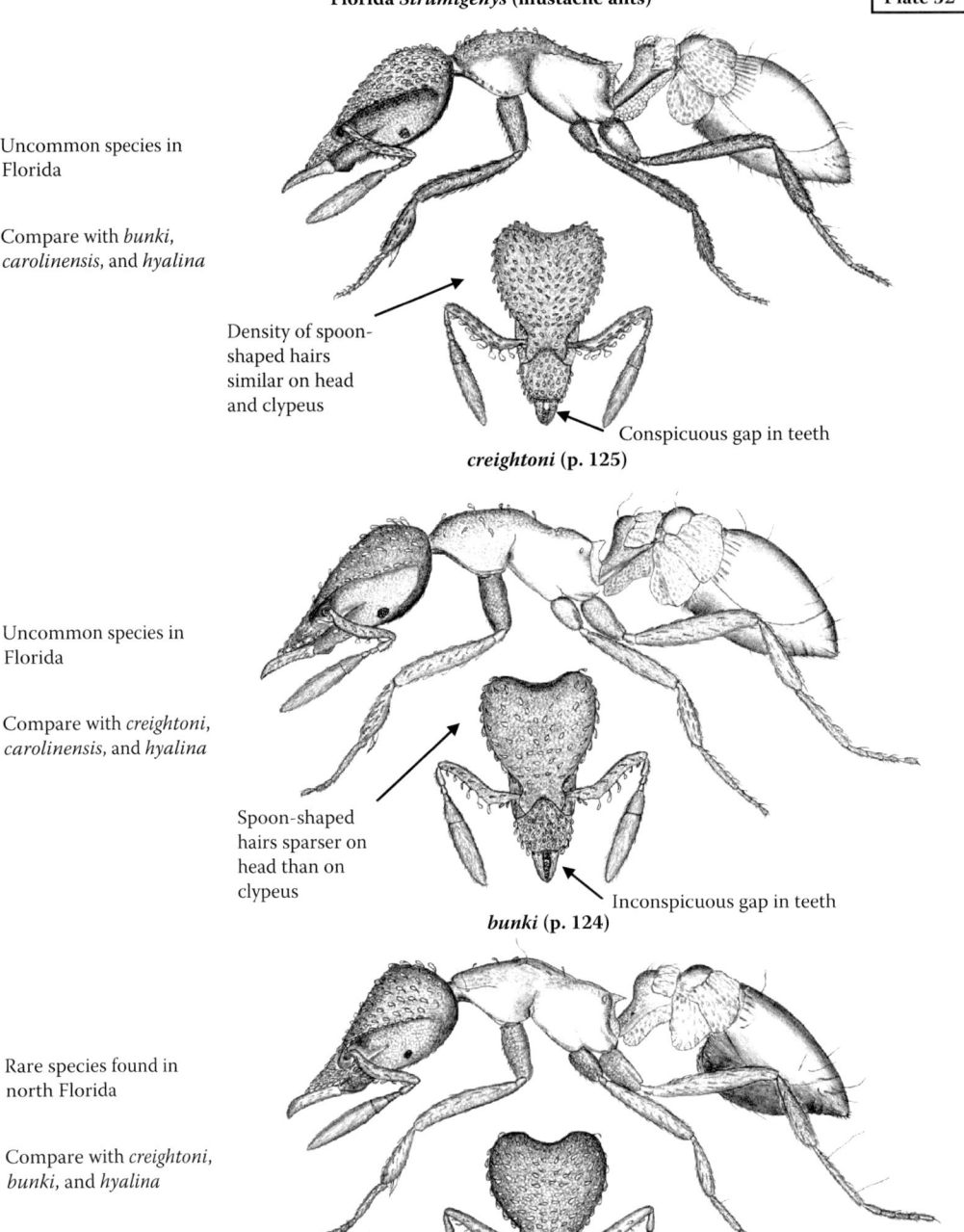

- Uncommon species in Florida

- Compare with *bunki*, *carolinensis*, and *hyalina*

Density of spoon-shaped hairs similar on head and clypeus

Conspicuous gap in teeth

creightoni (p. 125)

- Uncommon species in Florida

- Compare with *creightoni*, *carolinensis*, and *hyalina*

Spoon-shaped hairs sparser on head than on clypeus

Inconspicuous gap in teeth

bunki (p. 124)

- Rare species found in north Florida

- Compare with *creightoni*, *bunki*, and *hyalina*

Jaws thick with large teeth

carolinensis (p. 124)

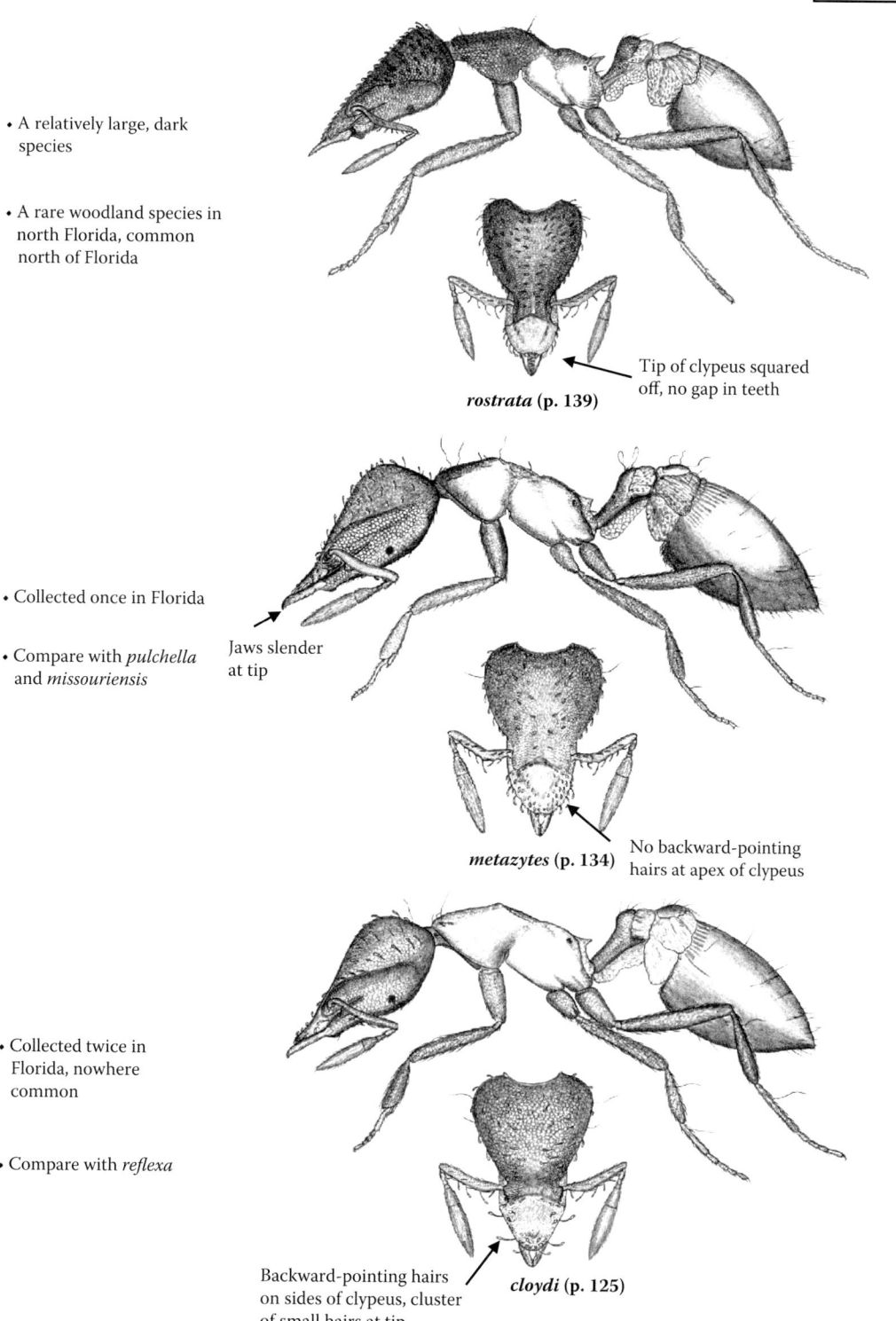

- A relatively large, dark species

- A rare woodland species in north Florida, common north of Florida

Tip of clypeus squared off, no gap in teeth

rostrata (p. 139)

- Collected once in Florida

- Compare with *pulchella* and *missouriensis*

Jaws slender at tip

metazytes (p. 134)

No backward-pointing hairs at apex of clypeus

- Collected twice in Florida, nowhere common

- Compare with *reflexa*

Backward-pointing hairs on sides of clypeus, cluster of small hairs at tip

cloydi (p. 125)

No membranous structure on postpetiole

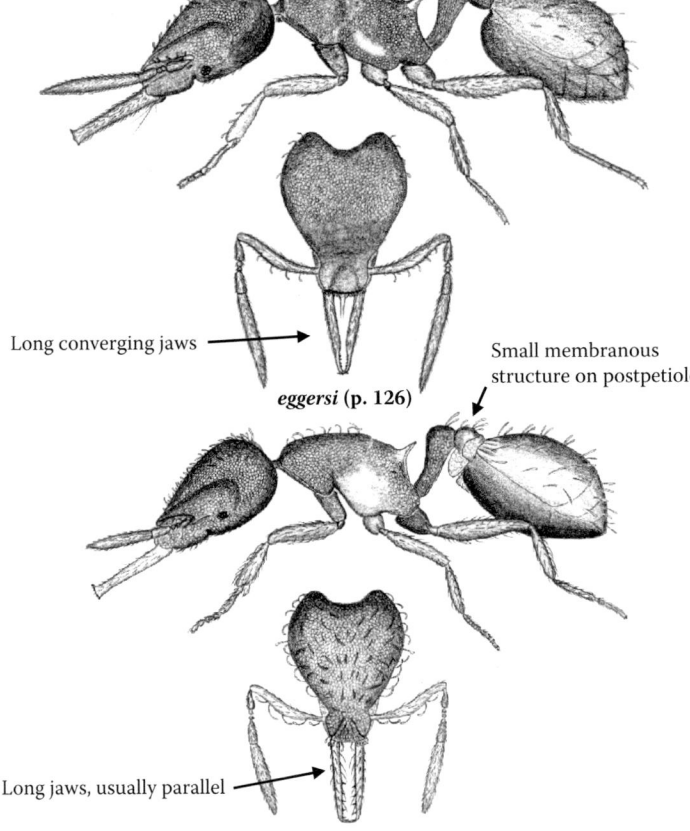

• Extremely common in south and central Florida in natural and disturbed areas

• Introduced from neotropics

Long converging jaws

Small membranous structure on postpetiole

eggersi (p. 126)

• Relatively rare in Florida, in southernmost Florida

• Introduced from neotropics

Long jaws, usually parallel

gundlachi (p. 129)

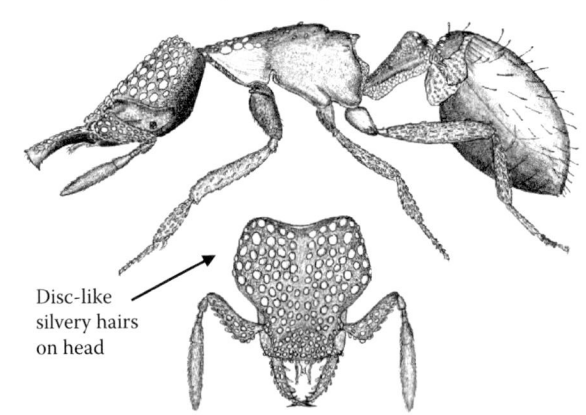

• Rare in Florida, commoner in Louisiana and Mississippi

• Introduced from Old World, probably Japan

Disc-like silvery hairs on head

hexamera (p. 130)

Plate 55

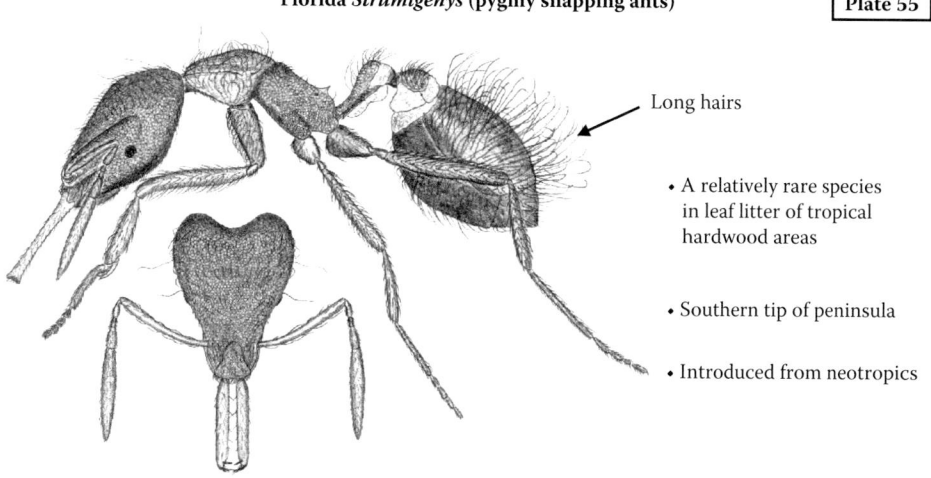

Long hairs

- A relatively rare species in leaf litter of tropical hardwood areas

- Southern tip of peninsula

- Introduced from neotropics

lanuginosa (p. 132)

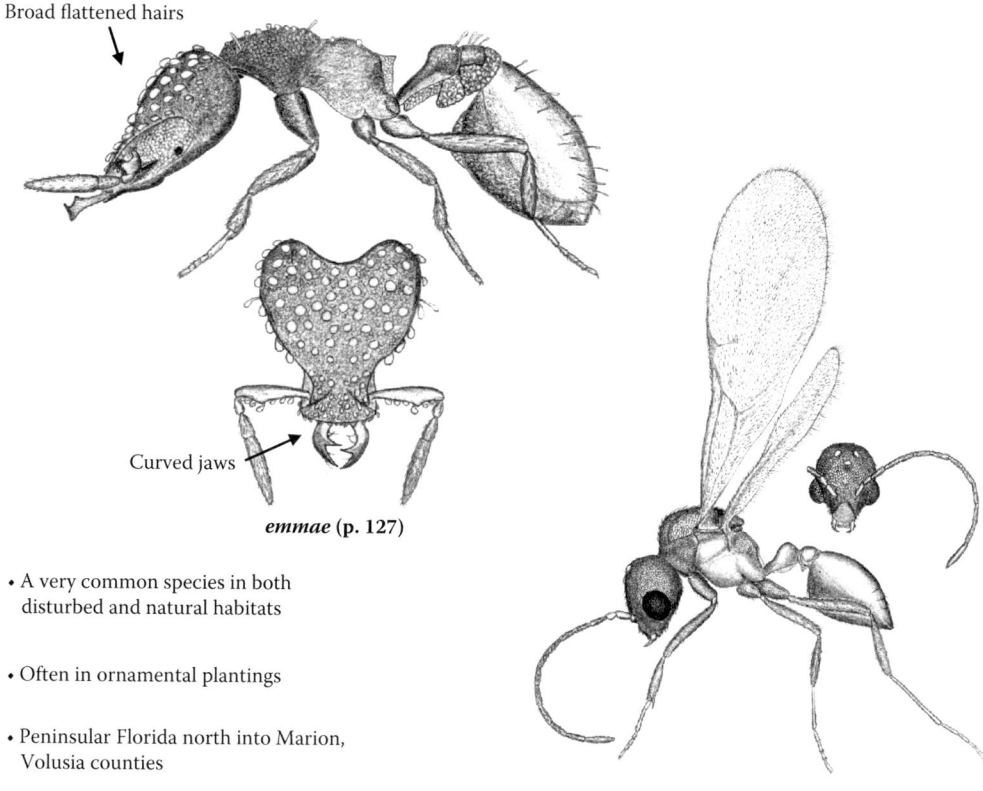

Broad flattened hairs

Curved jaws

emmae (p. 127)

- A very common species in both disturbed and natural habitats

- Often in ornamental plantings

- Peninsular Florida north into Marion, Volusia counties

- Introduced from Australian area

emmae male

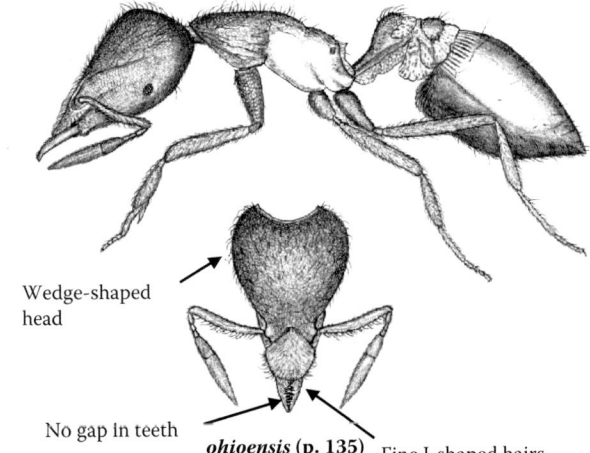

• A common northern species but rare in Florida

• Often in forest edges

Wedge-shaped head

No gap in teeth   Fine J-shaped hairs

***ohioensis* (p. 135)**

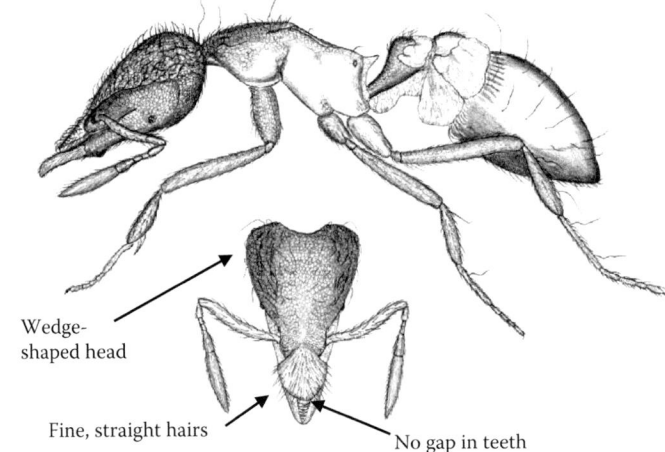

• A generally rare species, in Florida collected once, in Panhandle

• Compare with *pilinasis*

Wedge-shaped head

Fine, straight hairs   No gap in teeth

***laevinasis* (p. 131)**

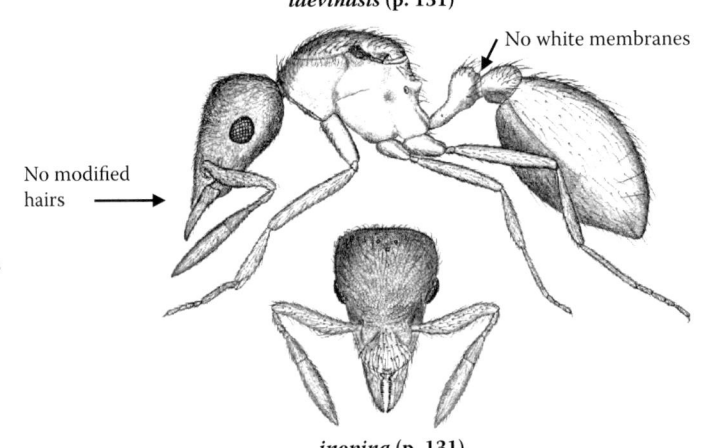

No white membranes

No modified hairs

• Known from three specimens from north-central Florida

• Lacks all modifications found in mustache ants

• Known from queens, possibly parasitic on other ants

inopina (p. 131)

Plate 57

Florida *Strumigenys* (pygmy snapping ants)

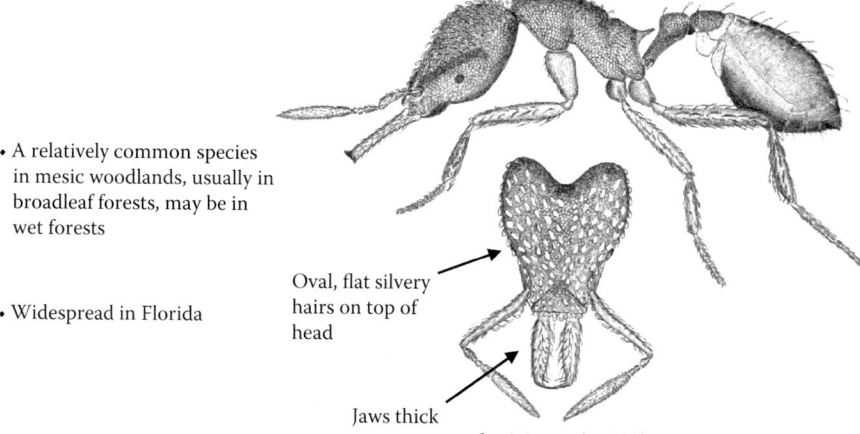

• A relatively common species in mesic woodlands, usually in broadleaf forests, may be in wet forests

• Widespread in Florida

Oval, flat silvery hairs on top of head

Jaws thick

louisianae (p. 132)

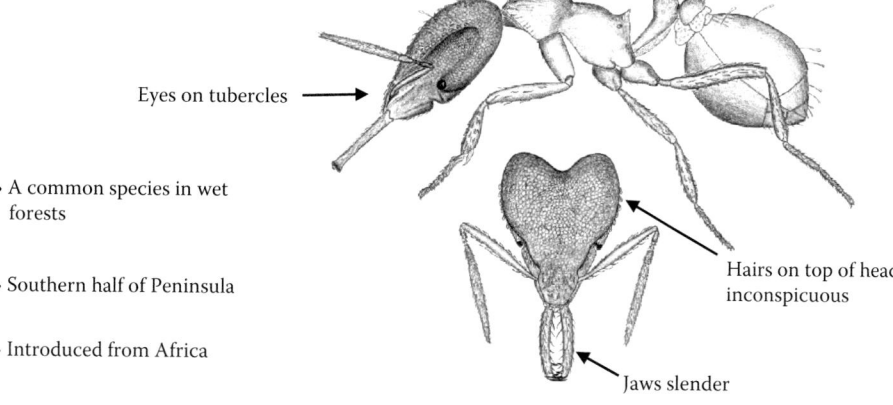

Eyes on tubercles

• A common species in wet forests

• Southern half of Peninsula

• Introduced from Africa

Hairs on top of head inconspicuous

Jaws slender

rogeri (p. 138)

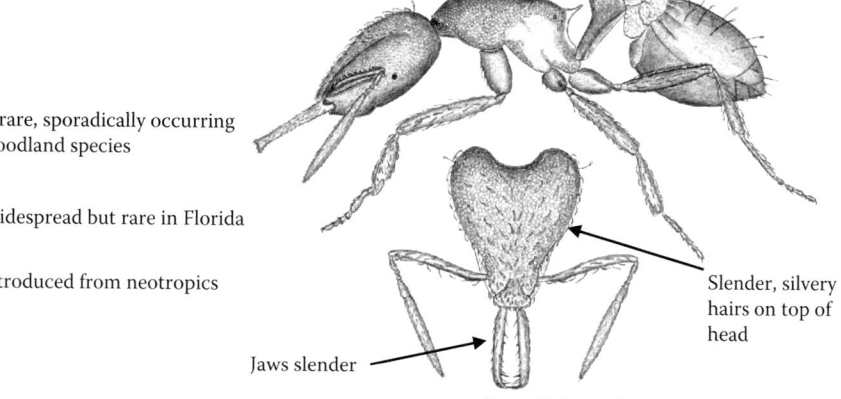

• A rare, sporadically occurring woodland species

• Widespread but rare in Florida

• Introduced from neotropics

Slender, silvery hairs on top of head

Jaws slender

silvestrii (p. 140)

- A forest species, rare in Florida, occurs north into Virginia

- Compare with *cloydi*

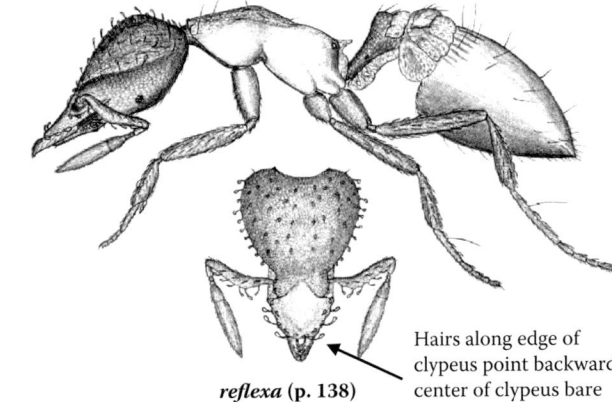

Hairs along edge of clypeus point backward, center of clypeus bare

reflexa (p. 138)

- A forest species, rare in Florida, occurs north into North Carolina

- Compare with *metazytes, rostrata,* and *pulchella*

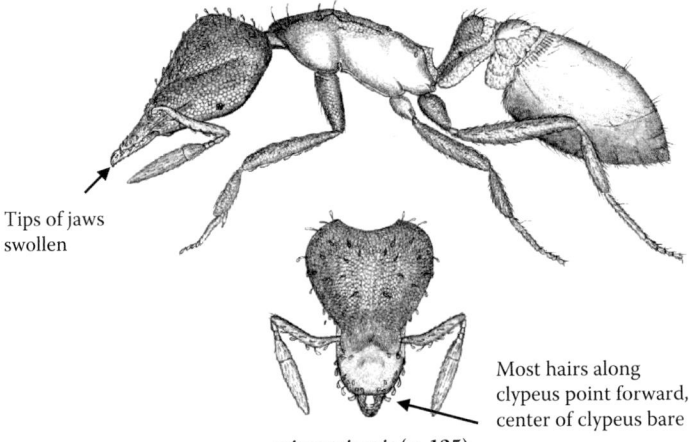

Tips of jaws swollen

Most hairs along clypeus point forward, center of clypeus bare

missouriensis (p. 135)

- A forest species, rare in Florida, occurs north into New York

- Compare with *metazytes* and *rostrata*

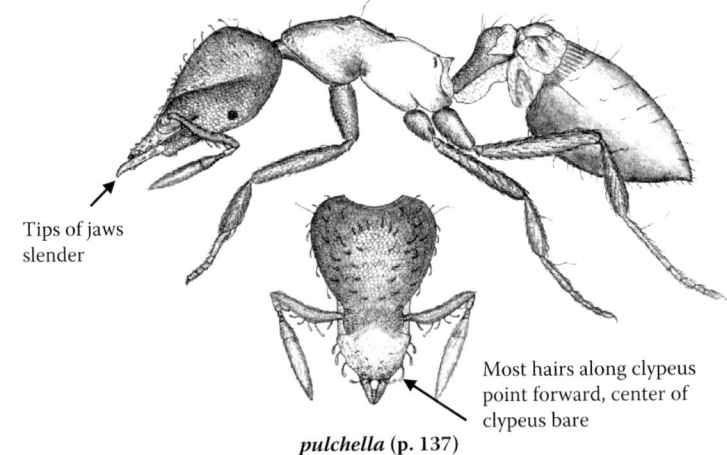

Tips of jaws slender

Most hairs along clypeus point forward, center of clypeus bare

pulchella (p. 137)

Reticulate sculpture

Long spines

- Dead twigs on live trees, old sawgrass culms

- Pale yellow to brownish yellow

- Often nocturnal, difficult to find

- Tropical Florida

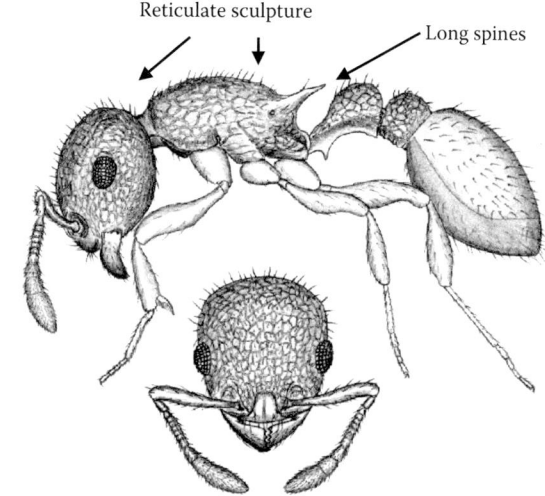

allardycei (p. 143)

- Nests in soil in open or semiopen well-drained sandy sites

- Blackish to dark brown, occasionally reddish brown

- Widespread in Florida

Head mostly shining, with fine ridges

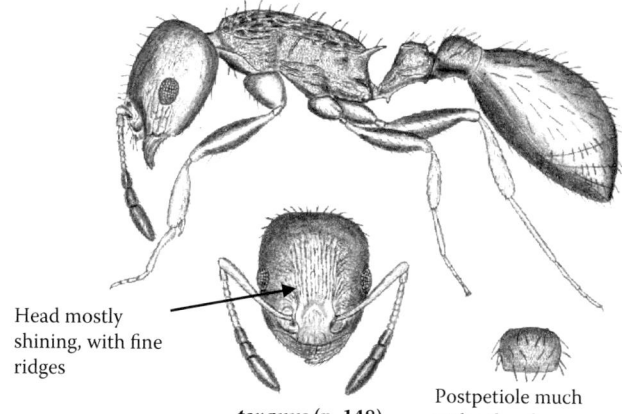

texanus (p. 148)

Postpetiole much wider than long

- Nests in soil in poorly drained sandy flatwoods

- Yellowish, head usually darker

- North Florida, has been collected four times

Head mostly shining, with fine ridges

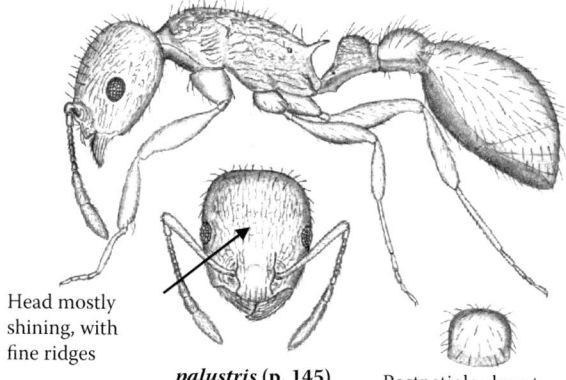

palustris (p. 145)

Postpetiole almost as long as wide

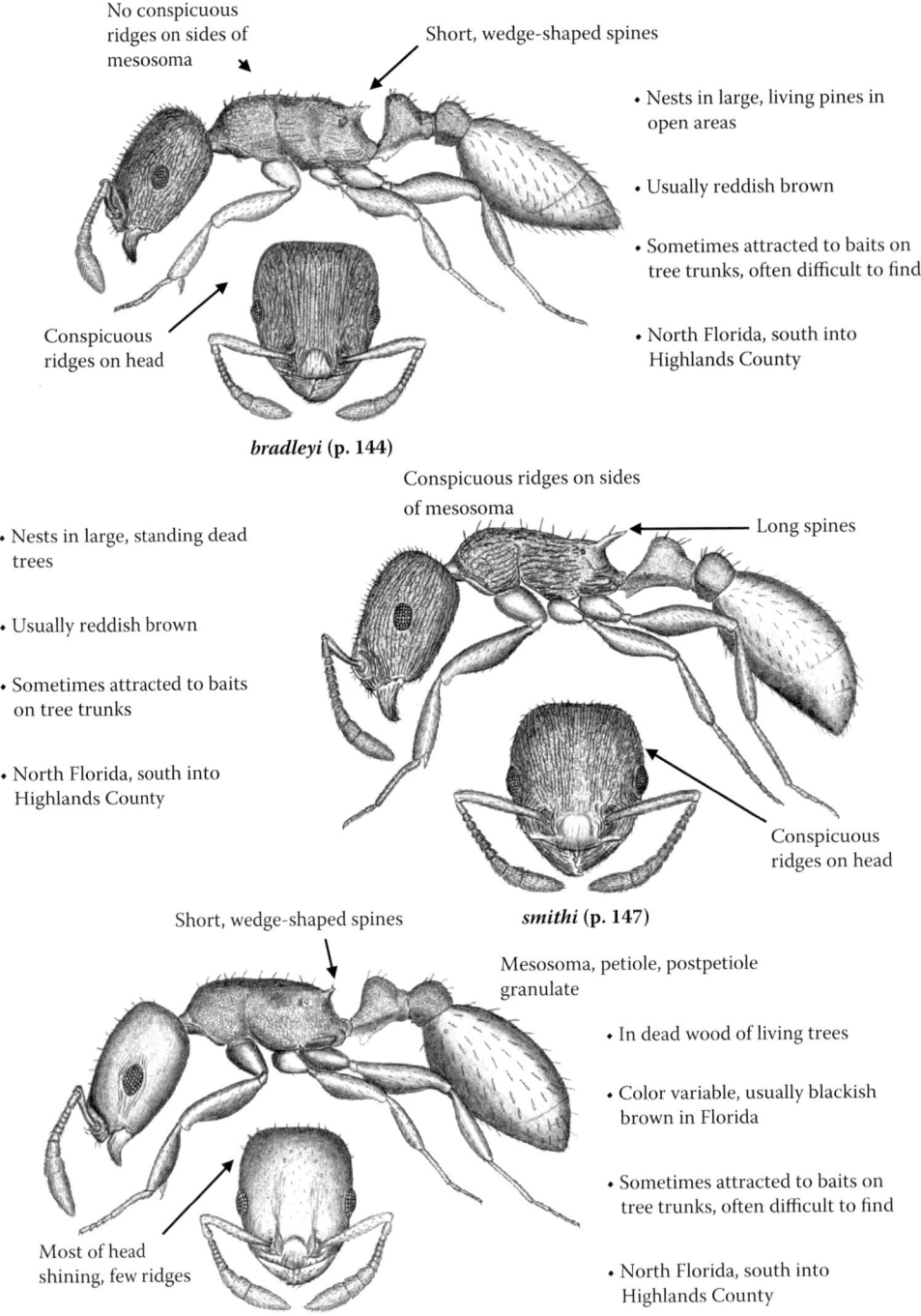

No conspicuous ridges on sides of mesosoma

Short, wedge-shaped spines

- Nests in large, living pines in open areas

- Usually reddish brown

- Sometimes attracted to baits on tree trunks, often difficult to find

- North Florida, south into Highlands County

Conspicuous ridges on head

bradleyi (p. 144)

Conspicuous ridges on sides of mesosoma

Long spines

- Nests in large, standing dead trees

- Usually reddish brown

- Sometimes attracted to baits on tree trunks

- North Florida, south into Highlands County

Conspicuous ridges on head

smithi (p. 147)

Short, wedge-shaped spines

Mesosoma, petiole, postpetiole granulate

- In dead wood of living trees

- Color variable, usually blackish brown in Florida

- Sometimes attracted to baits on tree trunks, often difficult to find

- North Florida, south into Highlands County

Most of head shining, few ridges

schaumii (p. 147)

Florida *Temnothorax* (creeper ants)

Plate 61

Mesosoma with conspicuous ridges

Long, curved spines

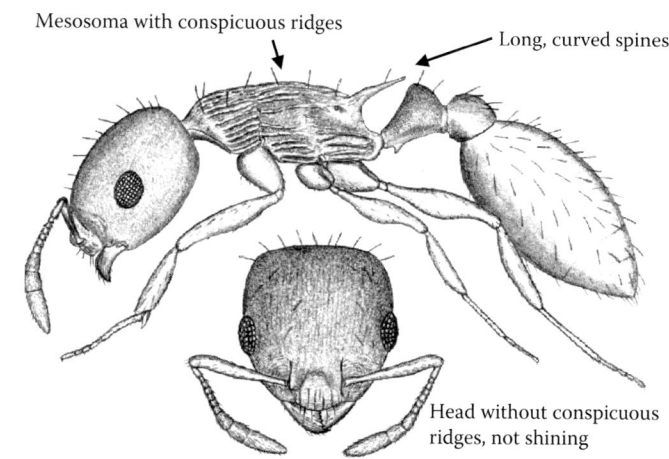

- Nests usually in hollow twigs, weed stems, etc., on ground

- Color yellowish

- Northern border of Florida

Head without conspicuous ridges, not shining

curvispinosus (p. 144)

Head and body not shining, no conspicuous ridges

Long spines

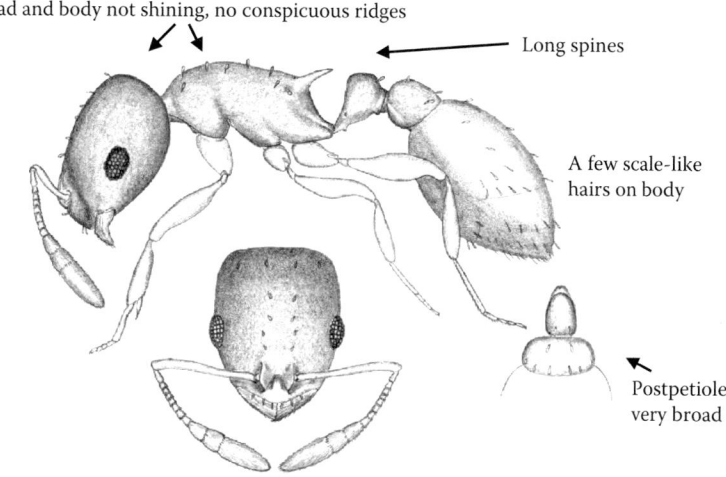

- Leaf litter in tropical hammocks

- Color pale yellow

- Usually collected by sifting or extraction

- Florida keys and coastal tropical hammocks

A few scale-like hairs on body

Postpetiole very broad

torrei (p. 149)

Head, most of mesosoma mostly smooth, shining

Mesosoma in profile with groove

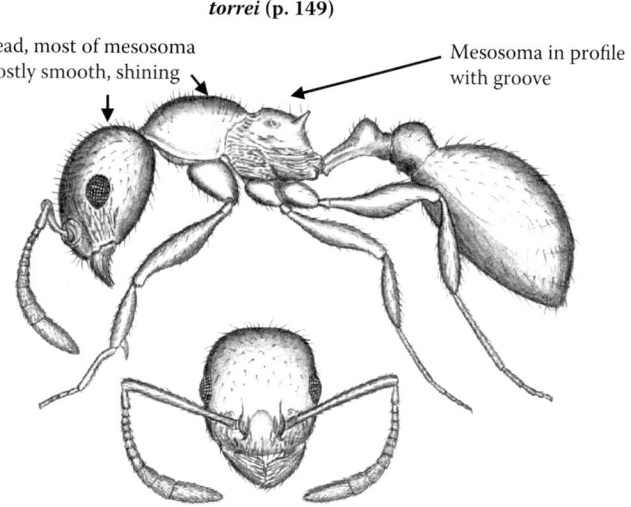

- Various habitats, from uplands to flatwoods and salt marshes

- Color blackish or dark brown, or yellow, occasionally bicolored

- Nests usually in hollow twigs or nuts on ground

- Throughout Florida

pergandei (p. 146)

Florida *Tetramorium* (pennant ants)

Plate 62

Frontal carinae strong at back of head

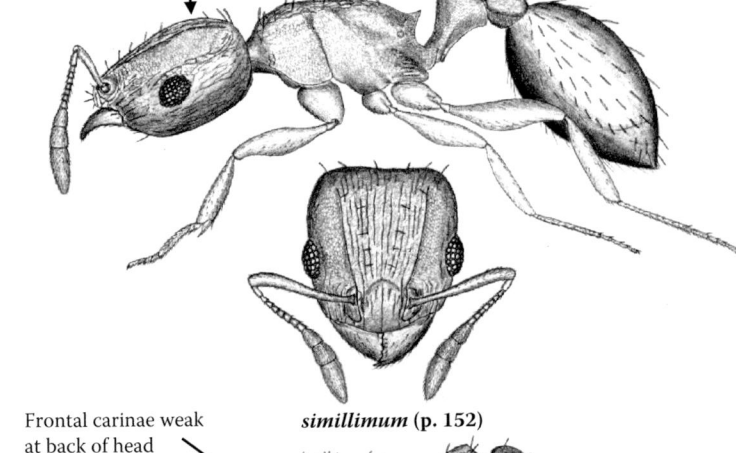

- In open grassy areas and beaches

- Yellow

- Length about 2 mm

- Introduced from Africa

simillimum (p. 152)

Frontal carinae weak at back of head

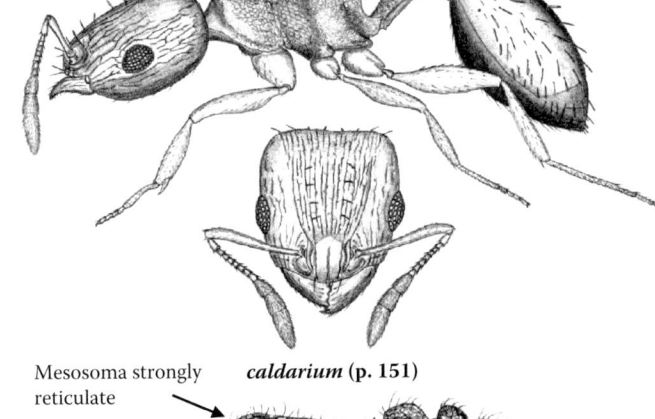

- Often outside buildings or on pavement

- Yellow

- Length about 2 mm

- Introduced from Africa

caldarium (p. 151)

Mesosoma strongly reticulate

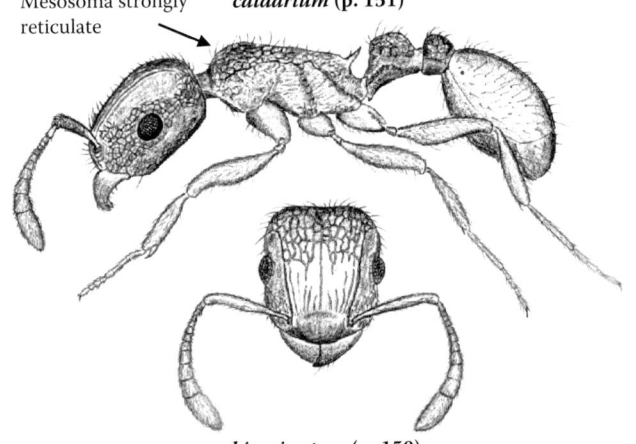

- Nests usually in rotten wood or dead plant stems

- Dark red–brown, gaster black

- Length more than 3 mm

- Introduced from Southeast Asia

bicarinatum (p. 150)

PLATES

**Florida *Tetramorium* (pennant ants)
and *Wasmannia* (little fire ants)**

Plate 63

Reticulate sculpture and dense short hairs

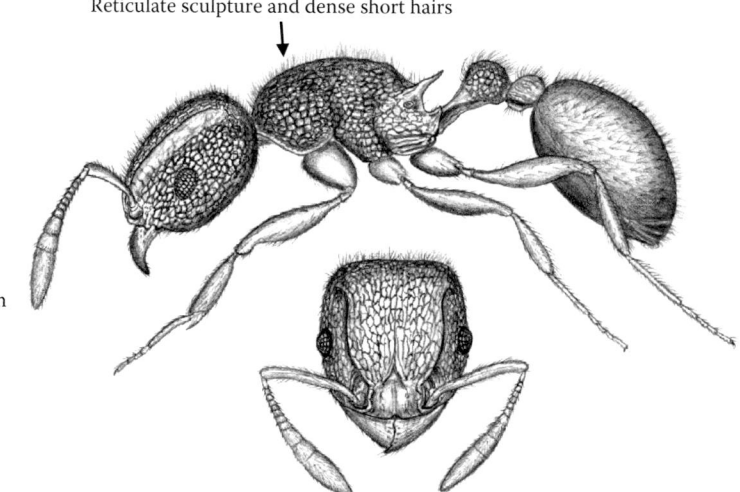

- Rare in Florida

- Dark reddish brown

- Probably introduced from
  Southeast Asia

Tetramorium lanuginosum (p. 152)

Long spines

Hatchet-shaped
petiole

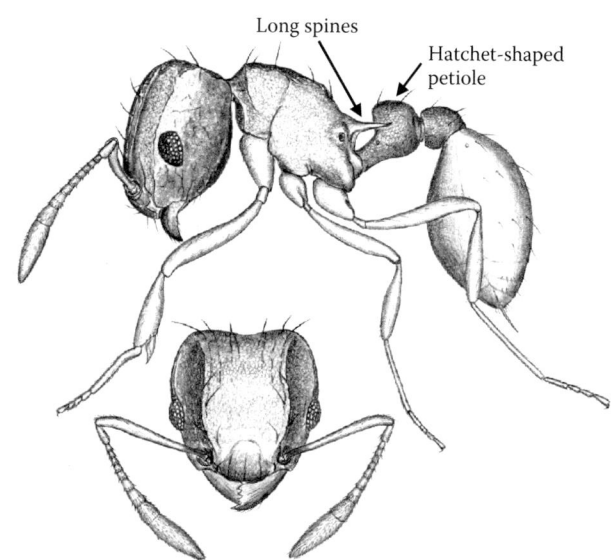

- Southern and central Florida

- Reddish yellow

- Worker about 1.5 mm long

- Nests on ground or in trees

- Amazingly powerful sting

- Introduced from mainland
  neotropics

Wasmannia auropunctata (p. 158)

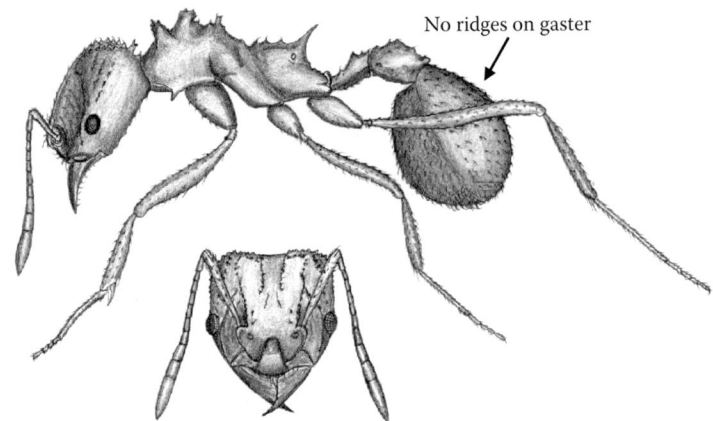

No ridges on gaster

- Lives in sandy uplands throughout Florida

- Length about 3 mm

- Crescent-shaped nest mound

***septentrionalis* (p. 155)**

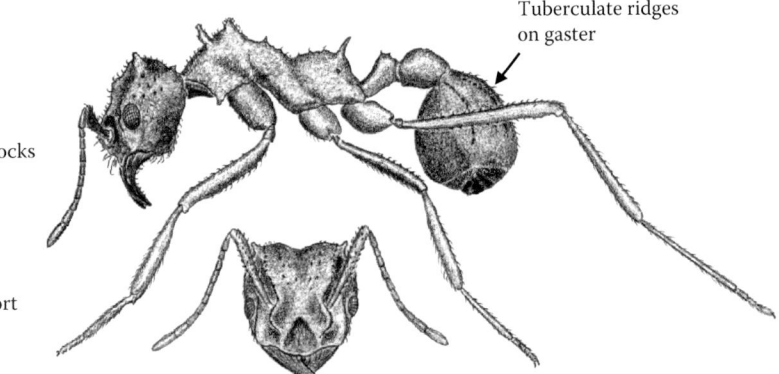

Tuberculate ridges on gaster

- Lives in tropical hammocks in keys

- Length about 4.5 mm

- Nest entrance has a short turret

***jamaicensis* (p. 154)**

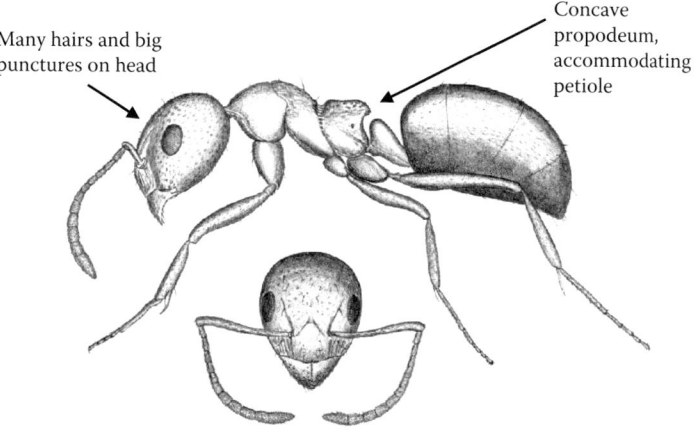

Many hairs and big
punctures on head

Concave
propodeum,
accommodating
petiole

- In Florida, lives in dead
  branches on live trees and
  shrubs

- Usually in wet areas

- Strongly resembles
  Camponotus impressus,
  lives in same sites

Dolichoderus pustulatus (**p. 163**)

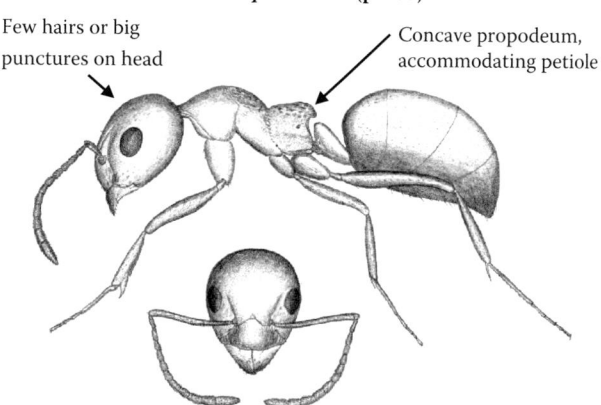

Few hairs or big
punctures on head

Concave propodeum,
accommodating petiole

- In Florida, lives in grass
  tussocks

- Rare in Florida: Leon and
  Taylor counties

- Colonies grow to be very
  large with many queens

Dolichoderus mariae (**p. 162**)

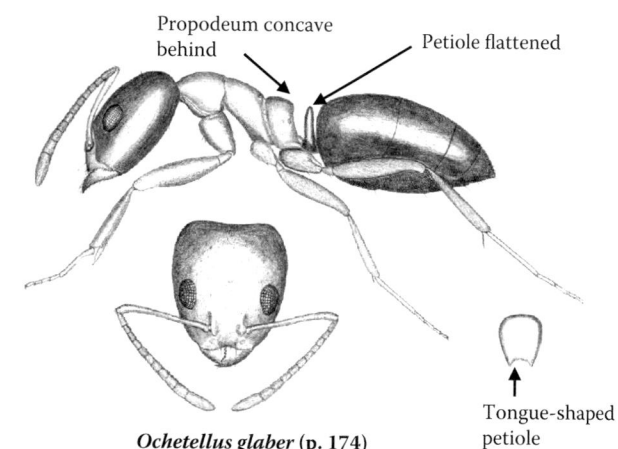

Propodeum concave
behind

Petiole flattened

- Nests in dead wood and in
  grass tussocks, often in wet
  areas

- Small, shiny, head and gaster
  black, mesosoma orange

- In Florida, known from a few
  sites in Orange County

- Introduced from Australia or
  New Guinea

Tongue-shaped
petiole

Ochetellus glaber (**p. 174**)

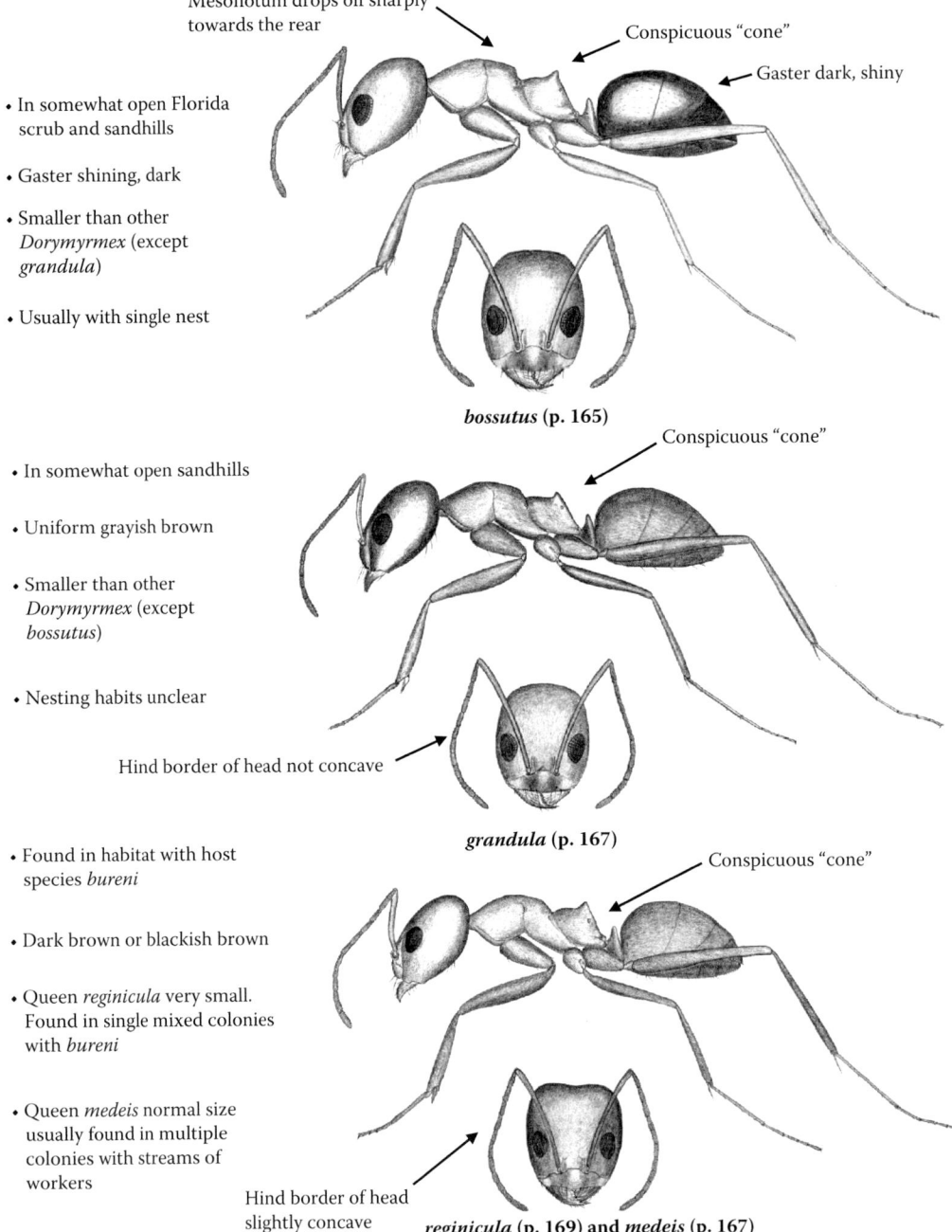

Mesonotum drops off sharply towards the rear

Conspicuous "cone"

Gaster dark, shiny

- In somewhat open Florida scrub and sandhills

- Gaster shining, dark

- Smaller than other *Dorymyrmex* (except *grandula*)

- Usually with single nest

***bossutus* (p. 165)**

Conspicuous "cone"

- In somewhat open sandhills

- Uniform grayish brown

- Smaller than other *Dorymyrmex* (except *bossutus*)

- Nesting habits unclear

Hind border of head not concave

***grandula* (p. 167)**

Conspicuous "cone"

- Found in habitat with host species *bureni*

- Dark brown or blackish brown

- Queen *reginicula* very small. Found in single mixed colonies with *bureni*

- Queen *medeis* normal size usually found in multiple colonies with streams of workers

Hind border of head slightly concave

***reginicula* (p. 169) and *medeis* (p. 167) (similar species)**

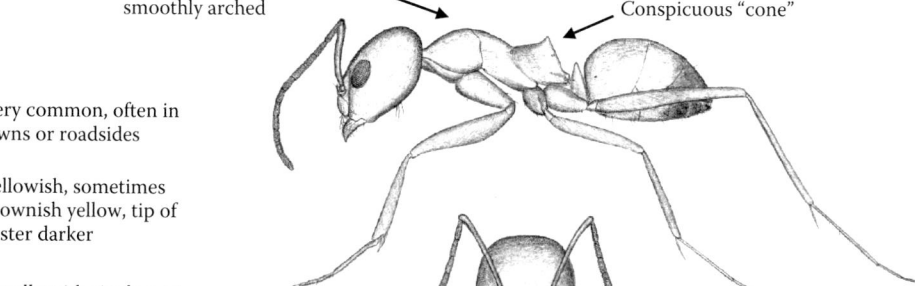

Pronotum and mesonotum smoothly arched

Conspicuous "cone"

- Very common, often in lawns or roadsides

- Yellowish, sometimes brownish yellow, tip of gaster darker

- Usually with single nest

bureni (p. 165)

Pronotum and mesonotum somewhat flattened

Small "cone"

- Rare, restricted to open sandy areas on Lake Wales and Brooksville Ridges

- Yellow

- Usually with single nest

elegans (p. 166)

Pronotum and mesonotum smoothly arched

Conspicuous "cone"

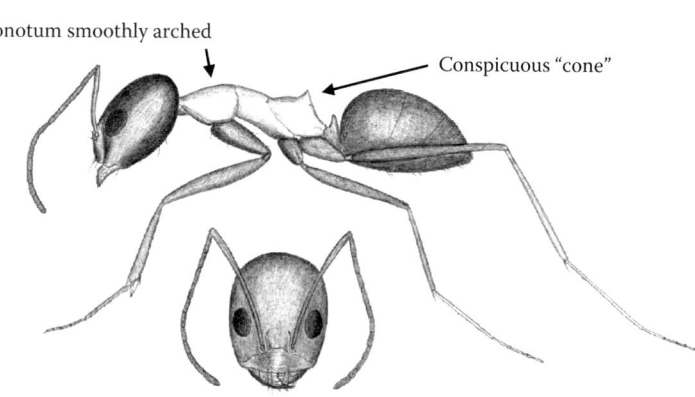

- Rare, restricted to open sites in Florida scrub

- Bicolored orange and black

- Usually with multiple nests

flavopectus (p. 166)

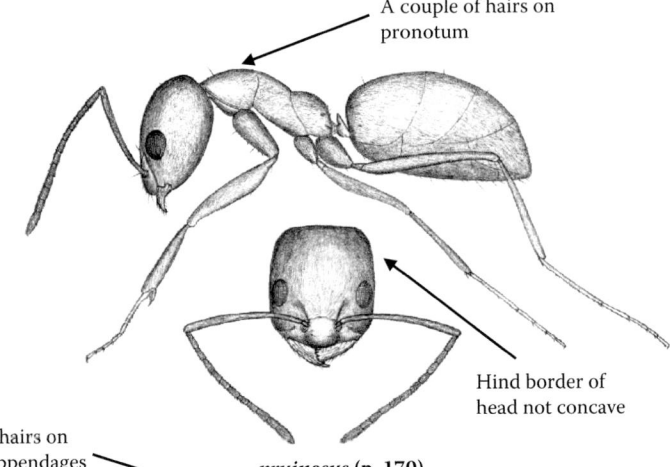

A couple of hairs on pronotum

Hind border of head not concave

- Found everywhere in Florida

- Silvery gray, very fast little ants in hot, open sites

pruinosus (p. 170)

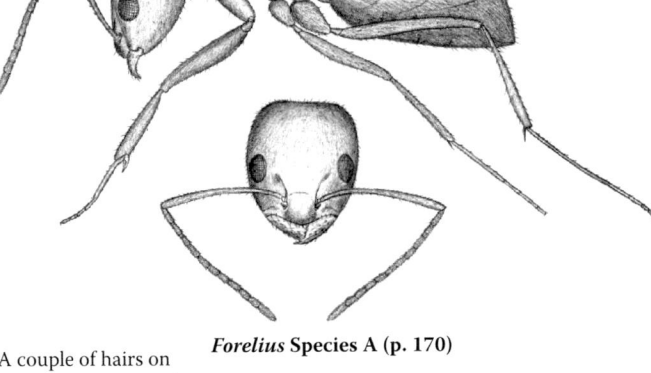

Many short erect hairs on head, body, and appendages

- Known from a few sites in northern Florida

- Silvery gray, very fast little ants in hot sandy sites

- Almost impossible to distinguish from *pruinosus* without high magnification

Forelius Species A (p. 170)

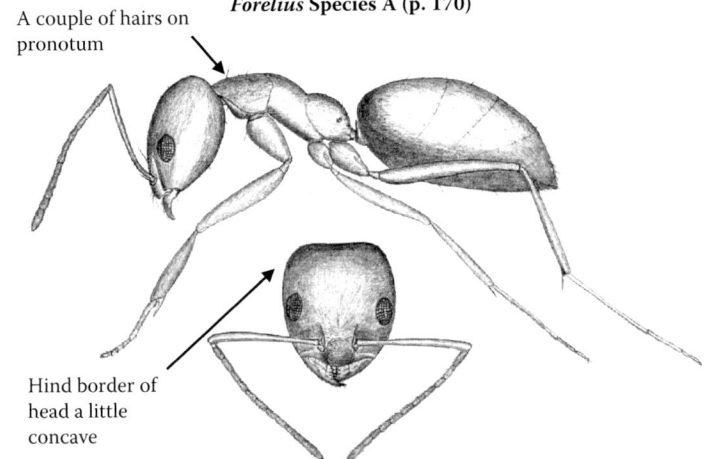

A couple of hairs on pronotum

Hind border of head a little concave

- Common in Southeast, collected once in Florida, look for it near Georgia border

- Distinctly yellowish gray, in hot, open sites, clay soils

Forelius Species B (p. 171)

Florida *Tapinoma* (odorous ants), *Linepithema* (Argentine ants), and a similar *Forelius* (asbestos ants)

Plate 69

No erect hairs on mesosoma or gaster →   ← Petiole flat, hidden

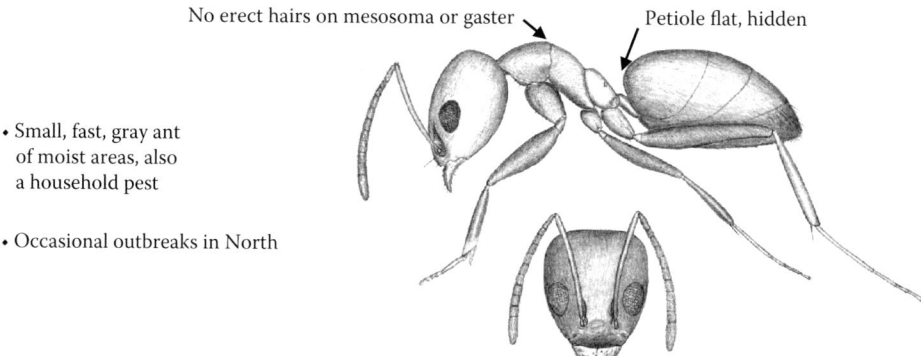

- Small, fast, gray ant of moist areas, also a household pest

- Occasional outbreaks in North

***Tapinoma sessile* (p. 177)**

No erect hairs on top of head or mesosoma →   ← Petiole scale-like

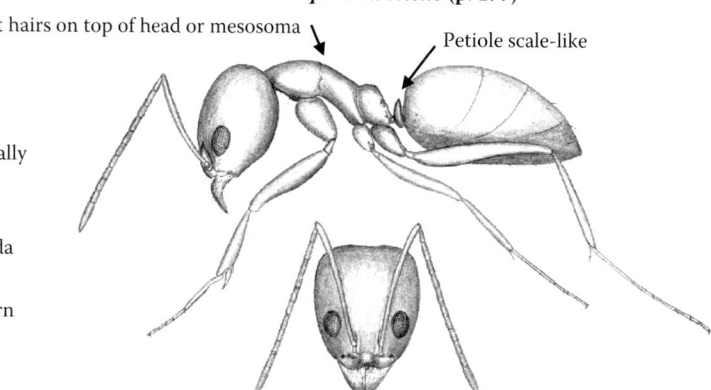

- Small, fast, gray ant, usually in disturbed sites

- Local outbreaks in Florida

- Introduced from southern South America

***Linepithema humile* (p. 172)**

Conspicuous pair of hairs on pronotum →   ← Petiole scale-like

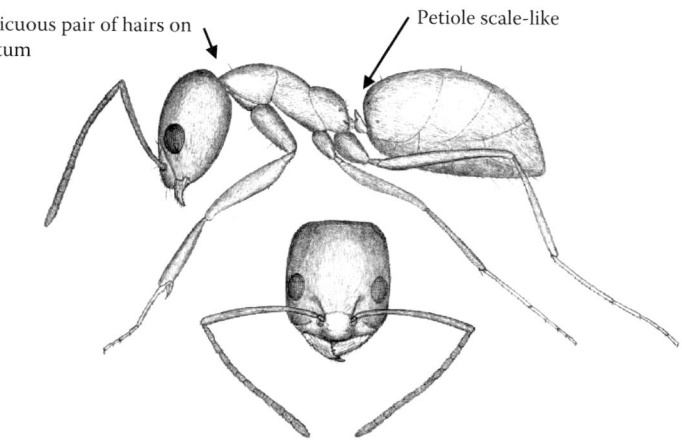

- Small, fast, gray ant of hot, open areas

- Not a household pest

***Forelius pruinosus* (p. 170)**

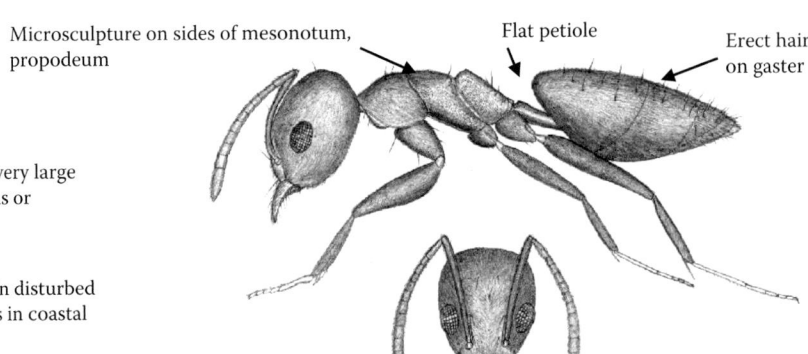

Microsculpture on sides of mesonotum, propodeum

Flat petiole

Erect hairs on gaster

- Sometimes has very large local populations or outbreaks

- Most common in disturbed sites, sometimes in coastal hammocks

- In field can be mistaken for *Nylanderia bourbonica*

***Technomyrmex difficilis* (p. 178)**

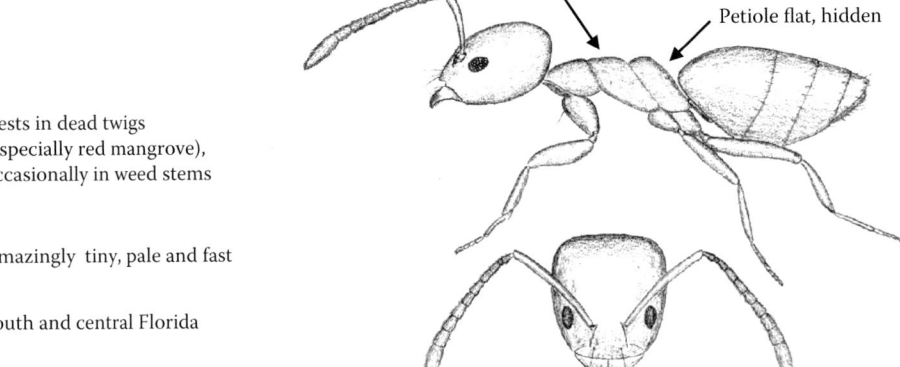

No standing hairs on head or mesosoma

Petiole flat, hidden

- Nests in dead twigs (especially red mangrove), occasionally in weed stems

- Amazingly tiny, pale and fast

- South and central Florida

***Tapinoma litorale* (p. 175)**

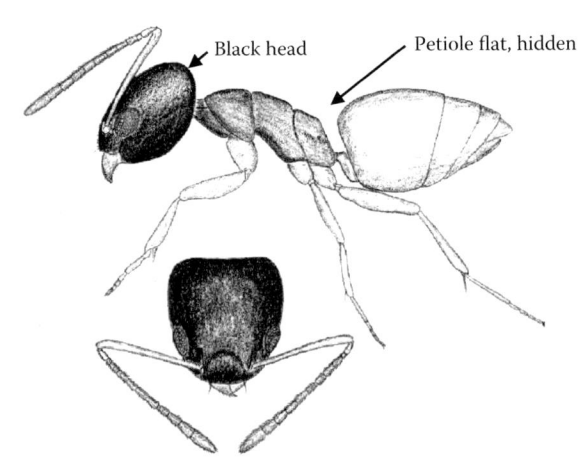

Black head

Petiole flat, hidden

- Nests in dry, dead twigs or branches or in dry piles of debris

- Very tiny and fast, but black head distinctive, look like running black dots

- Introduced, probably from India or Pacific Islands

***Tapinoma melanocephalum* (p. 176)**

No erect hairs on mesosoma

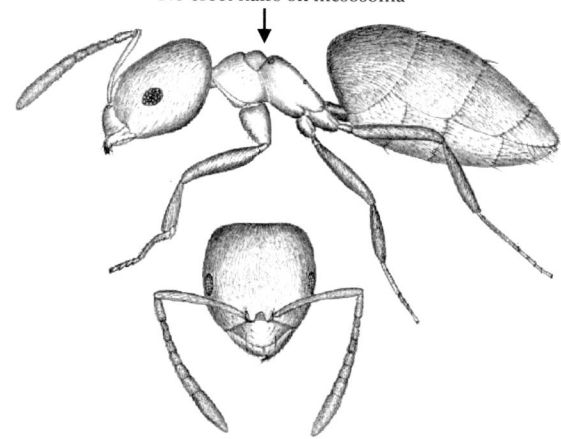

• Common throughout
  Florida

• Color yellow or
  brownish yellow

• Nests in ground or in
  rotten wood

depilis (p. 180)

A few erect hairs on mesosoma

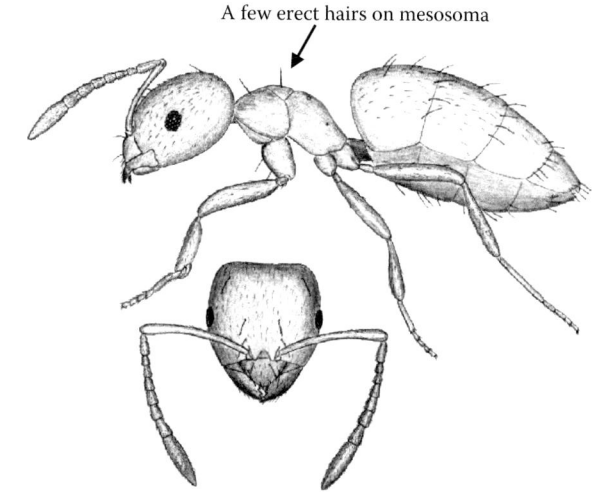

• Uncommon species, found in
  tropical Florida

• Color yellow

• Extracted from leaf litter

minutus (p. 181)

No erect hairs on mesosoma

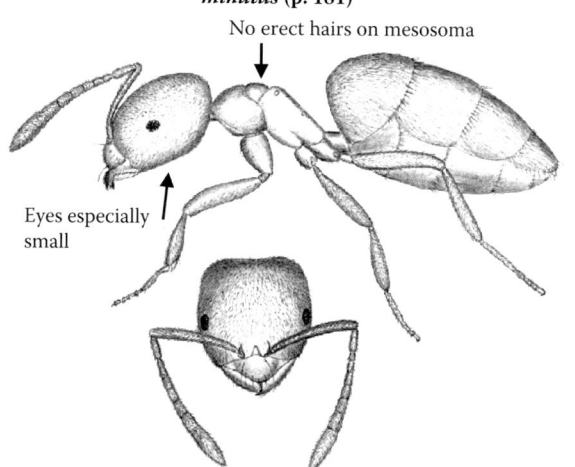

• Rarely collected, known only
  from Peninsular Florida

• Color yellow

• Subterranean, in dry sandy
  habitats

Eyes especially
small

Brachymyrmex Species A (p. 183)

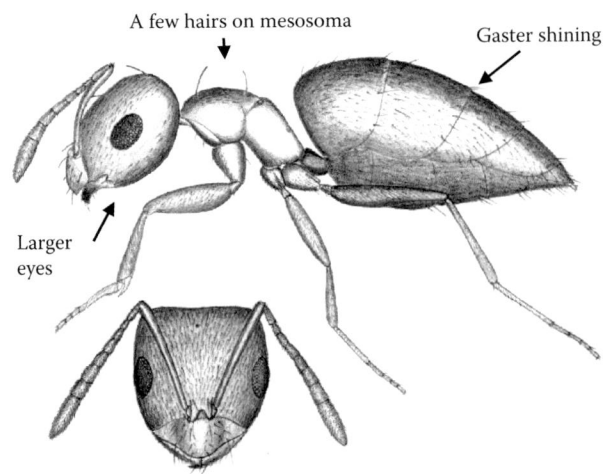

- Nests on ground, common throughout north and central Florida

- Dark brown

- Eyes longer than distance between eye and mandible

- Introduced from South America

A few hairs on mesosoma

Gaster shining

Larger eyes

patagonicus (p. 182)

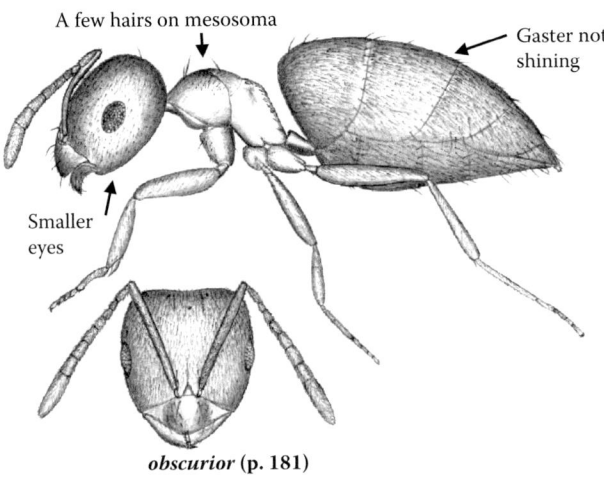

- Nests in ground, south and central Florida

- Medium brown to dark brown

- Eyes shorter than distance between eye and mandible

A few hairs on mesosoma

Gaster not shining

Smaller eyes

obscurior (p. 181)

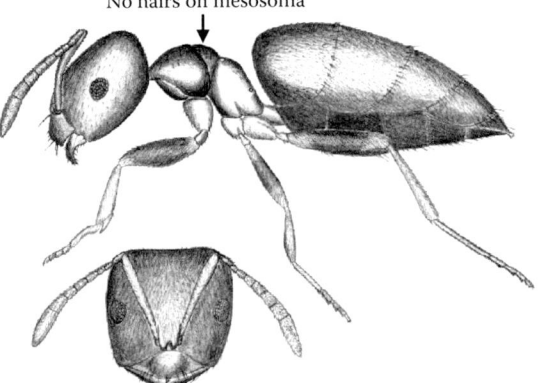

- Collected three times, north Florida

- Brown, legs pale, femora banded

- Nests in ground

No hairs on mesosoma

***Brachymyrmex* Species B (p. 184)**

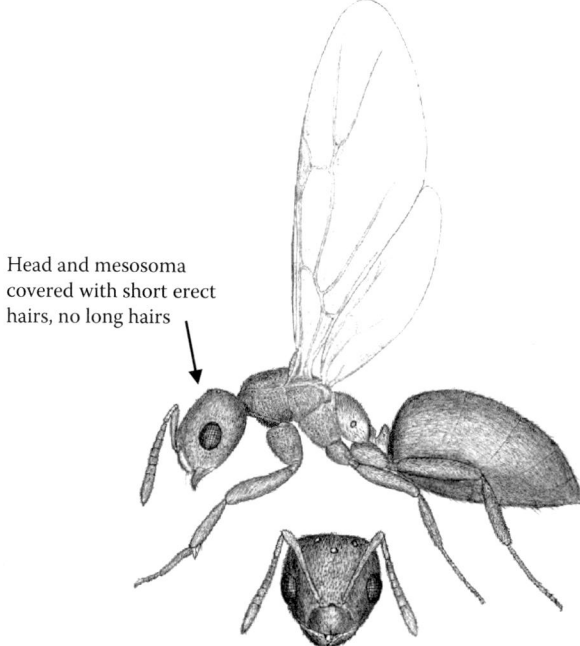

Head and mesosoma
covered with short erect
hairs, no long hairs

• About 3 mm long

• Uniformly dark brown

• Known from three
  queens, Highlands
  County

• Collected in flight
  traps, workers unknown

***Brachymyrmex* Species C (p. 184)**

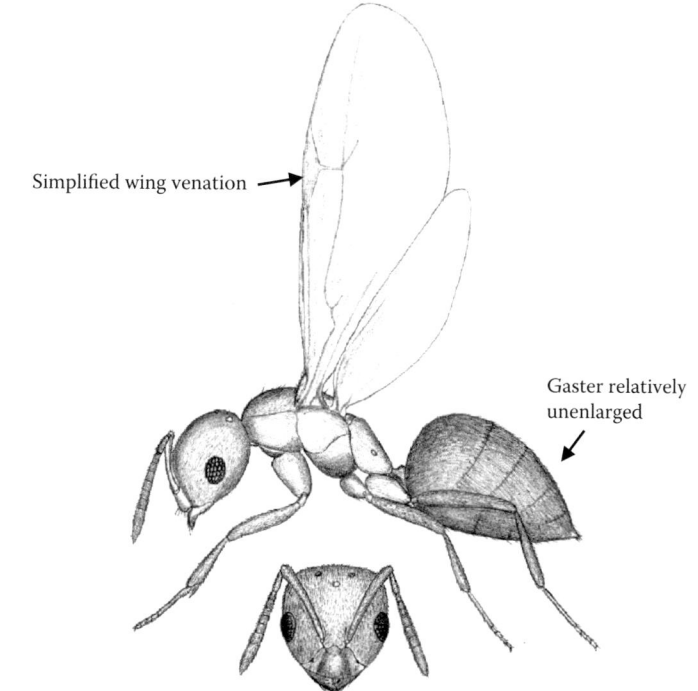

Simplified wing venation

• About 2 mm long

• Pale yellow

• Known from
  Highlands and Polk
  counties

• Collected in flight
  traps, probably
  parasitic

Gaster relatively
unenlarged

***Brachymyrmex* Species D (p. 184)**

- Large species, central Florida south through keys

- Red brown with black gaster

- Nests in natural and artificial cavities

- Compare with *floridanus*

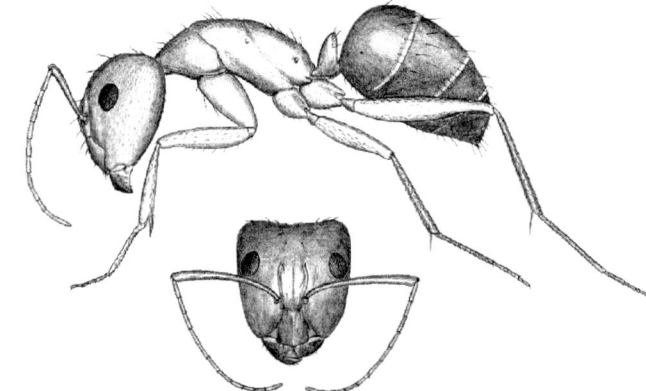

inaequalis (p. 193)

- Large species, widespread in Florida

- A bright chestnut brown species, head sometimes darker

- Nests in rotten wood or in ground

No big punctures on cheeks above jaws

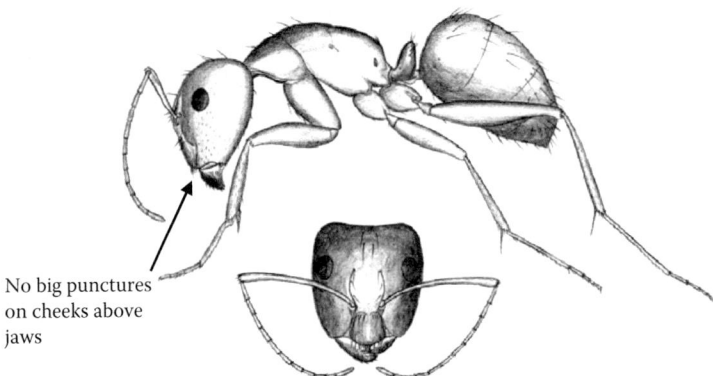

castaneus (p. 187)

- Large species, north Florida

- A chestnut brown species, head dark brown or black

- Nests underground, often below dead wood

Big punctures present on cheeks above jaws

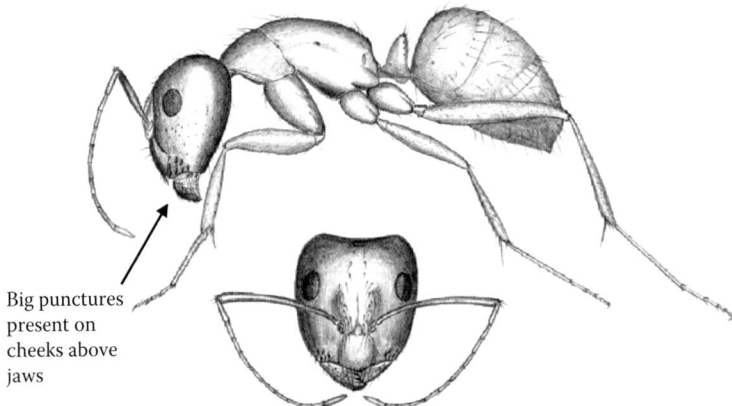

americanus (p. 186)

Plate 75

• *C. nearcticus*: an all-black species, in Florida usually found on large open-grown pines

• *C. decipiens*: a dark red species with black gaster, in dead branches on live trees

• *C. snellingi*: a dark red species, the gaster usually black, red at base, in dead branches on live trees

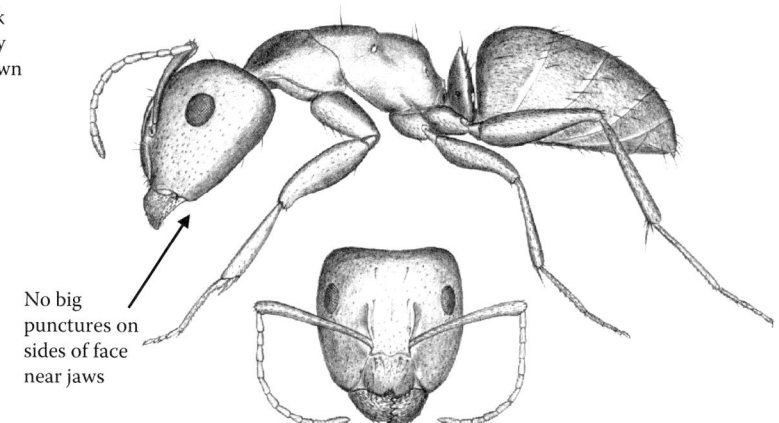

No big punctures on sides of face near jaws

nearcticus (p. 195) and decipiens (p. 188) and snellingi (p. 200) (similar species)

• *C. discolor*: a dark red species with black gaster, usually found on trunks of large oaks, uncommon in Florida, nests in dead branches

• *C. caryae*: an all black species in dead branches on hardwoods, rare in Florida (found twice)

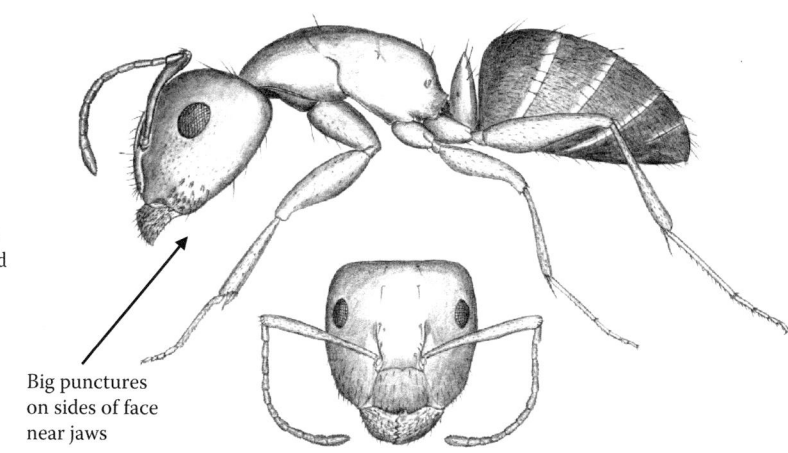

Big punctures on sides of face near jaws

discolor (p. 188) and caryae (p. 186) (similar species)

Plate 76

Dark red with black gaster

Very hairy

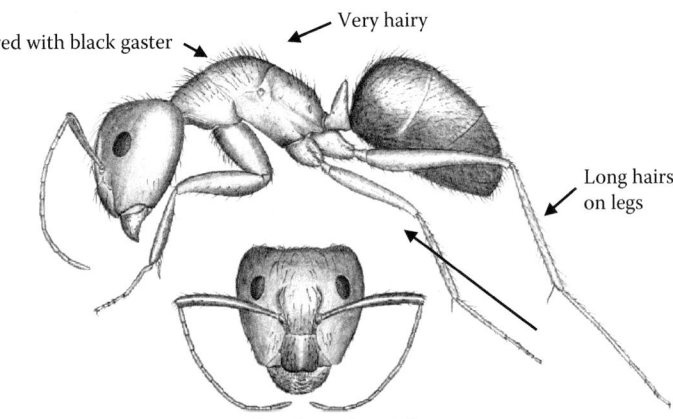

- A large species,
  widespread in Florida,
  commonest in south and
  central Florida

- Nests in rotten wood,
  under debris, in man-made
  structures or containers,
  sometimes in ground

Long hairs
on legs

floridanus (p. 189)

Completely black, very hairy

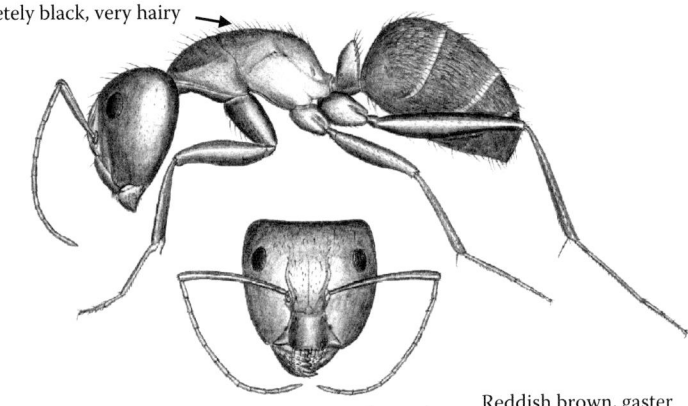

- A large species from north
  Florida

- In Florida usually nests
  in rotten portions of
  large hardwoods

pennsylvanicus (p. 196)

Reddish brown, gaster
black with yellowish
marking

Hairy

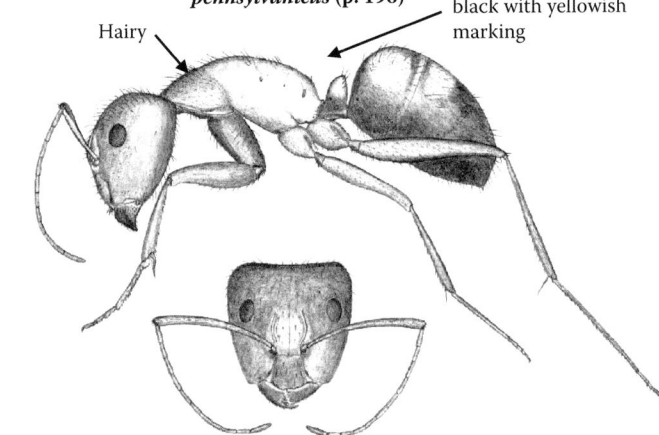

- A large species, more
  common in north Florida

- Deep subterranean nests
  in open sandhill habitat
  sometimes in scrub habitat

socius (p. 200)

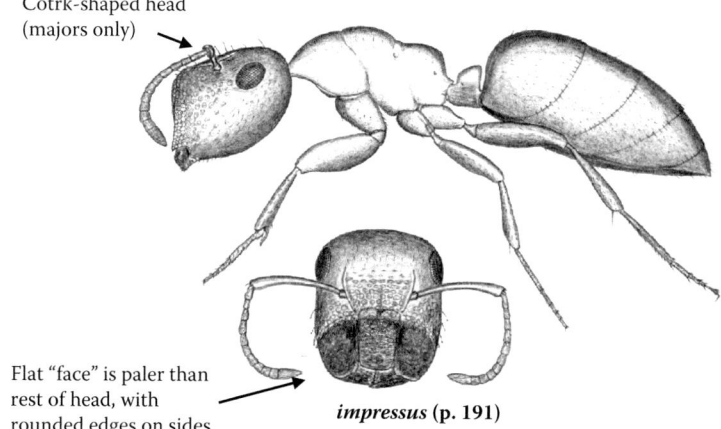

Cotrk-shaped head
(majors only)

• Common and widespread in
Florida

• Variable color, usually dark
reddish brown with pale
"face", sometimes pale spots
on gaster

• Nests in hollow twigs on live
trees and shrubs, often in
dead grape vines

Flat "face" is paler than
rest of head, with
rounded edges on sides

impressus (p. 191)

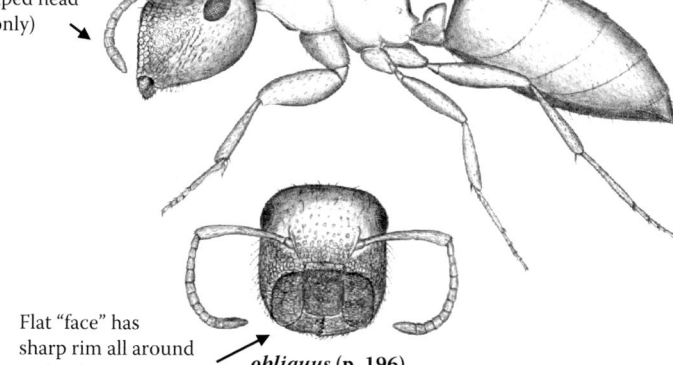

Cork-shaped head
(majors only)

• Apparently uncommon or
difficult to find

• No pale spots, as far as is
known

• Nests in dead twigs of trees,
often in wet areas

Flat "face" has
sharp rim all around
and is dark reddish
brown

obliquus (p. 196)

Strong notch, not found in
similar species

Cork-shaped head
(majors only)

• Tropical coastal areas

• Nests usually in dead twigs
of live red mangrove

• May occur in same areas as
impressus

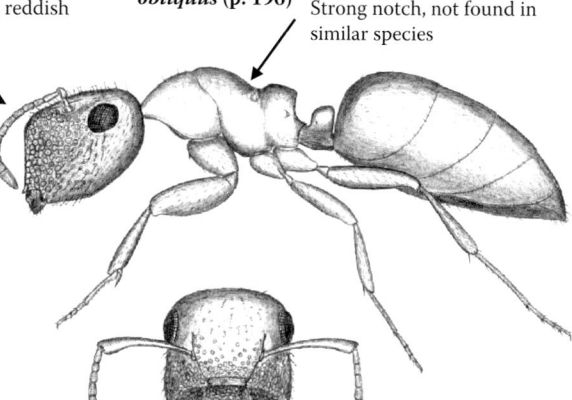

riehlii (p. 199)

Florida *Camponotus* (carpenter ants)

Plate 78

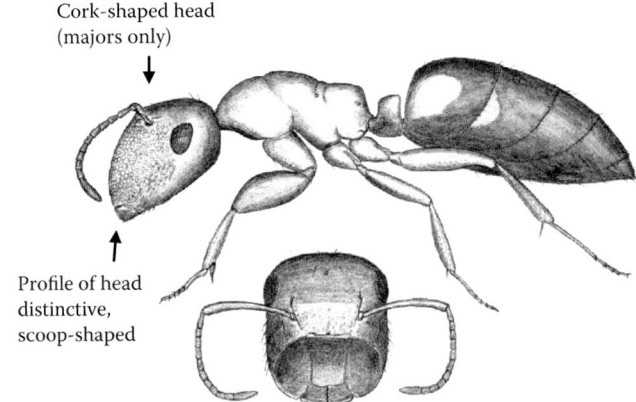

Cork-shaped head
(majors only)

Profile of head
distinctive,
scoop-shaped

- Nests in living twigs of white
  and green ash trees

- Rare in Florida, collected
  only once

mississippiensis (p. 194)

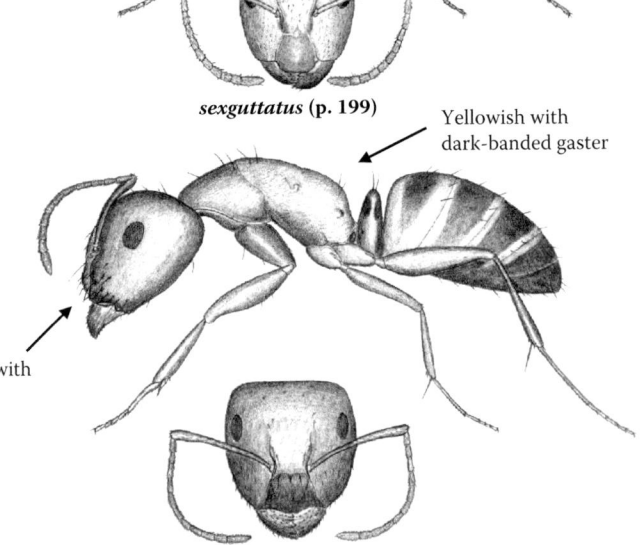

Only Florida *Camponotus* with
this notch (except *riehlii*)

Usually with
yellow spots
on gaster

- Has been found in tropical
  south Florida

- Nests in hollow branches or
  saw-grass stems

- Runs very fast

- Introduced from neotropics

sexguttatus (p. 199)

Yellowish with
dark-banded gaster

- Not reliably reported from
  Florida, but might occur near
  Georgia border

- Nests in hollow dead
  branches, usually on ground

Big punctures with
hairs near jaws

subbarbatus

Florida *Camponotus* (carpenter ants)

Plate 79

Head and body covered with fine erect hairs

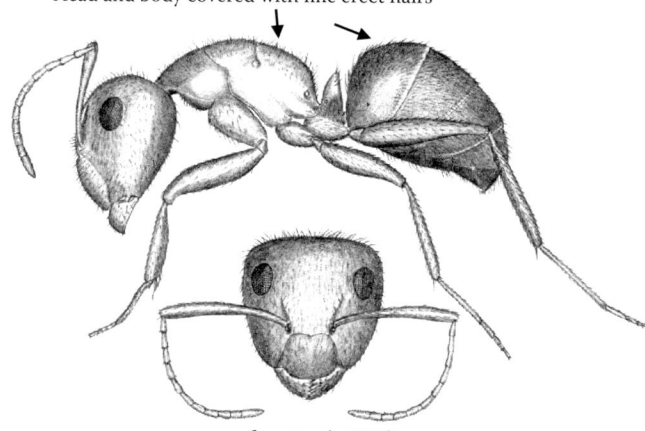

- A small, compact carpenter ant usually reddish with black gaster

- Nests in dead wood, sometimes in sawgrass stems

- Common and widespread in south and central Florida

- Introduced from neotropics

planatus (p. 198)

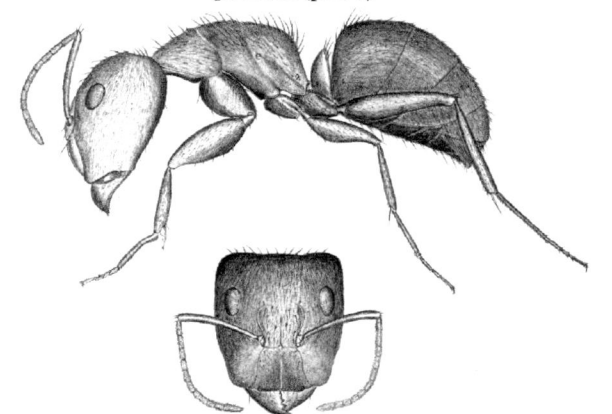

- A small, compact ant, in field resembles *planatus*, but black

- Nests in dead wood and hollow stems

- In Florida, known only from Koreshan State Park and Environs

- Introduced from neotropics

novogranadensis (p. 195)

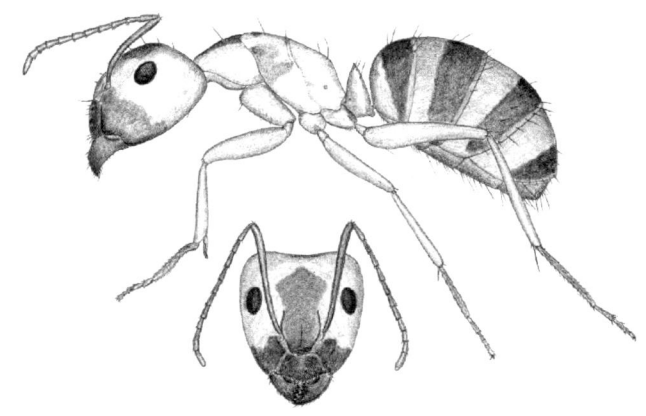

- Not known from Florida, occurs in nearby Bahamas, might appear in tropical Florida

- Yellow, with variable dark brown markings

- Nests in dead branches on live trees

ramulorum

**Florida *Formica* (fleet Formica ants)
and *Prenolepis* (fat-belly ants)**

Plate 81

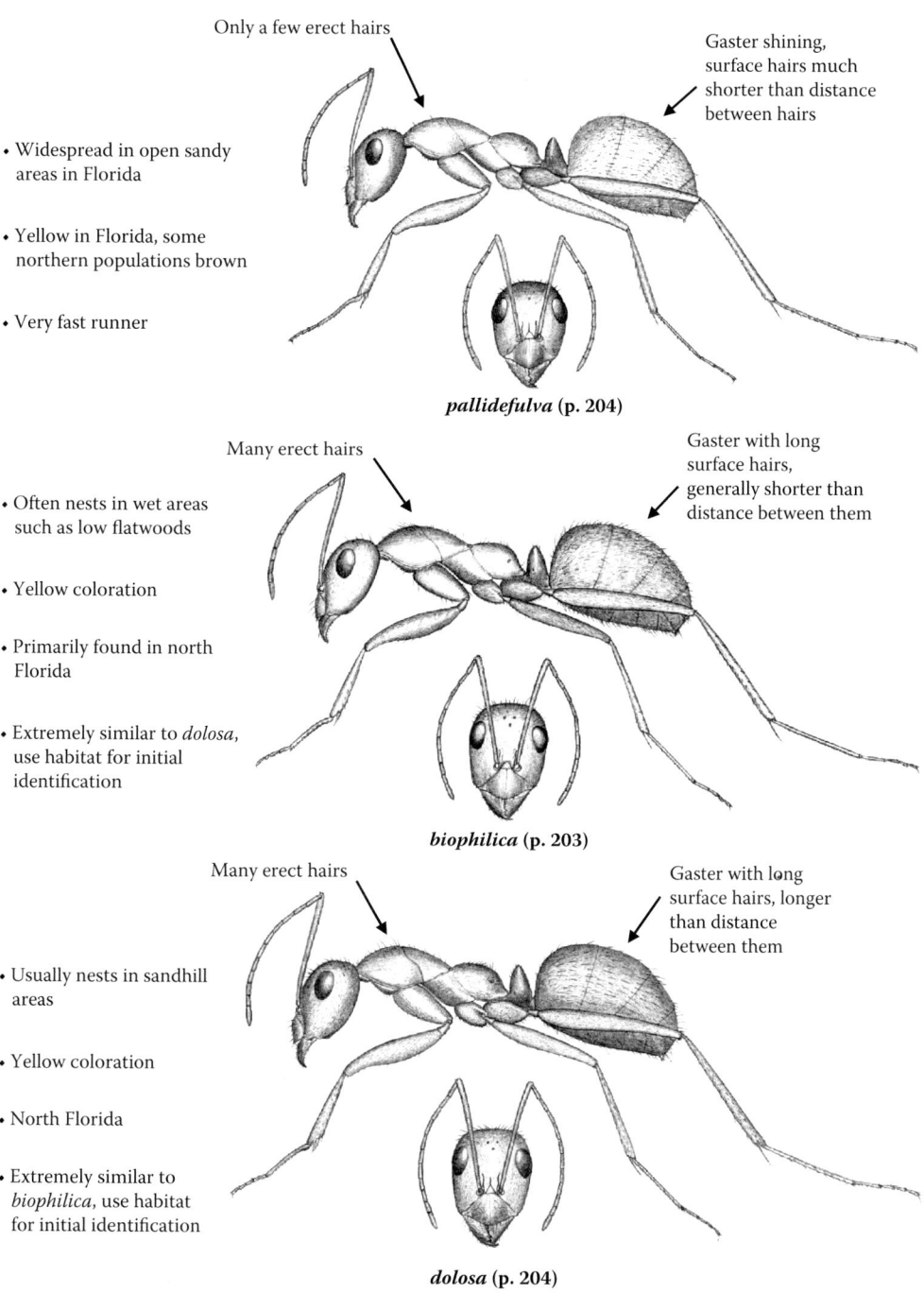

Only a few erect hairs

Gaster shining, surface hairs much shorter than distance between hairs

- Widespread in open sandy areas in Florida

- Yellow in Florida, some northern populations brown

- Very fast runner

pallidefulva (p. 204)

Many erect hairs

Gaster with long surface hairs, generally shorter than distance between them

- Often nests in wet areas such as low flatwoods

- Yellow coloration

- Primarily found in north Florida

- Extremely similar to *dolosa*, use habitat for initial identification

biophilica (p. 203)

Many erect hairs

Gaster with long surface hairs, longer than distance between them

- Usually nests in sandhill areas

- Yellow coloration

- North Florida

- Extremely similar to *biophilica*, use habitat for initial identification

dolosa (p. 204)

No erect hairs on antennal scapes

- Nests usually in rotten wood, often in swamp-forest

- Dark brown

- North and central Florida

No erect hairs on tibiae

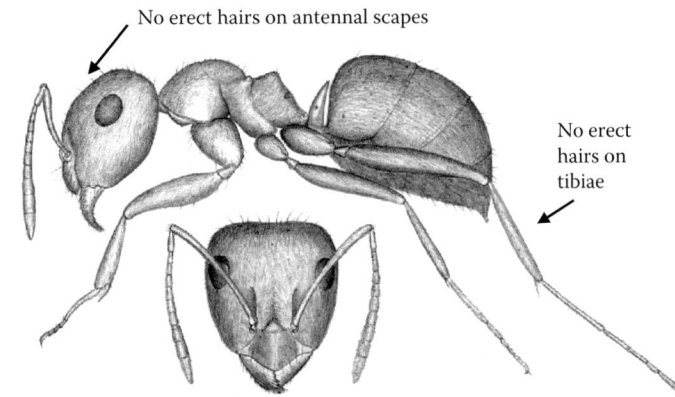

alienus (p. 206)

Erect hairs on antennal scapes

- Ground nester in fields

- Dark brown

- North Florida, uncommon

Erect hairs on tibiae

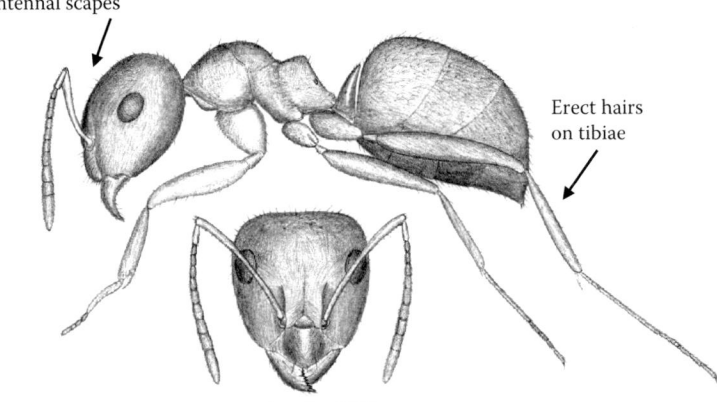

neoniger (p. 209)

Tufts of hair on mesosoma

- Temporary nest parasite of *alienus*

- Yellowish brown

- North Florida, uncommon

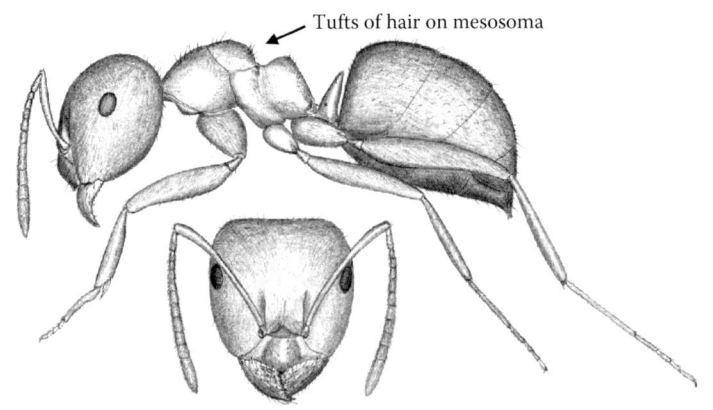

umbratus (p. 210)

Very small eyes

Erect hairs scattered on tergites

- Found once in
  Florida (Walton
  County)

- Yellow, shiny

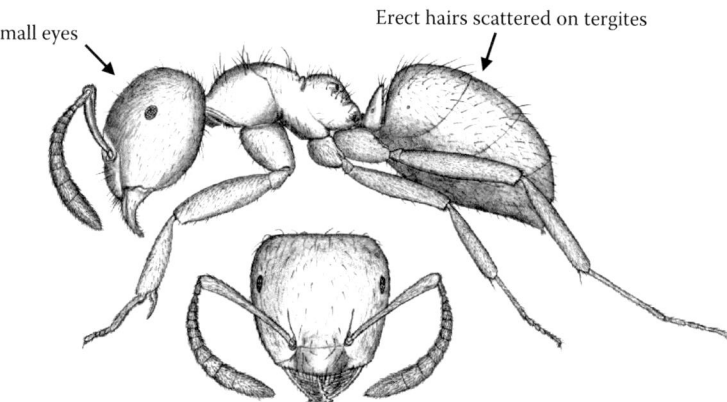

***Lasius claviger* (p. 207)**

Very small eyes

Erect hairs on edges of tergites

- Found once in Florida
  (Liberty County)

- Yellow, shiny

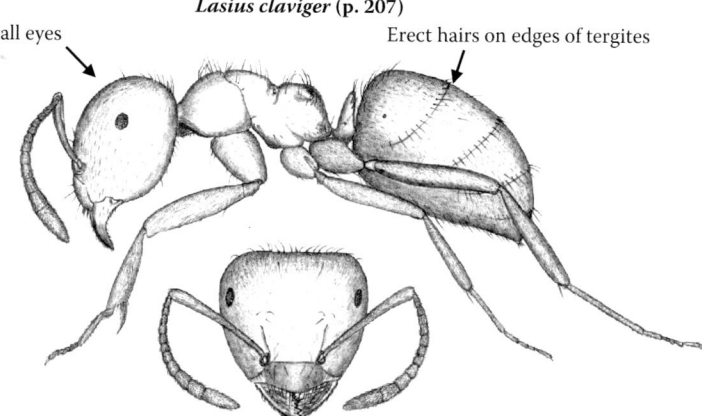

***Lasius interjectus* (p. 208)**

Very fine erect hairs

- Found once in Florida (Polk
  County), may have died out

- Head and gaster blackish,
  mesosoma orange

- Compare with *Ochetellus
  glaber*

- Introduced from neotropics

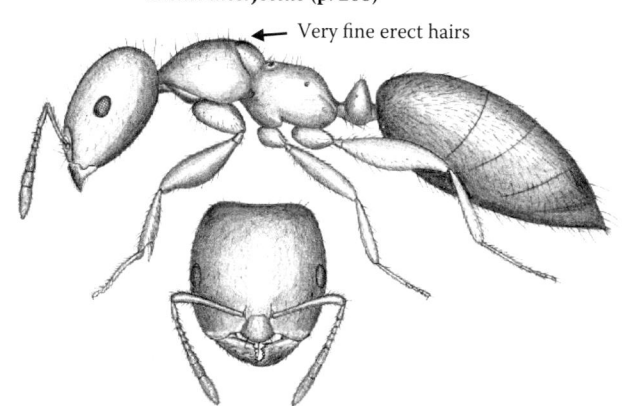

***Myrmelachista ramulorum* (p. 211)**

- Large hairs may be dark

- Yellowish, gaster usually slightly darker

- Nests in ground in open sand

- Usually in undisturbed scrub or sandhill habitats

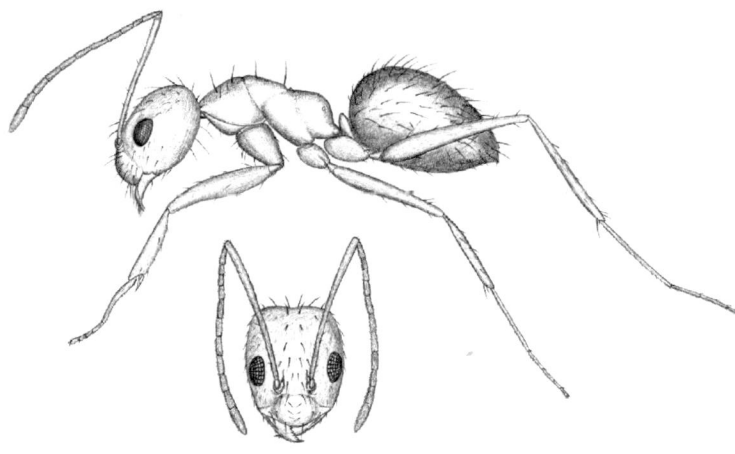

arenivaga (p. 213)

- Large hairs pale

- Yellowish or whitish

- Nests in ground in open sand

- In undisturbed scrub or sandhill habitats

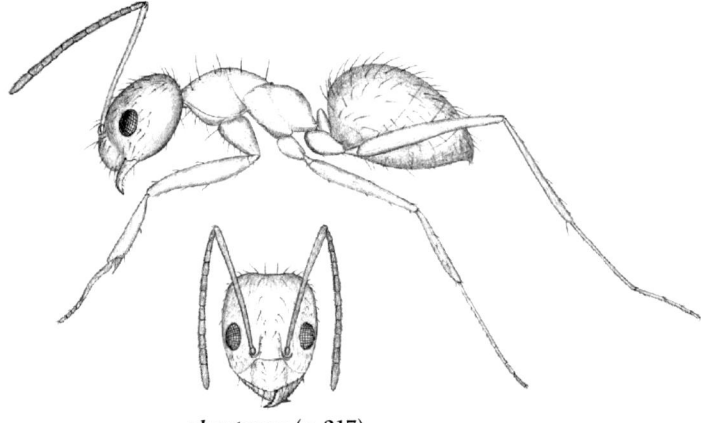

phantasma (p. 217)

Florida *Nylanderia* (crazy ants)

Plate 85

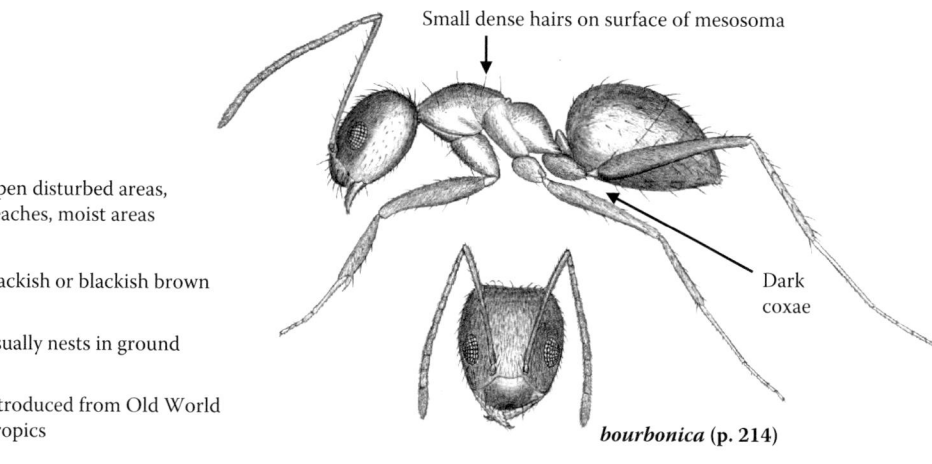

Small dense hairs on surface of mesosoma

- Open disturbed areas, beaches, moist areas

- Blackish or blackish brown

- Usually nests in ground

- Introduced from Old World Tropics

Dark coxae

bourbonica (p. 214)

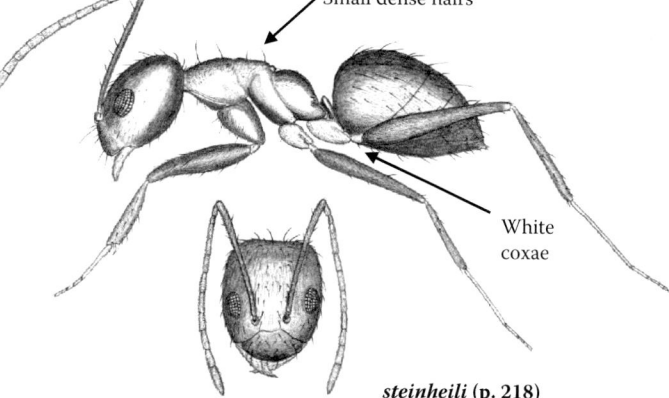

Small dense hairs

- Usually in shaded sites

- Blackish brown

- Southern and central Florida

- Nests usually in leaf litter or fallen hollow branches or twigs

- Introduced from West Indies

White coxae

steinheili (p. 218)

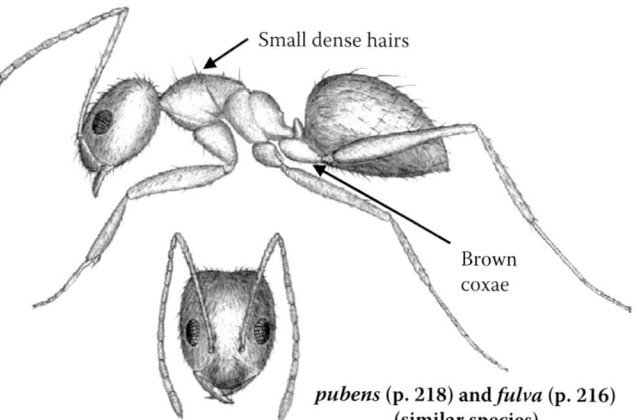

Small dense hairs

- Disturbed sites

- Medium or pale brown

- Nests usually in debris on or near ground

- No hairless area on head or body

- Introduced from neotropics

Brown coxae

pubens (p. 218) and *fulva* (p. 216)
(similar species)

Florida *Nylanderia* (crazy ants) Plate 86

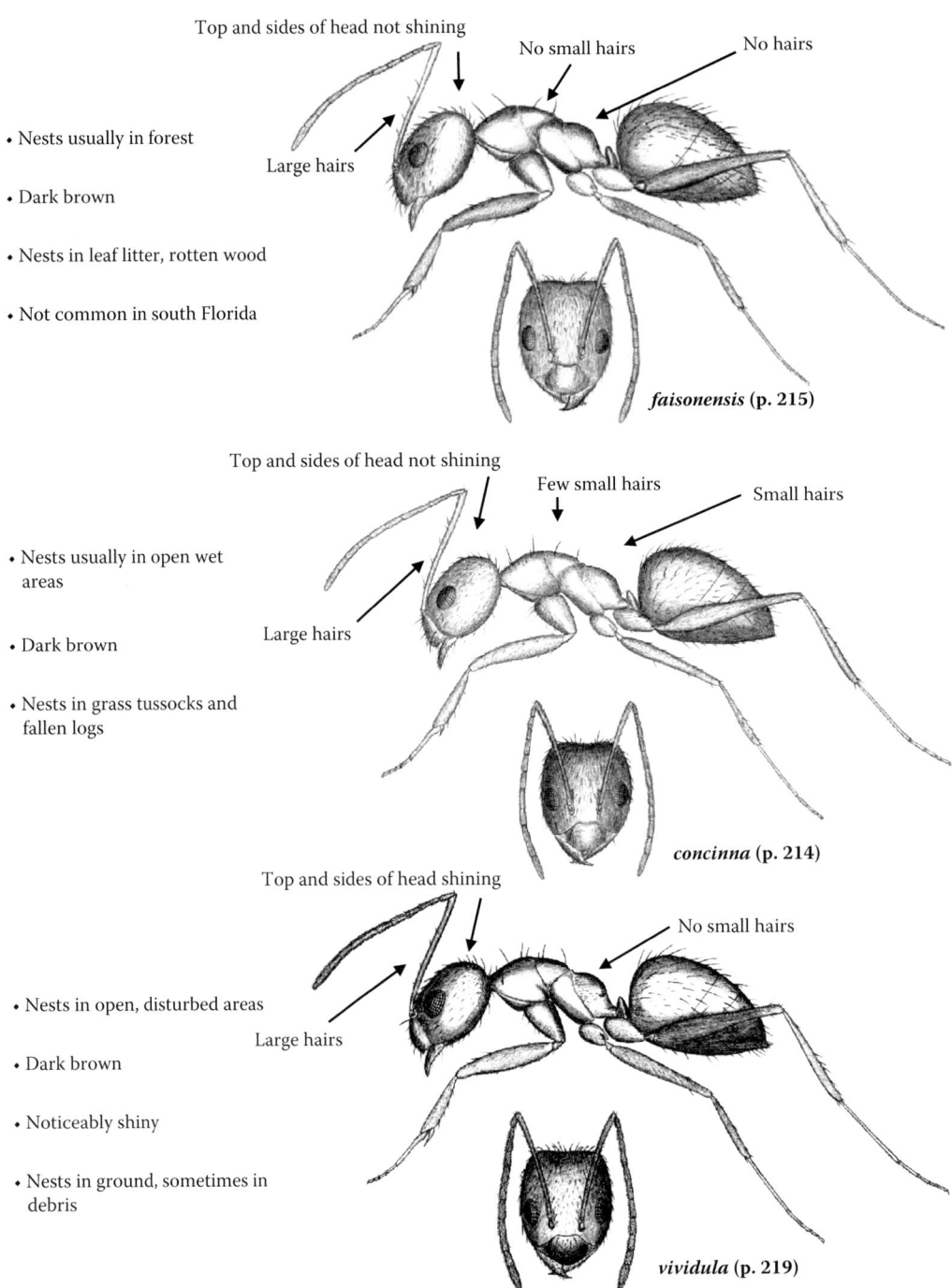

Top and sides of head not shining

No small hairs

No hairs

Large hairs

- Nests usually in forest

- Dark brown

- Nests in leaf litter, rotten wood

- Not common in south Florida

faisonensis (p. 215)

Top and sides of head not shining

Few small hairs

Small hairs

- Nests usually in open wet areas

- Dark brown

- Nests in grass tussocks and fallen logs

Large hairs

concinna (p. 214)

Top and sides of head shining

No small hairs

- Nests in open, disturbed areas

- Dark brown

- Noticeably shiny

- Nests in ground, sometimes in debris

Large hairs

vividula (p. 219)

PLATES

- Open pinelands, wet or dry

- Light brown, mesosoma usually lighter

- Small, less than 2 mm long

- Nests in logs, leaf litter, or bases of pines

No hairs or a couple on antennal scapes

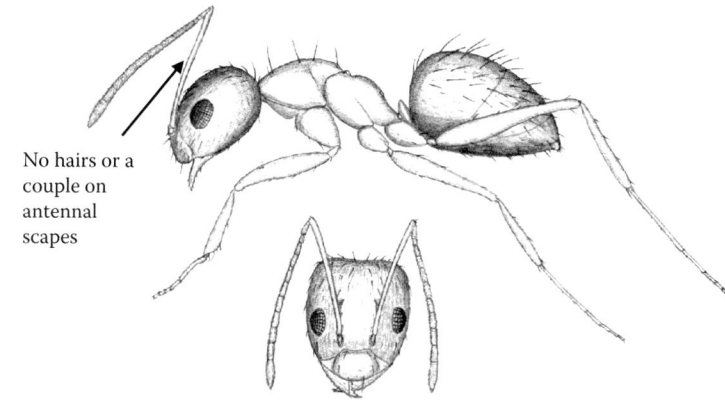

Nylanderia wojciki (p. 219)

- Semiopen areas

- Dark brown

- Nests may be in soil

- North and central Florida

No hairs on antennal scapes

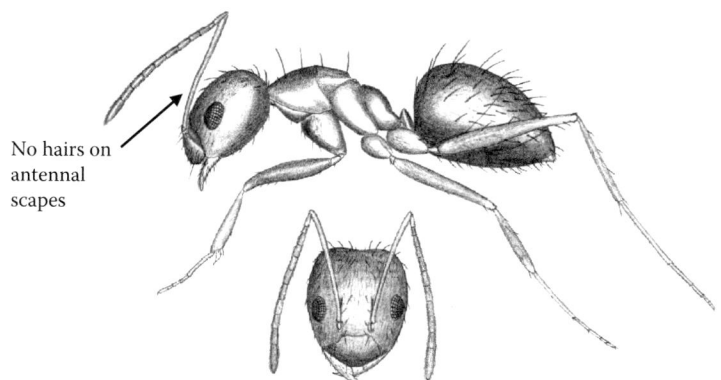

Nylanderia parvula (p. 217)

Very long antennae

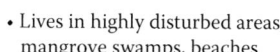

- Lives in highly disturbed areas, mangrove swamps, beaches

- Dark brown with slight bluish sheen in some lights

- Nests usually less than objects or debris on ground

- Introduced from Asia

Very long legs

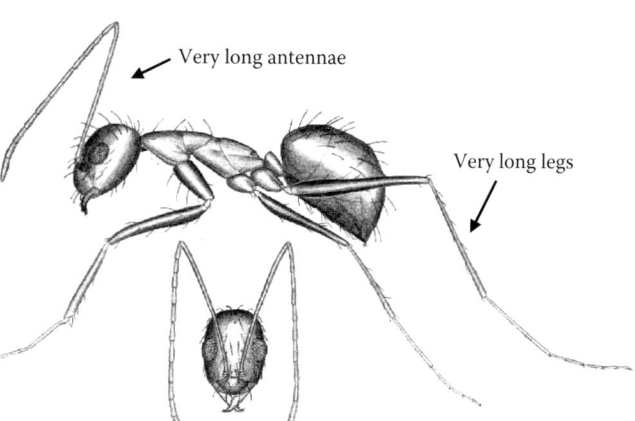

Paratrechina longicornis (p. 221)

- Found in fall in flight traps

- Somewhat like miniature queen of *wojciki*

- Known from Highlands County

- Parasite of *wojciki*

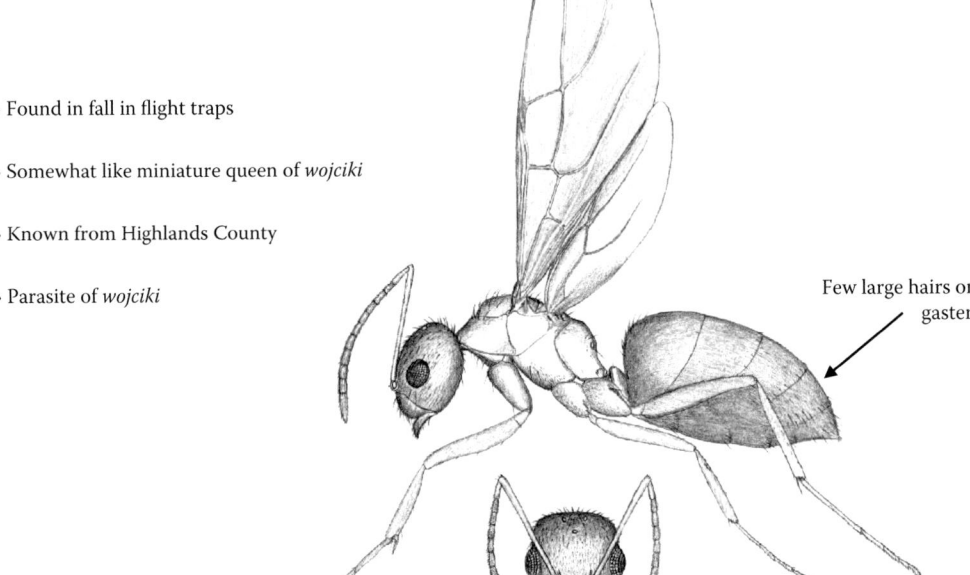

Few large hairs on gaster

Jaws with teeth

Nylanderia **Species A queen (p. 220)**

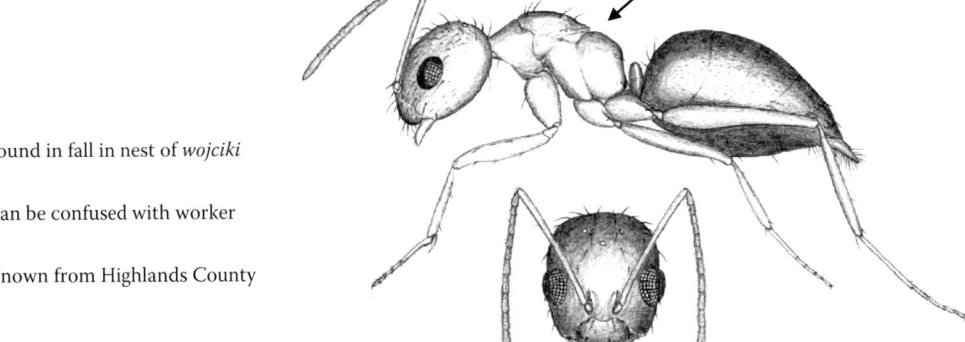

Flightless

- Found in fall in nest of *wojciki*

- Can be confused with worker

- Known from Highlands County

Nylanderia **Species A male**

PLATES

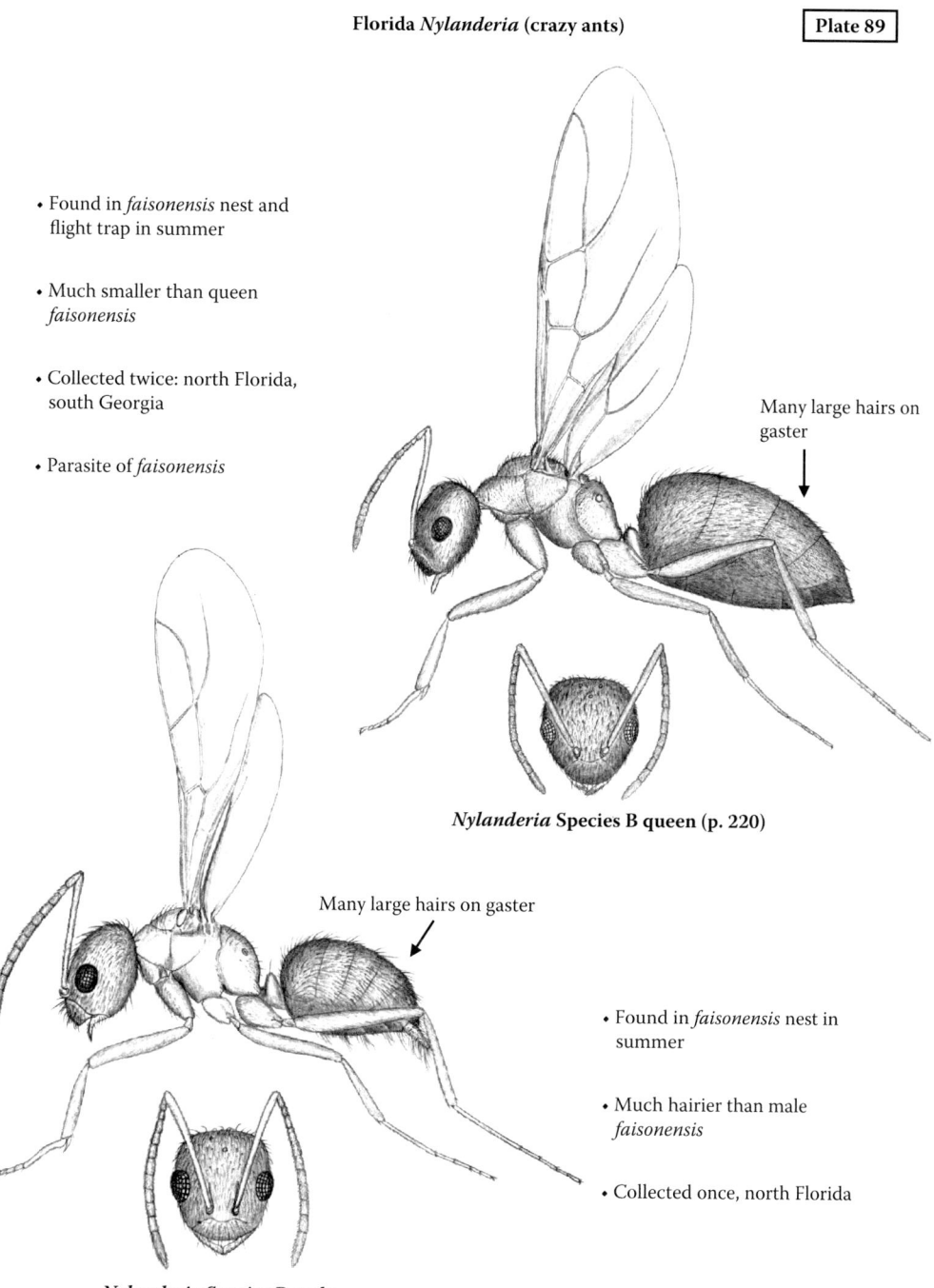

- Found in *faisonensis* nest and flight trap in summer

- Much smaller than queen *faisonensis*

- Collected twice: north Florida, south Georgia

- Parasite of *faisonensis*

Many large hairs on gaster

Nylanderia **Species B queen (p. 220)**

Many large hairs on gaster

- Found in *faisonensis* nest in summer

- Much hairier than male *faisonensis*

- Collected once, north Florida

Nylanderia **Species B male**

Plate 90

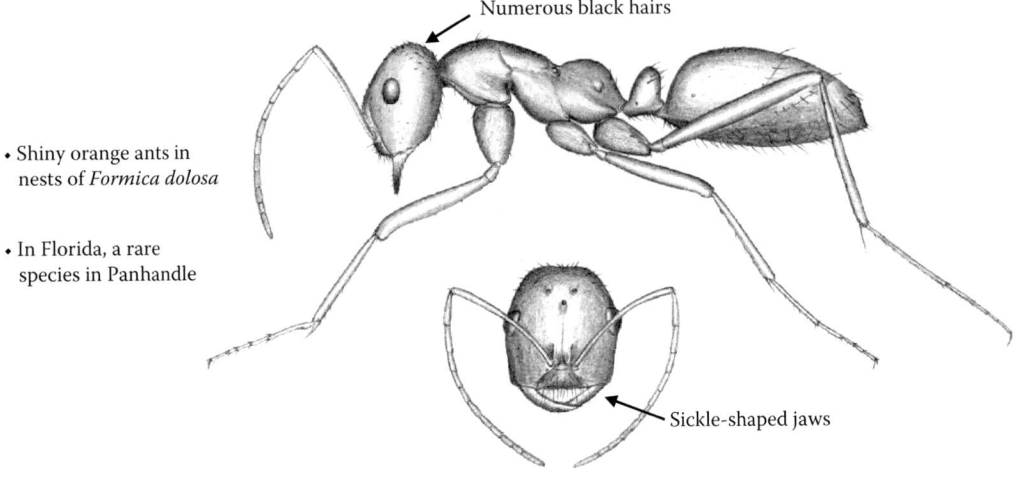

Numerous black hairs

Sickle-shaped jaws

• Shiny orange ants in nests of *Formica dolosa*

• In Florida, a rare species in Panhandle

longicornis (p. 225)

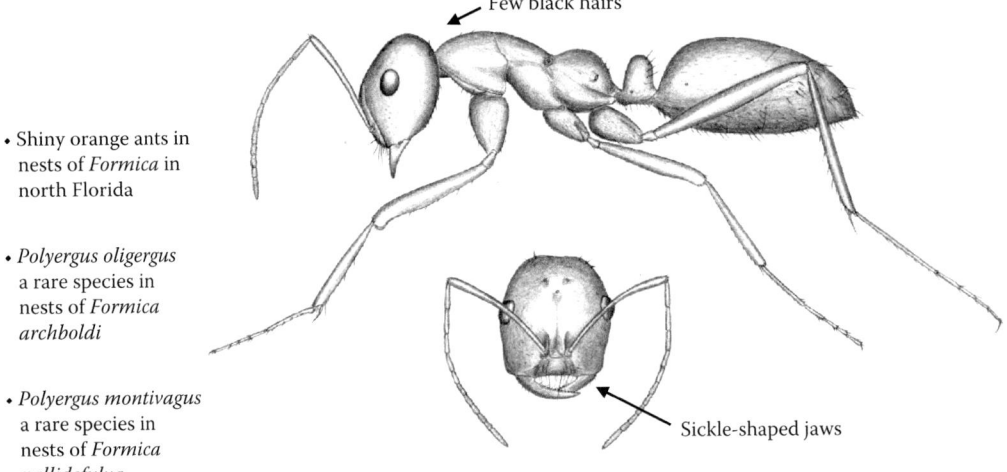

Few black hairs

Sickle-shaped jaws

• Shiny orange ants in nests of *Formica* in north Florida

• *Polyergus oligergus* a rare species in nests of *Formica archboldi*

• *Polyergus montivagus* a rare species in nests of *Formica pallidefulva*

montivagus (p. 225) and *oligergus* (p. 226)
(similar species)

DISTRIBUTION MAPS

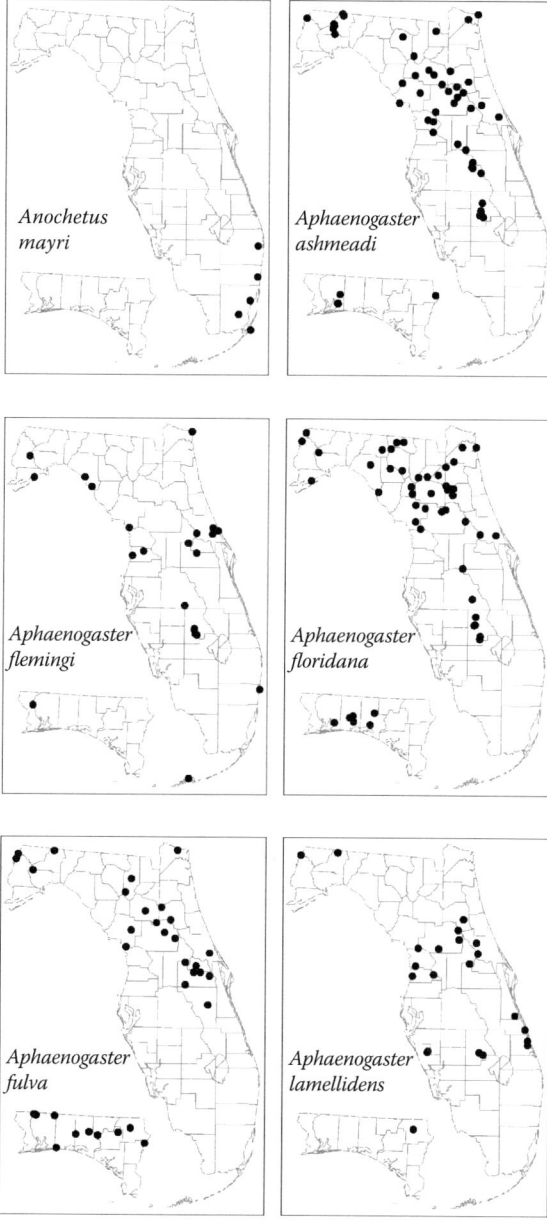

Anochetus
mayri

Aphaenogaster
ashmeadi

Aphaenogaster
flemingi

Aphaenogaster
floridana

Aphaenogaster
fulva

Aphaenogaster
lamellidens

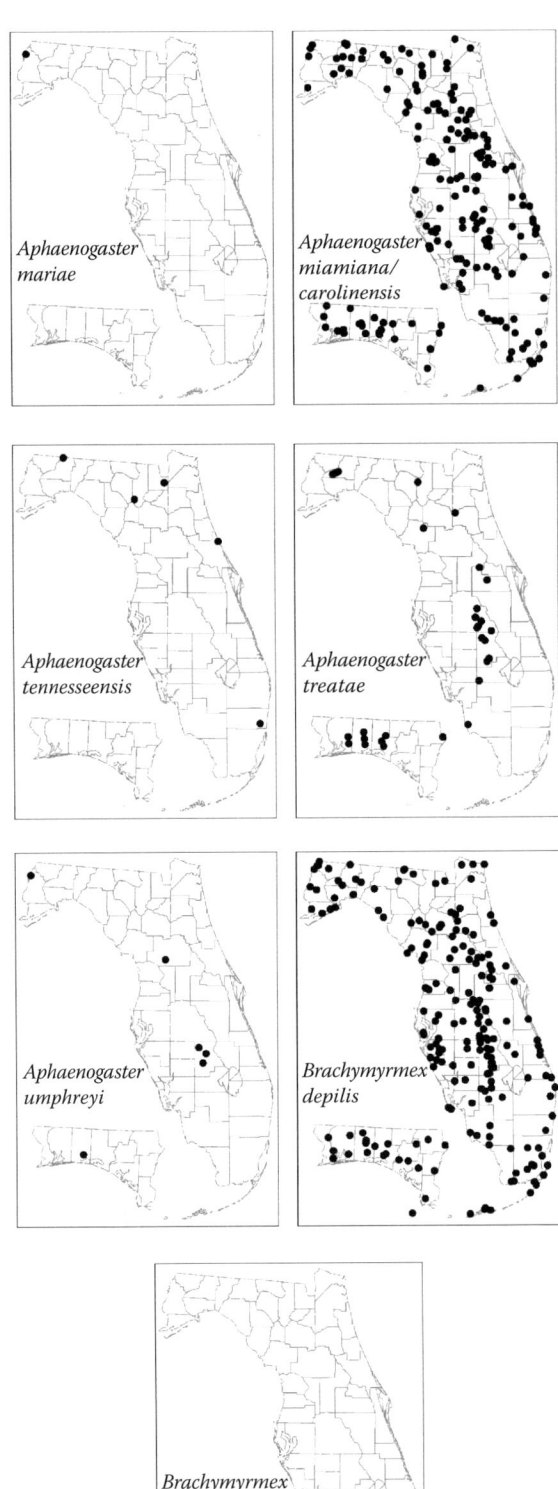

DISTRIBUTION MAPS

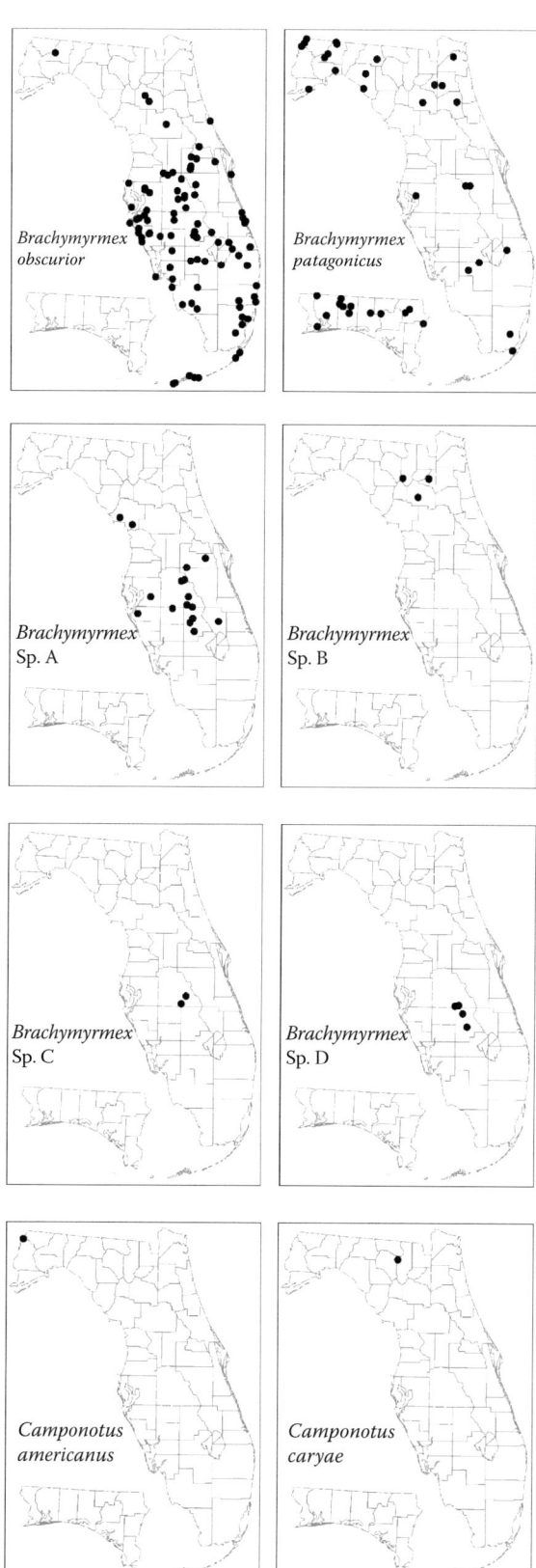

Brachymyrmex obscurior

Brachymyrmex patagonicus

Brachymyrmex Sp. A

Brachymyrmex Sp. B

Brachymyrmex Sp. C

Brachymyrmex Sp. D

Camponotus americanus

Camponotus caryae

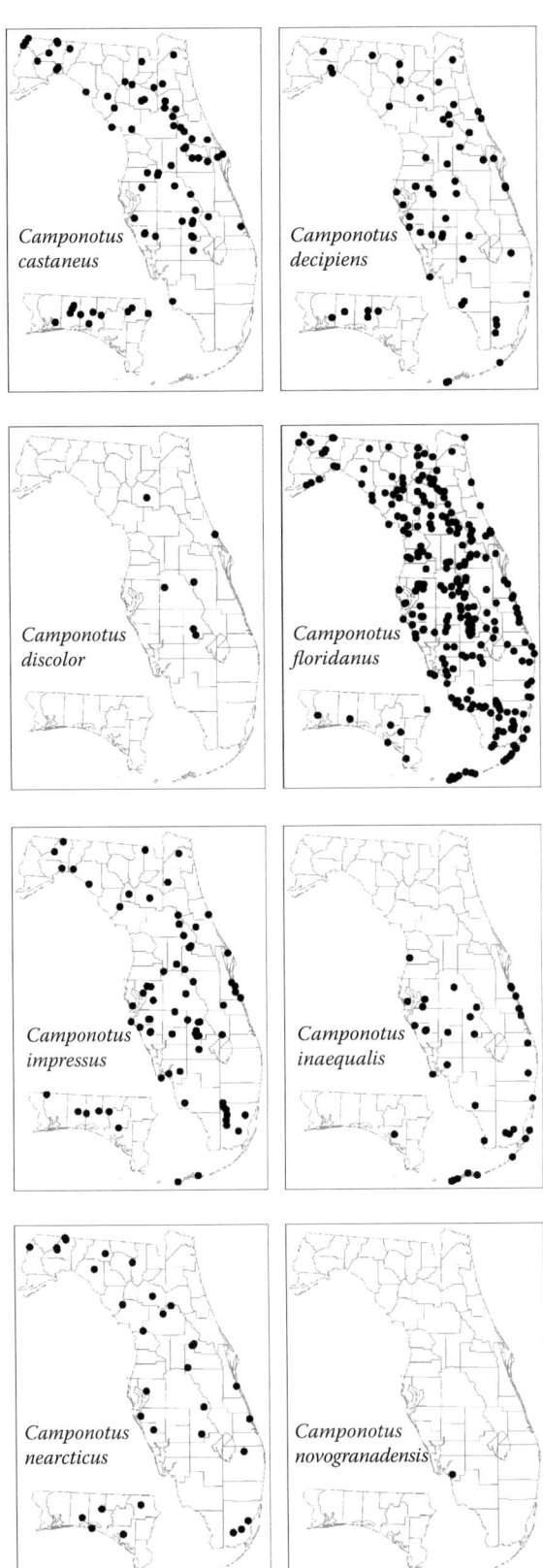

DISTRIBUTION MAPS

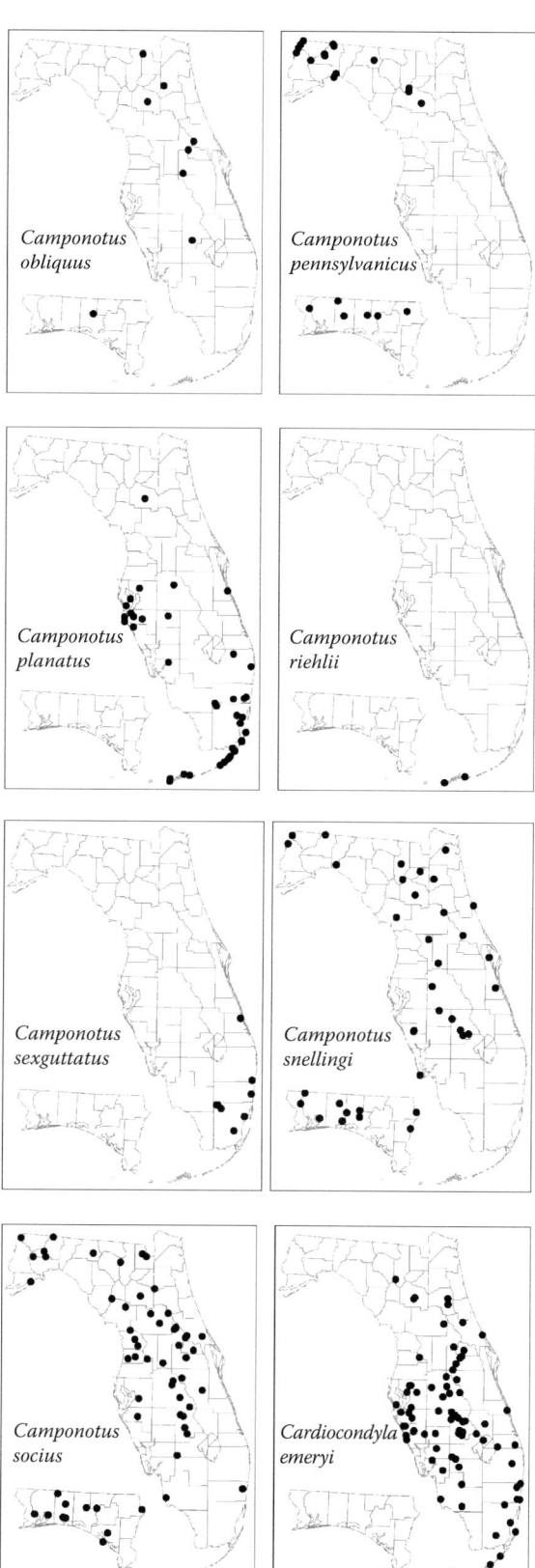

Camponotus
obliquus

Camponotus
pennsylvanicus

Camponotus
planatus

Camponotus
riehlii

Camponotus
sexguttatus

Camponotus
snellingi

Camponotus
socius

Cardiocondyla
emeryi

DISTRIBUTION MAPS

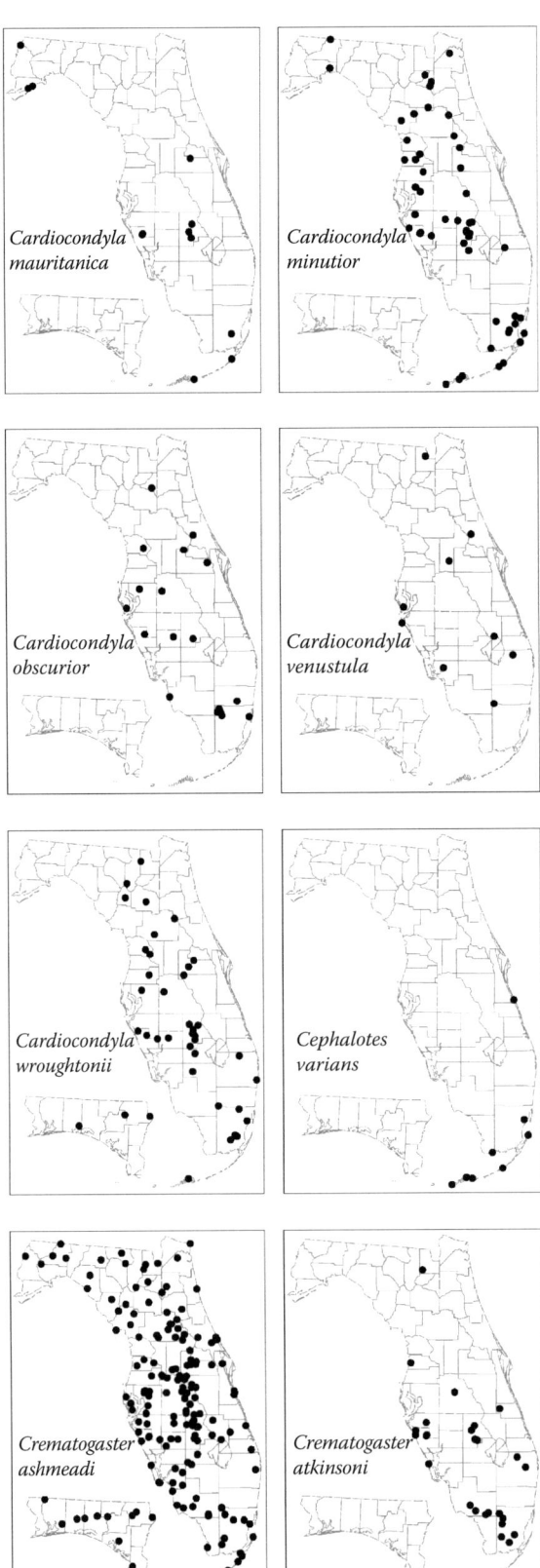

DISTRIBUTION MAPS

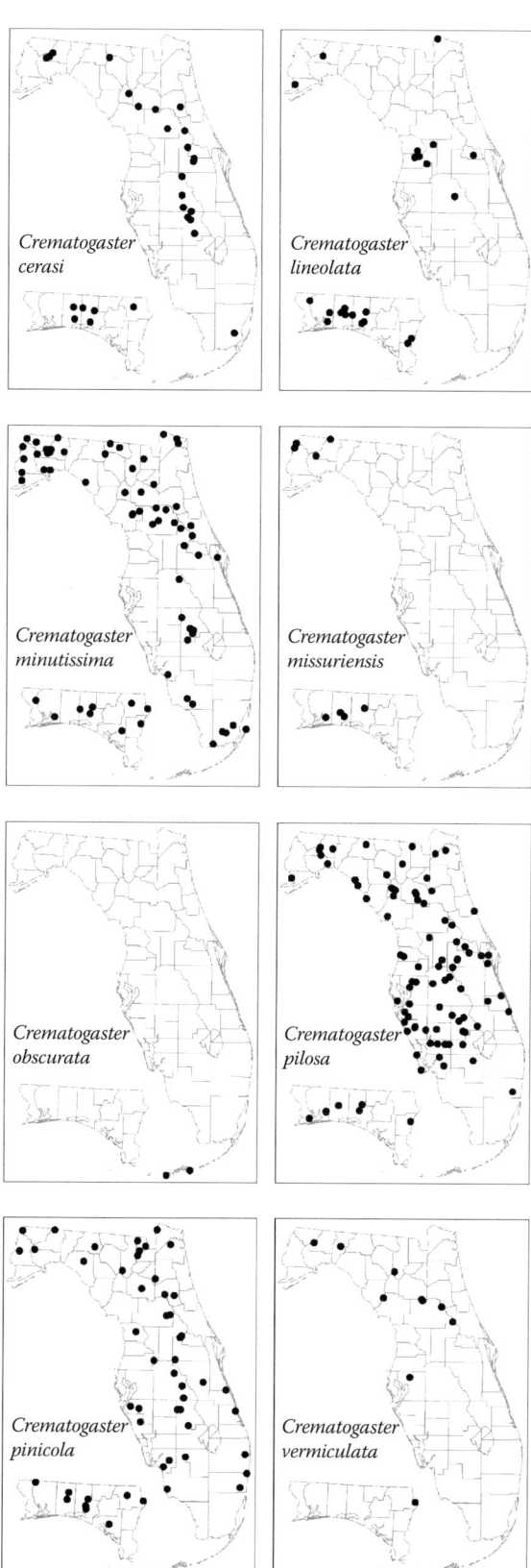

Crematogaster
cerasi

Crematogaster
lineolata

Crematogaster
minutissima

Crematogaster
missuriensis

Crematogaster
obscurata

Crematogaster
pilosa

Crematogaster
pinicola

Crematogaster
vermiculata

DISTRIBUTION MAPS

357

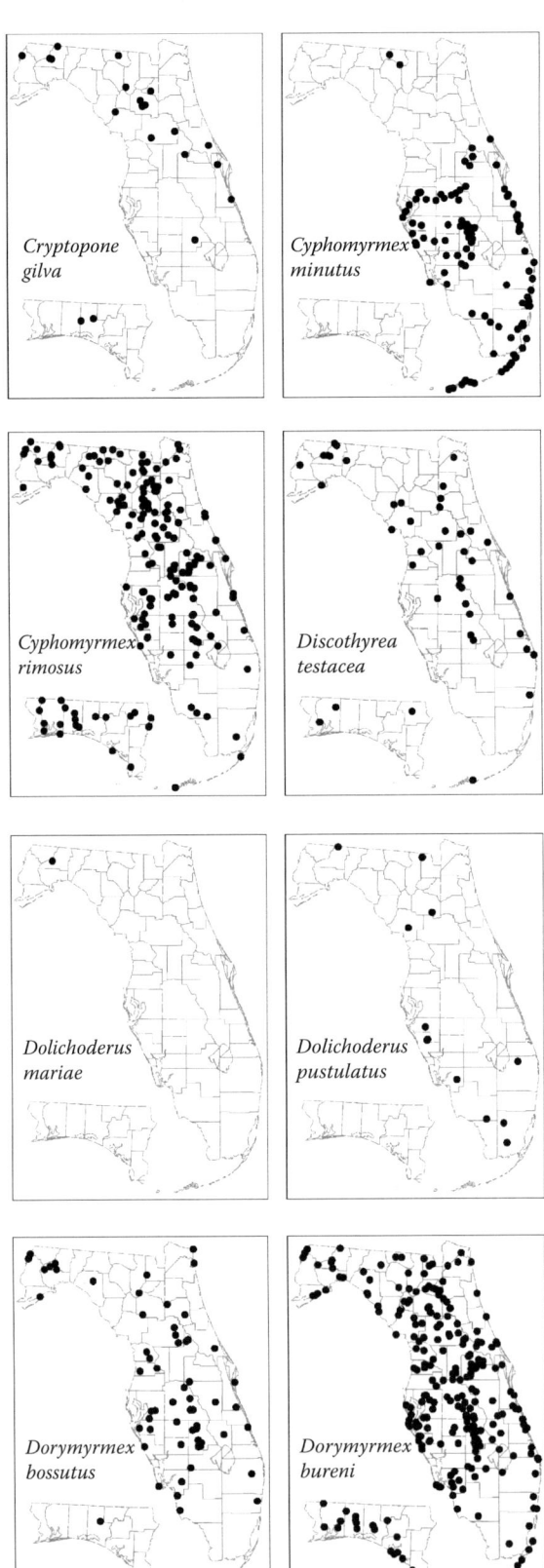

DISTRIBUTION MAPS

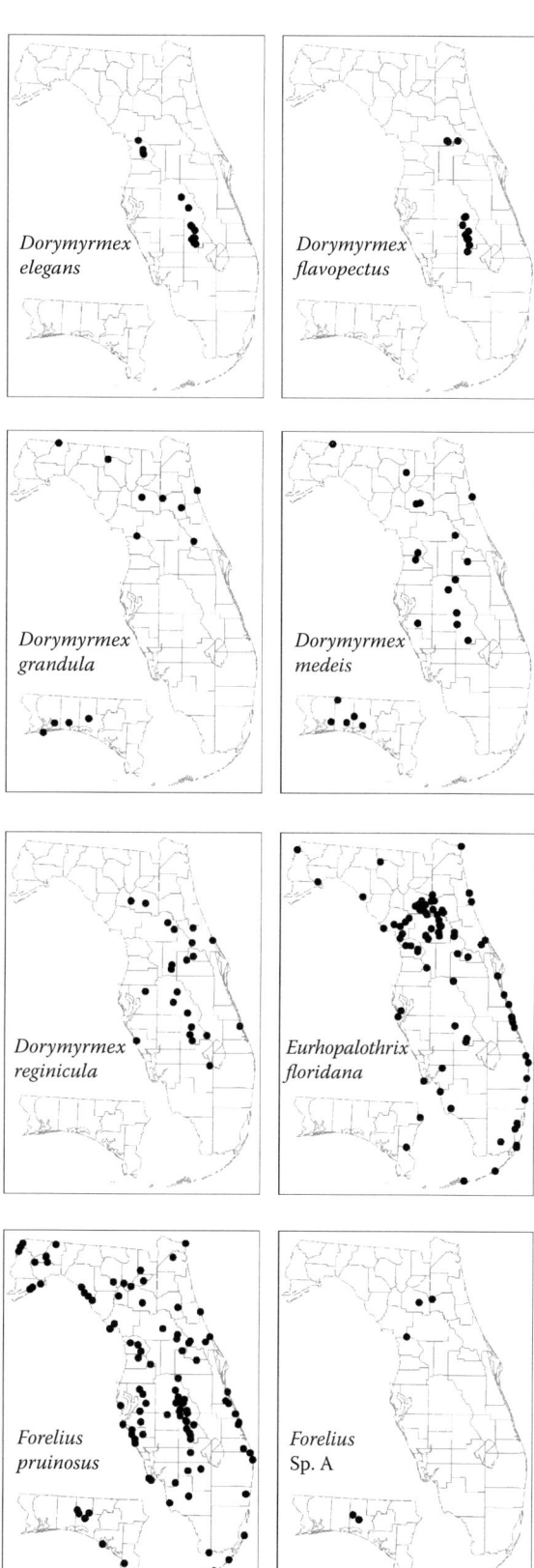

Dorymyrmex elegans

Dorymyrmex flavopectus

Dorymyrmex grandula

Dorymyrmex medeis

Dorymyrmex reginicula

Eurhopalothrix floridana

Forelius pruinosus

Forelius Sp. A

DISTRIBUTION MAPS

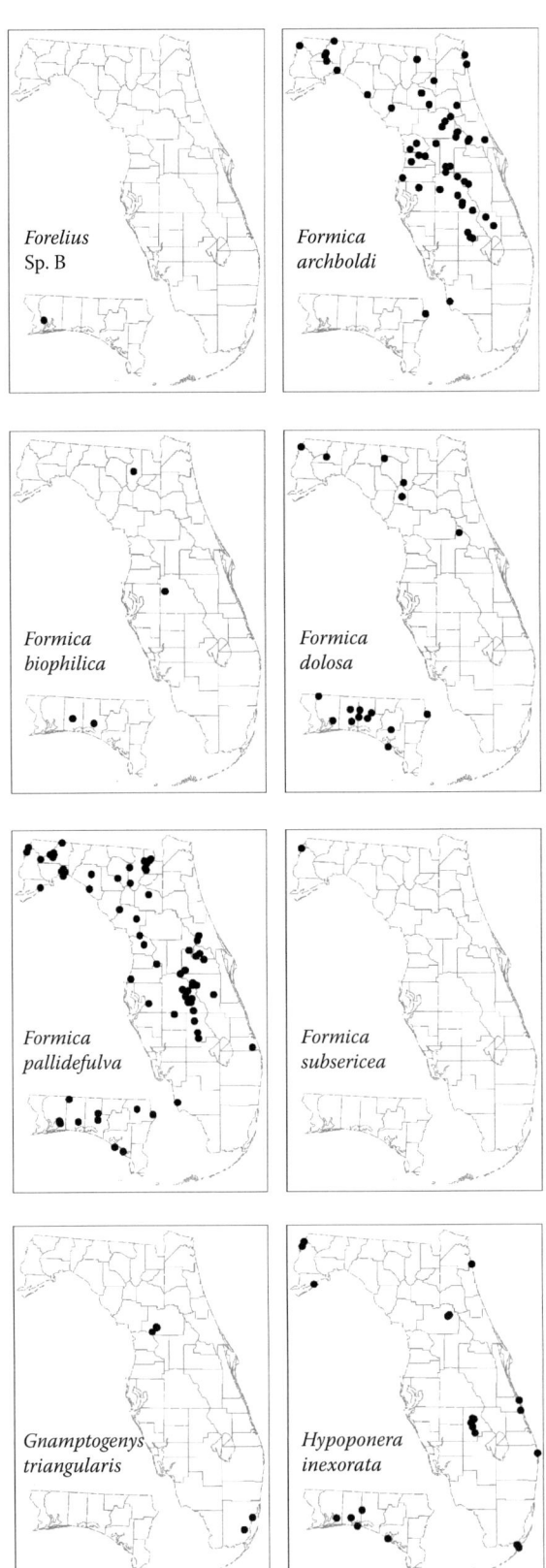

Forelius
Sp. B

Formica
archboldi

Formica
biophilica

Formica
dolosa

Formica
pallidefulva

Formica
subsericea

Gnamptogenys
triangularis

Hypoponera
inexorata

DISTRIBUTION MAPS

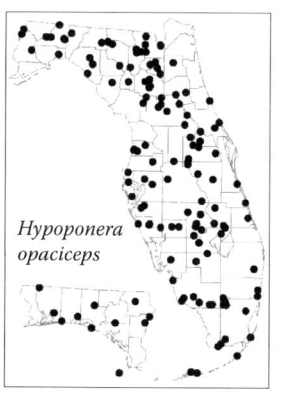

Hypoponera opaciceps

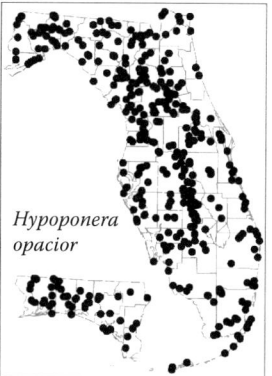

Hypoponera opacior

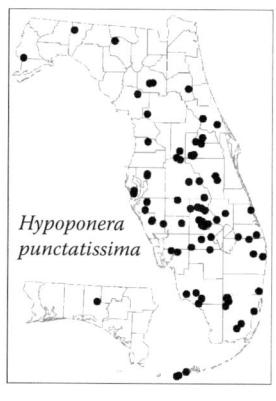

Hypoponera punctatissima

Lasius alienus

Lasius claviger

Lasius interjectus

Lasius neoniger

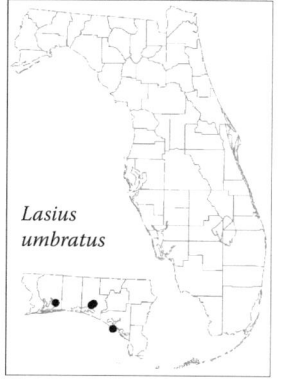

Lasius umbratus

DISTRIBUTION MAPS

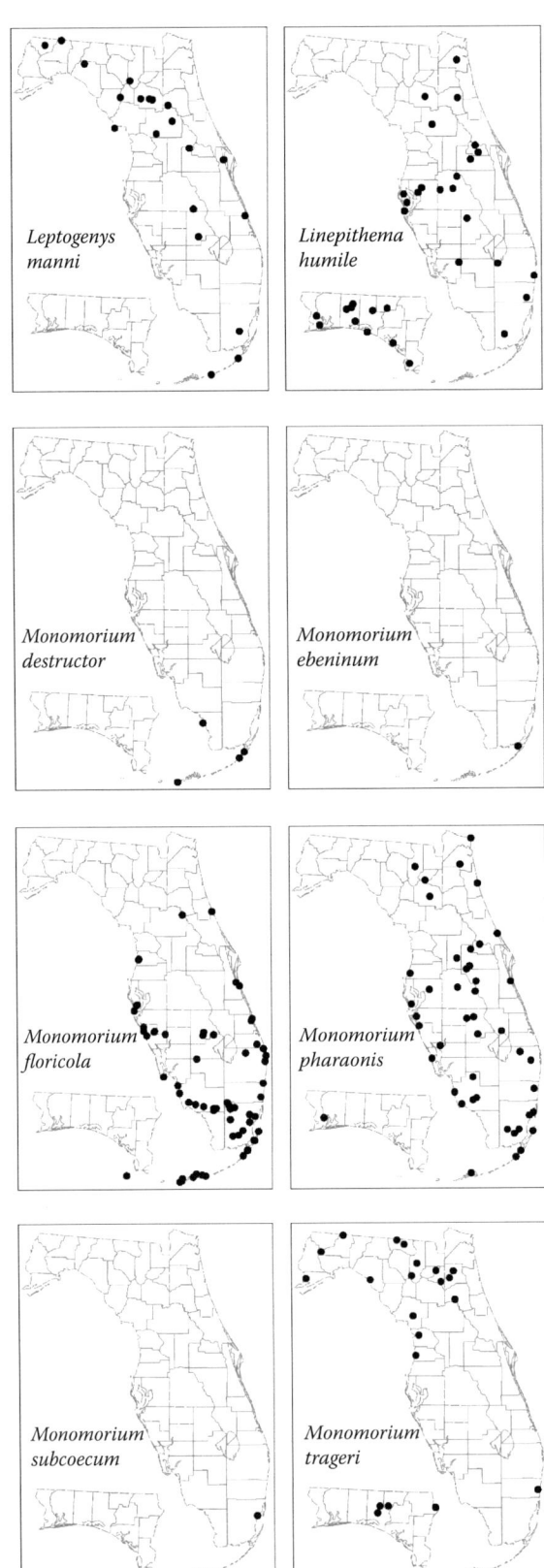

Leptogenys manni

Linepithema humile

Monomorium destructor

Monomorium ebeninum

Monomorium floricola

Monomorium pharaonis

Monomorium subcoecum

Monomorium trageri

DISTRIBUTION MAPS

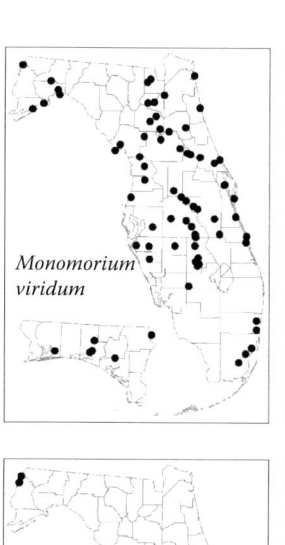

Monomorium viridum

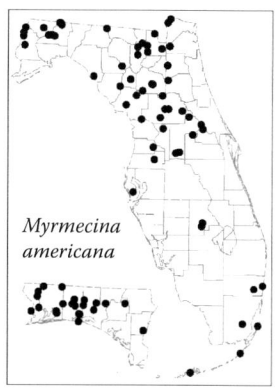

Myrmecina americana

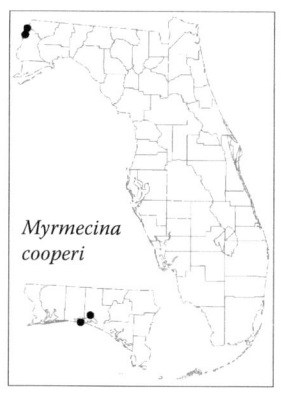

Myrmecina cooperi

Myrmecina Sp. A

Myrmecocystus ramulorum

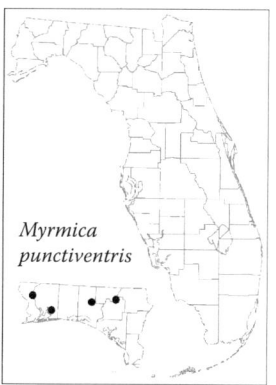

Myrmica punctiventris

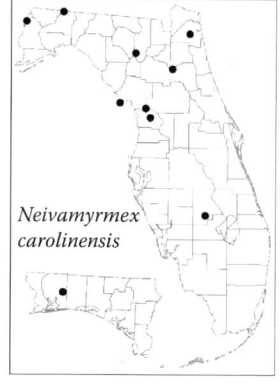

Neivamyrmex carolinensis

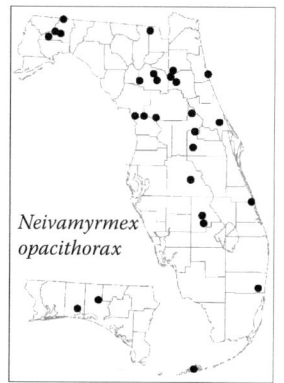

Neivamyrmex opacithorax

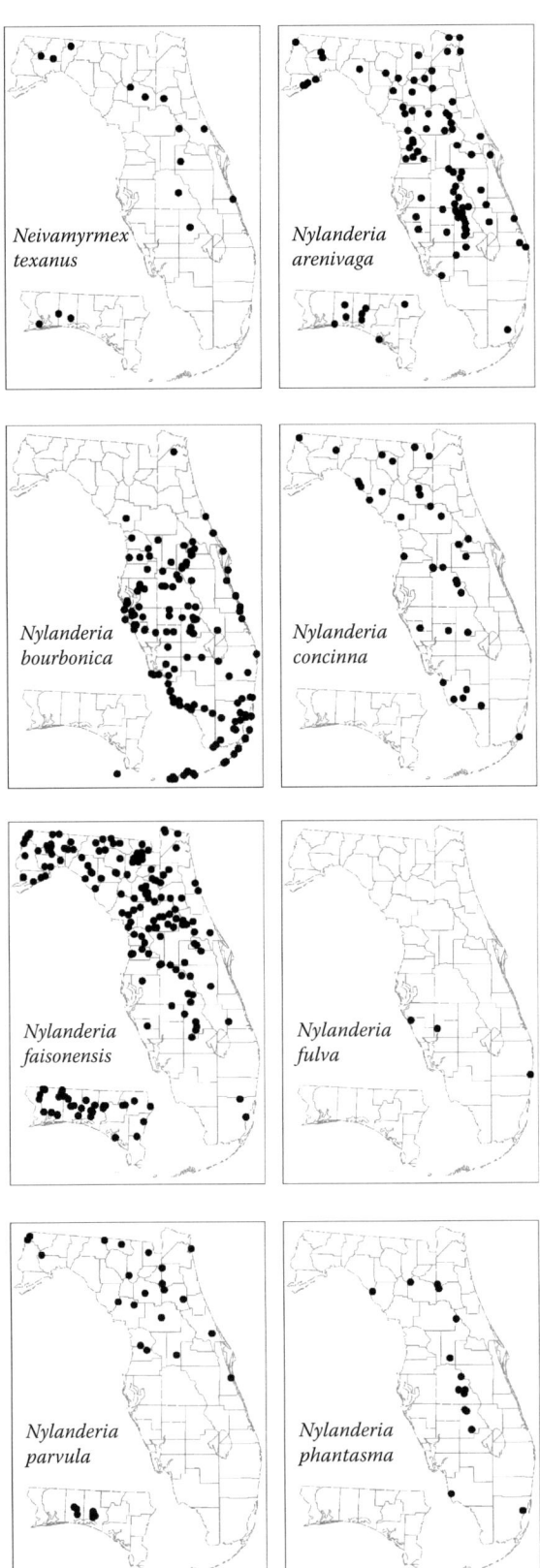

DISTRIBUTION MAPS

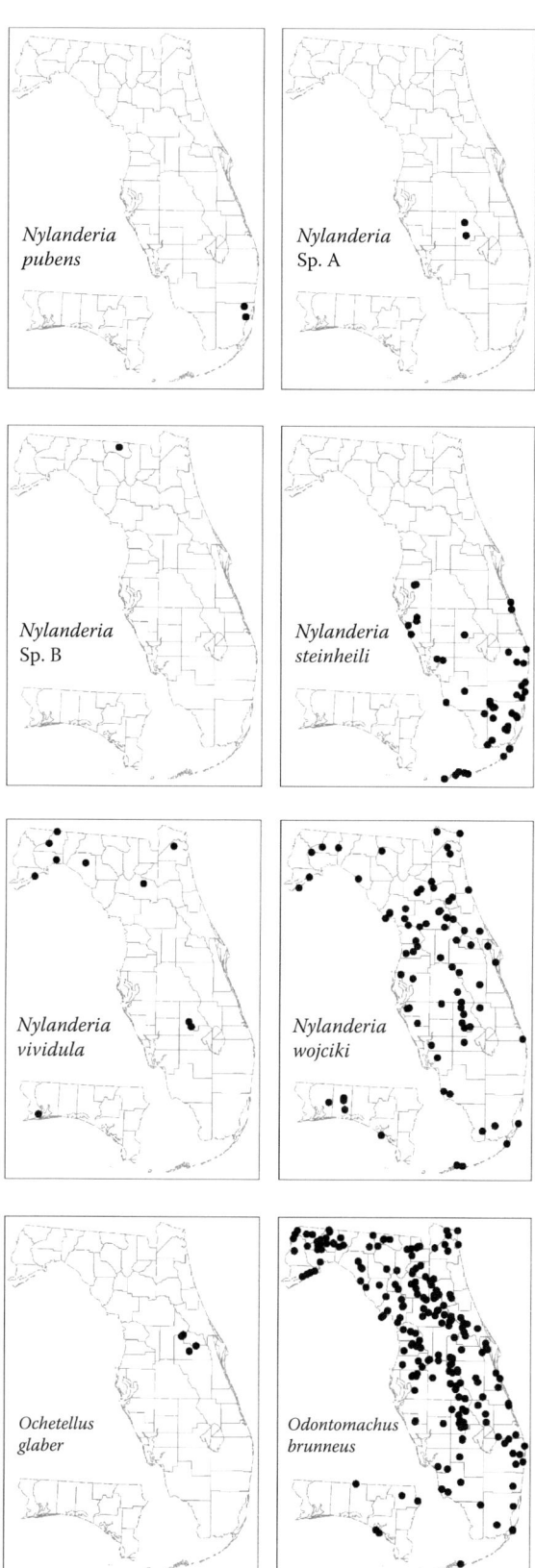

DISTRIBUTION MAPS

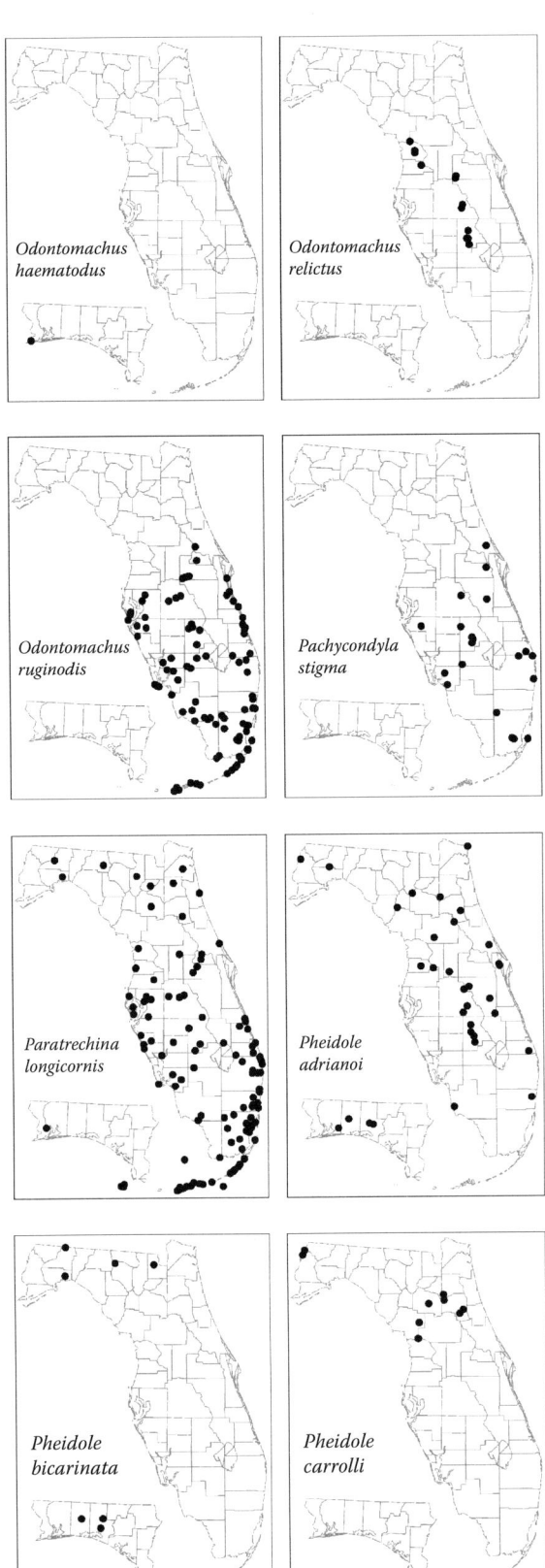

Odontomachus
haematodus

Odontomachus
relictus

Odontomachus
ruginodis

Pachycondyla
stigma

Paratrechina
longicornis

Pheidole
adrianoi

Pheidole
bicarinata

Pheidole
carrolli

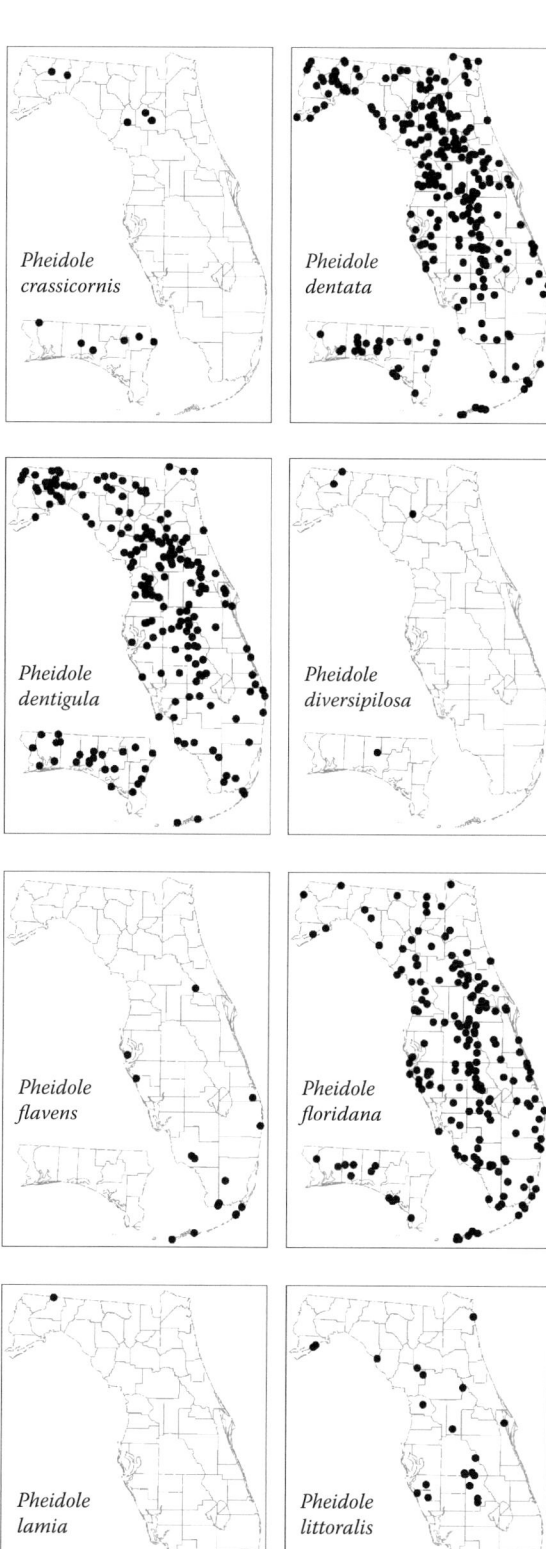

Pheidole crassicornis

Pheidole dentata

Pheidole dentigula

Pheidole diversipilosa

Pheidole flavens

Pheidole floridana

Pheidole lamia

Pheidole littoralis

DISTRIBUTION MAPS

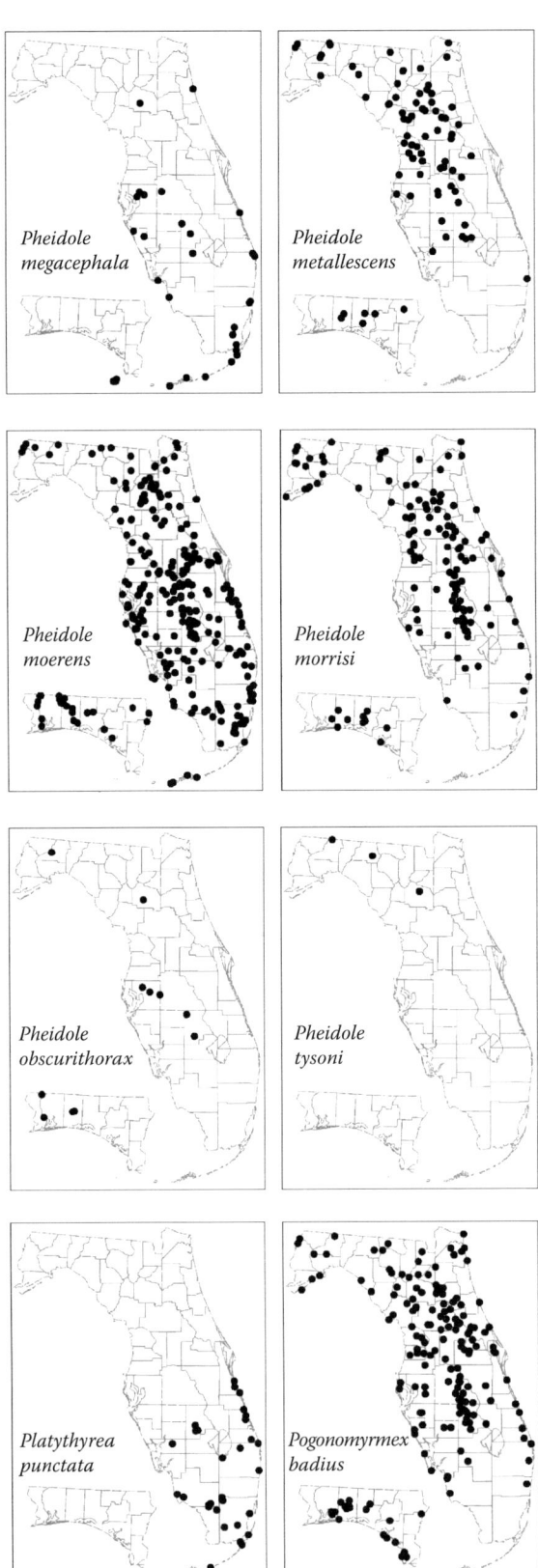

Pheidole
megacephala

Pheidole
metallescens

Pheidole
moerens

Pheidole
morrisi

Pheidole
obscurithorax

Pheidole
tysoni

Platythyrea
punctata

Pogonomyrmex
badius

DISTRIBUTION MAPS

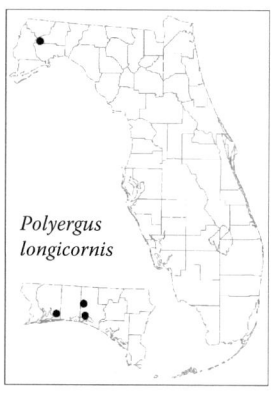

Polyergus longicornis

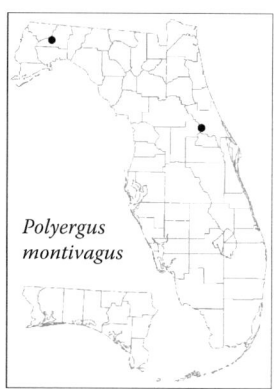

Polyergus montivagus

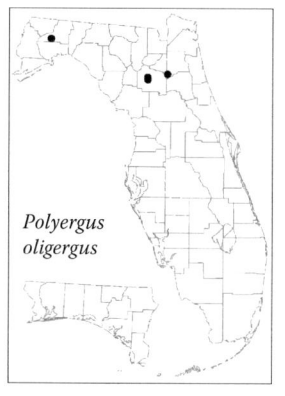

Polyergus oligergus

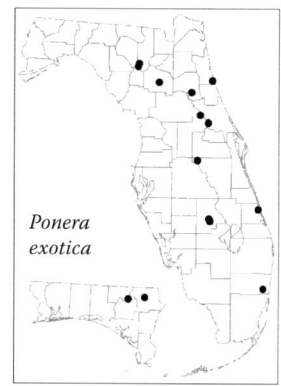

Ponera exotica

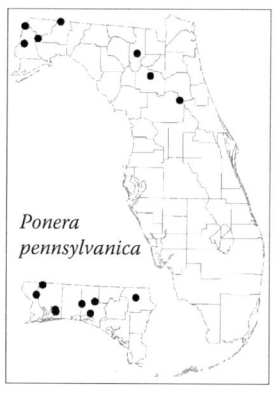

Ponera pennsylvanica

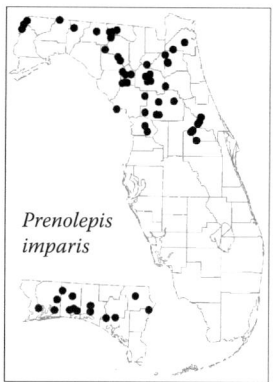

Prenolepis imparis

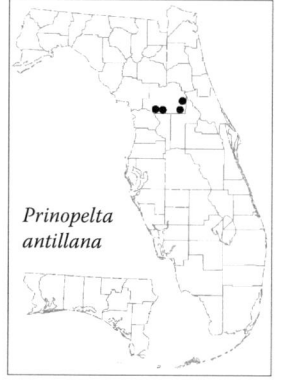

Prinopelta antillana

Proceratium chickasaw

DISTRIBUTION MAPS

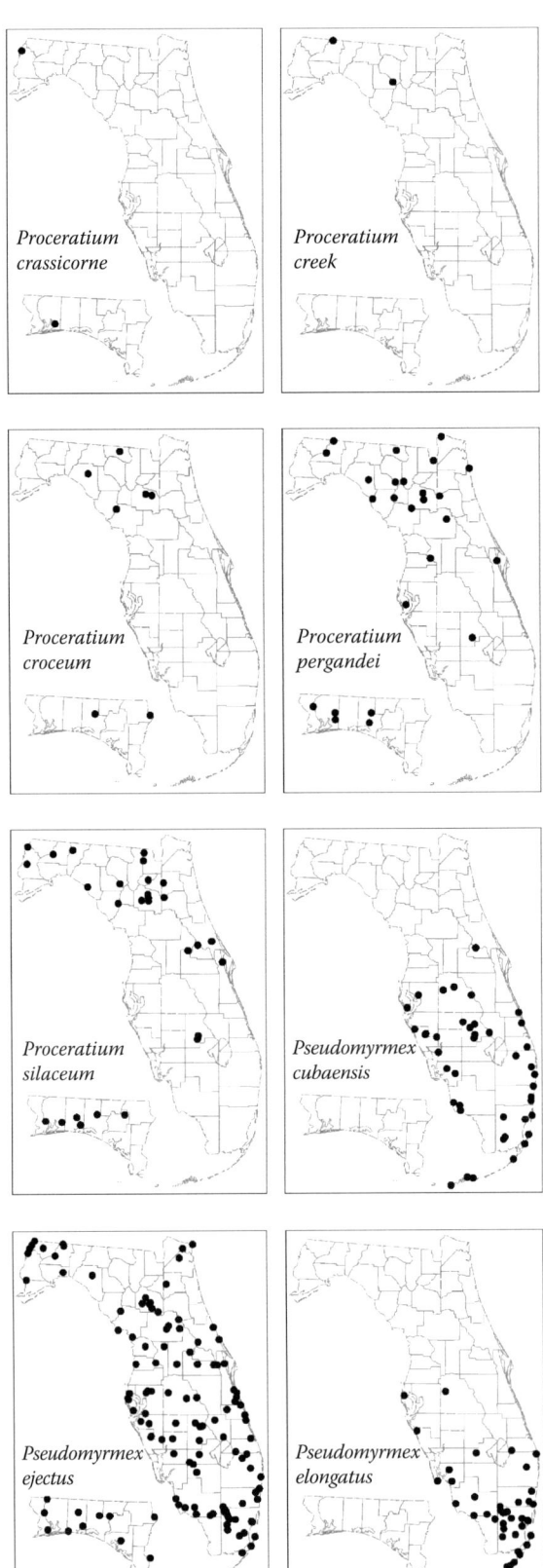

Proceratium
crassicorne

Proceratium
creek

Proceratium
croceum

Proceratium
pergandei

Proceratium
silaceum

Pseudomyrmex
cubaensis

Pseudomyrmex
ejectus

Pseudomyrmex
elongatus

DISTRIBUTION MAPS

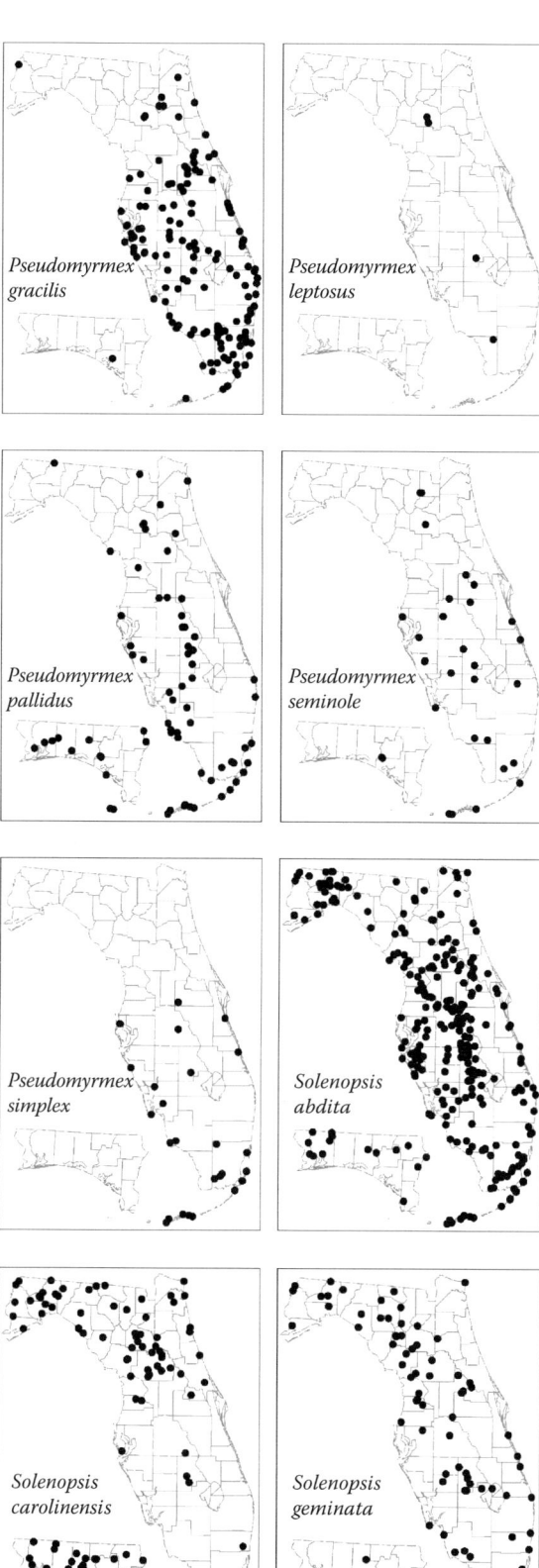

Pseudomyrmex gracilis

Pseudomyrmex leptosus

Pseudomyrmex pallidus

Pseudomyrmex seminole

Pseudomyrmex simplex

Solenopsis abdita

Solenopsis carolinensis

Solenopsis geminata

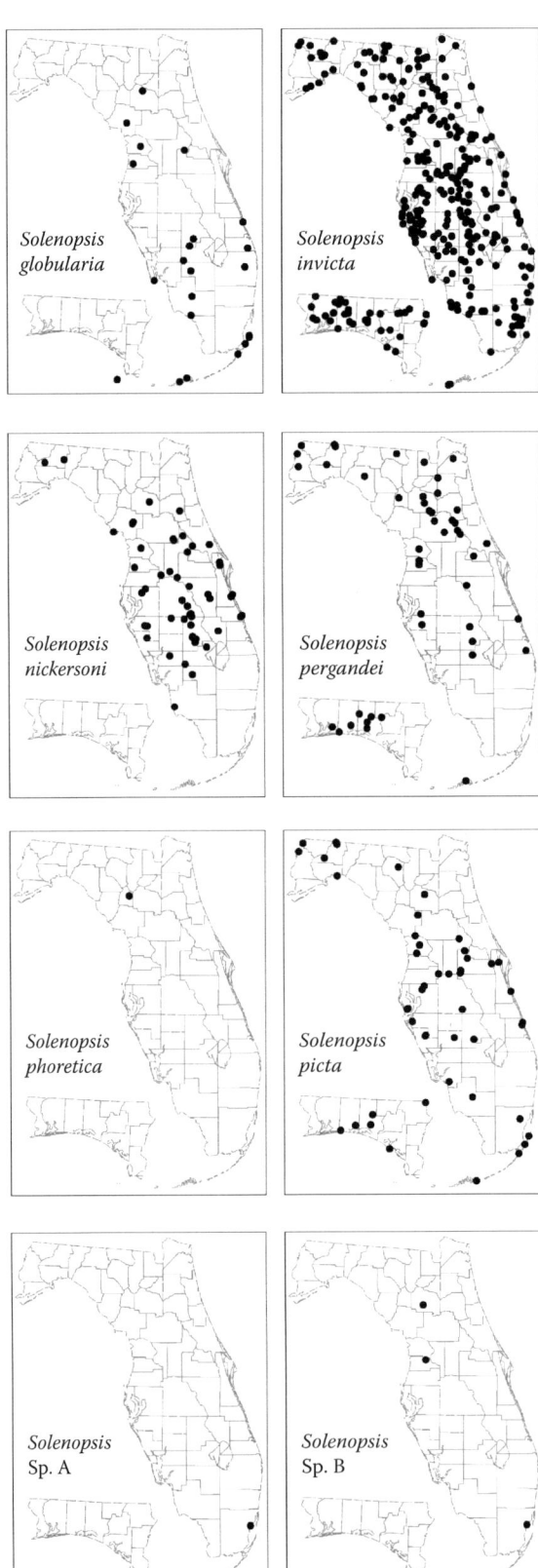

Solenopsis globularia

Solenopsis invicta

Solenopsis nickersoni

Solenopsis pergandei

Solenopsis phoretica

Solenopsis picta

Solenopsis Sp. A

Solenopsis Sp. B

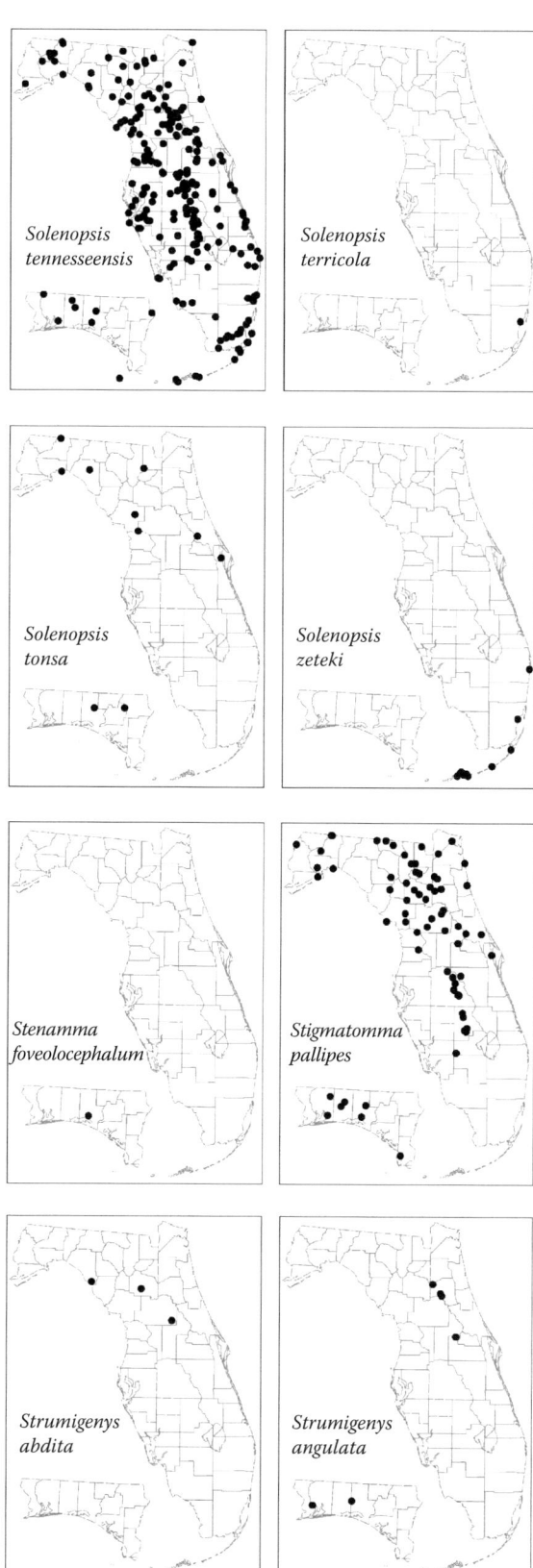

*Solenopsis
tennesseensis*

*Solenopsis
terricola*

*Solenopsis
tonsa*

*Solenopsis
zeteki*

*Stenamma
foveolocephalum*

*Stigmatomma
pallipes*

*Strumigenys
abdita*

*Strumigenys
angulata*

DISTRIBUTION MAPS

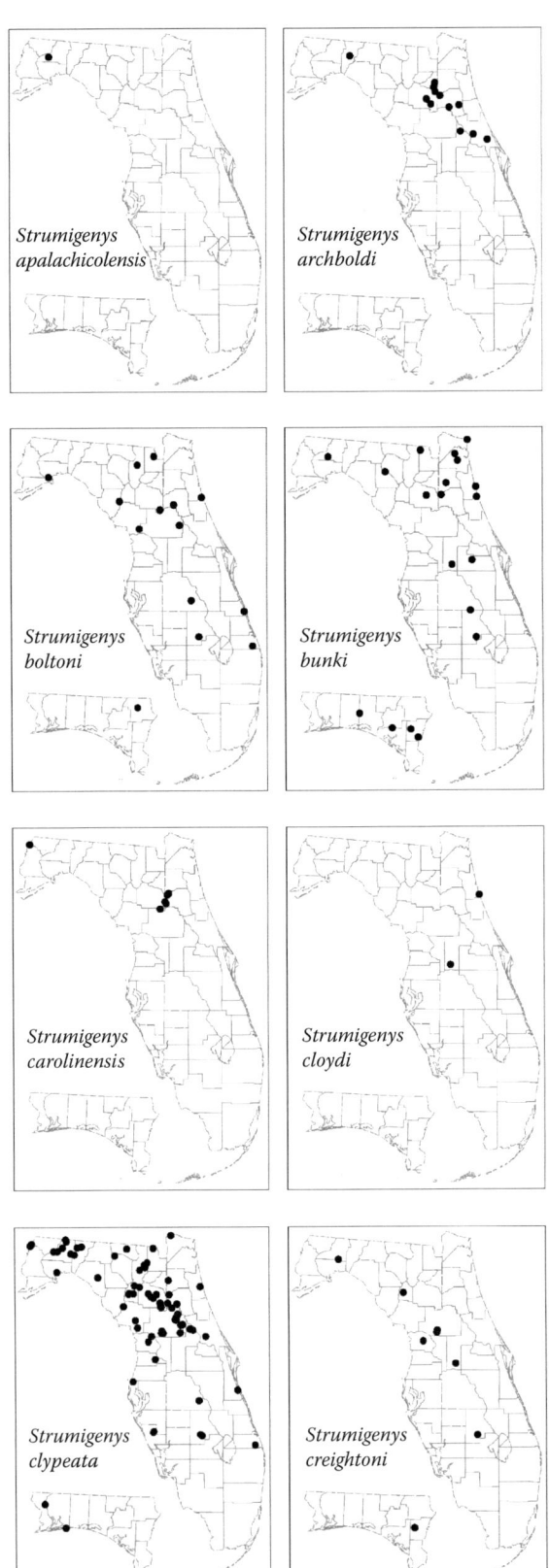

Strumigenys
apalachicolensis

Strumigenys
archboldi

Strumigenys
boltoni

Strumigenys
bunki

Strumigenys
carolinensis

Strumigenys
cloydi

Strumigenys
clypeata

Strumigenys
creightoni

DISTRIBUTION MAPS

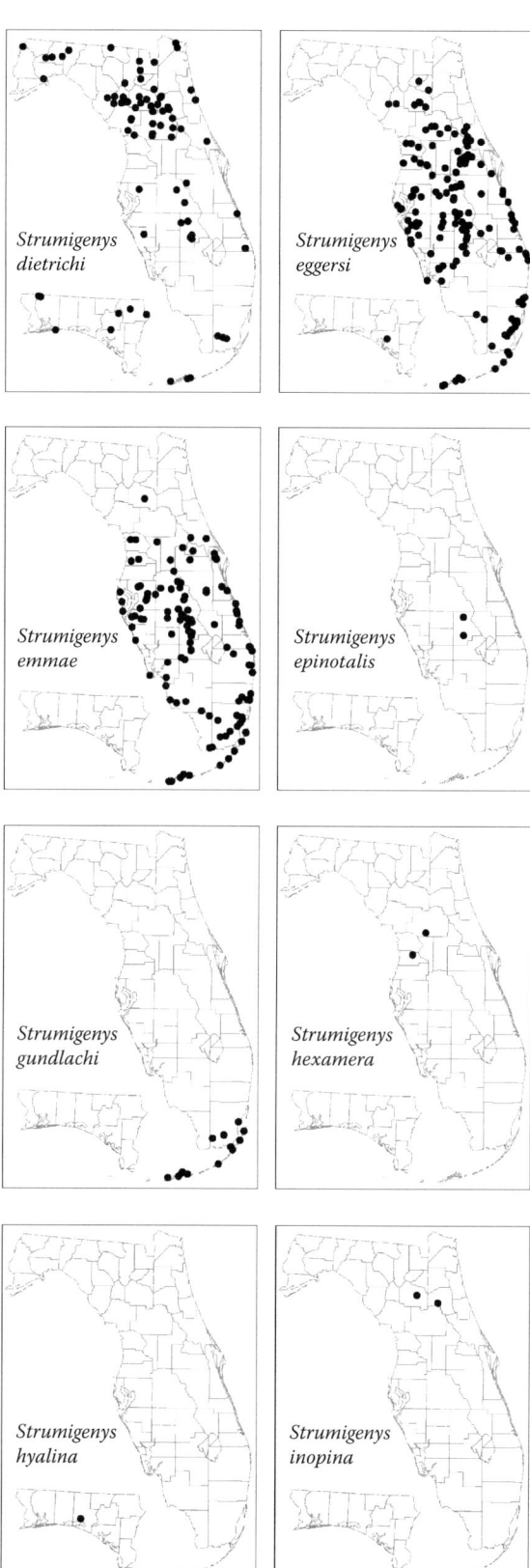

Strumigenys dietrichi

Strumigenys eggersi

Strumigenys emmae

Strumigenys epinotalis

Strumigenys gundlachi

Strumigenys hexamera

Strumigenys hyalina

Strumigenys inopina

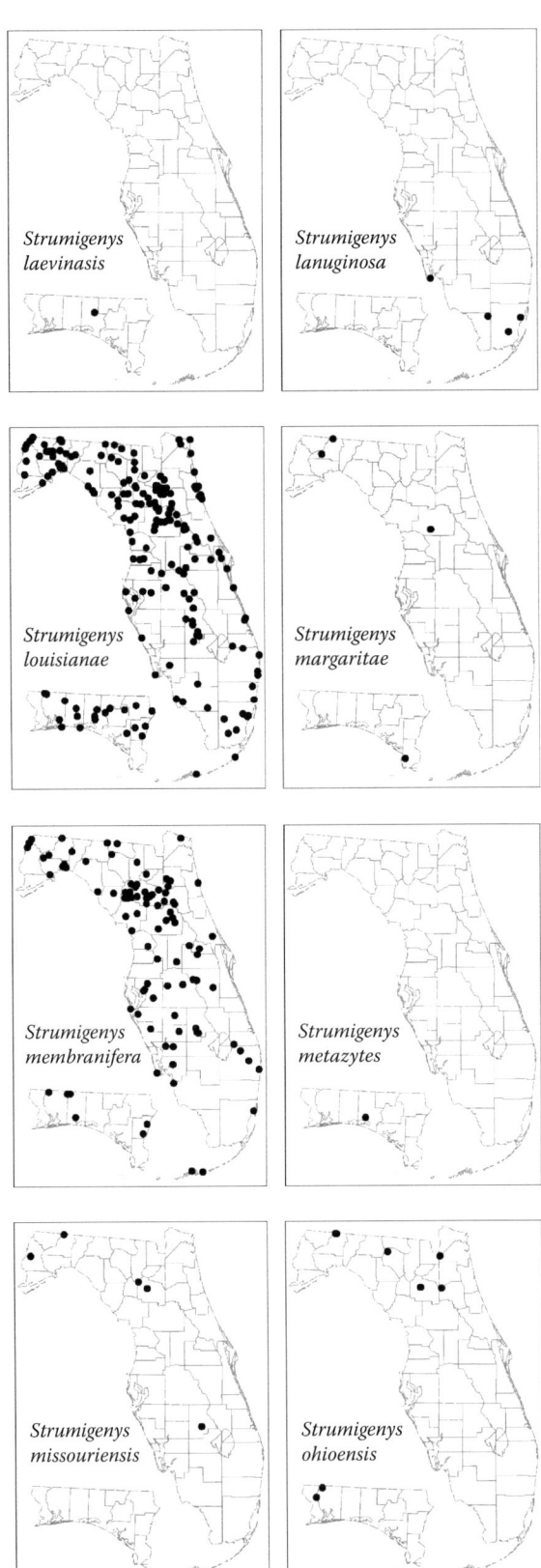

Strumigenys laevinasis

Strumigenys lanuginosa

Strumigenys louisianae

Strumigenys margaritae

Strumigenys membranifera

Strumigenys metazytes

Strumigenys missouriensis

Strumigenys ohioensis

DISTRIBUTION MAPS

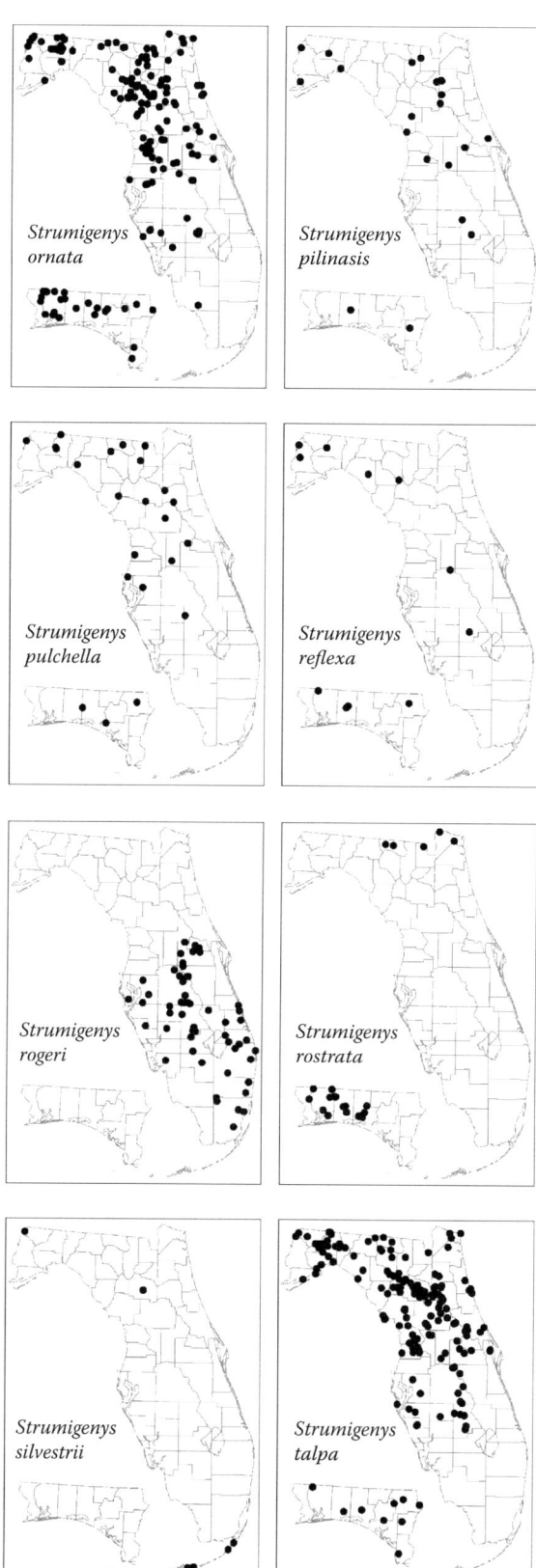

*Strumigenys
ornata*

*Strumigenys
pilinasis*

*Strumigenys
pulchella*

*Strumigenys
reflexa*

*Strumigenys
rogeri*

*Strumigenys
rostrata*

*Strumigenys
silvestrii*

*Strumigenys
talpa*

DISTRIBUTION MAPS

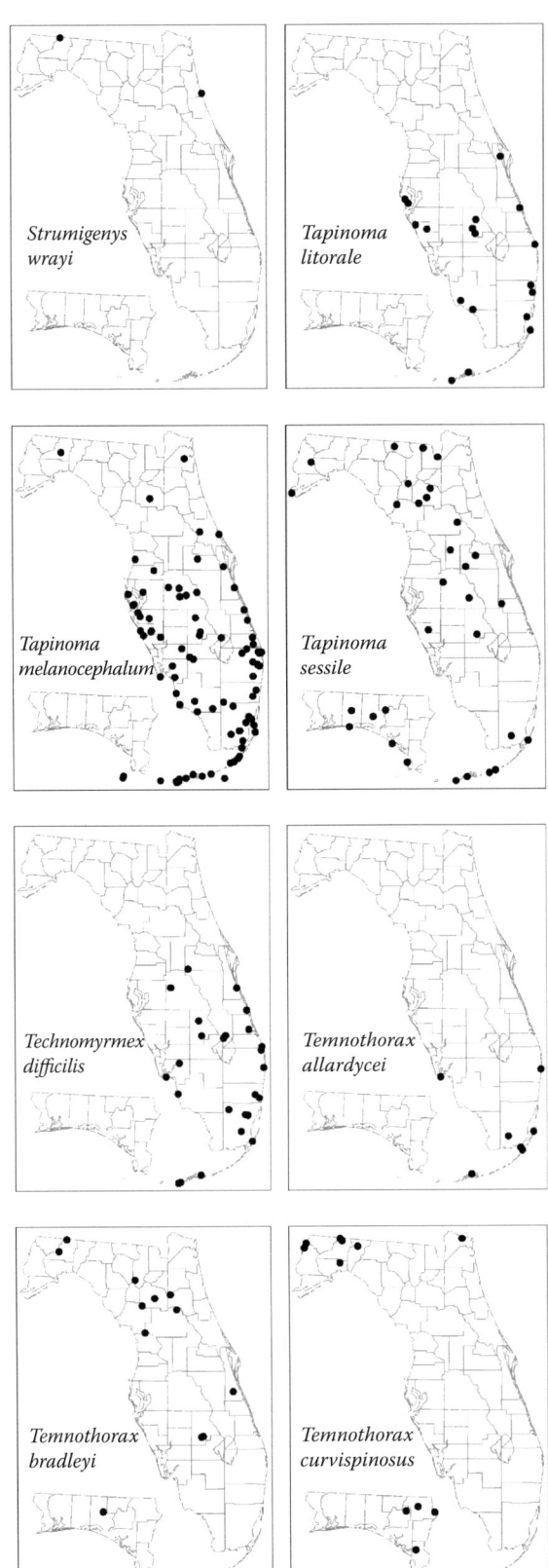

Strumigenys wrayi

Tapinoma litorale

Tapinoma melanocephalum

Tapinoma sessile

Technomyrmex difficilis

Temnothorax allardycei

Temnothorax bradleyi

Temnothorax curvispinosus

DISTRIBUTION MAPS

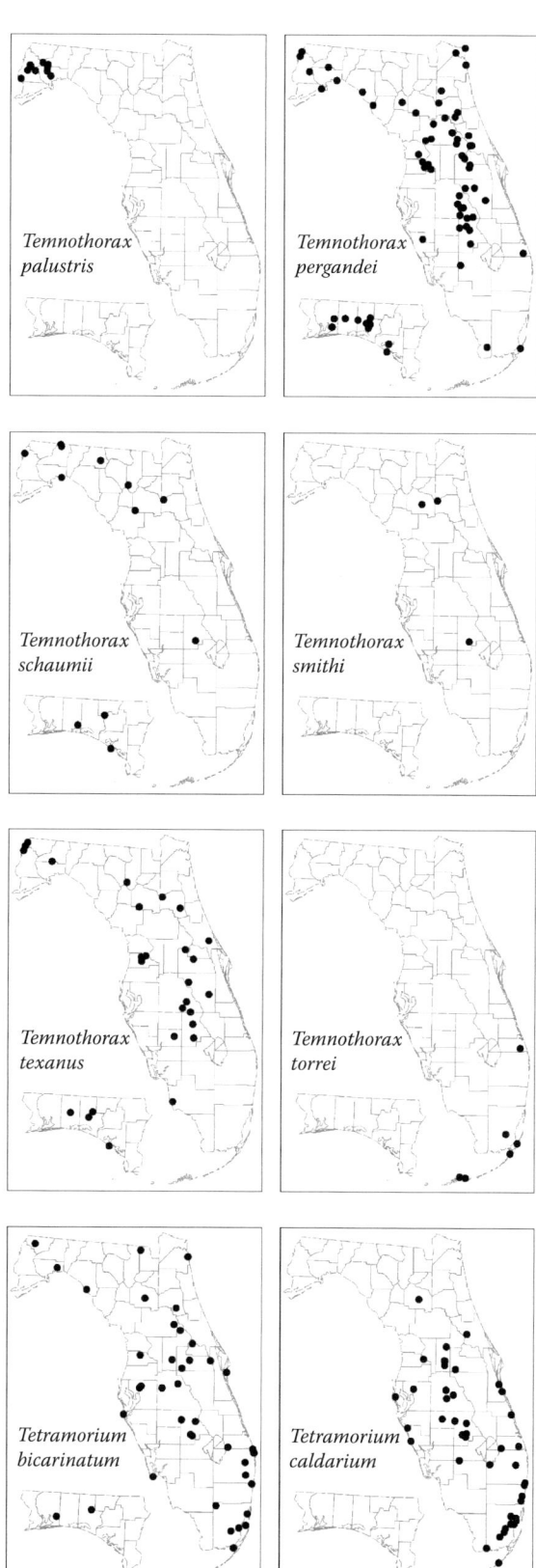

Temnothorax palustris

Temnothorax pergandei

Temnothorax schaumii

Temnothorax smithi

Temnothorax texanus

Temnothorax torrei

Tetramorium bicarinatum

Tetramorium caldarium

DISTRIBUTION MAPS

379

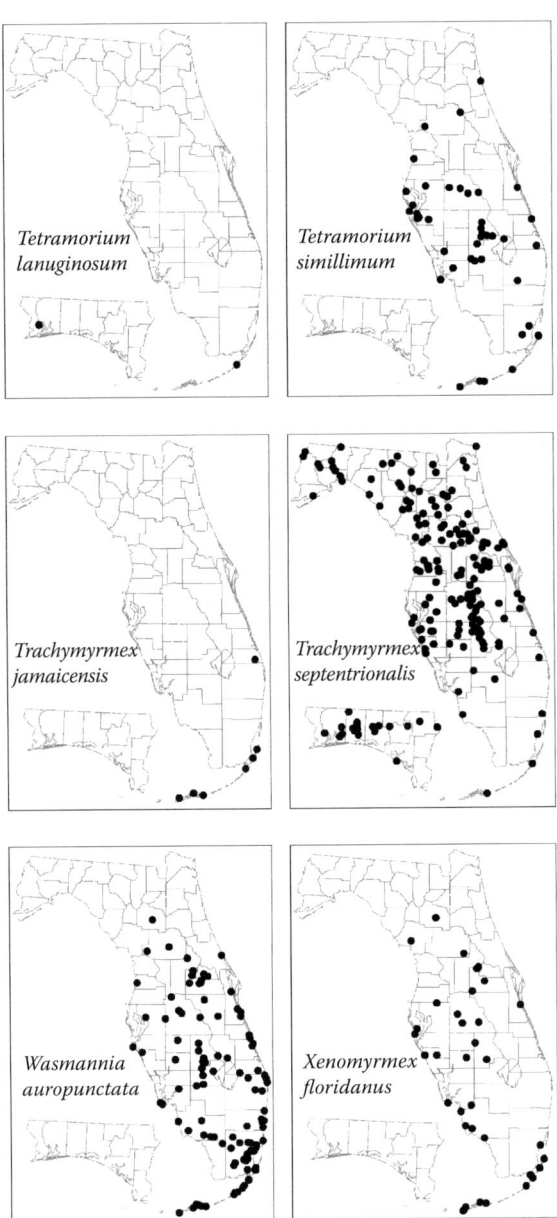

Tetramorium lanuginosum

Tetramorium simillimum

Trachymyrmex jamaicensis

Trachymyrmex septentrionalis

Wasmannia auropunctata

Xenomyrmex floridanus

DISTRIBUTION MAPS

INDEX

K

Kallal, Robert, 212
Kaufmann, S., 7
Kempf, W. W., 71
Kendra, Paul, 128
Kennedy, C. H., 126
Keularts, L. W., 216, 218
King, J. R., 92, 226
Klotz, J. H., 214
Koreshan State Park, 196
Kovarik, Peter, 18, 186
Kugler, C., 55, 58, 150
Kusnezov, Nicolas, 158

L

Laetilia sp, 5
Lake County, 125, 196, 204, 217
Lake Wales Ridge, 30, 47, 96, 98, 146, 165, 166, 167, 175, 213, 217, 220
LaPolla, John, 212
LaPolla, J. S., 212
Large imported big-headed ants, see *Pheidole obscurithorax* (Large imported big-headed ants)
Larger black sneaking ants, see *Cardiocondyla venustula* (Larger black sneaking ants)
Larger imported pennant ants, see *Tetramorium bicarinatum* (Larger imported pennant ants)
Largest genus, see *Strumigenys* spp (Mustache ants and Pygmy snapping ants)
Lasioglossum spp, 91
Lasius alienus (Woodland fuzzy ants), 206–207
 author, 206
 checklist, 231
 distribution, 206–207, 361
 name derivation, 207
 natural history, 207
 plates, 342
 taxonomy and similar species, 206
Lasius claviger (Hairy yellow underground ants), 207–208
 author, 207
 checklist, 231
 distribution, 207, 361
 name derivation, 208
 natural history, 207–208
 plates, 343
 taxonomy and similar species, 207
Lasius interjectus (Less hairy yellow underground ants), 208–209
 author, 208
 checklist, 231
 distribution, 208, 361
 name derivation, 208–209
 natural history, 208
 plates, 343
 taxonomy and similar species, 208
Lasius neoniger (Field fuzzy ants), 209–210
 author, 209
 checklist, 231
 distribution, 209, 361
 name derivation, 210
 natural history, 209–210
 plates, 342
 satellite nests, 177
 taxonomy and similar species, 209

Lasius niger americanus, see *Lasius neoniger* (Field fuzzy ants)
Lasius spp (Fuzzy ants and Yellow underground ants), 205–210
 description, 205–206
 most common and widespread, see *Lasius alienus* (Woodland fuzzy ants)
 name derivation, 206
 plates, 342–343
 as seed dispersers, 7
Lasius umbratus (Dusky fuzzy ants), 210
 author, 210
 checklist, 231
 distribution, 210, 361
 name derivation, 210
 natural history, 210
 plates, 342
 taxonomy and similar species, 210
Laskis, K. O., 162, 163
Latreille, Pierre-André, 205
Lattke, J. E., 21, 26
Lavigne, R. J., 212
Leaf miner bug, see *Eriocraniella* sp (Leaf miner bug)
Least fungus-gardener ants, see *Cyphomyrmex minutus* (Least fungus-gardener ants)
Least southeastern thief ants, see *Solenopsis tennesseensis* (Least southeastern thief ants)
Lebia spp, 91
Lechea deckertii (Deckert's pinweed), 90
Lee County, 186, 196
Leptogenys distinguenda, 25
Leptogenys elongata, 26
Leptogenys manni (Eastern isopod ants), 26
 author, 26
 checklist, 229
 distribution, 26, 362
 name derivation, 26
 natural history, 26
 plates, 272
 prey of, 6
 taxonomy and similar species, 2, 26
Leptogenys maxillosa, 25
Leptogenys spp (Beaked hunter ants), 25–26
 description, 25–26
 name derivation, 26
 plates, 272
Leptothorax spp, see *Temnothorax* spp (Creeper ants)
Lesser snapping ants, see *Anochetus* spp (Lesser snapping ants)
Less hairy yellow underground ants, see *Lasius interjectus* (Less hairy yellow underground ants)
Leuconotopicus borealis (Red-cockaded woodpeckers), 8, 195
Levings, S. C., 209, 210
Levy County, 18, 114, 183
Liberty County, 186, 205, 265, 340
Light stimulus, 9
Lignum Vitae Key Botanical State Park, 5
Lineolate acrobat ants, see *Crematogaster lineolata* (Lineolate acrobat ants)
Linepithema humile (Argentine ants), 172–174
 author, 172
 checklist, 231
 distribution, 172, 362
 name derivation, 174
 natural history, 172–174
 pest status, 9

N